Tumor Marker Protocols

METHODS IN MOLECULAR MEDICINE™

John M. Walker, *Series Editor*

14. **Tumor Marker Protocols,** edited by *Margaret Hanausek and Zbigniew Walaszek,* 1998
13. **Molecular Diagnosis of Infectious Diseases,** edited by *Udo Reischl,* 1998
12. **Diagnostic Virology Protocols,** edited by *John R. Stephenson and Alan Warnes,* 1998
11. **Therapeutic Application of Ribozymes,** edited by *Kevin J. Scanlon,* 1998
10. **Herpes Simplex Virus Protocols,** edited by *S. Moira Brown and Alasdair MacLean,* 1998
9. **Lectin Methods and Protocols,** edited by *Jonathan M. Rhodes and Jeremy D. Milton,* 1998
8. ***Helicobacter pylori* Protocols,** edited by *Christopher L. Clayton and Harry L. T. Mobley,* 1997
7. **Gene Therapy Protocols,** edited by *Paul D. Robbins,* 1997
6. **Molecular Diagnosis of Cancer,** edited by *Finbarr Cotter,* 1996
5. **Molecular Diagnosis of Genetic Diseases,** edited by *Rob Elles,* 1996
4. **Vaccine Protocols,** edited by *Andrew Robinson, Graham H. Farrar, and Christopher N. Wiblin,* 1996
3. **Prion Diseases,** edited by *Harry F. Baker and Rosalind M. Ridley,* 1996
2. **Human Cell Culture Protocols,** edited by *Gareth E. Jones,* 1996
1. **Antisense Therapeutics,** edited by *Sudhir Agrawal,* 1996

METHODS IN MOLECULAR MEDICINE™

Tumor Marker Protocols

Edited by

Margaret Hanausek

and

Zbigniew Walaszek

AMC Cancer Research Center, Denver, CO

Humana Press Totowa, New Jersey

999 Riverview Drive, Suite 208
Totowa, New Jersey 07512

This publication is printed on acid-free paper. ∞
ANSI Z39.48-1984 (American Standards Institute) Permanence of Paper for Printed Library Materials.

Cover illustration: Fig. 3A,B from Chapter 7, "Determination of Tumor Ferritin Concentration in Breast Cancer," by Jonathan F. Head and Robert L. Elliott.

Cover design by Patricia F. Cleary.

For additional copies, pricing for bulk purchases, and/or information about other Humana titles, contact Humana at the above address or at any of the following numbers: Tel.: 973-256-1699; Fax: 973-256-8341; E-mail: humana@humanapr.com; Website: http://humanapress.com

Printed in the United States of America. 10 9 8 7 6 5 4 3 2 1

Library of Congress Cataloging in Publication Data

Main entry under title:

Methods in molecular medicine™.

Tumor marker protocols/edited by Margaret Hanausek and Zbigniew Walaszek.
p. cm.—(Methods in molecular medicine;14)
Includes index.
ISBN 0-89603-380-5 (alk. paper)
1. Tumor markers. I. Hanausek, Margaret. II. Walaszek, Zbigniew. III. Series.
[DNLM: 1. Tumor Markers, Biological. 2. Neoplasms—diagnosis. QZ 241 T92532 1998]
RC270.3.T84T84 1998
616.99' 4075—dc21
DNLM/DLC 97-51263
for Library of Congress CIP

Preface

The development of cancer is a slow process and it very often takes a number of years before a tumor becomes clinically evident. No current technique is sensitive and specific enough to detect tumors at their earliest stage, i.e., when the tumor is smaller than one billion cells. Consequently, the disease is usually diagnosed in an advanced state, very often when it is already beyond the reach of therapeutic strategies. This is the main stumbling block to the secondary prevention that would reduce cancer mortality. There is hope, however, because in the last decade we have witnessed an explosion of reports dealing with tumor markers. In many instances, simple, noninvasive diagnostic tests are becoming available to detect the early signs of neoplasia. Interest in early detection of neoplasia is growing among those basic scientists, clinicians, and health professionals who realize that progress in reducing cancer mortality is dependent to a great degree on its early detection and prevention.

Some disappointment usually accompanies these discoveries, because what at first glance proves to be promising, is in many instances, applicable only to a limited number of cases. What is the cause of these failures? The answer may be that cancer is a multitude of neoplastic diseases in which endogenous and exogenous etiologic factors contribute either simultaneously or over a lifetime to the development of disease. Therefore, the reality is that no single test will be able to diagnose all neoplastic diseases. As scientists gain further knowledge of how tumors form, grow, and spread throughout the body, they hope to develop new markers that can identify a developing neoplasm, premalignant alterations, or even single circulating metastatic cells.

Throughout the world, both basic and clinical scientists involved in cancer research are trying to identify transformed phenotypes in order to develop their use as signs of neoplasia. However, despite great progress, the early diagnosis of cancer remains out of reach, with few exceptions. The aim of *Tumor Marker Protocols* is to bring together basic scientists, clinicians, and health professionals whose joint efforts may bring a breakthrough, or significant process in this field. This will happen when the biology of the tumor is fully elucidated, i.e., when our understanding of the complex biology of the early stages of neoplasia is better understood.

The scope of *Tumor Marker Protocols* is to thoroughly treat the major contemporary methods that have steadily broadened our understanding of early cancer formation, as well as to provide powerful routes to the detection or discovery of new cancer markers. The field of molecular tumor markers will be especially emphasized. An important related issue is how to develop methods to transfer technologies from basic scientists' hands to clinicians. The protocols in this book describe practical procedures for the broad range of protein and nucleic acid techniques frequently used by scientists in the fields of molecular biology, carcinogenesis, and cancer research. Each method is described by an author who has regularly used the technique in his or her own laboratory, and who describes all the practical steps necessary for carrying out such methods successfully.

Tumor Marker Protocols has been prepared in the same manner as previous volumes in the series of *Methods in Molecular Medicine,* published by Humana Press. The majority of the chapters are written as regular protocols, with notes at the end of each article commenting on critical points and issues. The book has been divided into parts in the following order: (I) significance of the multiple markers in the cancer marker field, with special emphasis on the statistical considerations in the analysis of tumor markers, (II) recent discoveries in the field of serum and tissue marker proteins, (III) cytogenetic markers, (IV) molecular markers, and finally (V) miscellaneous tumor markers and early detection of cancer. Part I of this book and a few chapters in Parts II, IV, and V bear the subtitle "Principles and Guidelines" in order to identify typical reviews. The book's many articles clearly illustrate a wide variety of promising opportunities. It is now the responsibility of cancer researchers to translate this promise into a comprehensive diagnostic strategy addressing the detection, therapy, and successful prevention of cancer.

In organizing *Tumor Marker Protocols,* every attempt has been made to cover the rapidly developing and changing field of molecular biology in cancer research. We hope that many readers will find the theoretical and applied chapters presented here a valuable source of information. It was a great privilege to be involved in the editing of this book and we are convinced that the scientific community will find it truly indispensable, not only for the beginners in the field of tumor markers, but also for experienced researchers. Finally, we want to express our sincere thanks to all those authors who contributed their expertise, experience, and hard work in preparing this book for publication.

Margaret Hanausek
Zbigniew Walaszek

Contents

Preface .... *v*
Contributors .... *xi*

PART I. MULTIPLE MARKERS: PRINCIPLES AND GUIDELINES

1 Integrating Multiple Clinical Tests to Increase Predictive Power
Harry B. Burke *3*

2 Statistical Considerations in the Analysis of Tumor Markers
Dennis A. Johnston *11*

3 Selection and Development of Biomarkers for Bladder Cancer
George P. Hemstreet, III, Robert E. Hurst, and Rebecca B. Bonner *37*

4 Clinical Application of Tissue and Serum Markers in Breast Cancer
Gordon F. Schwartz and Roland Schwarting *61*

PART II. IMMUNODIAGNOSIS OF CANCER: SERUM AND TISSUE MARKERS

5 Serum and Tissue Biomarkers in the Prognosis and Treatment of Breast Cancer
Tina J. Hieken and Rajeshwari R. Mehta *95*

6 The Oncofetal Protein p65 in Breast Cancer Detection
Margaret Hanausek, Zbigniew Walaszek, Ute Sherman, and Jerzy T. Klijanienko *103*

7 Determination of Tumor Ferritin Concentration in Breast Cancer
Jonathan F. Head and Robert L. Elliott *115*

8 Immunodiagnosis of Childhood Malignancies
David M. Parham and Hallie Holt *127*

9 Immunohistochemical Evaluation of Biomarkers in Prostatic and Colorectal Neoplasia: *Principles and Guidelines*
William E. Grizzle, Russell B. Myers, Upender Manne, and Sudhir Srivastava *143*

10 Factors Affecting Immunohistochemical Evaluation of Biomarker Expression in Neoplasia
William E. Grizzle, Russell B. Myers, Upender Manne, Cecil Rob Stockard, Lualhati E. Harkins, and Sudhir Srivastava *161*

11 Instrumentation, Accuracy, and Quality Control Issues in Development of Quantitative Fluorescence-Image Analysis (QFIA)
Rebecca B. Bonner, Robert E. Hurst, Jian Yu Rao, and George P. Hemstreet *181*

PART III. CYTOGENETIC MARKERS

12 Cytogenetics as a Diagnostic Aid for Childhood Hematologic Disorders: *Conventional Cytogenetic Techniques, Fluorescence* In Situ *Hybridization, and Comparative Genomic Hybridization*
Susanna C. Raimondi, Susan Mathew, and Ching-Hon Pui *209*

13 Preparation of Metaphase Chromosomes for Cytogenetic Analysis
Kevin A. Hahn and Penny K. Riggs *229*

14 Chromosome Staining and Banding Techniques
Kevin A. Hahn and Penny K. Riggs *239*

15 Fluorescence *In Situ* Hybridization to Chromosomes
Penny K. Riggs and Kevin A. Hahn *245*

16 Detection of Rearrangements in the *bcl-2* Gene Using the Polymerase Chain Reaction
Ruth W. Craig, Dorothy R. Belloni, Ernest S. Kawasaki, and Norman B. Levy *255*

PART IV. MOLECULAR MARKERS

17 Applications of Tissue Microdissection in Molecular Pathology: *Principles and Guidelines*
Michael R. Emmert-Buck, Irina A. Lubensky, Rodrigo F. Chuaqui, Larisa V. Debelenko, Cathy D. Vocke, Maria J. Merino, Paul H. Duray, W. Marston Linehan, Lance A. Liotta, and Zhengping Zhuang *269*

18 *EWS* Gene Fusions as Diagnostic Markers in Sarcomas: *Principles and Guidelines*
Marc Ladanyi *299*

19 p53 Detection in Breast Cancer
Penelope L. Davis and J. Dirk Iglehart *319*

20 Use of the Polymerase Chain Reaction Technique to Detect the t(14;18) Translocation in Lymphoid Tissue
Maryalice Stetler-Stevenson and Megan S. Lim *331*

21 Detection of *ras* Gene Mutations Using Oligonucleotide Ligation Technology
Faye A. Eggerding *341*

22 Detection of Prostate-Cancer Cells in Blood and Bone Marrow by RT-PCR
Robert L. Vessella and Eva Corey *357*

23 Quantitative, Competitive RT-PCR Analysis of Biomarkers in the Study of Neoplasia
Richard D. Hockett, Jr., Miriam D. Rogers, and Sudhir Srivastava *373*

24 Differential Display to Define Molecular Markers and Genes That Mediate Malignancy
Edward J. Pavlik, Katherine Nelson, Suseela Srinivasan, Thomas L. Johnson, and Paul D. DePriest *391*

Part V. Miscellaneous Markers and Early Detection of Cancer

25 Transforming Growth Factor Beta: *A Plasma Tumor Marker*
Feng-Ming Kong, Mitchell S. Anscher, and Randy L. Jirtle *417*

26 Anti-HMdU Autoantibodies in Human Sera as a Biomarker of Cancer Risk
Krystyna Frenkel and Jerzy Karkoszka *431*

27 A Scientific Basis for Cancer Prevention:*Defining the Role of Individual Cytosolic GST Isozyme*
Sanjay K. Srivastava and Sanjay Awasthi *447*

28 Aberrant Crypt Foci System to Study Cancer Preventative Agents in the Colon: *Principles and Guidelines*
Ranjana P. Bird *465*

29 Strain-Dependent Differences in the Expression of the Oncofetal Protein p65 in Mice Susceptible and Resistant to Chemical Carcinogenesis
Janusz Szemraj, Margaret Hanausek, Zbigniew Walaszek, and Alan K. Adams 475

30 D-Glucaric Acid as a Prospective Tumor Marker
Zbigniew Walaszek, Margaret Hanausek, Janusz Szemraj, and Alan K. Adams 487

Index 497

Contributors

ALAN K. ADAMS • *Department of Carcinogenesis, The University of Texas M.D. Anderson Cancer Center, Smithville, TX*

MITCHELL S. ANSCHER • *Department of Radiation Oncology, Duke University Medical Center, Durham, NC*

SANJAY AWASTHI • *University of Texas Medical Branch, Galveston, TX*

DOROTHY R. BELLONI • *Department of Pathology, Dartmouth Medical School, Hanover, NH*

RANJANA P. BIRD • *Department of Foods and Nutrition, University of Manitoba, Winnipeg, Manitoba, Canada*

REBECCA B. BONNER • *Department of Urology, University of Oklahoma Health Science Center, Oklahoma City, OK*

HARRY B. BURKE • *Department of Medicine, New York Medical College, Valhalla, NY*

RODRIGO F. CHUAQUI • *Laboratory of Pathology, National Cancer Institute, National Institutes of Health, Bethesda, MD*

EVA COREY • *Department of Urology, University of Washington School of Medicine, Seattle, WA*

RUTH CRAIG • *Department of Pharmacology and Toxicology, Dartmouth Medical School, Hanover, NH*

PENELOPE L. DAVIS • *Duke University Medical Center, Durham, NC*

LARISA V. DEBELENKO • *Laboratory of Pathology, National Cancer Institute, National Institutes of Health, Bethesda, MD*

PAUL D. DE PRIEST • *Department of Obstetrics and Gynecology, University of Kentucky College of Medicine, Lexington, KY*

PAUL H. DURAY • *Laboratory of Pathology, National Cancer Institute, National Institutes of Health, Bethesda, MD*

FAYE A. EGGERDING • *Molecular Oncology Laboratory, Huntington Medical Research Institutes, Pasadena, CA*

ROBERT L. ELLIOTT • *Mastology Research Institute, The Elliott Mastology Center, Baton Rouge, LA*

MICHAEL R. EMMERT-BUCK • *Laboratory of Pathology, National Cancer Institute, National Institutes of Health, Bethesda, MD*

KRYSTYNA FRENKEL • *Departments of Environmental Medicine and Pathology, and Kaplan Comprehensive Cancer Center, New York University Medical Center, New York, NY*

WILLIAM E. GRIZZLE • *Department of Pathology, University of Alabama, Birmingham, AL*

KEVIN A. HAHN • *Department of Comparative Medicine, College of Veterinary Medicine, University of Tennessee, TN*

MARGARET HANAUSEK • *AMC Cancer Research Center, Denver, CO*

LUALHATI E. HARKINS • *Veterans Administration Medical Center, Birmingham, AL*

JONATHAN F. HEAD • *Mastology Research Institute, The Elliott Mastology Center, Baton Rouge, LA*

GEORGE P. HEMSTREET, III • *Departments of Urology, Pathology, and Environmental Health Sciences, University of Oklahoma Health Science Center, Oklahoma City, OK*

TINA J. HIEKEN • *Rush Medical College, Rush North Shore Medical Center, Skokie, IL*

RICHARD D. HOCKETT JR. • *Department of Pathology, University of Alabama, Birmingham, AL*

HALLIE HOLT • *Department of Pathology and Laboratory Medicine, St. Jude Children's Hospital, Memphis, TN*

ROBERT E. HURST • *Departments of Urology, Biochemistry and Molecular Biology, and Environmental Health Sciences, University of Oklahoma Health Science Center, Oklahoma City, OK*

J. DIRK IGLEHART • *Duke University Medical Center, Durham, NC*

RANDY L. JIRTLE • *Department of Radiation Oncology, Duke University Medical Center, Durham, NC*

THOMAS L. JOHNSON • *Department of Obstetrics and Gynecology, University of Kentucky Medical Center, Lexington, KY*

DENNIS A. JOHNSTON • *Department of Biomathematics, The University of Texas M.D. Anderson Cancer Center, Houston, TX*

JERZY KARKOSZKA • *Department of Environmental Medicine, New York University Medical Center, New York, NY*

ERNEST S. KAWASAKI • *Procept, Inc., Cambridge, MA*

JERZY T. KLIJANIENKO • *Institut Curie, Paris, France*

FENG-MING KONG • *Department of Radiation Oncology, Duke University Medical Center, Durham, NC*

MARC LADANYI • *Departments of Pathology and Human Genetics, Memorial Sloan-Kettering Cancer Center, New York, NY*

Norman B. Levy • *Department of Pathology, Dartmouth Medical School, Hanover, NH*

Megan S. Lim • *Laboratory of Pathology, National Cancer Institute, National Institutes of Health, Bethesda, MD*

W. Marston Linehan • *Urologic Oncology Branch, National Cancer Institute, National Institutes of Health, Bethesda, MD*

Lance A. Liotta • *Laboratory of Pathology, National Cancer Institute, National Institutes of Health, Bethesda, MD*

Irina A. Lubensky • *Laboratory of Pathology, National Cancer Institute, National Institutes of Health, Bethesda, MD*

Upender Manne • *Department of Pathology, University of Alabama, Birmingham, AL*

Susan Mathew • *Department of Pathology and Laboratory Medicine, St. Jude Children's Research Hospital, Memphis, TN*

Rajeshwari R. Mehta • *Department of Surgical Oncology, University of Illinois at Chicago, Chicago, IL*

Maria J. Merino • *Laboratory of Pathology, National Cancer Institute, National Institutes of Health, Bethesda, MD*

Russel B. Myers • *Department of Pathology, University of Alabama, Birmingham, AL*

Katherine Nelson • *Department of Obstetrics and Gynecology, University of Kentucky Medical Center, Lexington, KY*

David M. Parham • *Department of Pediatric Pathology, Arkansas Children's Hospital, and Department of Pathology, University of Arkansas for Medical Sciences, Little Rock, AR*

Edward J. Pavlik • *Department of Obstetrics and Gynecology, University of Kentucky Medical Center, Lexington, KY*

Ching-Hon Pui • *Department of Pathology and Laboratory Medicine, St. Jude Children's Research Hospital, and Departments of Pathology and Pediatrics, University of Tennessee, Memphis, TN*

Susana C. Raimondi • *Department of Pathology and Laboratory Medicine, St. Jude Children's Research Hospital, and Department of Pathology, University of Tennessee, Memphis, TN*

Jian Yu Rao • *Department of Urology, University of Oklahoma Health Science Center, Oklahoma City, OK*

Penny K. Riggs • *Department of Comparative Medicine, College of Veterinary Medicine, University of Tennessee, TN*

Miriam D. Rogers • *Department of Pathology, University of Alabama, Birmingham, AL*

GORDON F. SCHWARTZ • *Department of Surgery, Jefferson Medical College, Philadelphia, PA*
ROLAND SCHWARTING • *Department of Pathology and Cellular Biology, efferson Medical College, Philadelphia, PA*
SUSEELA SRINIVASAN • *Department of Obstetrics and Gynecology, University of Kentucky Medical Center, Lexington, KY*
SANJAY K. SRIVASTAWA • *Department of Pathology, University of Alabama, Birmingham, A*L
SUDHIR SRIVASTAVA • *National Cancer Institute, National Institutes of Health, Bethesda, MD*
MARYALICE STETTLER-STEVENSON • *Laboratory of Pathology, National Cancer Institute, National Institutes of Health, Bethesda, MD*
CECIL ROB STOCKARD • *Department of Pathology, University of Alabama, Birmingham, AL*
UTE SHERMAN • *Department of Carcinogenesis, The University of Texas M.D. Anderson Cancer Center, Smithville, TX*
JANUSZ SZEMRAJ • *Department of Carcinogenesis, The University of Texas M.D. Anderson Cancer Center, Smithville, TX*
ROBERT L. VESSELLA • *Department of Urology, University of Washington School of Medicine, Seattle, WA*
ZBIGNIEW WALASZEK • *AMC Cancer Research Center, Denver, CO*
CATHY D. VOCKE • *Laboratory of Pathology, National Cancer Institute, National Institutes of Health, Bethesda, MD*
ZHENGPING ZHUANG • *Laboratory of Pathology, National Cancer Institute, National Institutes of Health, Bethesda, MD*

I

Multiple Markers: Principles and Guidelines

1

Integrating Multiple Clinical Tests to Increase Predictive Power

Harry B. Burke

1. Introduction

Clinical tests provide information that can be used by statistical methods to make patient outcome predictions. Outcomes are risk of disease, existence of disease, and prognosis. In this chapter we define and describe predictive factors and clinical prediction and explain how combining predictive factors can increase predictive accuracy, describe the advantages and disadvantages of commonly used statistical methods, and recommend an approach to the reporting of predictive factor research.

2. Predictive Factors

A predictive factor predicts an outcome (risk of disease, existence of disease, or prognosis) by virtue of its relationship with the disease process that causes the outcome. For example, the prognostic factor mutant p53 is associated with breast cancer because of its role in the regulation of apoptosis. Such terms as marker, biomarker, predictor, prognosticator, indicator, surrogate factor, and intermediate biomarker have been used to identify variables that are connected to medical outcomes. Their meanings overlap, and their undifferentiated use can cause confusion. All predictive factors are markers of disease (i.e., they are in some way associated with the disease process), but not all markers of disease have sufficient predictive power to be called predictive factors. We use the term factor to identify markers of disease that either are, or have the potential to be, predictive for a given outcome in a specified model.

Determining whether a marker is a predictive factor requires that:

1. The variable is measured in a defined population;
2. The population is followed until enough outcomes have occurred (i.e., deaths); and
3. The relationship between the variable and the outcome is determined.

From: *Methods in Molecular Medicine, Vol. 14: Tumor Marker Protocols*
Edited by: M. Hanausek and Z. Walaszek © Humana Press Inc., Totowa, NJ

If the variable predicts the outcome with "sufficient" accuracy (where "sufficient" varies with the question being addressed) in a specified model, it is called a predictive factor. If the predicted outcome always occurs, we say that the predictive factor and the outcome are 100% linked, i.e., the factor has a 100% predictive accuracy ***(1)***.

There are three types of predictive factors; risk, diagnostic, and prognostic ***(1)***. They differ in their outcomes and predictive power. "Risk" is an ambiguous term. We use "risk" to refer to "risk of disease." "Risk," when used in the context of "risk of recurrence" or "risk of death," is called "probability," as in "probability of recurrence" and "probability of death." Risk factor; the main outcome of interest is incidence of disease. The factor, either alone or in combination with other factors, is much less than 100% predictive of the disease occurring by a specified time in the future. Risk can be viewed as a propensity for the disease. Diagnostic factor; the main outcome of interest is also incidence of disease. The factor, either alone or in combination with other factors, is close to 100% predictive of disease. Prognostic factor; the main outcome of interest is death. A factor is rarely a strong predictor in isolation from other prognostic factors. There is domain overlap in that risk factors can be prognostic, but they cannot be diagnostic, and diagnostic factors can be prognostic, but they cannot be risk factors.

There are three subtypes of predictive factors: natural history, therapy-dependent, and post-therapy ***(1)***. Natural history predictive factors predict the future occurrence (risk), current existence (diagnosis), or course (prognostic) of a disease without an intervention. For risk and prognosis, natural history should the baseline against which all interventions are tested. Therapy-dependent predictive factors assume that there are effective therapies and predict whether the patient will respond to a particular intervention (for example, chemoprevention or chemotherapy). A natural history predictive factor may also be a therapy-dependent predictive factor. Post-therapy predictive factors require that patients respond to an intervention. They predict recurrence of the risk of disease or recurrence of the disease.

The predictive power of a factor depends on its intrinsic and extrinsic powers. The intrinsic predictive power of a factor is related to its "connectedness" to the disease process, i.e., its association to the disease process. The less connected the factor is, the less predictive it is. A direct connection means that the factor is an integral part of the disease process itself. An indirect connection means that it is not an integral part of the disease process but is related to the disease process, such as being a byproduct of it (i.e., a secondary infection). The extrinsic predictive power of the factor depends on the question being asked, i.e., the specific factor-outcome relationship being examined. For a specific disease process and outcome, the predictive accuracy of a factor depends on:

1. How closely connected the factor is to the disease process (individual factor power) and its relationship to the other factors (degree of predictive overlap);
2. How easy it is to collect and measure the factor; and
3. The degree to which the selected statistical method is able to capture the individual factor's predictive information and to integrate it with the information of other factors.

It is rarely the case that one factor is sufficiently predictive, i.e., that it is able to predict the outcome of interest with 100% accuracy. The usual strategy, when dealing with predictive factors, is to combine several in a predictive model. The most useful grouping of factors is one in which all of the factors are powerful and predictively orthogonal to each other, i.e., they index independent aspects of the disease process. If they represent aspects of the disease that are not independent of each other, then to the degree that their information overlaps is the degree to which one will not add predictive power. The statistical method employed must be able to capture the complexity of the disease process indexed by the predictive factors.

A predictive model for a specific outcome is the result of entering one or more predictive factors into a statistical method. The statistical method attempts to capture the relationship between the factors and the outcome. For example, the mathematical formula generated by the logistic regression statistical method relates the predictive factors (input variables), in terms of their β-coefficients, to a binary disease outcome (relapse, death, and so forth). It should be noted that the predictive power of a factor depends on the specific statistical method selected and on the other factors selected to be included in the model. The statistical model that results from the application of a statistical method, learning the relationship between the factors and the outcome, may or may not be the most efficient at capturing the predictive power of the factors.

Before discussing specific statistical methods, it is important to distinguish among significance, accuracy, and importance *(2)*. Model significance asks if the observed predictions are really different from those produced by another model or from those resulting from chance.

Significance is not accuracy. Accuracy is the association between the model's predictions and the known outcomes in a test population. The importance of a model or a factor is determined by whether the model or factor possesses sufficient accuracy to be useful in answering a particular clinical question. Finally, the assessment of model or factor significance, accuracy, and importance must be based on test data set results, not on training data set results.

3. Advantages and Disadvantages of Statistical Methods

Many methods can be used to combine predictive factors. In cancer, they include bins, stages, and indexes; decision trees; and regression methods, including logistic, proportional hazards, and artificial neural networks.

Bins are the result of the mutually exclusive and exhaustive partitioning of discrete variables. Each combination of variable values is a bin, and all patients are placed in the bin corresponding to their variable value combination *(2)*. An example is the TNM classification of breast cancer *(3)*. Tumor size (Tis, T1, T2, T3, T4), number of positive regional lymph nodes (N0, N1, N2, N3), and existence of metastases (M0, M1) produce 40 bins *(2)*.

Each patient in a bin receives the same prediction; namely, the most frequent outcome. If there are enough patients in each bin, it can be shown that the most frequent outcome is the best predictor of the true outcome. In other words, no prediction model can be more accurate than a bin model if the variables are discrete and the population is large. Problems with bin models *(2)* include:

1. Continuous variables must be cut up into discrete variables. This almost always results in a loss of predictive information and therefore a loss of accuracy.
2. As the number of discrete variables increases, the number of bins increases exponentially. In order to maintain accuracy, there must be a corresponding exponential increase in the size of the patient population.
3. The proliferation of bins reduces the ability to understand the phenomena. Bin proliferation negates the main advantage of a bin model; namely, its ease of understanding and ease of use.

Bin models are rarely used in situations in which there are more than two or three predictive factors or where each factor possesses more than a few strata.

A partial solution to the problems of a bin model is a stage model *(2)*. A stage model is the grouping of bins into super-bins. The justification for the grouping is the assumption that the factors selected represent "stages" of the disease process. For example, in breast cancer, the TNM staging system combines 40 TNM classification bins into six super-bins (TNM stages) based on decreasing survival ("stages of survival").

A small set of stages has the potential to maintain explanatory simplicity and ease of use. Problems with stage models include:

1. The combining of bins into super-bins/stages can substantially reduce predictive accuracy.
2. Stage systems do not overcome the exponential increase in bins and patients associated with adding a variable to the analysis: They just delay the problem at a cost in predictive accuracy. If the stages are held constant when variables (and their associated bins) are added to the staging system, the potential improvement

in accuracy associated with the additional bins will be small to nonexistent. But, if the stages are expanded to accommodate additional bins, the system loses its ease of understanding and usefulness. Thus, attempts to improve predictive accuracy by adding variables to a bin/stage model are rarely successful.

3. The problems of cutting up continuous variables, with the resulting loss in predictive accuracy, remains.
4. Finally, if a single staging system is used for more than one cancer site, the staging rules may be more applicable to some sites than to other sites. The sites to which they do not apply will experience major losses in predictive accuracy.

Indexes associate numerical scores (usually based on a bounded, linear scale) with bins or groups of bins. Each score is associated with one of a small number of disease stages (usually a severity of illness system). Each patient receives the prediction of the stage in which their score places them. Indexes offer some flexibility in the grouping of bins, but at the cost of further degradation in predictive accuracy because additional information is lost. The simplest example of an index is the Apgar. An example in breast cancer is the Nottingham Index ***(4)***.

The accuracy of different stratifications of a predictive factor(s) can be compared. For a specific site (i.e., breast) and predictor(s) (tumor size <2, 2–5, >5) any bin or group of bins, or stage (bin or index) or group of stages, can be compared, in terms of a specific outcome, with another stratification (tumor size <1, 1–<2, 2–<3, 3–<4, 4–<5, 5–>5). This contrast can be over a single time interval without respect to events within the interval (i.e., logistic regression) or with respect to the events within the interval ***(5,6)***. For a single interval without respect to events within the interval, accuracy has been assessed by several discriminative association approaches, including Goodman and Kruskall's Gamma ***(7)***, Kendall's Tau ***(8)***, or the area under the receiver operating characteristic ***(9)***.

The usual descriptive approach for contrasting predictive factors across a series of event time intervals is the Kaplan-Meier product-limit method ***(5)*** (inferential methods that can accommodate continuous variables, and that usually assume proportional hazards, will be discussed later when regression methods are presented). A Kaplan-Meier plot should always include confidence intervals for each stratum (i.e., each step function). A significant difference within a Kaplan-Meier stratification (tumor size <2, 2–5, >5) is usually assessed by a log-rank test ***(10)***. It is important to note that there is currently no method for comparing the accuracy of two different Kaplan-Meier plots (i.e., two different stratifications of the same predictive factors). It is incorrect to use the *p*-value of the log-rank test to select one stratification over another, because the log-rank test only determines whether a stratification is likely to have occurred by chance. An extreme stratification may result in smaller *p*-values, but it may also reduce predictive accuracy.

Decision trees split predictive factors to maximize predictive power using a loss function, such as the log-likelihood and a greedy search algorithm. A well-known decision tree approach is the Classification and Regression Trees (CART) recursive partitioning method ***(11)***. Empirically, we have not found CART, either pruned or shrunk, to be the most accurate statistical method when compared to regression methods. Its problems include the selection of the correct loss function, difficulty dealing with continuous variables, and overfitting when searching for the best predictors when there are more than two or three splits.

Univariate regression methods are not appropriate for determining whether a variable is a predictive factor. Univariate methods should not be used, because new variables must be assessed in the context of the known factors, and because some variables are only predictive when they interact with another variable.

Logistic regression assess the cumulative probability of a binary event occurring by a specific time. It uses a maximum likelihood loss function and a greedy search technique. It is a very efficient method for binary outcome problems (i.e., recurrence and survival). Its limitation is that it usually spans a large time interval and does not distinguish when events occur within the time interval. This limitation can be overcome if several sub-time intervals are created within the overall time interval. Logistic regression models can be created for each sub-time interval. Censoring can be accommodated by removing cases that are censored within the time interval that censoring occurs.

Proportional hazards methods include the Cox ***(6)*** and less commonly the Weibull or exponential ***(12)***. Proportional hazards methods assume that the hazard of each patient is proportional to the hazards of all the other patients, and that a patient's hazard is related to that patient's relative risk. The Cox model does not create survival curves. For Cox-related survival curves, a baseline hazard must be introduced (for example, Breslow-Cox estimates) ***(13)***. Some researchers incorrectly believe that the Cox is the only regression method that can deal with censoring (*see* paragraph on logistic regression above). Because, in cancer, the proportional hazards' assumption may be violated, researchers who use the Cox model must demonstrate that the proportional hazards assumption holds for their population.

Artificial neural networks are a general regression method ***(14,15)***. They can perform almost any regression task. In addition, three-layer artificial neural networks automatically capture nonlinearity and complex interactions. They can handle censoring in the same way that multi-interval logistic regression handles censoring. Artificial neural networks are as transparent as the phenomena contained in the data. For simple phenomena, artificial neural networks are easily understood; for complex phenomena they are complex and less easily understood. Artificial neural networks are especially recommended in the domain of complex systems (e.g., the molecular-genetic domain of cancer).

4. Reporting Predictive Factor Research Results

There is a great deal of variation in the reporting of predictive factor results. This variability makes it difficult to understand and compare results. The following is a recommended approach to reporting the discovery of a new predictive factor or the validation of an existing factor.

For a defined subset of patients with the ___a___ disease, ___b___ is a ___c___ predictive factor for ___d___ when assayed ___e___ by ___f___, for the ___g___ on a test data set with ___h__ characteristics, the ___i___ is significant at the ___j___ level using the ___k___ statistical method, which also incorporates ___l___ predictive factors, for ___m___ therapy. Using the ___n___ method to assess its accuracy, the ___k___ statistical model is ___o___ accurate on the test data set.

"Defined" means specification of collection method, inclusion and exclusion criteria, and so forth.

a: Name of disease.
b: Name of the predictive factor.
c: Type and subtype of predictive factor (i.e., risk, diagnosis, prognosis; natural history, therapy-dependent, post-therapy).
d: Outcome (i.e., 5-yr breast cancer-specific survival).
e: Time of assay (i.e., at discovery, prior to therapy, after therapy).
f: Specific laboratory method (i.e., immunohistochemistry).
g: If stratified, the specific range/cut-point/and so forth of the prognostic factor. If the variable value is based on rater judgment, then Cohen's κ should be reported.
h: Relevant characteristics of the data set, including:
   1. Data set size,
   2. Number of events, and
   3. Whether the therapy was randomized.
i: The value and confidence interval.
j: For example, $p < 0.05$ for one test of the data. If multiple tests of the data are performed, an adjustment may be required.
k: Type of multivariate statistical method (i.e., logistic regression, Cox).
l: Other relevant prognostic factors, if they are included in the multivariate model.
m: Specific type of surgery, chemotherapy, radiation therapy.
n: Area under the receiver operating characteristic (Az) R^2, χ-square, etc.
o: Numerical value and its range of possible values (i.e., Az = 0.75, 0.50, –1.0).

References

1. Burke, H. B. (1994) Increasing the power of surrogate endpoint biomarkers: aggregation of predictive factors. *J. Cell. Biochem.* **19,** 278–282.
2. Burke, H. B. and Henson, D. H. (1993) Criteria for prognostic factors and for an enhanced prognostic system. *Cancer* **72,** 3131–3135.

3. Beahrs, O. H., Henson, D. E., Hutter, R. V. P., and Kennedy, B. J. (1992) *Manual for Staging of Cancer,* 4th ed., Lippincott, Philadelphia, PA.
4. Haybittle, J. L., Blamey, R. W., Elston, C. W., Johnson, J., Doyle, P. J., Campbell, F. C., Nicholson, R. I., and Griffiths, K. (1982) A prognostic index in primary breast cancer. *Br. J. Cancer* **45,** 361–366.
5. Kaplan, E. L. and Meier, P. (1958) Nonparametric estimation from incomplete observations. *J. Am. Stat. Assoc.* **53,** 457–481.
6. Cox, D. R. (1972) Regression models and life-tables (with discussion). *J. Royal Stat. Soc. B.,* pp. 187–220.
7. Goodman, L. A. and Kruskal, W. H. (1954) Measures of association for cross classifications. *J. Am. Stat. Assoc.* **49,** 732–764.
8. Kendall, M. G. (1962) *Rank Correlation Methods.* Hafner, New York.
9. Bamber, D. (1975) The area above the ordinal dominance graph and the area below the receiver operating characteristic graph. *J. Math. Psy.* **12,** 387–415.
10. Mantel, N. (1966) Evaluation of survival data and two new rank order statistics arising in its consideration. *Cancer Chemother. Rep.* **50,** 163–170.
11. Breiman, L., Friedman, J. H., Olshen, R. A., and Stone, C. J. (1984) *Classification and Regression Trees.* Wadsworth and Brooks, Pacific Grove, CA.
12. Evans, M., Hastings, N., and Peacock, B. (1993) *Statistical Distributions,* 2nd ed., Wiley, New York.
13. Breslow, N. E. (1974) Covariance analysis of censored survival data. *Biometrics* **30,** 80–99.
14. Burke, H. B. (1994) Artificial neural networks for cancer research: outcome prediction. *Sem. Surg. Onc.* **10,** 73–79.
15. Burke, H. B., Rosen, D. B., and Goodman, P. H. (1995) Comparing the prediction accuracy of artificial neural networks and other statistical models for breast cancer survival, in *Advances in Neural Information Processing Systems*, vol. 7 (Tesauro, G., Touretzky, D. S., Leen, T. K., eds.), MIT Press, Cambridge, MA, pp. 1063–1067.

2

Statistical Considerations in the Analysis of Tumor Markers

Dennis A. Johnston

1. Introduction

In this chapter, we will lay out the basics of experimental design: how to organize a study and determine sample size, what data to use, how to set up the database for analysis, what statistics are necessary for the analysis of the study, and what statistical packages are available to analyze the study. These are considerations that should be included when the study is being planned. Consideration of statistical and data-collection requirements at the time of initial study planning will reduce missing data and prevent the collection of too few samples to ensure enough statistical power to be able to see the effects planned or too many samples, which wastes resources and time. The incorporation of data collection planned for the statistical analysis will streamline the data-collection process and minimize data coding errors, the amount of recoding, and poststudy data processing.

2. Experimental Design

The term "experimental design" encompasses all aspects of the design of the study, including both scientific considerations as well as the statistical aspects, from the mathematical structure of the study, the number of samples required, and the database structure to the statistical techniques required to analyze the study *(1–5)*. From the statistician's viewpoint the study can be broken down into several steps analogous to the scientific method. These are:

1. Define the study objectives as clear hypotheses;
2. Establish the type of trial;
3. Determine the statistical analyses required;
4. Determine the sample size;

From: *Methods in Molecular Medicine, Vol. 14: Tumor Marker Protocols*
Edited by: M. Hanausek and Z. Walaszek © Humana Press Inc., Totowa, NJ

5. Set up the database;
6. Conduct the trial; and
7. Analyze the data.

3. Objectives

These are the specific aims of the study. Often the specific aims are written in a narrative style in a general all-inclusive manner. The objectives are the specific testable hypotheses, which are derived from the specific aims and consist of the list of hypotheses about the attributes of the study population (treatment groups, markers—single or multiple, diagnosis, staging and disease extent, status over time or at a specific time, relapse, survival, and so forth). This is an iterative process best done during the time the specific aims are being defined.

3.1. Marker Data Types

Marker information on the subjects in a study falls into one of the following general categories:

1. The marker has two levels (binary): expressed or not expressed, and the subject or tumor either expresses the marker or not. Example: p53 expression ***(6)***.
2. The marker has two levels: expressed or not expressed, and a sample of cells from the subject or tumor has a proportion of cells expressing the marker. Example: polymerase chain reaction (PCR)-based detection system in leukemia ***(7,8)***.
3. The marker has several components each of which could be expressed or not. Examples: p53, MTS1, and others, where various exons and even nucleic acids may be changed ***(9)***. In p53 this means that any one of the 393 codons could be modified. One interest here is in associating frequency of particular modification(s) with other factors.
4. The marker has several levels or perhaps a continuous measure. Examples: prostate specific antigen (PSA), carcinoembryonic antigen (CEA), PCR-based detection system. These markers are measured using antigen-response assays or are a particular case of **item 2** above, respectively.

3.2. The Clinical Trial

The marker study often falls within a controlled clinical trial (protocol) ***(10,11)***. These trials in cancer treatment and research fall into one of five phases:

1. Phase 0: These are animal trials, which can be special-purpose trials to localize and identify specific markers or serve as a preliminary test of a treatment agent.
2. Phase I: Preliminary trials on human subjects designed primarily to establish dose limits and side effects with some effectiveness (tumor reduction) but on a limited number of patients.
3. Phase II: For specific doses of the treatment or treatment combination, to determine effectiveness of the treatment.

4. Phase III: To compare two or more treatment modalities to determine which is superior for a particular class of patient and disease.
5. Phase IV: The use of the treatment for the designed purpose in the general medical community.

This list is intended to be a general guideline. More and more, phases I and II are being combined into phase I/II to attempt to perform safety and fundamental effectiveness in one trial followed by a phase III trial for effectiveness in combination with other chemotherapies and other treatment modalities. More and more animal trials listed here as Phase 0 trials are being performed as Phase I through Phase III trials with animal subjects ***(12,13)***.

3.3. Marker Hypotheses

The hypotheses concerning markers also fall into groups:

1. The marker is equivalent to or opposite to another binary attribute. That is, both attributes are expressed under the same conditions, or one is expressed while the other is not expressed (i.e., lost).
2. The marker is more sensitive than the known binary attribute. Whereas the hypothesis is easy to state, defining sensitivity in this case and then developing a test is much more difficult.
3. The marker is a marker for a given stage of the disease. Another way to state the hypothesis is that the marker is prognostic for a given stage of the disease. As the cells undergo transformation to cancer cells, markers may be expressed differently and thus can be used to mark a stage of disease progression. In particular, early-stage markers indicating premalignant change or sensitive assays suitable for screening are particularly desirable. The problem here is that the marker may well be more sensitive or specific for early-stage disease than any currently known method, making testing very difficult.
4. The marker or its nonexpression is predictive for early relapse or other time to event (survival, time to metastases). This can either be the presence of the marker at diagnosis or change or presence of the marker at some time after diagnosis and initial treatment, such as an early predictor of eventual relapse, earlier than any other predictor. The first condition is simpler to analyze, whereas the latter requires either the development of a response model longitudinally or time-varying covariates in a prognostic factor model.
5. The marker is measured as a continuous or near-continuous variable, such as a proportion of cells expressing the marker out of a large number of total cells (for example, the number of cells with positive PCR in a bone marrow biopsy or aspirate in **ref.** *7*).

Once the hypotheses have been determined, they should be stratified into major and minor hypotheses. Major hypotheses are those that must be decided for the study to be a success, and thus drive the study design and the sample-size determination. Minor hypotheses are those that may be interesting and that may bear on future studies but are of less value.

Each hypothesis should consist of a null hypothesis, H_0, and one or more alternative hypotheses, H_A. The null hypothesis should be fully specified. For example, to compare a binary variable proportion A to another B, a typical null hypothesis would be equality (H_0: $A = B$), and the alternatives might be that A was less than B (H_{A1}: $A < B$), or greater (H_{A2}: $A > B$), or both (H_A: $A \neq B$). The hypotheses H_{A1} and H_{A2} are called one-sided alternatives, whereas H_A is a two-sided alternative. The null hypothesis should be simple, with all facets of the analysis defined. Generally the alternatives are compound as the ones above where the amount of difference between A and B is not specified, just that they are not equal. If the difference is great between A and B, there is a good chance to see a difference when a difference exists, and few subjects will be required to see difference. If the difference is small, then there is a small chance of seeing the difference, and many subjects will be required. In the next section, we describe the statistical tests available, and in the section on sample-size determination, we will again address the relationship between sample size, difference, and chance to see the difference, called power.

4. Statistical Analyses

The methods of analysis follow the hypotheses to be tested. In many cases there are alternative methods of testing, or the testing of a given hypothesis is complex and consists of several steps.

4.1. Binary Comparisons

In this case we have two attributes, call them A and B, each with two outcomes, call them expressed (E) and nonexpressed (NE). An example would be the comparison of a blast-colony assay (BCA) for detecting residual childhood leukemia vs a reverse transcription-polymerase chain reaction (RT-PCR) amplification of leukemia-specific rearrangements ***(7)***. Each subject would have both assays performed. The data can be summarized in a crosstabulation (contingency table) as in **Table 1**. In **Table 1**, a is the number of subjects that express both A and B; b is the number that express B but not A; c the number that express A but not B; and d the number that do not express either A or B. The total number of subjects in the study is n. The marginal totals are $R_1 = a + b$, $R_2 = c + d$ for total expressed and not expressed for B, respectively. The marginal totals of A are $C_1 = a + c$ and $C_2 = b + d$, respectively.

4.1.1. Binary Equivalence Studies

In this structure, there is a known or standard test with a binary outcome, A, which is the standard test or outcome (i.e., biopsy, BCA, radiological diagnosis) to which the marker test, B, is to be compared. Two statistics are used to describe this equivalence ***(2)***: the sensitivity, which is the number called posi-

Table 1
Crosstabulation of Binary Outcomes by Summing Similar Outcomes[a]

	Attribute A		
Attribute B	E	NE	Total
E	a	b	$a + b$
NE	c	d	$c + d$
Total	$a + c$	$b + d$	n

[a] *See* text for abbreviations.

tive by the marker B, as a proportion of the total actually positive by the standard A, and the specificity. In the notation of **Table 1**, with expressed being "positive," the sensitivity is expressed as $s = a/(a + c)$ and the specificity, the proportion called negative by the marker that are actually negative, as $f = d/(b + d)$. Sensitivity is also called the true-positive fraction (TPF), and the specificity the true-negative fraction (TNF). The other two proportions using the marginal totals of A are the false-negative fraction, FNF = $c/(a + c)$, and the false-positive fraction, FPF = $b/(b + d)$. If A and B are equivalent, s and f will approach 1.0 while FNF and FPF approach 0.0. Several statistical tests are used to confirm equivalence:

1. Independence: Use a standard chi-square analysis of the table to test independence of A and B

$$\chi_{\mathrm{I}}^2 = \frac{n(ad - bc)^2}{C_1 C_2 R_1 R_2} \tag{1}$$

which is distributed with a chi-square distribution with one degree of freedom. Thus, to test independence with a significance of 0.05, if $\chi_{\mathrm{I}}^2 \leq 3.841$, the chi-square critical value for significance 0.05 and one degree of freedom *(14)*, then accept the null hypothesis that A and B are independent of each other and thus are not related, much less equivalent. If $\chi_{\mathrm{I}}^2 > 3.841$, then A and B are not independent and are thus related to each other. For χ_{I}^2 to be large, then either the main diagonal product ad or the crossdiagonal product bc must be large in magnitude relative to the other product. If $\chi_{\mathrm{I}}^2 > 3.841$ and $ad > bc$, then we can say that A and B are positively related. If $\chi_{\mathrm{I}}^2 > 3.841$ and $ad < bc$, then we can say that A and B are inversely (or negatively) related. The degree to which they are related is in terms of the magnitude of the chi-square, but is usually expressed in terms of sensitivity and specificity.
2. Sensitivity and specificity: Often a minimum value for sensitivity and specificity to be sufficiently large so that A and B may be considered equivalent is given or customary in a scientific area. Often agencies, such as the FDA, will specify

minimums that are customary ***(4)***. If these are given, then the researcher need only compare to these known values. The researcher should use at least a chi-square test to verify that the hypothesis H_0:$s \geq s_{min}$ (i.e., that the specificity is at least as great as the minimum, s_{min}, desired) vs the alternative H_A:$s < s_{min}$ that the sensitivity is less than the minimum required:

$$\chi_s^2 = \frac{(a-\hat{a})^2}{\hat{a}} + \frac{(c-\hat{c})^2}{\hat{c}} \tag{2}$$

where

$$\hat{a} = (a+c)s_{min} \tag{3}$$

$$\hat{c} = (a+c)(1-s_{min}) \tag{4}$$

and χ_s^2 is chi-square with one degree of freedom. However, if $a > \hat{a}$, accept H_0:$s \geq s_{min}$; otherwise, if $\chi_s^2 > 2.706$, the chi-square critical value for 0.10, reject H_0:$s \geq s_{min}$ at significance 0.05 and accept that the sensitivity is below minimum. This test can be performed using the normal distribution approximation ***(14)*** to the binomial by using the *t*-test

$$t_s = \frac{s - s_{min}}{\sqrt{\frac{s_{min}(1-s_{min})}{a+c}}} \tag{5}$$

and if $t_s < -t_{0.05}(a+c)$, accept that the sensitivity is below minimum, where $t_{0.05}(a+c)$ is the critical value for the *t*-distribution at significance 0.05 and $a + c$ degrees of freedom. The one restriction to using either chi-square or *t*-test is that min $(\hat{a},\hat{c}) \geq 5$. If not, then an exact binomial must be calculated or tables must be used ***(15)***.

Specificity can be tested in a similar way by substituting (f, f_{min}, d, b) for (s, s_{min}, a, c) in the equations above. If both sensitivity and specificity are tested, it is best to correct for multiple testing by using the significance level

$$\alpha_{test} = 1 - \sqrt{(1-\alpha_{final})} \tag{6}$$

where α_{final} is the desired final significance (e.g., 0.05) and α_{test} is the level at which the individual tests are performed (i.e., ~0.025).

3. Tests of relationship: There are a whole family of tests of the strength of the relationship between A and B, generally called concordance ***(15–19)***. A relatively easy to use standard test is Cohen's kappa test, which calculates a statistic designated by the Greek κ, and calculated from **Table 1** as

$$\kappa = \frac{\theta_1 - \theta_2}{1 - \theta_2} \tag{7}$$

Table 2
Crosstabulation of Binary Outcomes Obtained from Table 1 by Dividing by the Number of Subjects[a]

	Attribute A		
Attribute B	E	NE	Total
E	p_{11}	p_{12}	$p_{1\cdot}$
NE	p_{21}	p_{22}	$p_{2\cdot}$
Total	$p_{\cdot 1}$	$p_{\cdot 2}$	1

[a] *See* text for abbreviations.

where

$$\theta_1 = \frac{1}{n}(a+d) \tag{8}$$

$$\theta_2 = \frac{1}{n^2}\{(a+c)(a+b)+(b+d)(c+d)\} \tag{9}$$

The quantity θ_1 is the sum of the main diagonal probabilities and θ_2 is the sum of the estimates of the diagonal probabilities. Kappa has a maximum of 1 when the off-diagonal elements are 0, and is 0 when the attributes are independent, and, in this way, is similar to the correlation coefficient of regression analysis. Kappa is somewhat easier to calculate and describe when we form a new table, **Table 2**, from **Table 1** by dividing each element by n and forming a table of estimated probabilities (i.e., $p_{11} = a/n$; $p_{\cdot 1} = (a + c)/n$) ***(17)***. In the notation of **Table 2**, $\theta_1 = \Sigma_i p_{ii}$ and $\theta_2 = \Sigma_i p_{i\cdot}\, p_{\cdot i}$. The asymptotic variance of κ is

$$\sigma_\kappa^2 = \frac{1}{n}\left\{\frac{\theta_1(1-\theta_1)}{(1-\theta_2)^2} + \frac{2(1-\theta_1)(2\theta_1\theta_2-\theta_3)}{(1-\theta_2)^3} + \frac{(1-\theta_1)^2(\theta_4-4\theta_2^2)}{(1-\theta_2)^4}\right\} \tag{10}$$

where

$$\theta_3 = \sum_i p_{ii}(p_{i\cdot} + p_{\cdot i}) \tag{11}$$

$$\theta_4 = \sum_i \sum_j p_{ij}(p_{j\cdot} + p_{\cdot i})^2 \tag{12}$$

so that we can test hypotheses such as $H_0{:}\kappa \geq \kappa_0$, where $\kappa_0 = 0.9$, say, by using a one-sided t-test. To test the above hypothesis, test if

$$t = \frac{(\kappa - \kappa_0)}{\sigma_\kappa} < -t_{0.05}(n) \tag{13}$$

where $t_{0.05}(n)$ is the critical value for the t-distribution with n degrees of freedom at significance level, as before. If the test is true, then κ is not at least κ_0, and the concordance is not as great as κ_0.

4. Tests of overcalls and undercalls: If the marker test (B) is determining more positives (E) than the standard test (A), then B is said to be overcalling A. Overcalling can be seen in **Table 1** in that the specificity of the test is low and the value of *b* is large. If fewer positives are determined by B than A, then the specificity is low and the value of *c* is large. The excess of overcalls to undercalls can be tested by testing $H_0{:}s = f$ and testing the equality of the sensitivity and specificity using a two-sample *t*-test or by using the McNemar test comparing *b* and *c* ***(14)***. First calculate the average misses $\hat{m} = (b + c)/2$ and then the chi-square directly:

$$\chi_m^2 = \frac{(b-\hat{m})^2}{\hat{m}} + \frac{(c-\hat{m})^2}{\hat{m}} \tag{14}$$

or use the calculation formula:

$$\chi_m^2 = \frac{(b-c)^2}{b+c} \tag{15}$$

which is chi-square with one degree of freedom, so that if $\chi_m^2 > 3.841$, then there are overcalls ($b > c$) or undercalls ($b < c$) at the 0.05 significance level.
5. Lod scores: Lod scores are used to test the relative frequency of a marker vs a standard percent. If the marker were due to Mendelian inheritance, we might expect that it would occur at the rate of 0.50 ***(20)***. If we know that the marker has a rate of, say, η in normal tissue, and we have examined *n* subjects and found *r* with the marker for a rate of $\hat{\theta} = r/n$, then the lod score is the logarithm (base 10) of the likelihood ratio:

$$\mathrm{lod}(\theta) = \log_{10}\left(\frac{\theta^r(1-\theta)^{n-r}}{\eta^r(1-\eta)^{n-r}}\right) \tag{16}$$

The lod score is often presented at a vector of values for θ as well as $\hat{\theta}$, and it is customary to consider a lod score greater than 3 to be significant ***(21)***. Since twice the logarithm (base *e*) of the likelihood is approximately chi-square with one degree of freedom, 3 is the equivalent of a chi-square of 13.8 ($p < 0.001$). A lod score equal to 0.834 is equivalent to a significance of $p = 0.05$. Thus, if the lod score exceeds the critical point, we would say that there is a significant difference from normal subjects.

4.1.2. Binary Improvement Studies

We would expect that as we develop more specific markers to a particular condition, rather than comparing our marker to see if it is as good as the "standard" marker available, we should be comparing it to see if the new marker is superior to the standard. This is easy to state, but the methods and experimental design considerations are complicated. In this section, we will set up a series of possible relationships that the standard has with the disease condition.

1. The condition is binary: Both the standard and the new marker are predicting the condition (i.e., cancer, relapse after remission). The trial will require that a determination that the condition occurs be made. This requires that time be allowed for the condition to develop or requires additional testing to verify the status of the subject as well as both standard and new marker status determined for each subject. Thus, each subject will have three determinations, the "true" subject condition, the standard status, and the new-marker status. The data may be analyzed using either log-linear models ***(22,23)*** or logistic regression, which is more common ***(24)***. In logistic regression, the standard and the new marker are used to predict the true condition:

$$\ln\left(\frac{\pi}{1-\pi}\right) = \beta_0 + \beta_1 A + \beta_2 B \tag{17}$$

where A is 1 if the standard is expressed and 0 otherwise, and B is 1 if the marker is expressed and 0 otherwise. The estimate π is the estimate of the true condition, which is compared to the true condition using log-likelihood. This is most easily done using a standard computer package discussed later. Briefly, the logistic analysis with a forward selection procedure will estimate π with neither A nor B included to establish a baseline log-likelihood, then attempt to include A or B singly as an improvement over baseline and then, once the more significant relationship of the two is included, both are included. In this way, we can see if either fit the true condition, and then, if each does, whether both are necessary to fit the true condition. An exploratory analysis examining the subjects missed by each and both is often helpful. A warning: when there is complete agreement with the true condition, the method is degenerate. If this occurs, check the tables, which is a good idea in any event.
2. The condition is time varying: An example of this is relapse in cancer therapy. In addition to predicting relapse accurately, the measure to predict relapse first is much more desirable in that a second course of therapy could begin sooner, perhaps with better results. For each subject, the time at which the true condition, standard, and new marker become expressed is recorded. At the time of analysis, each subject has three times (x,a,b, respectively). For each time there is a status variable indicating if the condition is expressed or not. Custom calls for recording that the condition has not yet been expressed with a "+" after the last time recorded for the variable. For the analysis, order each triple of times from smallest to largest. Because all three are on the same subject, the following combinations of status variables and codes are possible:
   a. All condition times are known;
   b. Two are known, one is unknown;
   c. One is known, two are unknown; or
   d. All three are unknown.

   The unknown times have the maximum time of the three; therefore, we can use a variant of Friedman's test. If any times are the same, average the orders of the identical times. If an unknown time is the same as a known time, the unknown

Table 3
Friedman's Method Applied to Comparison of Times to Expressed Markers

	Condition order		
Subject	New marker	Standard	True
1	0_{11}	0_{12}	0_{13}
2	0_{21}	0_{22}	0_{23}
—	—	—	—
n	0_{n1}	0_{n2}	0_{n3}
Rank sum	R_1	R_2	R_3

time has the higher order. Sum the orders over the subjects, as shown in **Table 3**. Apply Friedman's test ***(14)***:

$$\chi_F^2 = \frac{1}{n}\sum_{i=1}^{3} R_i^2 - 12n \tag{18}$$

which is chi-square with 3 – 1 = 2 degrees of freedom. If $\chi_F^2 > 5.991$, then the three conditions have different times at a significance of 0.05. To compare the standard and new marker, order only those two columns and calculate:

$$\chi_{AB}^2 = \frac{2(R_1^2 + R_2^2)}{n} - 9n \tag{19}$$

which is chi-square with one degree of freedom. If $\chi_{AB}^2 > 3.841$, then the standard and new marker have different times.

The method above does not include the differences in the times until the conditions are expressed. This analysis involves the analysis of prognostic factors with time-varying factors (*see* **Subheading 4.4.** for more information).

4.2. The Standard Has More Than Two Levels

In this section, we will consider the case in which the standard has *k* levels, such as a differential diagnosis. In this case, the data can be tabulated as in **Table 4**. The diagnosis can be either a differential diagnosis without strict order (i.e., nonproliferative breast tissue, fibroadenoma, ductal hyperplasia, atypical ductal hyperplasia, ductal carcinoma *in situ* [DCIS] invasive, DCIS only, carcinoma [CA] only) or in strict order (i.e., bladder cancer grades: normal, dysplasia grade 1, dysplasia grade 2, dysplasia grade 3, cancer *in situ*, invasive cancer).

4.2.1. Analysis Without Regard to Ordering

Two analyses are commonly used that are successful regardless of whether the diagnostic groups are ordered or not:

1. Chi-square tests of independence: For both ordered and nonordered diagnostic groups, a chi-square analysis can be performed to determine if any relationship is significant, as in **Table 4**:

Table 4
Comparison of a Marker with a Standard with *k* Levels

	Marker		
Standard	Expressed	Not expressed	Row total
Normal	f_{11}	f_{12}	$f_{1\cdot}$
D_1	f_{21}	f_{22}	$f_{2\cdot}$
—	—	—	—
D_{k-1}	f_{k1}	f_{k2}	$f_{k\cdot}$
Column total	$f_{\cdot 1}$	$f_{\cdot 2}$	n

$$\chi^2_{k\times 2} = \sum_{i=1}^{k}\sum_{j=1}^{2}\frac{(f_{ij}-\hat{f}_{ij})^2}{\hat{f}_{ij}} \quad \text{where } \hat{f}_{ij} = \frac{f_{i\cdot}f_{\cdot j}}{n} \tag{20}$$

which is chi-square with $k - 1$ degrees of freedom ***(14)***. If $\chi^2_{k\times 2} > \chi^2_{\alpha}(k-1)$, the chi-square critical value for significance a and $k - 1$ degrees of freedom, accept that there is some relationship between the marker and the diagnoses. Use subhypotheses and an examination of the individual chi-square terms to determine which diagnoses are more closely associated with the marker expression.

2. Lod scores: If we have an estimate of the marker expression frequency on normal tissue, η, we can use the lod score analysis of **Subheading 4.1.1.** ***(5)*** to test each diagnosis

$$\mathrm{lod}(\theta_i) = \log_{10}\left(\frac{\theta_i^{f_{i1}}(1-\theta_i)^{f_{i2}}}{\eta^{f_{i1}}(1-\eta)^{f_{i2}}}\right) \tag{21}$$

where θ_i is usually the maximum likelihood estimate of the probability of expression $f_{i1}/f_{i\cdot}$. The estimate of η can be obtained from the first row of **Table 4**. If the normal diagnosis is performed on tissue adjacent to the tumor, for example, this tissue may already reflect early transformation, which is expressed by the marker making row one inappropriate to estimate η. This may require that tissue samples either from noninvolved distant tissue be used or independent subjects be used to estimate η.

4.2.2. Analysis with an Ordered Diagnosis

If the diagnostic categories are ordered (not necessarily linearly or perfectly), there are analyses that can be performed in addition to those in the previous section. **Table 5** gives an example with 5 diagnostic levels, 10 subjects in each. Chi-square analysis yields a chi-square of 28 with 5 degrees of freedom which is highly significant ($p < 0.001$). All the expected frequencies are 5.0, providing the greatest deviations for the estimated quantities for levels 1 and 6 with partial chi-squares of 5 for each element of the table in those two rows. The differences between actual and estimated show the consistent shift from 0%

Table 5
Example of Uniform Change Through Five Levels

	Marker		
Standard	Expressed	Not expressed	Row total
1	0	10	10
2	2	8	10
3	4	6	10
4	6	4	10
5	8	2	10
6	10	0	10
Column total	30	30	60

expressed at level 1 to 100% expressed at level 6. Using a normal estimate of 0.05 and adjusting 0 counts to $0.5/f_{i\cdot} = 0.05$ in this example, to avoid indefinite values, the lod scores are 0.0, 0.6, 2.4, 5.0, 8.3, and 12.8, respectively for the six levels. Using a lod score of 3.0 as significant, levels 4–6 are significant individually. For the ordered analyses, form the five 2-by-2 tables (levels 1–5) shown in **Table 6** by dividing the 6-by-2 **Table 5** between each level and summing columns above and below the cutoff level.

1. Lod scores: Lod scores are calculated for the five cutoffs or higher providing scores of 24.9, 26.1, 24.8, 20.6, and 12.8, respectively, showing that the marker is a marker for the disease from cutoff 1.
2. Receiver operating characteristic curves (ROC): We could calculate the significance of all five cutoffs by calculating chi-squares for all five 2-by-2 tables (levels 1–5). A way to combine all five into a visual pattern in a single analysis is to use ROC analysis ***(25–27)***. Customarily, the plots are the true-positive fraction vs false-positive fraction for each cutoff. However, if we plot the false-negative fraction (FNF) vs true-negative fraction (TNF) assuming the marker to be the true value, then the cutoffs plot in increasing order from left to right. We use the marker as "truth" since the error in the marker analysis is considerably smaller than a pathologist's determination of a slide. In **Table 6**, take the first row of data for each level and divide by the column totals to get **Table 7**. The data in **Table 7** is plotted with (0.00, 0.00) appended before the data and (1.00, 1.00) after. These points correspond to the decisions that all subjects express and all subjects do not express the marker, respectively. Inputting the (FNF, TNF) pairs into the ROCFIT program of Metz ***(26)***, we can calculate an approximate area under the ROC curve and area standard deviation and compare the area to a known area (i.e., area if no relationship =0.5). **Figure 1** plots the data from **Table 7**, along with the so-called "guess" line of no relationship of area 0.5.

Table 6
Example of Uniform Change Through Five Levels as in Table 5 with Cutoff after Levels 1–5

Standard	Marker expressed	Marker not expressed	Row total
Level 1			
1	0	10	10
2–6	30	20	50
Column total	30	30	60
Level 2			
1–2	2	18	20
3–6	28	12	40
Column total	30	30	60
Level 3			
1–3	6	24	30
4–6	24	6	30
Column total	30	30	60
Level 4			
1–4	12	28	40
5–6	18	2	20
Column total	30	30	60
Level 5			
1–5	20	30	50
6	10	0	10
Column total	30	30	60

Table 7
Table of False-Negative Fraction vs True-Negative Fraction of Table 6

Cutoff	FNF	TNF
1	0.00	0.33
2	0.07	0.60
3	0.20	0.80
4	0.40	0.93
5	0.67	1.00

4.3. Analysis of Multiple Markers on the Same Subjects

Often a battery of related markers (i.e., p53, MTS, microsatellites) are applied to the same subjects and compared to diagnosis, grade, ploidy, and the other markers. **Table 8** shows a typical layout, which includes grade (i.e., 1, 2, 3),

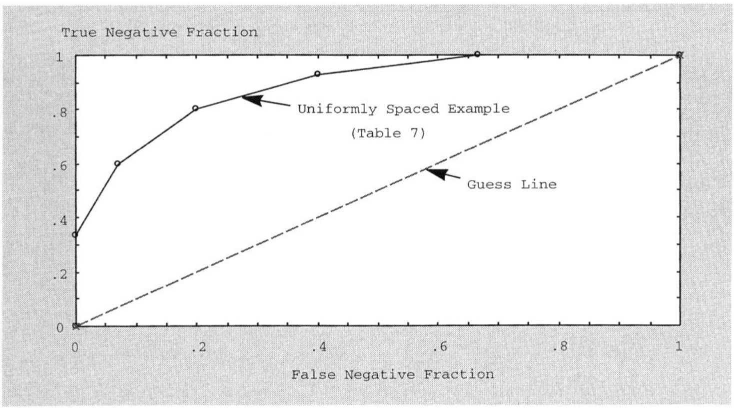

Fig. 1. ROC plot of the evenly spaced data shown in **Table 5**. The guess line is also plotted.

Table 8
Layout of Multivariable Marker Data

#	Gr	DNA	p53 EX5	—	EX9	MTS cosmid 1063.7	—	C1.B	Ki-67
1	g_1	d_1	x_{11}	—	x_{1p}	x_{1p+1}	—	x_{1p+m}	x_{1p+m+1}
2	g_2	d_2	x_{21}	—	x_{2p}	x_{2p+1}	—	x_{2p+m}	x_{2p+m+1}
—	—	—	—	—	—	—	—	—	—
n	g_n	d_n	x_{n1}	—	x_{np}	x_{np+1}	—	x_{np+m}	x_{np+m+1}

ploidy (i.e., diploid and aneuploid), p53 (i.e., exons 5–9), MTS (i.e., 1063.7–C1.B), and a continuous variable, the percent of cells expressing Ki-67. Each marker can be compared individually to the diagnosis (grade) using the methods described in **Subheading 4.2.** To compare the markers and develop a model of interaction in predicting the grade, a multivariate analysis must be performed.

4.3.1. Binary Diagnosis

If we wished to predict grade 3 versus grades 1 and 2, the reduced grade would be binary. Then we could use an extension of the logistic regression presented in **Subheading 4.1.2.** ***(1)***:

$$\ln\left(\frac{\pi_j}{1-\pi_j}\right) = \beta_0 + \delta d_j + \sum_{i=1}^{p+m+1} \beta_i x_{ji} \tag{22}$$

where π_j is the predictor of the jth subject grade, $j = 1,\ldots, n$. This model contains markers, continuous variables, and the other potential dependent variable, ploidy. The model can be specified as desired to test the significance of all or part of the markers and other variables obtained from the subject. The model can be constructed in a stepwise fashion, adding variables automatically one term at a time and testing the significance of the remaining variables to the model after entering the current model. In this way, a parsimonious model to predict grade or any other binary dependent variable can be constructed.

4.3.2. Multilevel Diagnosis

In this case, the diagnosis or grade of the disease is multinomial. In the example in **Subheading 4.3.1.**, if we wish to use all three grades to compare to the markers or if we have several nonordered diagnoses, such as the six diagnoses for breast cancer in **Subheading 4.2.**, we have a multilevel diagnosis. Because it is not binary, we use the more general method of log-linear analysis *(22,23)*. In this method an analysis of variance, such as the factorial model, is created using the log of counts, $n_{i_1 i_2 \mathrm{L}\, i_k}$, of the contingency table created by tabulating the levels of the k factors associated. In **Table 8**, there are the grade, the DNA, the p p53 factors, and the m MTS factors, as well as the Ki-67 continuous factor. With just the second-order interactions, this gives a factorial model:

$$\begin{aligned}\log(n_{i_1 \mathrm{K}\, i_k}) = \mu + \lambda_{i_1}^{g} + \lambda_{i_2}^{d} + \sum \lambda_{i_p}^{x_p} + \sum \lambda_{i_m}^{x_m} \\ +\sum\left(\lambda_{i_1}^{g} + \lambda_{i_2}^{d}\right)\lambda_{i_p}^{x_p} + \sum\left(\lambda_{i_1}^{g} + \lambda_{i_2}^{d}\right)\lambda_{i_m}^{x_m} \\ +\sum\sum \lambda_{i_p}^{x_p}\lambda_{i_m}^{x_m}\end{aligned} \qquad (23)$$

The model is more stable if built term by term, including the grade and/or ploidy or other dependent factors first and adding the markers one at a time or a group at a time rather than the entire model. The number of statistical cells into which a count is summed increases geometrically with added markers.

4.3.3. Dimensionality Reduction

When analyzing multidimensional tables, such as those of the previous sections, it is important to keep the expected number of subjects in each statistical cell of the table high enough to satisfy the chi-square rule of thumb that no more than 20–25% of the expected number of subjects be below 5 and none below 1. This rule of thumb is true for both chi-square and log-linear models. To do this, categories that have small expected frequencies should be combined, such as the combination of grade 1 and 2 into one grade to compare to grade 3 in **Subheading 4.3.1.**, which allows a logistic model. The logistic model is simpler, and there are better software tools for analysis than for general log-linear models.

4.4. Time to a Critical Event

Often the marker is not predicting another marker so much as a future event (i.e., cancer induction, remission, relapse, death). If just predicting the event itself, the methods of the previous parts of **Subheading 4.** will suffice. As often as predicting the occurrence of the event is the prediction of the time until the event. This requires survival analysis methods ***(11)***. In these methods, the time from some known point (start of study, diagnosis, surgery or other treatment) is to be estimated by the marker(s) along with other factors, such as grade of disease, age, gender, type of treatment (induction or therapy), other markers, and so forth. Standard parametric methods (t-tests, regression, analysis of variance) cannot be used because times are not normally (Gaussian) distributed. Also, nonparametric methods cannot be used since some subjects may not have completed the study (lost to follow-up: sacrifice or treatment toxicity in animal studies, death owing to other causes not related to the trial, failure to return for follow-up evaluation), or the study evaluation may have been performed before all subjects reached the critical event (withdrawn alive). To analyze the study without these subjects biases the result. For example, a very good treatment may have many subjects alive or free of disease at the end of the study, and to ignore them would reduce the median time to death (relapse), biasing the result.

The date of the last contact and status (censored or not censored, alive or dead, continued remission or relapse, and so forth) along with the values of the markers and other potential prognostic factors are needed for each subject. These are used to estimate the probability of surviving from the start of the study to the time of last follow-up (called the survivorship function) and is defined by the integral

$$\mathcal{F}(t) = \int_t^\infty f(u)du = 1 - F(t) \tag{24}$$

where f is the probability density function and $F(t)$ is the cumulative distribution function ***(11)***. Two methods are used to estimate $\mathcal{F}(t)$. For larger sample sizes, the life-table or Berkson–Gage method is used. For smaller sample sizes, a limiting form, called the Kaplan–Meier method, is used.

4.4.1. Comparing Survival Curves—Discrete Variables

If there is only one discrete variable (like a marker), or a discrete variable can be made from a continuous variable using a cutoff point, say, the survivorship functions calculated for each level of the variable can be compared directly using generalizations of the Mann–Whitney/Wilcoxon rank test (Gehan's test, Peto and Peto's logrank test, Peto and Peto's generalized Wilcoxon test) for two levels, and the Kruskal–Wallis test (Lee–Desu generalization in **ref. *28***) for k levels. Cox's F test and others are also used but assume the survivorship distribution to be similar to the exponential or Weibull distribution.

4.4.2. Comparing Survival Curves—the Regression Approach

Cox ***(11)*** developed a method to estimate the influence of potential prognostic factors on the survivorship function by estimating $h(t) = f(t)/\mathcal{F}(t)$, the hazard function (instantaneous failure rate, force of mortality), using the regression equation

$$h(t|\mathbf{x}) = h_0(t)\exp\left(\sum \beta_i x_i\right) \quad (25)$$

where the x_i's are the prognostic factors and $h_0(t)$ is the null hazard function. The model is usually written as

$$\ln\left(\frac{h(t|\mathbf{x})}{h_0(t)}\right) = \sum \beta_i x_i \quad (26)$$

which looks and is analyzed in a manner similar to log-linear models, substituting the hazard function for the odds ratio. The procedure to determine which potential prognostic factors are necessary to predict the hazard rate is referred to as proportional hazard regression, prognostic factor analysis, or Cox model regression.

4.5. Continuous Markers

Whereas the marker itself is a binary response of a biochemical probe to a cell in that it either interacts or does not interact, often the measurement is over a large number of cells so that the actual data is either a percentage or proportion of "marked" cells or a measurement proportional to the proportion (integrated optical density or intensity of stain or ng/ml concentration in the tissue). Practically, this data is continuous.

4.5.1. Calibration

When developing a continuous marker test, the first step often is the calibration of the continuous measure to the actual proportion of cells. A limiting dilution assay is most often used for this process. This is discussed in the classic paper by Taswell ***(29)***. Briefly, a known number (amount) of reactive cells (material) is diluted by known amounts of nonreactive material through the orders of magnitude that a typical unknown sample will contain. These known dilutions are subjected to the assay in the same manner in which an unknown sample would be processed. The result is a set of pairs (x_i, y_{ij}); $i = 1,\ldots, k$; $j = 1,\ldots, n$ where the x_i's are the k known concentrations of reactive material and the y_{ij}'s are the result of the assay with n replicates at each concentration. The

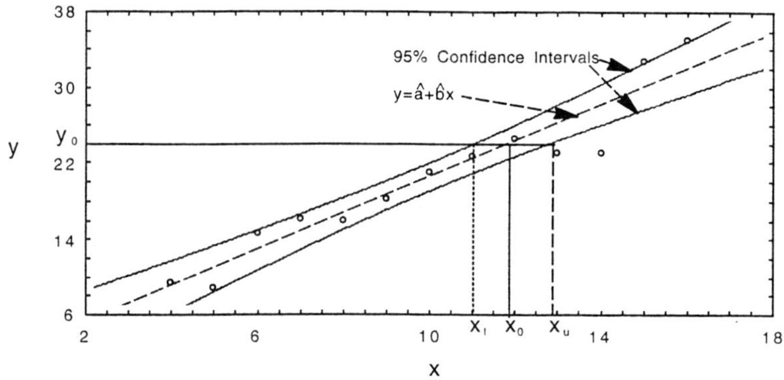

Fig. 2. Example of linear model fit $y = a + bx$ to sample data with the 95% confidence hyperbolae plotted. The inverse prediction given y_0 is shown as the vertical projection x_0 from the fitted line. The upper 95% fiducial estimate x_u is estimated by vertically projecting the intersection of y_0 and the rightmost 95% confidence hyperbola. The lower limit x_l is obtained projecting the leftmost hyperbola.

functional relationship between x_i and y_{ij} is to be expected to be linear $y = a + bx$ or log-linear $\log_{10}(y) = a + bx$ because of the dilutions in x_i, the proportional relationship of the assay, and the short-term culture often employed in the assay to increase the magnitude of y_{ij} to a measurable level. Since the models are linear in the coefficients (a and b), the coefficients can be estimated using regression methods or alternatives, such as those suggested by Taswell ***(29)***. Because regression methods are more readily available, they are generally used; *see* **Fig. 2** for an example constructed using Statistica (StatSoft, Tulsa, OK). Call the estimates $\hat{a}$ and $\hat{b}$. The model is applied by taking a sample of unknown concentration and using the assay to calculate the result y_0. The model is used by inverting the equation and calculating $x_0 = (y_0 - \hat{a})/\hat{b}$, which is a point estimate of the proportion of marked cells in the sample.

A problem arises when we need the distribution of x_0 or need to calculate confidence limits about x_0. Whereas the distribution of $\hat{a}$ and $\hat{b}$ are asymptotically Gaussian, the distribution of x_0 is Cauchy with a principal value for a mean and no variance to use to calculate an approximate Gaussian confidence interval. It has been customary to calculate an approximate variance by the method of moments, which is not a valid method here since there is no finite variance. Another approach has been simply to use $x = a + by$ or $x = a + b\log_{10}(y)$ and calculate the regression of x on y. Neither are appropriate models since they do not provide the same estimates as the other models, the x_i's are fixed known dilutions that are error-free relative to the error of the assay, and the y_{ij}'s are replications at the x_i's. Fieller ***(30)*** suggested that the confidence interval could be calculated using the confidence interval hyperbolae calculated on the linear model to estimate the confidence on x_0. For the example of **Fig. 2**, (x_l,x_u) is the 95% confidence interval, assuming y_0 is known and has no associated error. The interval will be symmetric about x_0 only when $x_0 = \bar{x}$. The formula to calculate (x_l,x_u) when y_0 is calculated is a tolerance interval and is calculated by

$$x_l, x_u = \bar{x} + \frac{\hat{b}(y_o - \bar{y})}{K} \pm \frac{t}{K}\sqrt{s_{y.x}^2\left[\frac{(y_0 - \bar{y})^2}{\sum(x_i - \bar{x})^2} + K(1 + 1/n)\right]} \quad (27)$$

where $K = \hat{b}^2 - t^2 s_b^2$, $s_{y.x}^2$ is the residual error, s_b^2 is the variance of $\hat{b}$, and t is the t-distribution confidence limit (two-sided) using the error degrees of freedom ***(14)***.

If the relationship between x and y is not linear or log-linear, two approaches are possible. The first is to isolate a portion of the data that lies in a linear or log-linear region, restrict the data analysis to just this region, and calculate a linear or log-linear regression model as above. This restriction works well when only a few data points must be lost in the analysis and where the slope of the relationship is relatively flat. Where the slope is steep or just a few data points remain after the restriction, nonlinear regression is necessary. This has been seen in analyses using PCR ***(7)***. In this analysis, only a few dilutions remained between maximum response and below measurable levels. Nonlinear models using all the data from the toe (below measurable levels) through the log-linear body of the function to the shoulder (maximum response) were applied to the data with good success. The difficulty is that the inverse calibration had to be simulated to obtain the inverse confidence limits.

5. Sample-Size Determination

Each statistical method discussed in **Subheading 4.** has a sample-size determination associated with the statistical method. There are general concepts that are applicable to all methods. At the conclusion of the experiment, we must decide if the null hypothesis is true or the alternative hypothesis is true when either the null hypothesis is really true or the alternative is really true. These four choices are depicted in **Table 9**. Two errors can be made:

1. We can decide that the alternative, H_A, is true when the null hypothesis, H_0, is really true. This is a Type I error, and we measure the size of the error by α, often called the significance of the test.
2. We can decide that H_0 is true when H_A is really true. This is Type II error, and we measure the size of the error by β. The probability of deciding that H_A is true when it really is called the power of the test and is equal to $1 - \beta$. Since H_A is usually a compound hypothesis, β is a compound function of the difference that the specific alternatives are from the null.

Table 9
Errors in Hypothesis Testing

	True state	
Decision	H_0	H_A
H_0	$1 - \alpha$	β
H_A	α	$1 - \beta$

As the sample size increases, the errors of making a decision decrease. The purpose of sample-size determination is to balance increasing sample size with its corresponding increase in cost to run the experiment (monetary, subject costs, and time) with the errors inherent in the experimental process. Before the experiment is begun, the researcher determines the criteria that constitute the major objectives and establishes the hypotheses to be tested, which are critical to the conclusions of the experiment. The Type I and II errors are specified with the Type II error defined for a given minimum difference that the alternative is from the null hypothesis. The presentation of all of the calculations is beyond the scope of this chapter. Further details are found in Mace ***(31)*** and Cohen ***(32)***. The parameters used in calculating the power will be presented as used in STPLAN ***(33)***, a public-domain sample-size program discussed in **Subheading 6.**

5.1. Sample Size with Binary Data

In many cases, the fundamental statistical comparison that the experiment must calculate is the binary marker with the binary standard. Additional comparisons may be made, but this is fundamental to the analysis. The other statistical tests mentioned in this chapter have more power to see difference than the binary test. Usually if the sample size is determined from the binary tests, all the other tests desired will be sufficiently powerful. The tests for comparisons of two-by-two tables that can be used depend upon the final comparisons:

1. Independence: Use the comparison of two proportions (Binomial, Fisher-Exact, matched pairs test). The two proportions are the sensitivity s (TPF) and the false-positive fraction, FPF. The researcher specifies the estimated sensitivity and the maximum FPF to be detected along with the significance and power. The sample size calculated will plan an experiment so that the two proportions s and FPF will be found statistically significantly different ($p \leq \alpha$) at least $(1 - \beta)$ 100% of the time when the true proportions are at least as different as the estimated parameters.
2. Sensitivity or specificity: Use the test of a proportion (one-sample exact binomial) for both sensitivity and specificity. The researcher specifies the hypothesized minimum sensitivity, say 0.95, and then the maximum alternate sensitivity

that must be detected with significance α and power (1 – β), say 0.65. The specificity would be planned the same way, If both are specified, then calculate the sample size for each and use the final significance calculated in **Subheading 4.1.2.** above, with the final number of samples being the higher of the two.

3. Association: Use a one-sample normal variate. The researcher specifies the minimum kappa hypothesized and then the maximum alternate kappa that must be detected with significance α and power (1 – β). Use a standard deviation of 1.0 for the estimate of the standard deviation

5.2. Other Comparisons

In most of the techniques that have been presented there is no simple closed-form method to calculate the sample size. The problem is simulated with a given null and specific alternative hypotheses producing a large number of simulated experiments with the postulated null and alternative conditions and tests performed with the specified significance. The power is calculated for systematically varied sample size to determine the interrelationship between sample size and power so that for a given power the sample size may be estimated and vice versa. Often the more sophisticated analyses contain a simple proportional test of interest, which can be used to provide a lower bound on the sample size needed and which will be sufficient for the purposes of study planning.

6. Computer Programs

There are programs available commercially and as public shareware that will perform most tasks in the analysis of biomarkers. This section lists some of those programs, with apologies for those programs not listed and some concern that this list is too temporal to be of lasting value. Please write or e-mail the author with suggestions for additions or corrections of omissions.

6.1. Data Entry/Management Programs

The most common method to enter and maintain experimental data has been the spreadsheet or small database program. To easily access most statistical systems, it is still better to use to a major spreadsheet rather than a database system since most major statistical systems have data import connections to spreadsheets and not to database systems. This is changing. The three spreadsheet programs with easy entry to statistical programs are:

1. Microsoft Excel (Microsoft, Redmond, WA): Current spreadsheet format that is imported by most major systems is 4.0. If you are using 5.0 or higher, check to see which is supported. It is an easy task to save worksheets in 5.0 as 4.0 worksheets.
2. Lotus 1-2-3 (Lotus, Cambridge, MA): Current spreadsheet format is for Lotus 3.0.
3. Quattro Pro (Novell, Orem, UT): Current spreadsheet format is for version 6.0.

The standard form for a spreadsheet to import into a statistical package is to have the first row provide the variable names for the variables collected on each subject. The variables are recorded in columns with the subjects occupying a row each.

There are many database programs available. The data entry is usually more difficult. For most there is no easy access to statistical systems, and it is necessary that a query be made in the database program to make a spreadsheet, tab-delimited, or comma-separated file for entry into the statistical system.

6.2. Statistical Systems

There are a number of statistical systems that will perform both the chi-square contingency table analysis as well as the log-linear, logistic, and survival analyses mentioned in **Subheading 4.** They expect the data in spreadsheet, tab-delimited, space-separated, or comma-separated files or, in most cases, will permit direct entry of the data into their own spreadsheet. In all cases, they expect the data with subjects on each row. If the data is already in tables, most have either manual table entry or tables functions for entry. All the systems mentioned in this section are general statistical systems:

1. SPSS (SPSS, Inc., Chicago, IL): This system comes in PC Windows, Macintosh, UNIX, and Mainframe versions. It works in both command-driven and pull-down menu versions.
2. Statistica (CSS, Inc., Tulsa, OK): This system comes in PC DOS and Windows as well as Macintosh versions. It has good integrated graphics.
3. SAS (SAS Institute, Inc., Cary, NC): This system comes in PC Windows, Macintosh, UNIX, and Mainframe versions. SAS has been primarily command driven and complicated to use because of its complexity. New interfaces are solving this problem.
4. Minitab (Minitab, Inc., State College, PA): This system is available in PC Windows and Macintosh versions. It operates in command-driven and pull-down-menu modes. Excellent improvements have been made to the graphics. It has easily used simulation facilities.
5. SYSTAT (SPSS, Chicago, IL): This system is available in PC DOS only. It has excellent graphics but a relatively difficult command interface.
6. Sigma Stat/Plot (Jandel, San Rafael, CA): Available in PC DOS only. These are actually two programs linked together. Sigma Stat does not have the breadth of statistics available in other packages.
7. BMDP (SPSS): Until recently, an independent Los Angeles company. The system is available in several versions. The main criticism has been the user interface. Many of the algorithms are the best available.

6.3. Other Programs

Programs available to calculate other statistics mentioned in this chapter, but not necessarily available in commercial program systems, include programs available through the U.T.M.D. Anderson Cancer Center, Department of Biomathematics FTP/Web server at www.odin.mdacc.tmc.edu:

1. CTA: Contingency table analysis program where the user enters the summary table rather than individual records. CTA calculates chi-square, kappa, and McNemar statistics.
2. STPLAN: Calculates a wide variety of sample size and power problems.
3. ROCFIT: A set of programs to do ROC analysis is available from Charles Metz ***(26)***.
4. LOD: Program to calculate a generalized lod score analysis.
5. GOFCHI: A chi-square goodness-of-fit program in which the user provides the actual data frequencies and the test frequencies.

7. Notes

The other chapters of this book provide excellent examples of the need for and use of statistics in the analysis of markers. These notes will use the data structure in several of the chapters to illustrate the methods presented.

1. The result of many of the marker analyses, especially as applied to pathologic tissue or isolated cells with fluorescence *in situ* hybridization (FISH) or conventionally stained markers (*see* Chapters 10, 13–15, and 19), comparative genomic hybridization (CGH) (*see* Chapter 12), and loss of heterozygosity (LOH) (*see* Chapter 17), is a determination for each patient that the marker is present or lost. This leads to a test of association to the disease as determined by other methods. It is important to note that LOH and other marker changes may occur earlier in the progression from normal cell to cancer cell than the other methods can detect the cancer and may be the cause or byproduct of an early transition. This speaks to the need for testing of the general population to develop negative controls and determine the prevalence of the marker in the general population. This will make the calculation of lod scores more accurate. The testing of nonaffected relatives constitutes an additional control but is not a substitute for the negative controls.
2. Once the marker has been established as a potential factor in the development of the cancer, it should be compared with other factors thought or proven to predict the cancer (diagnosis) or to predict the survival, relapse, complete remission, or other event in the progress of the disease (prognostic) (*see* Chapter 7). This is necessary to develop a panel of tests necessary to diagnose disease or prognosticate the potential outcome. For an overview of techniques for the analysis of prognostic factors, *see* Chapter 1. The Kaplan–Meier and the Berkson–Gage methods are methods of presenting time to critical-point analysis (time to relapse, death). As with other types of data, time to critical point may be analyzed by simple nonparametric univariate techniques or multivariate techniques (*see* **Subheading 4.4.**). Classification and regression trees (CART), which is a search technique to develop a hierarchical classification in disease states, for example, is also discussed. The technique usually develops cutoffs for continuous variables which "best" discriminate between the groups (i.e., diagnostic groups) according

to a criteria defined by the user to be "best." While CART may or may not produce the most efficient classification, often the tree describes the process well enough to be an easily understood classification of the data. We have found that CART often conforms to intuition regarding breakpoints in the classification and matches logistic regression and other multivariate techniques in the variables used and overall correct classification. Also discussed briefly in Chapter 1 is the use of artificial neural networks (NN). The user of this technique should be warned that there is a built-in criteria function for the classification of the data that can vary from least squares to logistic regression and, if possible, should be chosen to match the problem being analyzed. Also of note is that the NN contains several levels of parameters, depending on the choice of the user, and may overdetermine the data set if the data set is small. To properly use the NN technique, the user should divide the data set into a training set and a testing set, train the NN on the training set, and then evaluate the results on the testing set.

3. Markers are often interpreted from gels or gel panels (*see* Chapters 6, 16, 19, 22, and 27). Gels may be interpreted simply as a "spot" or "band" being present or absent. They may also be interpreted regarding the quantity of material present above background. This may be estimated by densitometry by calculating the height of the peak response above background or the area under the tracing above background. In some gels, an ellipsoidal- or teardrop-shaped spot may result. In this case the better measure of the amount of material is the integrated volume of the spot above background rather than the area along a line through the spot, as the spot has spread out in width and height along the gel. In Western blots (*see* Chapters 19 and 27) the controls can be used to estimate a smooth model of molecular weight over the length of the gel, which can be applied to the lanes for accurate molecular-weight estimation (see our web site for a version of NIH-Image that has a Western-blot analysis module).
4. ELISA is a common alternative to gels and other methods to quantitate the amount of response to a probe in a sample(s) (*see* Chapters 5, 23, 25, and 26). The 96 well plate permits standards to be run with every plate. The standards are included on each plate to ensure that the correspondence between the concentration and color intensity in the plate is accurate. The ELISA can be considered a dilution assay, as the standards follow a known concentration dilution. Thus, certain considerations are common to ELISA and other dilution assays (*see* Chapters 8–10, 16, and 22). The number of standards must be sufficiently large to permit estimation of the calibration function. Standards are provided as separate concentrations (dilutions) of the known, as well as negative, controls with replications (duplicate, triplicate) to estimate the variability in the assay as seen by the ELISA unit. The number of replications per concentration cannot substitute for the number of concentrations when estimating the calibration model. The number of parameters to be estimated in the calibration function determines the number of separate concentrations (negative standard is one concentration) required by the estimation. For example, if the model is linear, a minimum of three concentrations is required. For a cubic equation (*see* Chapter 25), the mini-

mum number is five, since the model $y = a + bx + cx^2 + dx^3$ estimates four parameters. A standard nonlinear exponential model $y = A(1 - e^{-at}) + b$ has a minimum of four. Michaelis/Menton: $y = (a + bx)/(1 + dx)$ has a minimum of four or three if $a = 0$. Logistic models require a minimum of five: $y = b + \{(a - b)/[1 + (x/c)^d]\}$ or four: $y = a/[1 + (x/b)^c]$ or more, depending on the number of parameters in the model. The standards and replications must be chosen to balance the need for concentrations to fit the model and the need to estimate the variability of the assay.

References

1. Johnston, D. A. (1980) Analysis of clinical trials. *Cancer Bull.* **32,** 216–221.
2. Feinstein, A. R. (1977) *Clinical Biostatistics.* Mosby Co., St. Louis, MO.
3. Wooding, W. M. (1994) *Planning Pharmaceutical Clinical Trials: Basic Statistical Principles.* Wiley-Interscience, New York.
4. Aziz, K. J. and Maxim, P. E. (1993) The FDA's perspective on the evaluation of tumor marker tests. *Clin. Chem.* **39,** 2439–2443.
5. Grizzle, W. E. (1994) Tissue resources in the detection and evaluation of markers, in *Early Detection of Cancer: Molecular Markers* (Srivastava, S., Lippman, S. M., Hong, W. K., and Mulshine, W. K., eds.), Futura Publishing, Armonk, NY, pp. 69–88.
6. Simms, W. W., Ordonez, N. G., Johnston, D. A., Ayala, A. G., and Czerniak, B. (1995) p53 expression in dedifferentiated chondrosarcoma. *Cancer* **76,** 223–227.
7. Ouspenskaia, M. V., Johnston, D. A., Roberts, W. M., Estrov, Z., and Zipf, T. F. (1995) Accurate quantitation of residual B-precursor acute lymphoblastic leukemia by limiting dilution and PCR-based detection system: a description of the method and principles involved. *Leukemia* **9,** 321–328.
8. Roberts, W. M., Estrov, Z., Ouspenskaia, M. A., Papusha, V. Z., Johnston, D. A., Harris, D., Vriesendorp, A., McClain, K. L., Pinkel, D. P., and Zipf, T. F. (1997) Measurement of treatment response during remission in childhood acute lymphoblastic leukemia. *N. Engl. J. Med.* **336,** 317–323.
9. Sell, S. (1993) Detection of cancer by tumor markers in the blood: a view to the future. *Crit. Rev. Oncogen.* **4,** 419–433.
10. Gehan, E. A. (1980) Planning clinical trials. *Cancer Bull.* **32,** 200–206.
11. Lee, E. T. (1992) *Statistical Methods for Survival Data Analysis.* Wiley-Interscience, New York.
12. Supplement to Cancer Research (1991) Vol. 51 (No. 23, Part 2), pp. 6407–6491.
13. Peto, R., Pike, M. C., Day, N. E., Gray, R. G., Lee, P. N., Parish, S., Peto, J., Richards, S., and Wahrendorf, J. (1980) Guidelines for simple sensitive significance tests for carcinogenic effects of long-term animal experiments. Annex to *Long-Term and Short-Term Screening Assays for Chemical Carcinogenesis: A Critical Appraisal.* IARC Monographs, Supplement 2, International Agency for Research on Cancer, Lyon, pp. 311–426.
14. Zar, J. H., Jr. (1996) Biostatistical Analysis, 3rd ed., Prentice-Hall, Englewood Cliffs, NJ.
15. Steel, R. G. D. and Torrie, J. H. (1980) *Principles and Procedures of Statistics,* 2nd ed., McGraw-Hill, New York.

16. Fleiss, J. L. (1981) *Statistical Methods for Rates and Proportions*, 2nd ed., Wiley, New York.
17. Bishop, T. M. M., Fienberg, S. E., and Holland, P. W. (1975) *Discrete Multivariate Analysis: Theory and Practice*. MIT Press, Cambridge, MA.
18. Simon, G. S. (1978) Efficacies of measures of association for ordinal contingency tables. *J. Am. Statist. Assn.* **73,** 545–551.
19. Agresti, A. (1990) *Categorical Data Analysis*. Wiley-Interscience, New York.
20. Ott, J. (1991) *Analysis of Human Genetic Linkage*, Rev. ed., Johns Hopkins University Press, Baltimore, MD.
21. Peltomaki, P., Aaltonen, L. A., Sistonen, P., Pylkkanen, L., Mecklin, J.-K., Jarvinen, H., Green, J. S., Jass, J. R., Hamilton, S. R., de la Chapelle, A., and Volgelstein, B. (1993) Genetic mapping of a locus predisposing to human colorectal cancer. *Science* **260,** 810–816.
22. Tanner, M. A. and Young, M. A. (1985) Modeling agreement among raters. *J. Am. Statist. Assn.* **80,** 175–180.
23. Agresti, A. (1988) A model for agreement between ratings on an ordinal scale. *Biometrics* **44,** 539–548.
24. Hosmer, D. W. and Lemeshow, S. (1989) *Applied Logistic Regression*. Wiley-Interscience, New York.
25. Egan, J. P. (1975) Signal Detection Theory and ROC Analysis. Academic, New York.
26. Metz, C. E. (1989) Some practical issues of experimental design and data analysis in Radiological ROC studies. *Invest. Radiol.* **24,** 234–245.
27. Chaturvedi, V., Johnston, D. A., Ro, J. Y., Logothetis, C., von Eschenbach, A. C., Batsakis, J. G., and Czerniak, B. (1997) Superimposed histologic and genetic mapping of chromosome 17 alterations in human urinary bladder cancer. *Oncogene*, **14,** 2059–2070.
28. Lee, E. T. and Desu, M. M. (1972) A computer program for comparing k-samples with right censored data. *Comp. Progr. Biomed.* **2,** 315–321.
29. Taswell, C. (1981) Limiting dilution assays for the determination of immunocompetent cell frequencies I. Data analysis. *J. Immunol.* **126,** 1614–1619.
30. Finney, D. J. (1978) *Statistical Method in Biological Assay*, 3rd ed., Charles Griffin, London.
31. Mace, A. E. (1974) *Sample Size Determination*. Krieger, Huntington, NY.
32. Cohen, J. (1988) *Statistical Power Analysis for the Behavioral Sciences*, 2nd ed., Lawrence Erlbaum Associates, Hillsdale, NJ.
33. Brown, B. W. and Herson, J. (1981) STPLAN: An interactive study planning package. *Am. Statist.* **35,** 164.

3

Selection and Development of Biomarkers for Bladder Cancer

George P. Hemstreet, III, Robert E. Hurst, and Rebecca B. Bonner

1. Introduction

Bladder cancer attacked approx 50,500 Americans in 1995 and killed about 11,200 *(1)*. Bladder cancer appears to develop along two main tracks: a deeply invasive, high-grade form that rapidly becomes life-threatening, and a much less dangerous low-grade form *(2–5)*. Although low-grade tumors are usually cured readily, by simple resection if detected early or by Bacille Calmette-Guerin (BCG) therapy in the case of multiple tumors, the detection of low-grade tumors is pressing because approx 15% of patients with these tumors progress to dangerous disease *(6)*. Given this tendency to progress, even though approx 70% of bladder cancers are low grade on initial diagnosis, the number of deaths caused by bladder cancer is almost equally divided between those with aggressive disease upon presentation and those who progress from low-grade disease. Thus, the ability to detect a group at high risk for progression, or to detect progression early, is crucial to decreasing the death toll from bladder cancer, particularly if detection could be based on noninvasive techniques that quantitate biochemical changes in exfoliated cells found in urine. Conventional cytologic methods have poor sensitivity to low-grade tumors *(2,3)*, though the addition of DNA ploidy by image analysis, which only detects the limited class of low-grade tumors with aberrant ploidy or bladders with field disease, improves the sensitivity 15–20% compared to Papanicolaou cytology *(7–9)*.

The normal bladder, or any other solid organ, represents a complex ecosystem of interacting epithelial and stromal cells whose growth is highly regulated, and the progressive subversion of proliferation, death, and differentiation controls *(10–15)* leads to emergence of cells with tumorigenic phenotypes.

From: *Methods in Molecular Medicine, Vol. 14: Tumor Marker Protocols*
Edited by: M. Hanausek and Z. Walaszek © Humana Press Inc., Totowa, NJ

Some of these altered phenotypes result from genotypic changes or from altered differentiation arising from a changed cytokine and stromal environment ***(16,17)*** whereas others result from epigenetic effects driven by exogenous agents ***(18–22)***. Bladder cancer seems to develop typically by a process of "field disease," frequently involving widespread histopathologic or biochemical changes ***(23)*** and carrying increased risk for progression and recurrence ***(3,6,24)***. The potential for recurrence seemingly is directly related to the number of partially transformed cells remaining after therapy and to the continuance of epigenetic events promoting further carcinogenesis. Increasing understanding of the process of tumorigenesis at the molecular genetic ***(25)*** or biochemically defined phenotypic ***(23)*** level shows that many detectable changes are often present in the absence of morphologic changes ***(23,26,27)***.

Bladder cancer apparently develops along distinct high- and low-grade pathways ***(4,5,28,29)***. The high-grade pathway results in a distinct series of morphologically evident premalignant changes ***(28)***, but the low-grade pathway does not. The high-grade pathway involves mutations in the p53 suppressor gene and possibly other loci on chromosome 17p as an apparently late step ***(4,5)***, whereas loss of a tumor-suppressor function on chromosome 9q seems to be an early, obligate step in bladder carcinogenesis of both types ***(30–41)***. There is considerable evidence for the possibility of more than a single tumor-suppressor gene on chromosome 9 ***(30,36,37,40)***. However, there is probably considerably more complexity to these pathways than is currently appreciated, and some of the results may be clouded by genetic instability. Further study of genotypic and phenotypic changes in the bladder-cancer field may clarify the important genotypic and phenotypic alterations and the network of genetic alterations in these pathways.

Biomarkers are of great interest because they can: serve to identify pathological processes well before they become symptomatic, identify individuals who are susceptible to disease, and provide prognostic information on individuals identified with disease processes ***(9)***. Even a cursory review of the literature on biomarkers shows that there are potentially thousands available, and for bladder cancer, the number of potential markers reported in the literature probably exceeds 100, as has been summarized ***(42)***. Clearly, each biomarker cannot be evaluated in a 5-yr randomized clinical trial. In this chapter, we distinguish markers of potential clinical use from those with only scientific interest, and demonstrate that relatively simple and straightforward approaches can quickly sift through the multiplicity of markers to identify those with potential clinical interest. In particular, we relate examples of biomarkers assayed by quantitative fluorescence image analysis (QFIA).

2. Classification of Markers

Markers can be classified by several logical approaches ***(43,44)***. Markers can be either "genotypic" or "phenotypic." If the former, they usually represent probes of DNA; if the latter, usually protein. Probes of mRNA can be either. Markers can also reflect the relationship to the development of disease. "Markers of exposure" simply detect whether or not organisms have been exposed to a particular agent that may promote or retard tumor development, without regard to any biological effect, such as induced mutations in affected sequences. Exogenous exposures, either promotional (carcinogens) or preventive (nutritional), are underappreciated, because they are frequently difficult to reconstruct in the complexities of genetic polymorphism. "Markers of effect" show some biological effect, which may or may not be relevant to the development of disease. For example, DNA adducts represent both markers of exposure and of effect, since they show binding to DNA, but it is not clear that they bind to specific gene sequences or induce mutations in those sequences. "Markers of disease" reflect the presence of disease, whatever the origins. It is logical to focus on biomarkers of effect, provided their functional role or specific sequence of tumor genesis can be established. "Markers of susceptibility" determine whether an individual is susceptible to disease resulting from a particular exposure, and in conjunction with markers of effect or exposure, can be powerful tools in risk assessment. "Markers of detection" are used to identify the presence of disease, while "markers of prognosis" are used to predict the patient's future risk from the disease, including predicting the risk for having already suffered metastasis, the susceptibility to therapy, or the likelihood of progression. The above definitions are all somewhat arbitrary, and at some point grade into each other. Aberrant DNA ploidy is, for example, either a marker of detection or a prognostic marker, depending on the context. Most diseases are a result of subtle functional disregulation, and all begin in the cell. Biomarkers may be detected in cells quantitatively, and in fact, West proposed that under appropriate conditions stoichiometric determinations of biomarkers could be obtained at the single-cell level ***(45)***. This approach is analogous to that with soluble biomarkers quantitatively determined in a test tube.

In the case of carcinogenesis and detection of premalignant disease, it is logical to study biomarkers as single cells rather than as soluble cellular products because of the dilutional effects of urine, serum, or other body fluids, such as semen. Thus, one of the powers of quantitative fluorescence image analysis is quantitation at the single-cell level, associating the biomarker change with a specific cell type, or analysis in a specific cellular compartment (i.e., nucleus vs cytoplasm) ***(42)***. These concepts relate specifically to selected sample types and methods of analysis. Several important aspects of fluorescence should be emphasized here, particularly as they relate to quantitation and to more conventional immunocytochemistry. Understanding subtleties and methodology is important if one is to optimally select biomarkers, optimize receiver operator curve (ROC) plots, and then combine biomarkers in profiles, effectively setting thresholds for biomarker combination in single cells or in cell populations. Historically, fluorescence was employed routinely for cytologic evaluation because of rapid staining conditions. The cost of microscopic instrumentation, the lack of quantitation, the inability to achieve single-cell suspensions, and the inability to archive specimens led to the broad applications of Papanicolaou cytology for routine application.

However, for years fluorescence maintained a viable position in most laboratories, and rediscovery of the quantitative power of fluorescence was appreciated with the introduction of flow cytometry. The technology has now received wide acceptance, although absolute standardization and rare-event detection remain a problem. Advantages of image analysis and flow cytometry have been reviewed, but few appreciate the power of combining quantitative fluorescence with image analysis ***(46,47)***. The ultimate advantage resides in the increased biomarker resolution, which may convert a marker of marginal utility to a highly useful marker. A clear-cut example of quantitative resolution is related later in the text. In a small blinded study, the visual resolution of the tumor-associated antigen M344 as measured by fluorescence was compared to the resolution of conventional immunohistochemistry. A discordance between the two methods was apparent in 8 of 13 samples. The poor performance for detecting M344 antigen in cells with conventional peroxidase immunohistochemistry compared to those previously reported by fluorescence ***(48)*** was recently confirmed by Fradet in a blinded study ***(49)***. Logically, the method selected for biomarker analysis is equally as important as the selection of the biomarker. Selection of a biomarker must thus consider the method and the reagents available for optimizing the assay. **Table 1** summarizes the sensitivity of various assays currently in use, whereas **Table 2** illustrates the general principles germane to the selection of a specific biomarker. Unfortunately, most biomarkers evaluated today are not subjected to the rigors of scientific logic or scientific method. All too frequently, a new biomarker is identified and integrated into a study without checking whether the anticipated result would lead to a strong biomarker or a biomarker profile that will positively alter the clinical management.

3. The Importance of Strong Markers

In order to be useful clinically, the results of biomarker tests must be definitive enough to select or alter an individual patient's treatment. This means that selected markers must be relatively clear-cut in providing individual risk assessment, regardless of whether the marker is for diagnosis or prognosis. The requirement that markers be strong pays an important statistical benefit in

Table 1
Methods of Detecting Molecules and Mutations with Approximate Lower Limits of Sensitivity

Marker molecule and method	Lower limit of detection
Nucleic acid/autoradiography	3×10^4 molecules
DNA/southern blot/autoradiography	1×10^6 molecules
DNA/PCR	1–10 molecules
RNA/cDNA/PCR	100 molecules
Mutations by DNA/PCR/SSCP[a]	1/100 molecules, relative abundance
Proteins in solution by	
ELISA or RIA	pg–ng (10^7 to 10^{10} molecules)[b]
Electrophoresis/autoradiography	1–10 pg (10^7 to 10^8 molecules)
Electrophoresis/blot/immunochemistry	10–100 pg (10^8 to 10^9 molecules)
Photometric	μ*M* range
in 1 mL	1 nmol (6×10^{14} molecules)
in 1 μL	1 pmol (6×10^{11} molecules)
Proteins in cells by	
Fluorescence immunofluorescence	300 molecules
Absorption immunochemistry	Not quantitative

[a]SSCP, single-strand conformational polymorphism.
[b]Assuming 60,000 molecular weight in relating weight and molecular units.

Table 2
Criteria for Biomarker Selection

- Clinical utility
- Strong biomarker
  - Sensitivity
  - Specificity
  - Negative predictive value
  - Positive predictive value
- Functional role
- Sequence in oncogenesis
- Assay considerations
  - Stability of reagent
  - Cost of reagent
  - Fixation requirements
  - Reproducibility of the assay
  - Machine-sensible parameters
- Contribution to biomarker profile
- Adaptability to automation

that their efficacy can be demonstrated in small studies. Indeed, if their efficacy is not demonstrable in a small study, the marker cannot be strong, and therefore will not be clinically useful. Moreover, there are also a number of currently used markers, and for a new marker to provide any additional information, it should provide an improvement over what is currently available.

The selection and evaluation of markers does not proceed within a vacuum and is driven by the clinical problem being solved and the effectiveness of alternative approaches. The standard for monitoring for bladder cancer recurrence and progression is cystoscopy, and any new approaches need to be measured against the standard of effectiveness of cystoscopy, even though the technique is not without false negative results ***(50)***. Biomarkers can be used as adjuncts to cystoscopy to discover clues to the existence of, for example, upper tract disease or cryptic disease. Potentially, biomarkers might be used to replace cystoscopy, at least for certain subsets of patients. However, any stratification by biomarkers needs to be carefully designed to minimize the possibility that dangerous disease that would normally be detected by cystoscopy would be missed by the biomarker.

The relative costs of false negatives and false positives are crucial considerations as well. For example, a test with high sensitivity that also had relatively poor specificity would not be a problem for monitoring patients for recurrence because at worst, a patient would be subjected to cystoscopy, a procedure that would be routinely used were the marker not available. On the other hand, detecting cancer in an asymptomatic population requires careful balancing of both false positives and false negatives. The usual factor limiting the performance of any marker is its prevalence in individuals without disease, and all things being equal, a marker having a low positive prevalence in the nondisease population will bc morc powerful than one having a background prevalence. This is true whether or not the marker is being used for prognosis or detection.

Stratification of patients on the basis of biomarker measurements needs to reflect that biomarkers actually assess risk and do not diagnose cancer, which requires a pathologic diagnosis. Although traditionally laboratory tests are forced into a binary decision of positive and negative, in actuality three results are usually achieved. If the breakpoint for blood glucose is 120 mg/dL, a person with a value of 119 mg/dL will not be automatically considered to be well, and one of 121 mg/dL will not automatically be considered as diabetic. A person with a blood glucose of 160 mg/dL will be classified as a diabetic with a high degree of confidence, whereas one with a value of 90 mg/dL will be considered normal, also with a high degree of confidence. The three results achieved in practice are, in fact, positive, negative, and more information is required. Having two thresholds facilitates this kind of decision-making and is illustrated by the use of two thresholds in classification of results achieved with the M344 antibody in detection of bladder cancer ***(48)***. The lower limit of two M344-positive cells per 10,000 bladder cells was drawn to maximize sensitivity, and the upper limit of 10/10,000 to maximize specificity. The majority of individuals without disease or at low risk fell below the first threshold, whereas about 50% of tumor cases and virtually no individuals without cancer or cancer risk fell above the higher limit. Thus, the assignments of high and low risk could be made with high confidence, leaving a group in the middle composed of some individuals with bladder cancer, some with premalignant disease, and some with confounding diagnoses such as bladder outlet obstruction. Additional information is required to assess correctly the status of individuals in the middle category, and can include other markers, such as aberrant DNA ploidy in the cited study, or clinical examinations.

Researchers often are seduced into believing that a marker can classify all aspects of a disease, and rarely is this belief justified. Again, the clinical problem that needs to be solved should guide judgment and study design. Rarely is detecting advanced disease a problem. The more common problem is to detect small, recurrent tumors, and any study, from the very beginning, should incorporate this spectrum of cases. Often it is more productive to concentrate on a subset of patients, rather than trying to capture the entire range of variation from stage Ta N0M0 to T4 with metastases. In bladder cancer, the main problems are to identify T1 tumors with significant potential for metastasis and to detect patients at high risk for recurrence and those with significant risk of progression or recurrence. For example, a study of aberrant ploidy in low-grade tumors demonstrated that it was a significant risk factor. In 62 patients followed for at least 15 yr, 43 suffered recurrences and 13 died. The most significant risk factors for death and recurrence were stem-line aneuploidy and the presence of cells with greater than 5C DNA, respectively ***(2)***. This important finding would have been diluted out and likely missed in a large study of all grades, particularly since the association between ploidy and high grade was well known at that time. A screening test for bladder cancer applicable to high-risk groups (smokers, persons over 50 with other risk factors, or workers exposed to carcinogens) would also be an effective tool in control of bladder cancer ***(51)***.

How markers are selected for evaluation is worthy of some discussion. Strong markers are likely to reflect primary biochemical events involved in carcinogenesis or are characteristics intimately associated with the general malignant phenotype. There are many changes in the biochemistry of cancer cells, and each of the changes has the potential to serve as a marker. However, most are probably secondary and unlikely to be strong. Aberrant DNA ploidy arising from genomic instability is one of the most powerful markers yet developed for prognosis, regardless of whether it is assessed by the central

Table 3
Sample Size and Power of Biomarker Measurements

Test result	Disease positive	Disease negative	Disease negative
Test positive	9	0	1
Test negative	3	12	11
		$\chi^2 = 0.0007$	$\chi^2 = 0.005$

tendency of cell populations by flow cytometry ***(52)*** or the appearance of rare, aberrant cells as determined by image analysis ***(7,48)***. Neoantigens form another well-used group, but with these, positivity in the population without disease is always a major consideration, as well as what fraction of tumors express the marker. Studies of model systems, for example, cultured cells, can be powerful in identifying potential markers, an example being the demonstration that levels of actin reflected differentiation and dedifferentiation ***(23,53)***.

Because many components of, for example, signaling pathways, share common intracellular biochemical components, the possibility of finding markers that reflect alterations in any one of several possible systems would provide higher sensitivity than would using individual signaling-pathway components as markers (i.e., growth-factor receptors). An example is the use of alterations of the cytoskeleton on the path to carcinogenesis, which has proven to be a powerful marker for assessment of carcinogenic risk ***(23,53–56)***. In comparing the efficacy of genotypic and phenotypic markers, similar considerations hold in that many genotypes, for example mutations of different codons on the p53 gene, may share a common phenotype. In the case of cancer, it is important to ascertain whether the marker is altered because of genetic instability or if it is a driving event in the carcinogenic process.

4. Practical Study Designs for Pilot Marker Investigations

Selection of markers for random clinical trials should only be made after pilot studies have demonstrated that they are likely to offer improvements over existing markers and sufficient preliminary data has been obtained to support study design. Over the years, several study designs have been found to be valuable in the evaluation of biomarkers ***(43,44)***. These proceed in a logical order, and at each stage markers that are not useful are eliminated from further study.

4.1. The "Quick and Dirty" Pilot Test

This test uses about a dozen normal and a dozen abnormal samples and derives its usefulness from the requirement that clinical markers be strong. Consider the data shown in **Table 3** as a 2 × 2 contingency table. Analysis of the data by χ^2 yields a value of $p = 0.0007$ with no false positives, and even

Table 4
Illustration of Stratified Risk Model for G-actin in Bladder-Cancer Patients and Controls[a]

Risk group	Hematuria	Biopsy result	QFIA cytology	Previous history of bladder cancer	Abnormal fraction (%)
A	NR	Positive	NR	NR	18/19 (95)
B	NR	ND	Positive	NR	46/51 (90)
C	NR	ND	Intermediate	NR	18/24 (75)
D	Yes	ND	Negative	Yes	34/52 (66)
E	Yes	ND	Negative	No	13/36 (36)
Control	No	ND	Negative	No	3/38 (7)

[a]ND, not done; NR, not relevant.

with a single false-positive, the result is $p = 0.005$. In fact, a marker would need to be about this effective in categorizing patients in order to be useful. It is clear that ineffective markers can be eliminated quickly with small studies.

4.2. The "Stratified Risk" Study

This represents a variation of the cross-sectional study design in which several groups of patients are stratified by conventional clinical criteria and laboratory results into groups at different relative risk ***(54)***. The candidate marker is now measured in this population, and, depending upon the selection of groups, one can determine whether a marker becomes abnormal early or late in the carcinogenic process. A marker such as altered actin will show a distribution of abnormal results throughout the risk stratum, but one that is associated with active disease will be restricted to the top risk groups. **Table 4** illustrates the use of this design to investigate abnormal F-actin content as a risk factor for bladder cancer.

4.3. The "Simple" Trial

This study is modeled on the "simple" clinical trial model currently being evaluated to test drugs and uses three groups: known bladder cancer cases, individuals attending the urology clinic who do not have cancer, and asymptomatic controls such as laboratory workers and individuals attending other clinics, such as the orthopedic clinic ***(48)***. Individuals fill out a short questionnaire to assess age, occupation (to assess potential occupational exposures), smoking history, and a brief medical history. The purpose of including the two control groups is that rarely is one attempting to diagnose cancer in an asymptomatic population, but instead selected markers are more likely to be used to evaluate symptomatic individuals, including those who attend a urology clinic.

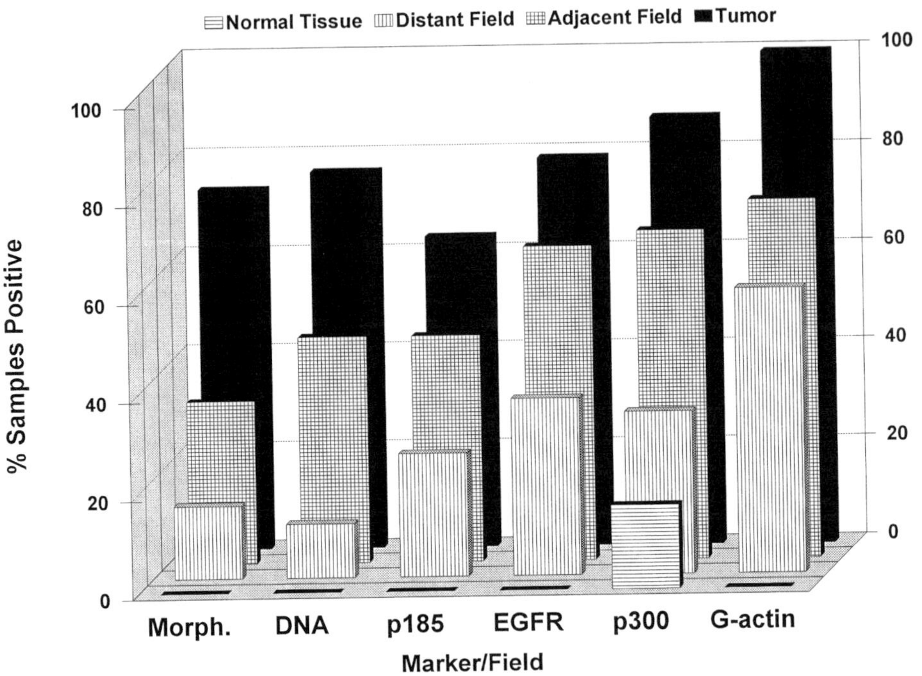

Fig. 1. Progression of biomarkers from distant field to adjacent field to cancer. Normal tissue was obtained from separate noncancer patients. Values are means for several cancer patients. Samples were obtained from "touch preps" made in the operating room. EGFR is epidermal growth-factor receptor, p185 is the product of the HER2/*neu* oncogene, and p300 is the antigen that reacts with M344 antibody.

The completely asymptomatic controls provide a "normal–normal" group to identify potentially confounding conditions. This design is very effective in identifying confounding variables, such as outlet obstruction, a confounder for the M344 antibody against a low-grade tumor antigen ***(36)***.

4.4. The Field Disease Model

This model is well suited to investigating progressive changes that occur during neoplasia and in identifying markers that are independent or dependent on each other ***(23)***. The idea is to follow the progression of markers by taking advantage of the cross-section in space that recapitulates the longitudinal development of cancer. Samples are obtained at surgery, by touch-prep or other techniques, from the tumor, the bladder epithelium immediately adjacent to the tumor, and the bladder epithelium at least 2 cm away from the tumor. This model can also provide valuable information with which to characterize how a given marker relates to progression. An example of such a study is shown in **Fig. 1**, which presents the fraction of samples that were abnormal for a given biomarker in the tumor, adjacent, and distant epithelial fields.

4.5. Study of Biomarkers in Patients Undergoing Tumor Progression or Regression

Selection of a biomarker is enhanced by an appreciation of its functional role and a knowledge of its temporal expression in the cascade of tumorigenesis. Sequential monitoring of patients at high risk for developing bladder cancer (i.e., occupationally exposed cohorts, patients with previous tumors) establishes an association of a biomarker with known risk factors and when it is expressed in tumorigenesis. Because multiple genotypic alterations may lead to fewer phenotypic changes, the mounting evidence that a single genetic alteration may dramatically affect the expression of multiple gene products justifies a focus on the functional protein products. This is not to de-emphasize the importance of biomarkers of susceptibility and deregulation of messages, but quantitative relations are difficult to assure with these biomarkers, particularly since posttranscriptional and posttranslational modifications of the gene products may occur. A quick assessment of biomarkers is possible by studying biomarkers expressed in patients with a tumor and in those in which a tumor has been resected. Provided that all the tumor has been resected, markers re-expressed in the patients with previous tumors, in all probability, are related to those identified in the bladder cancer field ***(56,57)***. The low false positive for DD23, a tumor-associated antigen, in patients with previous tumors indicates that this biomarker is expressed late. Treatment with BCG in one pilot study eliminated cells expressing M344 and aberrant DNA in 68 and 89% of cases, respectively ***(56)***. However, minimal effect was noted on the expression of G-actin. The administration of intravesical DMSO, a known differentiation agent, corrected the G-actin marker in 91% of the cases ***(56)***. Thus, BCG corrected the later markers, aberrant DNA ploidy, and M344, whereas G-actin, an early marker, was corrected by the differentiation-inducing agent, DMSO. It is also logical that following biomarkers for disease recurrence is another model for defining the successive expression of a phenotypic biomarker.

5. Quantitation and Standardization of Cellular Biomarkers

The principal assumptions in quantitation are: that the fluorescence signal is proportional to the content of biomarker; and sample collection and processing do not obscure the relationship of quantitation of biomarker to disease. Cellular components are usually assayed with a fluorescent-labeled affinity probe, which is defined as a labeled molecule or combination of molecules exhibiting a specific and strong affinity for the target biomolecule and carrying a fluorescent

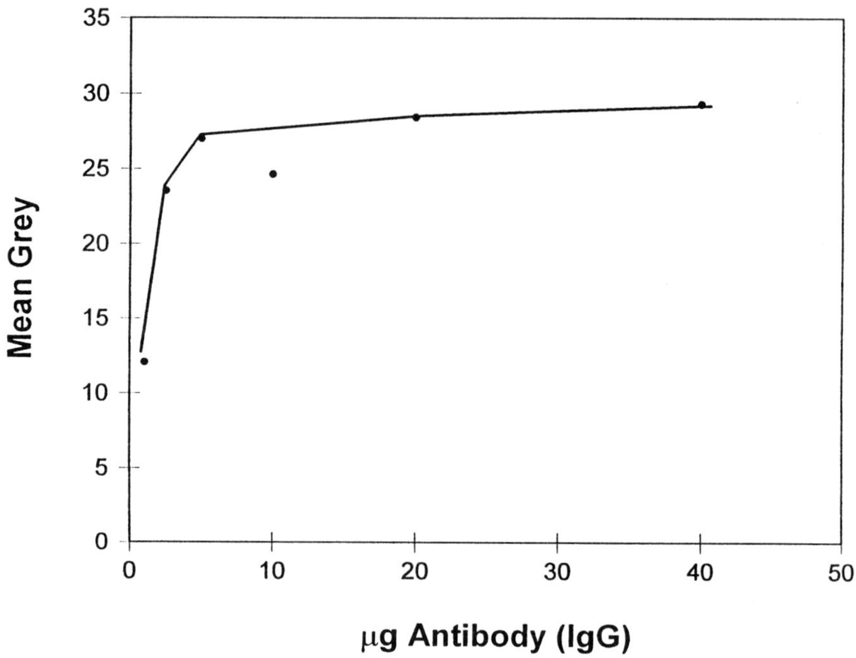

Fig. 2. Titration of transglutaminase system with antibody in a secondary system. A prostate cancer cell line (PC-3) expressing high amounts of transglutaminase was titrated with different amounts of primary antibody and a fixed excess of secondary reagents (biotin-labeled goat antimouse and Texas red-labeled avidin).

label. Such probes can be antibody reagents, enzymes, cofactors or inhibitors, peptide ligands for receptors, oligonucleotides, cDNA or gene sequences, or specific dye molecules. In a direct affinity system, quantitation is easily established because the covalently labeled probe binds in a fixed stoichiometry, and the opportunity for nonspecific interactions is usually less than is seen with indirect systems. Direct probes are also easier to combine in multiple-marker combinations than are indirect probes, in which careful consideration must be given to antibody crossreactivity. Indirect probes, on the other hand, offer signal amplification and a single detection system useful with several primary antibodies, though not simultaneously. In an indirect system, each component must be separately titrated. Moreover, each time a new reagent is obtained, it is necessary to retitrate the reagent because of concentration and activity differences in different lots from the same manufacturer. The titration of antibody reagents follows the same principles as were proposed earlier with dyes ***(58–60)***. **Figure 2** illustrates the titration of the transglutaminase protein and shows clearly the saturation of binding sites.

Standardization, accuracy, and quality control are crucial considerations in obtaining and maintaining accurate results with markers used clinically. The comments below are directed mainly at image analysis, which offers several advantages over flow cytometry derived from the availability of the image of a fluorochrome-labeled cell for further image processing. Many markers are restricted to specific areas of the cell, such as the nucleus, cytoplasm, or cell membrane. Image processing can often be used to electronically isolate the signal from the desired compartment of interest while rejecting signals from outside that area. This method has recently been illustrated for quantitative analysis of G-actin in the nucleus compared to the cytoplasm ***(55)***. Results in these studies confirmed the value of nuclear actin as a late transformation marker in the bladder cancer. Image processing can also be used to identify and substantially reduce errors resulting from cellular autofluorescence ***(48)***.

Quantitation in absolute terms is possible when one or more fixed points are known. With individual cells, the mean fluorescence of 100–200 cells will equate with the mean content of biomarker measured by an independent biochemical analysis, such as ELISA or other techniques ***(61)***. Cultured cells can provide a fixed reference point in that prepared slides are usually stable when stored frozen and can therefore be used over time as a common standard for quantitation.

Figure 3 shows the linearity in response achieved with the transglutaminase system titrated above. Previous studies with p185 analysis of a series of *neu*-transfected cell lines expressing different levels of the protein showed similar linearity ***(61)***, establishing that QFIA methods are accurate methods for analysis of cellular proteins.

In order to obtain the highest accuracy, our laboratory uses one cell line for standardization and a second, independent cell line as a quality control standard. The effectiveness of quality control is illustrated in **Fig. 4**, which shows the stability of quality control samples over approximately a year in G-actin analyses of a worker cohort being monitored for bladder cancer over a several-year period. Stability of results is obviously of crucial importance in such a study.

6. Establishing Thresholds

Establishing thresholds is a complex matter involving several considerations. The first consideration is whether the biomarker is a qualitative or a quantitative marker. Qualitative markers are simple "count" markers in which a cell usually either expresses the marker or does not. In this case, the only threshold that must be established is the threshold for the number of positive cells. If the marker is quantitative, then one must ask whether it is quantitative with respect to individual cells, that is, a threshold of positive can be established for each cell, and cells above the threshold can be considered as posi-

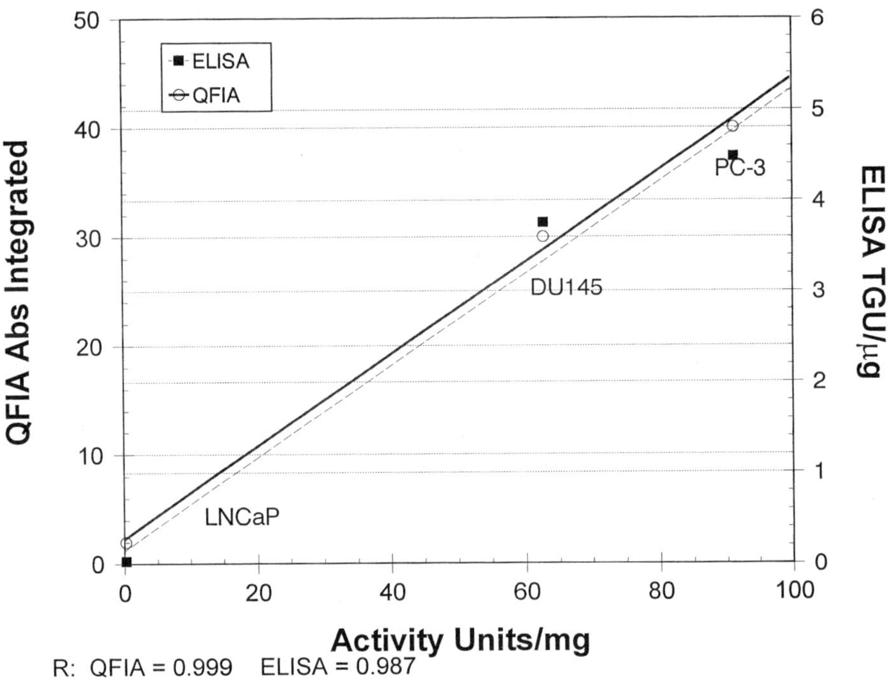

Fig. 3. Linearity of response for transglutaminase system. The same batches of prostate cancer cell lines were analyzed in parallel for mean fluorescence intensity by QFIA and mean content of transglutaminase per cell (ELISA, dotted line).

tive, or whether the marker appears in many cells, field cells as well as cancer cells, for example, and the threshold must be established for the population of cells as a whole. In general, the distributions of marker quantities in cells must be examined in order to determine what is the most effective means of analyzing each particular marker.

Figure 5 illustrates the conversion of a quantitative marker to a "count" marker. Epidermal growth-factor receptor (EGFR) is apparently not down-regulated in high-grade tumor cells, and drawing the threshold at the higher

---

Fig. 4. *(opposite page)* Reproducibility of G-actin assay demonstrating the ratio of two batch controls over time with approx 150 independent batches. The shaded band represents acceptable assays.

Fig. 5. *(opposite page)* Illustration of a system for converting quantitative markers to a positive–negative system. Two thresholds are illustrated. Threshold 1 represents the threshold for all normal cells, whereas Threshold 2 is designed to label high-grade tumor cells as positive.

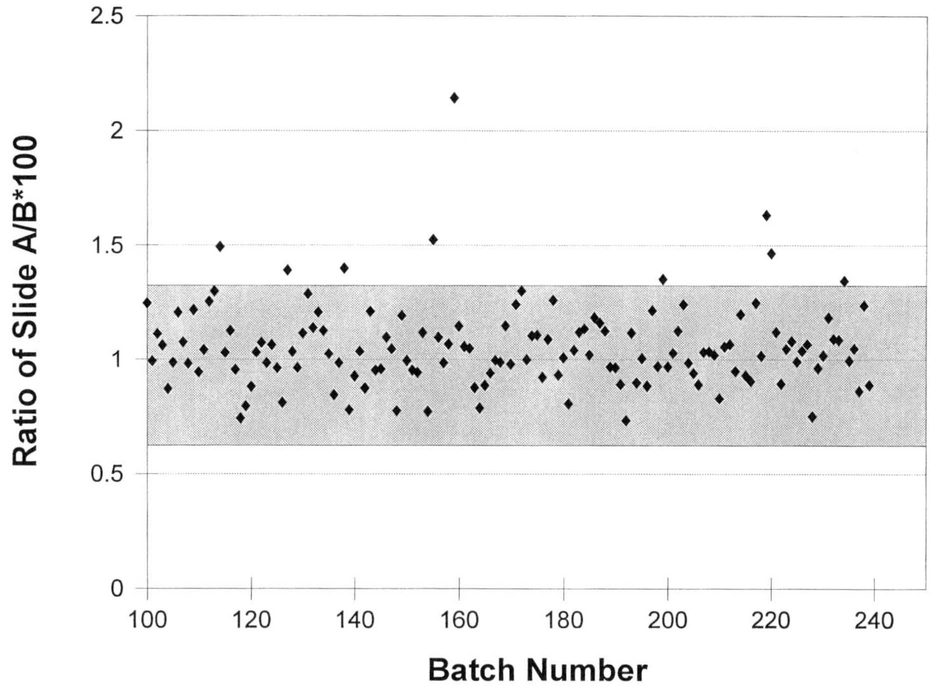

2.5
2
1.5
1
0.5
0
Ratio of Slide A/B*100
100
120
140
160
180
200
220
240
Batch Number

Fig. 4.

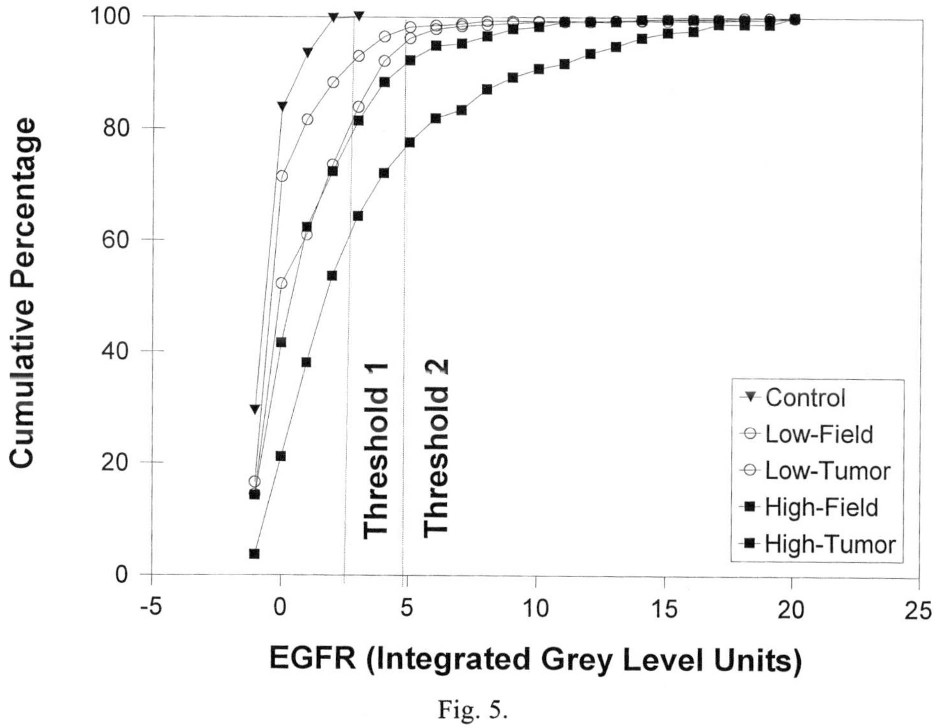

100
80
60
40
20
0
Cumulative Percentage
-5
0
5
10
15
20
25
EGFR (Integrated Grey Level Units)
Threshold 1
Threshold 2
Control
Low-Field
Low-Tumor
High-Field
High-Tumor

Fig. 5.

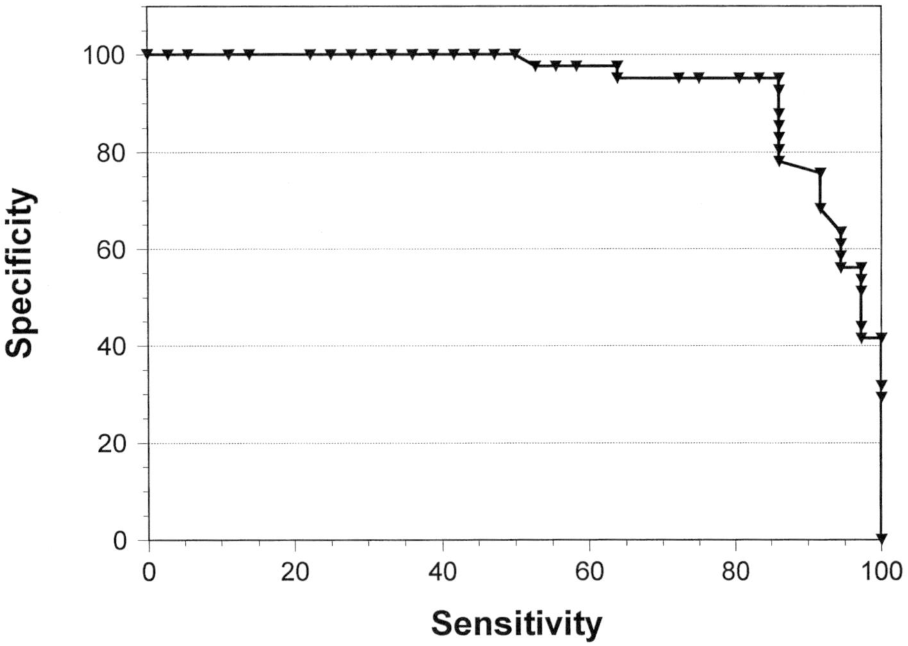

Fig. 6. ROC plots of sensitivity as a function of specificity for the bladder cancer marker DD23. The threshold for a positive cell was 90 DD23 units, as determined.

value essentially flags high-grade tumor cells, which can be counted separately. The lower threshold flags some percentage of low-grade and field cells as well, but does not flag normal cells. Quantitative markers have advantages over qualitative markers in that they are reproducible and referable to independent assays, such as ELISA. This is a very distinct advantage when a marker is likely to be used clinically in multiple laboratories.

A direct comparison of the same marker being employed in quantitative and count modes was performed for DD23, a tumor-related antigen that is expressed in tumor cells as well as apparently normal cells in a tumor-containing bladder. **Figure 6** illustrates cumulative frequency and ROC plots for the marker used in bladder-cancer detection using the mean content of the marker in exfoliated bladder-cell samples. Under these conditions, the marker achieved approx 95% specificity and 87% sensitivity. The same DD23 marker can be used as a "count" marker as well, as shown in **Fig. 7**. The first task is to define a threshold to define a "positive" cell. Reasoning that the specificity should not be less than was achieved with the marker in a quantitative mode, the sensitivity was determined as a function of the threshold by reading off the family of cumulative frequency curves generated at different thresholds for cell-positive, keeping the sensitivity constant at 95%. The results of this measurement, shown in **Fig. 7**, demonstrate that the sensitivity shows an optimum and then drops off

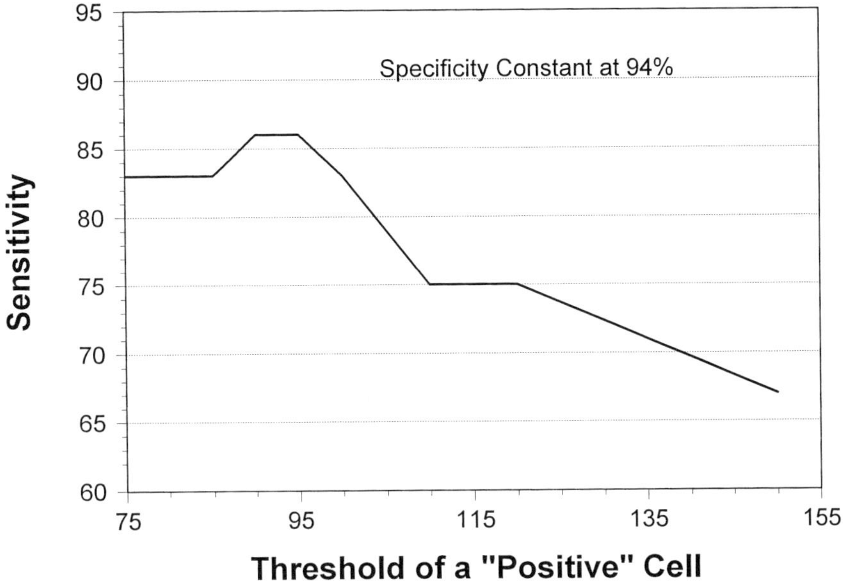

Fig. 7. Selection of threshold defining a positive cell using DD23 as a "count" marker. The specificity was maintained at a constant 94% and the effect on sensitivity of varying the threshold for defining a positive cell is shown. The sensitivity was maximal at 90 DD23 units for a positive cell.

gradually as the threshold for cell-positive is raised. Interestingly, the maximum sensitivity occurs at close to the intensity at which cells become discernible by fluorescence. At a higher intensity, corresponding to what might be easily read by immunocytochemistry, the sensitivity is decreased to about 76%.

7. Biomarker Panels

Because each tumor is unique, and several pathways apparently exist by which cells can become cancerous, individual biomarkers may not detect all cancers and will therefore have a decreased sensitivity. One possible solution to this problem is to select independent biomarkers that reflect different potential pathways or phenotypes. A major consideration is the statistical independence of markers. If two markers are highly correlated, then one provides much the same information as the other, and one is superfluous. The technique of cluster analysis can be used to identify which biomarkers cluster together. This clustering can be used to identify markers that cluster together, and are therefore redundant, as well as a set of independent markers ***(23)***. For example, in a study involving five markers; G-actin, EGFR, and p185 *(HER2/neu)*, cells with >5C DNA (a measure of genomic instability), M344 antigen, G-actin, and

M344 were independent, while cells with >5C DNA, EGFR, and p185 formed a cluster. Consequently the minimum set of independent markers with the least overlap consisted of G-actin, M344, and cells with >5C DNA,chosen because it is technically easier than either of the antibody-based techniques for EGFR and p185.

Prognostic markers represent a more complex situation. Given that metastasis is relatively rare, even though circulating tumor cells are relatively common, the metastatic phenotype is likely to represent a minority of cells within a tumor ***(62)***. Metastatic cells must also contain a number of independent traits, such as weak cell–cell adhesion (allowing them to break free of the tumor), the ability to survive in circulation, the ability to adhere to and subsequently penetrate a capillary bed, which implies both mobility on the part of the cancer cell and the ability either to degrade intracellular matrix or stimulate normal cells to degrade matrix, and the ability for autocrine growth or to use growth signals at the metastatic site ***(62–71)***. Molecular investigations at either the gene or gene-product level are identifying the molecular bases of these traits, and it is widely believed that understanding the molecular basis of metastasis will make it possible to predict metastatic potential. This belief may not be warranted because cells lacking any one of these traits are unlikely to be metastatic, though measurement of any single trait is likely to be positively correlated with metastasis. The situation is further complicated by the likelihood that each trait may be acquired by different molecular pathways, for example, by the activation of any one of several matrix-degrading proteases. Formally, if there are i traits and j ways to achieve each trait, and if each has an associated correlation, p, with metastasis, then

$$\sum_{ij} p_{ij} = 1.0$$

Because the risk is partitioned among all the possible means of achieving the metastatic phenotype, no single marker will be strong in the sense discussed above. It is for this reason that no single biomarkers predict risk better than pathologic stage and grade.

Analysis of the problem suggests several possible solutions. The first simplification is to assume that prediction is not necessarily advantageous for all stages. T3–T4 tumors are most likely at least locally metastatic and must be treated as if they were metastatic. T_a tumors, on the other hand, are very unlikely to be metastatic because they have not penetrated the underlying connective tissue and muscle. Only for T1 and T2 tumors is metastatic potential of particular importance. When considered in this restricted way, many markers are capable of subdividing T1–T2 tumors in survival studies, even though the above equation must still hold. A second approach would be to search for mark-

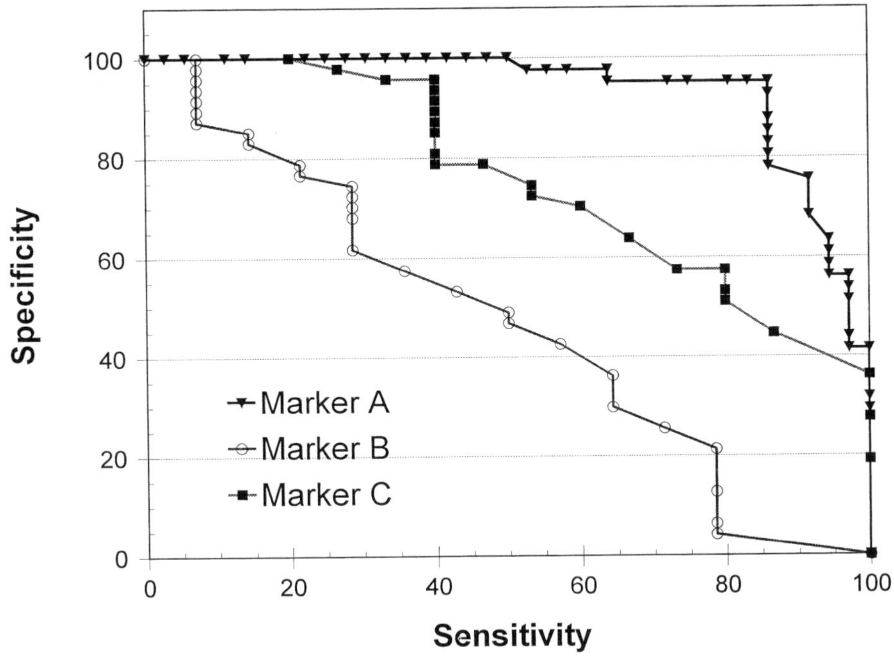

Fig. 8. Examples of ROC Plots. Marker A shows excellent specificity and sensitivity in detection of bladder cancer, whereas markers B and C are less effective. Marker B is virtually useless when used alone as a marker.

ers of the metastatic phenotype, rather than individual molecular markers. In other words, is there a single marker that identifies the phenotype or a small set of markers that identifies the individual subphenotypes (i.e., a single marker for the capability of autocrine growth)? Finally, consideration must be given as to how such rare cells will be detected. A key to selecting biomarkers for inclusion into a biomarker panel is a comparison of ROC plots. **Figure 8** shows the comparison of three different biomarkers assayed in the same clinical samples with biopsy-proven bladder cancer as the gold standard. Although the sequential determination of the marker B and marker C may be useful for monitoring patients for bladder-cancer recurrence as determined by the QFIA assay, it is clearly a poor marker for cancer detection. The biomarker C is clearly an improvement over the biomarker B, but marker A reflects an optimum sensitivity and specificity.

As shown, when assayed alone, both biomarkers B and C were poor markers, but when considered in combination in the same cell, the sensitivity improved to approx 75% with a specificity of 75%. These results do not approach the sensitivity and specificity observed with biomarker-A series of patients.

8. Summary

The selection and development of biomarkers is driven by the clinical question and the need to select strong markers that will impact clinical management. Sample type and treatment and development of optimal assays are critical to achieving the desired sensitivity and specificity. Because all disease begins in the cell and because in cancer biochemical changes occur prior to morphometric alterations, it is logical to study precancerous alterations, at the biochemical and immunological level. In this chapter we have related the general principles of biomarker development as they relate to quantitative fluorescence image analysis and bladder cancer. Biomarkers may be assayed at the gene, message, or protein level. The selected method depends on the assay sensitivity, the class of marker to be studied, and a knowledge of the functional role and when in tumorigenesis (i.e., early vs late) the biomarker is expressed. Early biochemical cellular alterations of effect are detectable in cells derived from the cancer field prior to the development of overt malignancy. A study of biomarkers in the field eliminates many of the problems associated with tumor heterogeneity because the system is not perturbed by genetic instability, which drives heterogeneity. Not all precancerous lesions progress to malignancy; thus a study of biomarkers of susceptibility and exposure in relation to early biomarkers of effect should enhance the power of future epidemiological studies that incorporate intermediate end-point markers of effect as correlative end points ***(42)***. These recent developments in biomarker research and changes in health-care delivery systems make possible strategic cost effective approaches for cancer prevention.

References

1. Parker, S. L., Tong, T., Bolden, S., and Wingo, P. A. (1996) Cancer statistics. *Cancer J. Clin.* **46,** 5–27.
2. Farrow, G. M. (1990) Urine cytology in the detection of bladder cancer: a critical approach. *J. Occup. Med.* **32,** 817–821.
3. Koss, L. G. (1979) Tumors of the urinary tract and prostate, in *Diagnostic Cytology and Its Histologic Basis* (Koss, L. G., ed.), Lippincott, Philadelphia, PA, pp. 749–811.
4. Presti, J. C., Jr., Reuter, V. E., Galan, T., Fair, W. R., and Cordon-Cardo, C. (1991) Molecular genetic alterations in superficial and locally advanced human bladder cancer. *Cancer Res.* **51,** 5405–5409.
5. Spruck, C. H., III, Ohneseit, P. F., Gonzalez-Zulueta, M., Esrig, D., Miyao, N., Tsai, Y. C., Lerner, S. P., Schmütte, C., Yang, A. S., Cote, R., Dubeau, L., Nichols, P. W., Hermann, G. G., Steven, K., Horn, T., Skinner, D. G., and Jones, P. A. (1994) Two molecular pathways to transitional cell carcinoma of the bladder. *Cancer Res.* **54,** 784–788.
6. Heney, N. M., Ahmed, S., Flanagan, M. J., Frable, W., Corder, M. P., Hafermann, M. D., and Hawkins, I. R. (1983) Superficial bladder cancer: progression and recurrence. *J. Urol.* **130,** 1083–1086.

7. Bass, R. A., Hemstreet, G. P., Honker, N. A., Hurst, R. E., and Doggett, R. S. (1987) DNA cytometry and cytology by quantitative fluorescence image analysis in symptomatic bladder cancer patients. *Int. J. Cancer* **40,** 698–705.
8. Amberson, J. and Laino, J. (1993) Image cytometric deoxyribonucleic acid analysis of urine specimens as an adjunct to visual cytology in the detection of urothelial cell carcinoma. *J. Urol.* **149,** 42–45.
9. Hemstreet, G. P., Bonner, R. B., Hurst, R. E., and O'Dowd, G. A. (1996) Cytology of Bladder Cancer, in *Comprehensive Textbook of Genitourinary Oncology* (Vogelzang, N. J., Scardino, P. T., Shipley, W. U., and Coffey, D. S., eds.), Williams and Wilkins, Baltimore, MD, pp. 338–350.
10. Weinberg, R. (1989) Oncogenes, antioncogenes, and the molecular bases of multistep carcinogenesis. *Cancer Res.* **49,** 3713–3721.
11. Pienta, K., Partin, A., and Coffey, D. S. (1989) Cancer as a disease of DNA organization and dynamic cell structure. *Cancer Res.* **49,** 2525–2532.
12. Tzen, C., Estervig, D. N., Minoo, P., Filipak, M., Maercklein, P., Hoerl, B., and Scott, R. (1988) Differentiation, cancer, and anticancer activity. *Biochem. Cell Biol.* **66,** 478–489.
13. Heldin, C., Betscholz, C., Claesson-Welsh, L., and Westermark, B. (1987) Subversion of growth regulatory pathways in malignant transformation. *Biochim. Biophys. Acta* **907,** 219–244.
14. Couture, J. and Hansen, M. (1991) Recessive genes in tumorigenesis. *Cancer Bull.* **43,** 41–50.
15. Kastan, M. B., Onyekwere, O., Sidransky, D., Vogelstein, B., and Craig, R. W. (1991) Participation of p53 protein in the cellular response to DNA damage. *Cancer Res.* **51,** 6304–6311.
16. Ruoslahti, E. and Yamaguchi, Y. (1991) Proteoglycans as modulators of growth factors. *Cell* **64,** 867–869.
17. Nathan, C. and Sporn, M. (1991) Cytokines in context. *J. Cell Biol.* **113,** 981–986.
18. Harris, C. C. (1991) Chemical and physical carcinogenesis: advances and perspectives for the 1990s. *Cancer Res.* **51,** 5023s–5044s.
19. Trosko, J. E., Chang, C. C., Madhukar, B. V., and Oh, S. Y. (1990) Modulators of gap junction function: the scientific basis of epigenetic toxicology. *In Vitro Toxicol.* **3,** 9–26.
20. Cuthill, S. (1994) Cellular epigenetics and the origin of cancer. *BioEssays* **16,** 393,394.
21. Hemstreet, G. P., III, Rao, J. Y., Hurst, R. E., Bonner, R. B., Jones, P. L., Vaidya, A. M., Fradet, Y., Moon, R. C., and Kelloff, G. J. (1992) Intermediate endpoint biomarkers for chemoprevention. *J. Cell. Biochem. Suppl.* **16I,** 93–110.
22. Prehn, R. T. (1994) Cancers beget mutations versus mutations beget cancers. *Cancer Res.* **54,** 5296–5300.
23. Rao, J. Y., Hemstreet, G. P., Hurst, R. E., Bonner, R. B., Jones, P. L., Min, K. W., and Fradet, Y. (1993) Alterations in phenotypic biochemical markers in bladder epithelium during tumorigenesis. *Proc. Natl. Acad. Sci. USA* **90,** 8287–8291.
24. Norming, U., Nyman, C., and Tribukait, B. (1989) Comparative flow and cytometric deoxyribonucleic acid studies on exophytic tumor and random mucosal biopsies in untreated carcinoma of the bladder. *J. Urol.* **142,** 1442–1447.

25. Vogelstein, B., Fearon, E., Hamilton, S., Kern, S., Preisinger, A. C., Leppert, M., et al. (1988) Genetic alterations during colorectal tumor development. *N. Engl. J. Med.* **319,** 525–532.
26. Sidransky, D., von Eschenbach, A., Tsai, Y. C., Jones, P., Summerhayes, I., Marshall, F., Paul, M., Green, P., Hamilton, S. R., Frost, P., et al. (1991) Identification of p53 gene mutations in bladder cancers and urine samples. *Science* **252,** 706–709.
27. Sidransky, D., Frost, P., von Eschenbach, A. C., Dyasu, R., Preisinger, A. C., and Vogelstein, B. (1992) Clonal origin bladder cancer. *N. Engl. J. Med.* **326,** 759–761.
28. Pagano, F., Pegoraro, V., Prayer-Galetti, T., Pizzarella, M., Milani, C., and Garbeglio, A. (1987) Prognosis of bladder cancer. II. The fate of patients with T1b transitional cell bladder cancer. *Eur. Urol.* **13,** 305–309.
29. Dalbagni, G., Presti, J., Reuter, V., Fair, W. R., and Cordon-Cardo, C. (1993) Genetic alterations in bladder cancer. *Lancet* **342,** 469–471.
30. Tsai, Y. C., Nichols, P. W., Skinner, D. G., and Jones, P. A. (1990) Allelic losses of chromosomes 9, 11, and 17 in human bladder cancer. *Cancer Res.* **50,** 44–47.
31. Hopman, A., Moesker, O., Smeets, A., Pauwels, R., Vooijs, G., and Ramaekers, F. C. S. (1991) Numerical chromosome 1, 7, 9, and 11 aberrations in bladder cancer detected by in situ hybridization. *Cancer Res.* **51,** 644–651.
32. Borland, R., Brendler, C., and Isaacs, W. B. (1992) Molecular biology of bladder cancer. *Hematol. Oncol. Clin. North Am.* **6,** 31–39.
33. Cairns, P., Shaw, M. E., and Knowles, M. A. (1993) Initiation of bladder cancer may involve deletion of a tumour-suppressor gene on chromosome 9. *Oncogene* **8,** 1083–1085.
34. Linnenbach, A. J., Pressler, L. B., Seng, B. A., Kimmel, B. S., Tomaszewski, J. E., and Malkowicz, S. B. (1993) Characterization of chromosome 9 deletions in transitional cell carcinoma by microsatellite assay. *Human Mol. Genet.* **2,** 1407–1411.
35. Miyao, N., Tsai, Y. C., Lerner, S. P., Olumi, A. F., Spruck, C. H., III, Gonzalez-Zulucta, M., Nichols, P. W., Skinner, D. G., and Jones, P. A. (1993) Role of chromosome 9 in human bladder cancer. *Cancer Res.* **53,** 4066–4070.
36. Ruppert, J. M., Tokino, K., and Sidransky, D. (1993) Evidence for two bladder cancer suppressor loci on human chromosome 9. *Cancer Res.* **53,** 5093–5095.
37. Keen, A. J. and Knowles, M. A. (1994) Definition of two regions of deletion on chromosome 9 in carcinoma of the bladder. *Oncogene* **9,** 2083–2088.
38. Orlow, I., Lianes, P., Lacombe, L., Dalbagni, G., Reuter, V. E., and Cordon-Cardo, C. (1994) Chromosome 9 allelic losses and microsatellite alterations in human bladder tumors. *Cancer Res.* **54,** 2848–2851.
39. Wheeless, L. L., Reeder, J. E., Han, R., O'Connell, M. J., Frank, I. N., Cockett, A. T., and Hopman, A. H. (1994) Bladder irrigation specimens assayed by fluorescence in situ hybridization to interphase nuclei. *Cytometry* **17,** 319–326.
40. Habuchi, T., Devlin, J., Elder, P. A., and Knowles, M. A. (1995) Detailed deletion mapping of chromosome 9q in bladder cancer: evidence for two tumour suppressor loci. *Oncogene* **11,** 1671–1674.

41. Sauter, G., Moch, H., Carroll, P., Kerschmann, R., Mihatsch, M. J., and Waldman, F. M. (1995) Chromosome-9 loss detected by fluorescence in situ hybridization in bladder cancer. *Int. J. Cancer* **64,** 99–103.
42. Finn, W. and Hemstreet, G. (1995) Biological Markers in Urinary Toxicology, in National Research Council (Helmstreet, G. P., ed.), National Academy Press, Washington, DC, pp. 81–152.
43. Schatzkin, A., Freedman, L., Schiffman, M., and Dawsey, S. M. (1990) Validation of intermediate end points in cancer research. *J. Natl. Cancer Inst.* **82,** 1746–1752.
44. Schulte, P. A., Ringen, K., Hemstreet, G. P., and Ward, E. (1987) Occupational cancer of the urinary tract, in *Occupational Cancer and Carcinogenesis* (Rauf, P. B., ed.), Hanley and Belfus, Philadelphia, PA, pp. 85–107.
45. Granados, E., de la Torre, P., and Palou, J. (1991) Echography and cystoscopy: 2 diagnostic means in bladder tumor (1) [Spanish]. *Actas Urol. Espanol.* **15,** 540–542.
46. Parry, W. and Hemstreet, G. P. (1988) Cancer detection by quantitative fluorescence image analysis. *J. Urol.* **139,** 270–274.
47. Koss, L. G. and Czerniak, B. (1992) Image analysis and flow cytometry of tumors of prostate and bladder; with a comment on molecular biology of urothelial tumors. *Monographs Pathol.* **34,** 112–128.
48. Bonner, R. B., Hemstreet, G. P., Fradet, Y., Rao, J. Y., Min, K. W., and Hurst, R. E. (1993) Bladder cancer risk assessment with quantitative fluorescence image analysis of tumor markers in exfoliated bladder cells. *Cancer* **72,** 2461–2469.
49. Fradet, Y., Veltri, R., Simard, P., Blumenstein, B., O'Dowd, G., Johnson, K., and Miller, C. (1996) Improved detection of bladder cancer by immunocytology with monoclonal antibodies M344 and 19A211. *Canadian J. Urol.* **Suppl. 3,** A40.
50. Devonec, M., Darzynkiewicz, Z., Kostyrka-Claps, M. L., Collste, L., Whitmore, W. F., Jr., and Melamed, M. R. (1982) Flow cytometry of low stage bladder tumors: correlation with cytologic and cystoscopic diagnosis. *Cancer* **49,** 109–118.
51. Bi, W., Rao, J., Hemstreet, G. P., Fang, P., Asal, N. R., Zang, M., Min, K. W., Ma, Z., Lee, E., Li, G., Hurst, R. E., Bonner, R. B., Weng, Y., Fradet, Y., and Yin, S. (1993) Field molecular epidemiology. Feasibility of monitoring for the malignant bladder cell phenotype in a benzidine-exposed occupational cohort. *J. Occup. Med.* **35,** 20–27.
52. Wheeless, L. L., Badalament, R. A., DeVere White, R. W., Fradet, Y., and Tribukait, B. (1993) Consensus review of the clinical utility of DNA cytometry in bladder cancer. *Cytometry* **14,** 478–481.
53. Rao, J. Y., Hurst, R. E., Bales, W. D., Jones, P. L., Bass, R. A., Archer, L. T., and Hemstreet, G. P. (1990) Cellular F-actin levels as a marker for cellular transformation: relationship to cell division and differentiation. *Cancer Res.* **50,** 2215–2220.
54. Rao, J. Y., Hemstreet, G. P., Hurst, R. E., Bonner, R. B., Min, K. W., and Jones, P. L. (1991) Cellular F-actin levels as a marker for cellular transformation: correlation with bladder cancer risk. *Cancer Res.* **51,** 2762–2767.

55. Rao, J. Y., Bonner, R. B., Hurst, R. E., Qiu, W. R., Reznikoff, C. A., and Hemstreet, G. P. (1996) Quantitative changes in cytoskeletal and nuclear actin levels during cellular transformation. *Int. J. Cancer* **70,** 423–429.
56. Hemstreet, G. P., Rao, J. Y., Hurst, R. E., Bonner, R. B., Waliszewski, P., Grossman, H. B., Liebert, M., and Bane, B. L. (1996) G-actin as a risk factor and modulatable endpoint for cancer chemoprevention trials. *J. Cell. Biochem. Suppl.,* **255,** 197–204.
57. Carter, H., Amberson, J., Bander, N., Badalament, R. A., Gorelick, J., Vaughan, E., and Whitmore, A. (1987) Newer diagnostic techniques for bladder cancer. *Urol. Clin. North Am.* **14,** 763–769.
58. Nakamura, N., Hurst, R. E., West, S. S., Menter, J. M., Golden, J. F., Corliss, D. A., and Jones, D. D. (1980) Biophysical cytochemical investigations of intracellular heparin in neoplastic mast cells. *J. Histochem. Cytochem.* **28,** 223–230.
59. West, S. S., Hemstreet, G. P., Hurst, R. E., Bass, R. A., Doggett, R. S., and Schulte, P. A. (1987) Detection of DNA aneuploidy by quantitative fluorescence image analysis: potential in screening for occupational bladder cancer, in *Biological Monitoring of Exposure to Chemicals* (Dillon, K. and Ho, M., eds.), Wiley, New York, pp. 327–341.
60. McGowan, P., Hurst, R. E., Bass, R. E., Hemstreet, G. P., and Postier, R. (1988) Equilibrium binding of Hoechst 33258 and Hoechst 33342 fluorochromes with rat colorectal cells. *J. Histochem. Cytochem.* **36,** 757–762.
61. Jones, P. L., O'Hare, C., Bass, R. A., Rao, J. Y., Hemstreet, G. P., and Hurst, R. E. (1990) Quantitative immunofluorescence, anti-ras p21 antibody specificity and cellular oncoprotein levels. *Biochem. Biophys. Res. Commun.* **167,** 464–470.
62. Fidler, I. J. (1991) The biology of human cancer metastasis. *Acta Oncologica* **30,** 669–675.
63. Aznavoorian, S., Murphy, A. N., Steller-Stevenson, W. G., and Liotta, L. A. (1993) Molecular aspects of tumor cell invasion and metastasis. *Cancer* **71,** 1368–1383.
64. Ichikawa, T., Nihei, N., Kuramochi, J., Kawana, Y., Killary, A. M., Rinker-Schaeffer, C. W., Barrett, J. C., Isaacs, J. T., Kugoh, H., Oshimura, M., and Shimazaki, J. (1996) Metastasis suppressor genes for prostate cancer. *Prostate* **6,** 31–35.
65. Kerbel, R. (1989) Towards an understanding of the molecular basis of the metastatic phenotype. *Inv. Metast.* **9,** 329–337.
66. Klienman, H. K. and Kibbey, M. C. (1991) Basement membrane regulation of tumor growth and metastasis. *J. NIH Res.* **3,** 63,64.
67. Lu, C. and Kerbel, R. S. (1994) Cytokines, growth factors and the loss of negative growth controls in the progression of human cutaneous malignant melanoma. *Current Opinion Oncol.* **6,** 212–220.
68. Raz, A. (1988) Actin organization, cell motility, and metastasis. *Adv. Exp. Med. Biol.* **233,** 227–233.
69. van den Hooff, A. (1991) The role of stromal cells in tumor metastasis: a new link. *Cancer Cells* **3,** 186,187.
70. Ware, J. L. (1993) Growth factors and their receptors as determinants in the proliferation and metastasis of human prostate cancer. *Cancer Metast. Rev.* **12,** 287–301.
71. Yokozaki, H. and Tahara, E. (1994) Metastasis-related genes. [Japanese]. *Gan to Kagaku Ryoho* [Japanese Journal of Cancer and Chemotherapy], **21,** 2541–2548.

4

Clinical Application of Tissue and Serum Markers in Breast Cancer

Gordon F. Schwartz and Roland Schwarting

1. Introduction

How different the practice of oncology would become if physicians could predict which of their patients will develop cancer, and, if the disease does occur, then determine who might remain disease-free. Programs stressing prevention, early detection, and prompt treatment could be aimed at the appropriate population, relieving anxiety in those destined to remain disease-free, and targeting treatment to improve what might have been the predicted outcome. Not only would our patients benefit, but expensive resources could be allocated more effectively.

Although that millennium has not yet arrived, the pursuit of biological markers shared by patients with malignant disease, and the use of these markers to differentiate between classes of risk for occurrence and recurrence of cancer, have already led to significant changes in physicians' recommendations for cancer therapy. These observations are particularly noteworthy in the treatment of women with carcinoma of the breast, especially as the genetics of this disease are being unraveled, and this disease may be an appropriate paradigm to demonstrate the emerging use of these markers in clinical medicine.

Breast cancer is the most prevalent malignancy in women who live in what is commonly called the "Western" world. Until recently overtaken by lung cancer, it had also been the most common cause of death from cancer in this group of women. Between the early 1970s and the 1990s, an American woman's lifetime threat of breast cancer increased from one chance in twenty to one chance in eight or nine. Moreover, as life expectancy, in general, increases further, as it has already increased over the past 50 yr, there will be more cases of breast cancer, even if the incidence remains the same. The impact

From: *Methods in Molecular Medicine, Vol. 14: Tumor Marker Protocols*
Edited by: M. Hanausek and Z. Walaszek © Humana Press Inc., Totowa, NJ

of breast cancer will continue to become an even greater problem. Society must address not only the emotional toll of breast cancer on its victims and their families, but also, since health care has assumed such a large proportion of any country's gross national product (GNP), the cost of detection and treatment for this burgeoning group.

Until breast cancer can be prevented, would it not be highly desirable to have available a single marker or group of markers that would reliably distinguish which women are at greater risk than others to develop breast cancer? This information would permit innovative screening programs that would identify these women and focus detection programs where they would be most efficient.

2. General Aspects of Tumor Markers

The transformation of the normal cell into a malignant one is a complex process involving multiple steps, culminating in a group of cells that become autonomous. It is assumed that abnormalities in the genetic composition of these cells permit their multiplication, uninhibited by the host's intrinsic mechanisms of defense. The abnormal genes for various malignancies carried within the genomes and probably responsible for the expression of clinical cancer are known as oncogenes. These are probably altered or derived versions of proto-oncogenes, the genes that regulate normal cell growth and differentiation. By some mechanism, the proto-oncogene undergoes somatic mutation that alters its structure or its expression, and the resultant gene product no longer exerts the same regulatory activity as its predecessor, thereby promoting carcinogenesis.

Conversely, there also exists a group of "tumor-suppressor" genes that apparently function in a manner contrary to that of oncogenes, i.e., as inhibitors of cellular growth. Thwarting the expression of these tumor-suppressor genes then becomes a necessary, perhaps critical, step in the initiation of carcinogenesis. The detection of altered oncogenes or tumor-suppressor genes, therefore, has major clinical significance. If carcinogenesis may be divided into three phases—initiation, promotion, and progression—the best marker would be one that identifies the individual at risk for the initiation of the malignant process, such as the identification of a specific oncogene or its product. Those markers that may be detected later in the natural history of the disease, i.e., during progression or thereafter, are perhaps important for prognostic information or to influence treatment decisions, but are already too late to permit the interruption of the evolution of the neoplastic process altogether.

It is simplistic to think of a single oncogene, *per se*, occurring in a breast epithelial cell, for example, as the inciting agent of breast cancer. It is more likely that the evolution of clinical cancer requires several genetic changes acting in concert to orchestrate the full neoplastic phenotype, at least for solid tumors. The multistep phases of carcinogenesis may be the manifestation of this serial acquisition of these genetic changes.

The quest currently has been for a single feature or combination of features that precisely distinguishes a normal cell from a malignant cell, to permit the diagnosis of a specific cancer with absolute accuracy. Whatever these factors may be, they reside in the malignant cell and not in normal cells, or, conversely, their absence from a malignant cell differentiates them from the adjacent normal cells. Carried further, however, this pursuit should strive to identify genetic markers residing within each cell of the host individual that might be used to identify those people who are today free from disease, but who are at risk to develop that malignancy because of these genetic traits, and the identification of tumor-associated markers that determine the prognosis or outcome of the disease when it does occur. Tests that search for genomic changes in the tumor itself require access to the tumor material, whereas if the tumor releases a protein into the systemic circulation, serologic assay may be employed using a serum probe. Because many markers currently investigated do not enter the circulation, assays must be employed on a portion of the tumor itself.

Additionally, if the apparent familial occurrence ("genetic predisposition") of specific cancers, including breast, ovarian, and colon carcinomas, implies a specific genomic identity among affected relatives, an additional mechanism other than the somatic mutations that occur sporadically and activate oncogenes must be invoked. These so-called cancer-predisposing genes acquired at conception must have differing mechanisms of action. Perhaps they affect the host's ability to resist environmental carcinogens or to regulate cellular proliferation. They might even adversely affect the ability of the immune system to recognize and destroy aberrant cells as they arise. Except for retinoblastoma, a malignant retinal tumor occurring in young children that has been mapped to a single mutation on chromosome 13, there have been few descriptions of these unique cancer-predisposing genes. In retinoblastoma, loss of function of the tumor-suppressor gene (anti-oncogene) at this single locus leads to the disease in all of the individuals with this mutation *(1–3)*.

The historical prototypes of biological markers have been serum proteins released from tumor cells used to monitor the course of malignant disease, after the disease has already been detected and treated. Among these are α-fetoprotein (AFP), whose demonstration in serum has been associated with hepatocellular and germ-cell tumors, and carcinoembryonic antigen (CEA). Increases in the serum concentrations of CEA have been associated with progression of cancers of the gastrointestinal tract, lung, and breast. Neither of these, however, is a specific marker for a well-defined malignancy that pinpoints the occurrence or recurrence of that particular disease, and they may be elevated in nonmalignant conditions, as well.

2.1. The Accuracy of Tumor Markers

Many so-called markers currently in vogue do not discriminate between benign and malignant cells well enough. Specificity and sensitivity are terms that are often used somewhat pedantically to describe the value of any test according to whether it creates too many false positives or too many false negatives. Those of us who do not use these terms daily, frequently use them incorrectly. Nevertheless, the concepts that they convey are readily apparent to clinical practitioners. For example, in screening women for breast cancer by mammography, if the radiologist calls every mammogram "positive," no cancers will be missed, although too many women may undergo biopsy for benign disease. Relating these terms to markers, melanoma-associated antigen (MEL) is also expressed in normal melanocytes; prostate-specific antigen (PSA) may be expressed not only in carcinoma but also in normal prostatic tissue. Cytokeratin may be expressed in other epithelial cells as well as in carcinomas. Thus, in these cases, the markers define the lineage of the cell rather than discriminating between benign and malignant.

The accuracy of markers may be enhanced by using a combination of them. A battery of markers may offer information that confirms a diagnosis that none alone would define. For example, an "epithelial-looking" skin lesion that lacks cytokeratin but expresses positivity for vimentin and S-100 protein highly suggests malignant melanoma. The absence of a marker may also offer information. In the above example, because the skin lesion lacks cytokeratin expression, a marker universally seen in epithelial tumors, carcinoma is excluded.

Positive and negative predictive values (PPV and NPV) are perhaps better terms to use when describing any marker that attempts to discriminate between normal individuals and those afflicted by any disease, because this ratio includes not only the accuracy of the test but also addresses the prevalence of the disease within the population studied. The positive predictive value is the number of patients with a disease who test positively divided by the number of all subjects who test positively; conversely, the negative predictive value is the number of normal people who test negatively, divided by the total number who test negatively. If a test had 100% sensitivity and 100% specificity, so that only people with a disease test positively for it, the PPV would be unrelated to the frequency with which the disease might be encountered. A test might be unsuitable, even if all of those affected were to test positively for it, if the disease were rare and if too many "normals" also had a positive result. Using these guidelines, for example, if mammographies were employed only in women whose mothers or sisters had breast cancers, its PPV would increase because of the increased prevalence of breast cancer in these women. Unfortunately, the prevalence of this disease in the female population is too high to overlook those women affected who do not have this family history.

2.2. Diagnostic vs Prognostic Value of Markers

In breast carcinoma as for other malignancies, tumor markers may be used to identify a particular tumor as originating from breast epithelium, thus differentiating it from other neoplasms. Theoretically, this is particularly useful in defining the origin of metastatic disease with an unknown primary source. Tumor markers or the expression of a combination of them may direct the clinician to the probable site of origin.

Markers that are used in this context are considered diagnostic markers. An axillary lymph node metastasis expressing both vimentin and S-100, but lacking cytokeratin expression, is in favor of metastatic melanoma, virtually excluding metastatic breast carcinoma. An anaplastic large cell neoplasm in the skin expressing leukocyte common antigen (LCA, CD45) is diagnostic of a hematologic process, again making a primary breast carcinoma unlikely. Conversely, an axillary lymph node metastasis with strong expression for hormone receptors favors primary breast carcinoma, even though the expression of estrogen and progesterone receptors is shared by some nonmammary neoplasms. In the latter example, the presence of hormone receptors contributes to the diagnosis of breast cancer. In addition, hormone-receptor expression implies a more favorable prognosis than one would generally expect in hormone-receptor negative tumors. Hence, in this one instance, assessment of hormone-receptor status may be of both diagnostic and prognostic value.

Different from this "hybrid" status of hormone receptors, other markers may be of prognostic value only. The mutated gene product of the anti-oncogene p53 is overexpressed in a variety of tumors. Tumors may arise when a mutation of p53 renders its usual tumor-suppressor behavior nonoperative. While detection of nuclear mutated p53 in significant amounts is an ominous prognostic indicator, its expression is not of diagnostic value.

A foolproof diagnostic marker to detect an inevitable predisposition for or the presence of a malignancy would be a remarkable clinical tool. Despite their extremely limited current availability and/or lack of precision, it is likely that their more accurate definition and clinical use are within the grasp of this generation of scientists. Additionally, however, biological markers that predict prognosis once a cancer has occurred are of great importance, since they may influence major therapeutic recommendations. At least for breast cancer, these tools have become part of contemporary clinical practice. Infrequently, the same marker may have both diagnostic and prognostic implications. The expression of estrogen receptors, for example, clearly delimits the origin of the cell, and presence or absence of these receptors affects both therapy and implied outcome. Demonstration of these nuclear steroid hormone receptors in the tumor is currently a general requirement for antihormonal therapy, such as tamoxifen. Because resting cells may not respond to radiation or chemotherapy, a marker that determines the proportion of proliferating tumor cells, such as Ki-67 (vide infra), may also have major importance in determining treatment as well as indicating prognosis.

2.3. Circulating Tumor Markers

In a generic sense, a circulating tumor marker is any substance that may be detected in human serum that separates those with a disease from those without it. The perfect marker would be highly accurate as well as inexpensive to perform, so that large populations could be screened effectively and economically. False positives would be more tolerable than false negatives, since no one with the disease would be overlooked, even if some truly negative patients might be unduly alarmed, so long as the subsequent differentiation of those with from those without the disease is not unduly tedious or dangerous.

The detection of primary disease would be the major but not the only use for a circulating tumor marker. A prognostic marker, if available, could predict the likelihood of recurrent disease and influence therapeutic recommendations. Markers for metastasis could detect the presence of disease in asymptomatic patients, and if the serum level of this marker varied with the "burden" of disease, what a foolproof way it would be to monitor a patient's response to treatment.

Unfortunately, there is no magic marker currently available! Today's dilemma concerning serum markers is their lack of both sensitivity and specificity. Although a cancer may have occurred, its "products" may not yet have been released into the systemic circulation and cannot, therefore, be detected by serologic studies. The detection system may not be sensitive enough to measure subtle increases in serum concentration of these markers, or too many patients without the disease test positively for the markers. Therefore, combinations of serologic markers that are theoretically independent of one another are often used to screen patients, in the hope that one or more of the markers in the combination will detect the patients with the disease. The known markers are generally divided into categories, depending upon their origin, including tumor-associated antigens, hormones, enzymes, and products of known biochemical sequences or pathways.

As new markers are evaluated, there are several questions that must be considered to determine their incremental value in patient assessment. Not only should the marker be produced by only tumor cells and not normal cells, it should be detectable early in the natural history of the disease, while the cancer is localized to the site of origin. A serum marker would be the easiest marker to employ, requiring venipuncture only for determination. The measurable serum level of the ideal marker would fall to nil after successful treatment, and be detectable again only if metastasis occurred, so that its serial measurement might predict outcome of treatment. Its circulating value should be proportional to tumor burden, so that regression after treatment could also be measured simply. Finally, responding to the exigencies of contemporary healthcare concerns, the marker must be inexpensive. Each of these criteria is difficult enough to achieve alone; the successful combination of them is even more formidable! The benefits to the practice of clinical medicine, however, would be inestimable.

2.4. Invasive Cancer vs Noninvasive Cancer

It is now accepted that a majority of ductal carcinomas *in situ* (DCIS) may never progress to invasion and, therefore, they may require less drastic therapeutic (surgical) intervention than infiltrating carcinoma ***(4)***. Not always is the distinction between invasive and noninvasive carcinoma easy. Invasive cribriform ductal carcinoma may deceptively resemble cribriform ductal carcinoma *in situ*. Moreover, localized invasion (microinvasion) may be hard to distinguish from lobular cancerization (involvement of terminal ducts and lobules by ductal carcinoma). The unifying feature of all *in situ* carcinomas is an intact basement membrane. A major basement-membrane component is collagen IV. Antibodies to collagen IV may visualize basement membrane components by immunohistochemistry and identify gaps in the basement membrane where so-called microinvasion is suspected. This has proven an effective way to distinguish between invasive and *in situ* carcinoma ***(5,6)***.

2.5. Quantitative DNA Analysis

Many malignancies exhibit chromosomal abnormalities. They may not demonstrate a diploid or tetraploid DNA content. These uneven DNA peaks that differ from the normal population are termed aneuploidy. It is commonly accepted that tumors with aneuploid cell populations are less favorable than those whose cells are diploid. Although perhaps true, such a large majority of breast cancers are aneuploid that this finding is not sufficiently discriminating. From the DNA distribution curves obtained by flow cytometry or image analysis of cell suspensions, the cells that are in the S-phase of division can be estimated, and this percentage has been correlated inversely with prognosis (the higher the S-phase fraction, the worse the prognosis). This parameter currently appears to be more reliable at predicting outcome than measuring ploidy alone ***(7–9)***.

2.6. Cytogenetics

DNA analysis by flow cytometry is a relatively crude assessment of the nature of chromosomal abnormalities in tumor cells. More illuminating is visualization of the chromosomes themselves using cytogenetic techniques that are beyond the scope of this chapter. These techniques, including *in situ* hybridization, may reveal numerical aberrations using chromosome-specific probes.

2.7. Proto-Oncogenes, Oncogenes, and Translocations

The difference between a proto-oncogene and an oncogene may be a qualitative one or a quantitative one. A qualitative change may be as little as a point mutation on a chromosome. Quantitatively, if an excess amount of its product is present, the proto-oncogene is then considered to have become an oncogene. Amplification is the production of an excess amount of this gene or gene product due to mutation or rearrangement within the regulatory region of the gene. The identification of nonrandom gene translocations is an important technologic achievement in cancer diagnostics. By nonrandom, it is implied that a particular type of malignancy is associated with a specific genomic change, such as the association of the "Philadelphia chromosome" with chronic myelogenous leukemia as a translocation between chromosomes 9 and 22. Other examples occur in Burkitt's lymphoma and in follicular B-cell lymphomas. When they occur, these are highly specific tumor markers for unique clinical entities. Unfortunately, unlike these well-defined markers, random chromosomal abnormalities occur that are not associated with a particular morphological change, and these then give rise to clinical cancer ***(10–12)***.

2.8. Gene Products as Tumor Markers

Gene sequences may be transcribed into messenger RNA, which may then be translated into proteins. In principle, both specific messenger RNA and proteins may be detected. Historically and technically, the most commonly used markers to date represent serum proteins, such as α-fetoprotein or CEA. These are not tumor specific, although other markers may be at least tissue specific, such as PSA for prostate. The identification of markers in a metastasis, for example, may help to identify the origin of the disease. Additionally, the concentration of these markers in the serum may offer information about the likelihood of recurrent cancer, since there may be quantitative differences in concentrations of these markers in patients in remission and those experiencing recurrence.

Immunohistochemical techniques identify the protein products of an oncogene. An antibody specific to the protein stains a slide containing the appropriate cells for examination, usually from the tumor itself, and the quantitative determination of the antibody present indicates an accurate estimate of the oncogene product. Since these antibodies stain the product of the oncogene itself, adjacent normal cells that may contain the proto-oncogene remain unstained. Current techniques to isolate, produce, quantify, and compare the multitude of antibodies to these gene products are controversial as they evolve, since they are of such great biological (and commercial) importance. Because of the multitude of complicated events that culminate in the clinical manifestation of breast cancer, the identification of new gene products that may affect prognosis and, therefore, influence treatment recommendations has become a burgeoning industry itself, worthy of its own publications (such as *Oncogene* and *Oncogene Research*).

Intermediate filament (IF) analysis has been used to aid the identification of some cancers. Intermediate filaments provide a communication network between the extracellular matrix and cytoplasmic structures, and neoplastic cells generally retain the IF of the cell of origin. Cytokeratins are present in squamous (complex) and simple epithelium and the malignancies derived from these tissues (carcinomas). When dealing with a poorly differentiated cancer, the expression of cytokeratins characterizes the tumor as a carcinoma. An additional IF is vimentin, historically an IF regarded as a marker of mesenchymal cells. Most carcinomas are negative for vimentin. Another IF is desmin, most closely associated with tumors derived from muscle cells. Similarly, there are other markers that define cell origin often used to suggest possible diagnoses in tumors of questionable morphologic appearance. Hematologists use a variety of these markers to differentiate the hematologic/lymphopoietic malignancies from one another.

3. Tumor Markers Utilized for Breast Cancer

3.1. Circulating vs Tissue Markers

3.1.1. Tumor-Associated Antigens (TAA)

Concerning breast cancer, there are few markers available that have any relevance to contemporary clinical practice. Although there are a host of putative markers, each with its ardent proponents (usually its discoverer), probably the only ones currently used with any degree of enthusiasm outside the research community are CEA and CA 15-3.

CEA, at its best, may correlate with stage of disease at diagnosis, whether localized or disseminated, and it is commonly used currently to monitor patients after initial treatment. An elevation in the CEA level is interpreted as a possible indication of recurrence, even in the absence of overt evidence of metastasis. Whether this so-called lead-time, between the detection of as yet subclinical metastasis and its subsequently clear manifestations, contributes to any gain in length or quality of survival has not been established. Moreover, it is the presence of visceral or osseous metastasis rather than soft tissue metastasis that is related to CEA elevations. Despite its ubiquitous application, because CEA levels may be elevated in patients without disease and may overlook as many as half the patients with metastasis, it is of questionable efficacy in the care of breast cancer patients. It is certainly not cost effective.

CA 15-3 is another, recently described, breast cancer-associated antigen. As with other such markers, progression of disease has been associated with increases in serum levels ***(13,14)***. However, CA 15-3 has engendered enough controversy to make regulatory agencies insist that clinicians who request this test affirm that the results of the determination will not be used to influence treatment without additional (unspecified) information. Whether appropriate as a criterion of use, many insurance carriers have thus far refused reimbursement for it without further tedious documentation. This requirement alone has probably minimized its use by clinical oncologists.

More recently, another serum marker has entered the commercial marker market, termed "TRUQUANT®BR™RIA," or CA27-29. First studies seem to indicate that this marker is more sensitive and specific than CA15-3 ***(15)***. This marker has been granted premarket approval by the FDA for patients previously treated for stages II and III breast cancer ***(16)***. However, all the above-mentioned reservations with regard to serum markers also apply to CA27-29.

Other serum markers are of even less documented value, despite what are often exciting predictions of success as they are announced. As a plethora of these TAA appear, it becomes tempting to consider their use as an accompaniment to one another, as though their use in combination might detect recurrent disease even earlier. Even if true, it is yet to be documented that earlier diagnosis of recurrence improves outcome. However, as the treatment of women with metastatic disease advances, now using autologous bone marrow transplantation, for example, to permit more intensive chemotherapy, and as new chemotherapy agents and protocols are implemented, it is likely that the size of the tumor burden will be a major factor in treatment decisions. At that time, the consequences of the earlier detection of recurrence will justify the more widespread use of multiple markers, and the isolation of these markers will become even more important.

3.1.2. Traditional Findings and Clinical Assessment

Over the course of the past decade, a multitude of tumor markers has evolved, and almost as many have been forgotten. Notwithstanding the advances in the identification of genomic markers that may someday be used to predict who is destined to develop carcinoma of the breast, it is probably fair to state that the standard against which prognostic markers must be judged is the careful evaluation of the patient by a capable clinician along with the appropriately fixed, stained, and mounted microscopic sections of the tumor (and the regional, i.e., axillary, nodes) studied by an equally skilled pathologist.

The assessment of the clinical stage of the tumor when detected is the *sine qua non* of the care of the patient with breast cancer. Most of the therapeutic recommendations and the estimations of outcome are based upon this evaluation. This includes a carefully performed history and physical examination, review of mammograms and other diagnostic studies, and finally the microscopic study of the tumor sections. *The size of the tumor and status of the axillary lymph nodes are the most important factors that predict the patient's outcome.* Other variables are of secondary importance. Now that quantifiable prognostic markers are available, and, because histologic grading is somewhat subjective, its value has been often deprecated; details of tumor morphology are often ignored. Although few studies directly correlate tumor morphology with these other markers, the results of the currently available marker assays are often quite accurately predicted by study of the morphology of the tumor itself. Whether one or another grading system is used is less important than the need to document the architectural arrangement of cells, the degree of nuclear differentiation, and the rate of mitosis ***(17,18)***.

3.1.3. Proliferation Markers

It is almost intuitive that a rapidly dividing tumor would be more aggressive than one that proliferates more slowly. The ability of tumor cells to divide does not itself predict the likelihood of metastasis; the many mitoses seen in typical medullary carcinomas of the breast, yet the apparently more favorable prognosis of this tumor, bears testimony to this observation. In general, however, it is an oncologic axiom that proliferation rate varies inversely with outcome. An attempt to quantify tumor kinetics has become part of the study of virtually all patients with breast cancer. For example, the thymidine-labeling index (TLI) is a highly sensitive and accurate technique to measure dividing cells, and is predictive of both recurrence and death from breast cancer. It is, unfortunately labor intensive and time consuming, and its accuracy burdened by many technical problems. From a purely pragmatic perspective, TLI is unlikely to be adopted as a technique for clinical use outside the confines of a purely research environment.

3.1.4. Mitotic Index

Traditionally, as they look at microscopic sections of tumors, pathologists gain an excellent impression of the proliferative potential of tumor cells by counting mitotic figures. This is often expressed as "mitoses per high power field." Quantitative evaluation is difficult, however, and the proliferative potential of the tumor may be underrated because of a low number of cells in the mitotic phase. The aim of proliferative assessment is the capability of detecting active (DNA-replicating) cells by DNA analysis, either by flow cytometry or image analysis.

Parenthetically, there has been a resurgence of interest in the quantification of the morphologic features of breast cancers. The so-called morphometric prognostic index (MPI), which includes the mitotic activity index, tumor size, and lymph node status, has been correlated strongly with outcome and is stated to be reproducibly assessed in routine histologic sections. If this technique is as reliable and reproducible as suggested, perhaps the need for more expensive markers would be obviated.

3.1.5. DNA Analysis

The analysis of DNA within breast tumor cells may be accomplished by studies of isolated cells by flow cytometry or by touch imprints from fresh tumor material. Both techniques use dyes that combine with DNA; in flow cytometry, fluorescent dyes are used that are detected by laser excitation; for image analysis, traditional Feulgen staining is used and evaluated in an appropriate optical system. Both techniques have disadvantages; in flow cytometry, the tissue is disrupted to gain access to single nuclei. When the nuclear suspension is evaluated, there is no distinction between tumor and normal nuclei. For example, an increase in inflammatory cells in the neighborhood of the tumor increases the number of diploid cells in the population studied. Similarly, if there is a small volume of tumor in an abundant stroma, sampling errors are common. The advantage of flow cytometry is the ability to evaluate a large number of nuclei for DNA content rapidly. Image analysis is more labor intensive and requires computer-assisted microscopy. Intact nuclei are required. DNA analysis offers a glimpse into the proportion of cells in the DNA-synthesis phase of the cell cycle, the S-phase, and reveals uneven DNA content (aneuploidy). Because tumor cells may have differing numbers of chromosomes, the DNA distribution curves may be abnormal. Although the evaluation of ploidy itself appears to be of little merit alone in determining the prognosis of breast cancer, there is general agreement that the proportion of cells in the S-phase of the cycle is a predictor of outcome, and the greater this number, the worse the prognosis ***(19,20)***. The observations about S-phase and DNA content have led to the search for other markers of cellular proliferation and information about their significance, individually and collectively.

3.1.6. Ki-67

Monoclonal antibody (MAb) Ki-67 is specific for an antigen associated with cell nuclei, expressed in the G1, G2, and M-phases of the cell cycle, as well as those cells in S-phase ***(21,22)***. Although the function of this antigen is unknown, the available evidence indicates that it is an accurate marker of cellular proliferation, correlating well with TLI ***(23)***. This marker, therefore, may be used to determine the proportion of actively dividing cells in a given tumor. In addition to the presumption of metastatic potential related to cellular proliferation, this marker may also help predict which tumors are more likely to be sensitive to radiation or chemotherapy. Ki-67 antibody staining required fresh (frozen) tissue until very recently, and long enough follow-up information was not available to assess its usefulness as a separate prognostic indicator. The few observations cited suggest an inverse relationship between Ki-67 positivity and disease-free survival, a "positive result" generally defined as 15% or more of tumor cells examined staining for this marker.

Newly reported MAbs termed MIB-1-3 detect an epitope of Ki-67 antigen that survives formalin fixation ***(24)***. This means that tissues already fixed and embedded may be retrospectively examined for Ki-67 expression. This marker may be compared with other information already extant to examine its role in predicting recurrence. We have begun to retrieve paraffin blocks in our patients with DCIS treated by excision and surveillance alone to see if we can predict which patients are at highest risk for recurrence by determining the proportion of tumor cells that stain for Ki-67 ***(25)***. The small population of tumor cells in patients with DCIS and other clinically occult cancers detected by mammography previously limited the use of many of the biological markers. These restrictions have now been removed, thanks to the use of immunochemical techniques. The Ki-67 determination, because of its ability to tag all proliferating cells, not just those in a single phase of the cell cycle, may prove to be one of the most important prognostic indicators in breast cancer.

3.1.7. Proliferating Cell Nuclear Antigen (PCNA)

Another nuclear antigen related to cellular growth is the PCNA. Like Ki-67, it is a cell cycle-regulated nuclear protein and is directly involved in DNA synthesis ***(26)***. Monoclonal antibodies to this marker that may be used in conventionally fixed histologic material have been described, so that this marker does not suffer from the limitations imposed by the need for fresh tissue ***(27)***. Unfortunately, there is more controversy about the value of PCNA than about other markers. While some correlation has been observed that would indicate similar information from the PCNA index as from measurements of the S-phase fraction and other clinicopathologic variables, there is enough debate about its value to question its use, at least currently, as an independent predictor of outcome in breast cancer ***(28)***. Its major advantage was its ability to be detected in paraffin-embedded fixed tissue. Now that antibodies against Ki-67 antigens that also survive fixation are available, PCNA is obsolete, until and unless a different function for this marker can be discovered.

3.2. Oncogene Products

*3.2.1. HER-2/*neu *(c-*erb*B-2)*

Amplification of the proto-oncogene HER-2/*neu* (c-*erb*B-2 is a synonym) and overexpression of its protein product has been implicated as an indicator of poor prognosis in breast cancer ***(29)***. After an initial flurry of excitement about its prognostic significance, controversy has arisen about its independent importance *(30)*. Whether the overexpression of this proto-oncogene is of consequence in both node-negative and node-positive women is an additional topic of debate. Its protein product may be amplified in as many as one-third of breast cancers, and antibodies to it are measurable in fixed tissue. An interesting observation about the HER-2/*neu* antibody is its overexpression in women who have nursed, without regard to length of lactation *(31)*. As of this date, it is probably reasonable to question whether measurement of the amplification of this proto-oncogene will become a useful prognostic variable in breast cancer, i.e., one that might itself influence a treatment recommendation. Currently, it should be considered as one among many markers that require further study before making categorical comments about their value.

Exciting but not germane to this discussion is a new interest in this particular marker as a target for immunotherapy in women with metastatic breast cancer. Currently, abundant research in this area of immunotherapy is under way, using antibodies that attack tumor cells that overexpress this marker *(32)*.

3.2.2. p53 Tumor-Suppressor Gene

So-called tumor-suppressor genes perform multiple, but incompletely understood, functions relating to the growth of normal cells. When these particular genes are inactivated, their regulatory function is impaired, and uninhibited neoplastic transformation may occur. p53 is a human nuclear protein, and mutation of its gene may occur in association with the development of malignancies. There have been reports of p53 alterations in breast cancers, suggesting that overexpression of this gene product, detectable immunohistochemically, correlates with absence of estrogen-receptor activity and high nuclear grade; it may have importance as an independent indicator of prognosis in both node-negative and node-positive patients *(33,34)*. Formalin-fixed, paraffin-embedded tissue may be used for immunohistochemical analysis. The data implying the importance of this marker in breast cancer continue to accumulate, but the suggestion that p53 protein accumulation is second only to lymph node status as an indicator of outcome demands its further intense investigation.

3.2.3. p21

p21 (WAV-1/CIP-1) has been recently identified as a binding partner that interacts with p53, differentiated between the wild (normal) type and mutated p53 *(35–38)*. WAV-1 mediates tumor-suppressor activity of p53 and vice versa. In normal breast tissue, it is apparently found in high amounts, whereas in breast carcinoma with mutated p53, the values for this marker are low. Monoclonal antibodies against WAV-1 exist and are currently experimentally used for WAV-1 detection in breast carcinomas. Since the protective effect of tumor-suppressor gene p53 is probably mediated through WAV-1, its expression should implicate favorable prognosis.

3.2.4. Epidermal Growth Factor Receptor (EGF-R)

Epidermal growth factor (EGF) is a polypeptide that influences cellular differentiation in different cell lines and may play a role in carcinogenesis ***(39)***. Overexpression of its cell membrane receptor (EGF-R) gene product has been associated with a number of malignancies, including breast cancer. Receptors for EGF have been documented on both primary tumors and metastases, and data suggest an inverse relationship between estrogen receptors and EGF receptors ***(40)***, which, if valid, would imply a correlation between the expression of EGF-R and allegedly unfavorable prognostic indicators.

3.2.5. c-myc *and* ras *Gene Products*

The proto-oncogene, c-*myc*, has been shown to be amplified in some invasive ductal carcinomas ***(41,42)***. This gene may be involved in the development of neoplasia and has been a new marker receiving increasing attention. Similarly, amplification of genes of the *ras* family have also been implicated in the progression of breast cancer. Thus far, these markers are of interest but are not yet clinically useful.

The excitement generated by investigators as new gene products associated with breast cancer are identified is not yet proportional to their value in detection, treatment, or predictions of prognosis. Thus far, the only oncogene that seems to have some relevance to current clinical concerns is c-*erb*B-2.

3.2.6. bcl-2

A protein that is expressed in the majority of breast carcinomas, bcl-2 has attracted much interest recently because of its involvement in the regulation of programmed cell death (apoptosis) ***(43–47)***. High levels of bcl-2 inhibit apoptosis, whereas low levels are conducive to programmed cell death. Interestingly, in our own experience, metastatic breast carcinoma has been almost exclusively positive for bcl-2 ***(45)***. Therefore, strong positivity for this marker may indicate tumors that are more likely to metastasize. If true, the presence of this marker, even in patients with small, node-negative tumors, might indicate the usefulness of adjuvant chemotherapy, since these tumors would be considered among the subsets at highest risk for recurrence.

3.2.7. Cathepsin D

Cathepsin D is an estrogen-induced glycoprotein that has both growth-promoting and proteolytic activity. Initial studies using monoclonal antibodies against this enzyme indicated that, at least for node-negative patients, a higher level of cathepsin D in the tumor was associated with a shorter disease-free interval *(48,49)*. There was no apparent difference noted in node-positive patients. For this reason, consideration has been given to an elevated level of cathepsin D as one criterion in the selection of node-negative patients for adjuvant chemotherapy. Recently, the usefulness of this marker has been challenged.

3.2.8. Factor VIII-Related Antigens (FAVIIIRA), CD31 and CD34

FAVIIIRAs, CD31 and CD34 are endothelial markers widely used for detection of vessels by immunohistochemistry. Many attempts have been made to correlate the extent of vascularization with metastatic potential *(50–57)*. It is probably fair to state that at this time quantitative evaluation of vascularization is of limited value to predict metastatic behavior. In addition, quantification by image analysis is cumbersome, and burdened with technical problems.

3.2.9. NM23

Metastasis suppressor gene (NM23) has recently been discovered in murine melanoma *(58–60)*. In humans, two forms of this gene have been identified: NM23.H1 and NM23.H2. These two genes code for the A and B subunit of nucleoside diphosphate kinase (NDPK). It has been suggested that the gene product of NM23.H1 acts as a metastasis suppressor gene *(58–74)*. Presence of NM23.H1 has been frequently associated with the inability of a tumor to become invasive and metastasize. More recently, it has been shown that expression, or overexpression, of NM23.H1 stimulates the assembly of basement membrane components *(67)*. Therefore, expression of NM23 may be involved in surveillance of basement membrane integrity. Looking at NM23 expression in DCIS may be particularly helpful to separate DCIS destined to remain indolent from DCIS that may predict the subsequent development of invasive cancer.

3.3. Steroid Hormone Receptors

The relationship between estrogen and breast cancer has been known for almost 100 years, since Beatson's observation that castration could produce remission *(75)*. In the 1950s, oophorectomy became a widely used accompaniment to radical mastectomy, until it became recognized that the overall length of survival was not affected by this procedure. Oophorectomy apparently delayed the emergence of recurrence in some women, and also was noted to induce regression after recurrence in others. How hormone dependency could be predicted or measured, and which women would have their course favorably affected by castration and/or other hormonal manipulations, however, was unknown until the discovery of an estradiol-binding protein in the rat uterus and subsequent research that identified the way in which estrogens interact with human breast cancer cells to alter their growth *(76)*.

The first report that correlated outcome with the level of steroid receptors was published in 1977 ***(77)***. The body of knowledge about estrogen (ER) and progesterone (PR) binding proteins in breast cancer has exploded, with reliable and reproducible methods of assay for these receptors in tumor tissue. Currently, the most commonly used technique of analysis is a biochemical dextran-coated charcoal (DCC) binding assay, and this is the "reference" standard for estrogen and progesterone receptor determinations. However, inaccuracies of measurement are still inherent in this technique, especially as mammography has detected smaller and smaller tumors. It is necessary to freeze the specimen intended for analysis as soon as it is removed from the patient. As little as a 15-min delay may render the test inaccurate. Sampling errors may occur if the specimen does not contain enough tumor, or if there is a significant enough desmoplastic response within the contiguous tissues to make the ratio of malignant cells to other breast cells (stromal or epithelial) too low. It is virtually impossible to measure the receptor content of *in situ* carcinomas (DCIS) using this technique. Another error may be introduced if the patient is either taking exogenous hormones or is producing endogenous estrogen in sufficient quantity to bind to the available receptor sites. These errors are all in one direction, producing false negative rather than false positive results. The optimal specimen for an accurate DCC assay is 1.0 g of tumor (approximate 1.0 cm^3 in volume), although as little as 0.1 g may be used.

Fortunately, within the past few years, immunocytochemical assay of ER and PR has been accomplished, and the concentration of these receptors that are bound to tumor nuclei can be counted by staining the receptors with a monoclonal antibody. This is reported as the proportion (percent) of cells that stain positively for the receptor antibody. This technique avoids sampling errors, since the pathologist can determine if the receptor is expressed on a normal or a malignant cell, and extraneous tissue does not affect the result. The additional advantages of the immunochemical assay are its applicability to formalin-fixed, paraffin-embedded tissue, and the ability to perform these analyses on the smallest of tumors, even the nonuniform sections that are the usual findings in intraductal carcinomas (DCIS).

The vast bulk of data that relate both treatment and outcome to measured levels of estrogen and progesterone receptors in breast cancer indicate a direct correlation between them, i.e., the greater the expression of these receptors, the more likely the tumor to respond to hormonal therapy and the better the outcome ***(78)***. Low levels of hormone receptors are more often associated with recurrence, and when metastasis occurs, response to treatment correlates with receptor activity.

A small minority of patients (<20%) with negative receptors will still respond to hormonal manipulation. Because the majority of the clinical studies that correlate the expression of hormone receptors and patient outcome have been based on the biochemical (DCC) determination of the presence of receptors, it is interesting to speculate that these patients who respond to hormonal manipulation despite low levels of receptors may be patients in whom sampling errors produced the negative results. Had the receptor assay been performed by the immunocytochemical technique, would the results have been the same?

3.4. Gene Deletions

3.4.1. Loss of Heterozygosity (LOH)

Extensive allelotyping of breast cancer for gene deletions of loci on multiple chromosomes has been reported ***(79–100)***. Deletion of genomic material is important, because the lost segment of DNA may contain tumor suppressor activity. Gene deletions are discovered by polymerase chain reaction (PCR) using microsatellite probes to various chromosomes and sites. Tumor-suppressor genes thought to play a role in breast cancer include p53 at 17p13.1, Rb at 13q, colorectal carcinoma gene DCC on 18q, and Brush-1 (proximal to Rb on 13q) in close proximity to the inherited early onset breast cancer gene BRCA2. Several genes located on 17q are implicated in breast cancer oncogenesis, such as the recently cloned BRCA1 gene at 17q21 and the metastasis-suppressor gene NM23 (distal to BRCA1). In addition, a plethora of allelic losses with more or less significant breast carcinoma associations on virtually all chromosomes have been reported.

3.5. Cancer Susceptibility Genes

Presuming that the genetics of breast cancer will be ascertained, so that those individuals "scheduled" to develop the disease can be identified almost from conception, a whole new approach to this disease will evolve, one of prevention rather than detection or treatment. Whether such a discussion is appropriate in this chapter is moot, but since such a gene would certainly be considered the ultimate "marker," how can it not be addressed? With the 1990 publication of the apparent localization of a gene for inherited susceptibility to breast (and ovarian) cancer to chromosome 17q21 by King and her colleagues, christened BRCA1, this era of breast cancer genetic investigation truly began to explode ***(101)***. This gene was precisely identified in 1994 by a team at the University of Utah ***(102)***. A second gene, BRCA2, traced to chromosome 13q12-q13, was identified by Wooster and colleagues, but this gene, unlike BRCA1, played little role in the development of ovarian cancer in these women ***(103)***. It was, however, apparently more closely linked to the families that had a male relative who had developed breast cancer.

It is estimated that by the age of 70, as many as 90% of the women with BRCA1 will develop breast cancer. While in women, BRCA1 also predisposes to ovarian cancer, men carrying the BRCA1 gene are at increased risk to develop prostate cancer, and perhaps also colon cancer.

The detection of these susceptibility genes has led to further detection of germline mutations of them that identify populations of women (and men) at even greater risk for the lifetime development of breast and other cancers. A single mutation of BRCA1, 185delAG, has been noted in approx 20% of Ashkenazi Jewish women with early onset breast cancer, and in almost 1% of all Ashkenazi Jewish women ***(104,105)***. Recently, other mutations on BRCA1 and a mutation on BRCA2 have been identified as well, also among Ashkenazim ***(106)***. All of these current findings suggest that the risk of breast cancer before age 40 in this population of Jewish women with these mutations on BRCA1 is more than 30 times the risk of breast cancer in the general population; the mutation on BRCA2 implies a fourfold risk. Thus far, no other mutations have been identified that affect a specific ethnic or racial group. Fortunately, because most breast cancers are not familial, the overall risk of breast cancer in these women is not as high as these data imply when taken out of the overall context of breast cancer incidence.

4. Discussion

Prognostic variables associated with outcome in breast cancer have been addressed at several Consensus Conferences sponsored by the National Cancer Institute (NCI). In 1985, adjuvant chemotherapy was noted as "standard care" for premenopausal women with positive axillary lymph nodes. At the same time, because the effectiveness of endocrine manipulation "closely correlates with the measured receptor levels," obtaining this information about receptors was implicitly, if not explicitly, recommended. This conference recognized the prognostic significance of tumor size, hormone receptors, cell differentiation, TLI, and aneuploidy as well as lymph node status, and it was recommended that, for "certain high-risk patients in this (premenopausal, node negative) group, adjuvant chemotherapy should be considered." "High risk" was not defined further, and there was no additional discussion of biological markers, either as independent indicators of prognosis or as affecting treatment recommendations.

In 1988, the oncologic community was stunned by the highly publicized and controversial National Cancer Institute's "Clinical Alert on Breast Cancer" concerning recommendations for adjuvant chemotherapy in node-negative patients ***(107)***. This brief summary was mailed to about 13,000 physicians and released to the media in advance of publication of results in any peer-reviewed journal, based upon "persuasive" information that was not published until months later and still remains controversial. However, since then, oncologists have sought the appropriate algorithm to discriminate among node-negative patients and offer treatment to those deemed at greatest risk for recurrence. Determination of estrogen (and progesterone) receptors has become one of the cornerstones of this decision tree, since the absence of receptors is one of the criteria often used to justify adjuvant chemotherapy in this group of patients ***(20,108)***. Although the data currently available may not indicate the use of receptor status alone to influence treatment in node-negative patients, the level of estrogen/progesterone receptors does predict tumor response to tamoxifen. The importance of knowing this information is well-established enough that failure to attempt to measure estrogen/progesterone receptors on a breast cancer specimen is probably a culpable error of omission.

The adoption of prognostic factors as a consideration was publicly addressed by the National Cancer Institute in its 1990 consensus statement concerning early-stage breast cancer ***(109)***. In order to be clinically useful, the consensus suggested that any prognostic factor have an independent predictive value and have therapeutic importance, as well as being widely available and reproducible. Cost was not addressed, but this issue has become increasingly more important as the fraction of the GNP devoted to healthcare has reached double digits. Mentioned in the NCI consensus statement were tumor size, estrogen and progesterone receptors, nuclear grade, histology, proliferative rate, and "other factors." It was not recommended to treat node-negative patients with tumors less than 1 cm in diameter, but for patients with larger tumors, it was suggested that the various factors putatively associated with prognosis should be weighed and individual patient recommendations considered. Listed first among the various directions for future research was the definition of prognostic factors and their relationship to one another.

The establishment of tissue banks and the ability to evaluate each emerging prognostic factor as discovered will certainly permit increasing accuracy in predicting outcome, thereby influencing therapeutic recommendations significantly. Nevertheless, we cannot wait for tomorrow's science to benefit the 120,000 or more node-negative American women who will be diagnosed with breast cancer this year. What, therefore, are reasonable guidelines that may be used currently to integrate the information currently available about these various factors with more traditional, "standard" treatment considerations?

There are charts available that assign node-negative patients to "good" and "high" risk categories based upon these factors ***(110)***. The criteria for assignment to one or the other category usually include tumor size, presence or absence of hormone receptors, and some measures of tumor proliferation, by nuclear grade or tumor histology, or measurement of DNA content or S-phase fraction. Although these various factors do correlate with one another, it is tacitly assumed that the various factors carry equal weight and that they correlate well enough that they will all be aligned in one column or the other. If that were indeed true, then measurement of one alone would be sufficient to affect treatment. If, however, as suspected, the factors are different although altogether related, which is the most important to consider? By confining the discussion to node-negative patients, it is already assumed that lymph-node status is the most important prognostic factor, and this assumption is currently uncontested. Nuclear grading is subjective and therefore less reproducible than more quantifiable markers. However, this information is so readily available from review of microscopic slides that major efforts should be made to standardize this system. A variation of the morphometric prognostic index or mitotic activity index, both mentioned previously, for node-negative patients might be the logical extension of this attempt.

The presence or absence of hormone receptors is thus far the only quantifiable tumor marker that is not currently controversial, the vast majority of investigators conceding that the absence of receptors predicts a greater likelihood for recurrent cancer than does their presence. There has been a measurable difference between these two groups of node-negative patients with respect to recurrence, albeit only about 8–10% ***(111)***. Additionally, receptor information predicts the response to hormonal treatment, an added advantage that other markers do not possess, and the response to hormonal manipulation is probably directly related to the magnitude of this marker.

Using proliferation markers as criteria for adjuvant chemotherapy, and categorizing patients into "high" and "low" risk groups based upon arbitrary cutoff values is tempting, but perhaps not yet completely justified by the available information. The logical premise that rapidly proliferating tumors are more dangerous than those that divide slowly is difficult to challenge, but the assertion that measurement of DNA content and/or S-phase fraction is enough to make this distinction is questionable. At the ends of the scale, for example, S-phase fractions above 20%, or less than 2%, decisions are easier than when they are in the mid-range. As information about Ki-67 and p53 accumulate, it is our impression that they will be more helpful than ploidy or S-phase alone at predicting prognosis. The cells in S-phase fraction are included with those measured by Ki-67, so that the overall information obtained by Ki-67 is more likely to be accurate than S-phase alone. p53 overexpression may be an additional, independent marker of the propensity of a breast cancer to metastasize. Both may be retrieved from fixed tissue so they are not time-dependent.

As noted, when all of the prognostic factors, irrespective of their relative importance (known or speculated), are aligned on one side of the "good risk/bad risk" chart, therapeutic decisions are relatively straightforward. If complications of therapy are minimal, and they generally are, adjuvant treatment is justifiable for the "average" patient. However, when these factors are scattered, some on one side and some on the other, presuming that they do measure independent variables, or if the patient has other major medical problems, the decisions are more difficult to make, since the order of importance of these several factors is another controversy. There is still no substitute for clinical judgment in the care of patients with breast cancer, but it is reasonable to assume that biological markers will provide information that will be incorporated into clinical staging systems for breast and other cancers, to codify the diagnosis as well to define therapy.

Finally, the exponential growth of information about breast cancer susceptibility genes has ushered in the newest era of "marker" studies. Although neither BRCA1 nor BRCA2, nor their mutations, indicate risk precisely, it is crucial to consider the implications of these markers, even as they continue to evolve. Gene alterations may themselves not be the final determinants of risk. What, if any, roles do environmental, hormonally or other inherited factors play in the development of breast cancer? As news about BRCA1 and BRCA2 has been widely publicized, many women with family histories of breast cancer have bombarded their physicians with inquiries about undergoing these studies, most of them without first considering how that information might be used, and by whom. For example, would the presence of one of these genes be considered a "pre-existing condition" when a woman applies for health insurance? Would it make an individual uninsurable? Several states have begun to enact legislation that would prevent discrimination based upon genetic information. The psychological impact of genetic testing for breast cancer has not been studied well enough yet to predict its effects on so many individuals. Does prophylactic mastectomy make any sense, and even if it does, when should it be performed? Unfortunately, it is currently possible to arrange for genetic testing for breast cancer on the Internet. The commercial exploitation of genetic testing for breast cancer has just begun. It is certainly destined to be a multimillion dollar business.

Each patient must understand the difficulty of using an incomplete, evolving body of information about these various biological and genetic markers to influence contemporary therapy. Until we can truly modify the course of breast cancer so that it can be prevented, we are committed to the continual revision of our recommendations based upon the best available information, subject to daily change.

Carcinoma of the breast is arguably the best-studied solid tumor, and the scientific literature concerning the molecular biology of this disease continues to expand exponentially. It is an excellent example of the use of biological markers to learn more about predilection, incidence, diagnosis, treatment, and outcome. Information gathered about breast cancer will be applicable to the study of other solid tumors and to the study of malignancies in general. The next generation of oncologists will almost certainly use genetic information widely to screen large populations for risk factors to determine who is destined to develop these diseases, perhaps even when, and thus target programs of prevention, earlier detection, and treatment to individuals identified by these disease-specific probes. The millennium that currently eludes us is perhaps not so far away!

References

1. Hansen, M. F., Koufos, A., Gallie, B. L., Phillips, R. A., Fodstad, O., Brogger, A., Gedde-Dahl, T., and Cavenee, W. K. (1985) Osteosarcoma and retinoblastoma: a shared chromosomal mechanism revealing recessive predisposition. *Proc. Natl. Acad. Sci. USA* **82,** 6216–6220.
2. Cavenee, W. K., Hansen, M. F., Nordenskjold, M., Kock, E., Maumenee, I., Squire, J. A., Phillips, R. A., and Gallie, B. L. (1985) Genetic origin of mutations predisposing to retinoblastoma. *Science* **228,** 501–503.
3. Cavenee, W. K., Dryja,T. P., Phillips, R. A., Benedict, W. F., Godbout, R., Gallie, B. L., Murphree, A. L., Strong, L. C., and White, R. L. (1983) Expression of recessive alleles by chromosomal mechanisms in retinoblastoma. *Nature* **305,** 779–784.
4. Schwartz, G. F., Schwarting, R., Cornfield, D. B., Shen, R., and McDade, T. (1996) Subclinical duct carcinoma of the breast (DCIS): Treatment by local excision and surveillance alone. *Proc Am. Soc. Clin. Oncol.* **199,** 15.
5. Charpin, C. L., Andrac, B., Devictor, M. C., Habib, H., Vacheret, L., Xerri, M., Lavaut, N., and Toga, M. (1989) Type IV collagen immunostaining and computerized image analysis (SAMBA) in breast and endometrial disorders. *Histopathology* **14,** 47–60.
6. Lanzafame, S. and Puzzo, L. (1988) Morpho-functional changes in the basement membrane in benign disease and in preinvasive and microinvasive carcinoma of the breast: immunohistochemical study with anti-collagen IV monoclonal antibody . [Italian]. *Pathologica.* **80,** 309–318.
7. Fisher, B., Gunduz, N., Costantino, J., Fisher, E. R., Redmond, C., Mamounas, E. P., and Siderits, R. (1991) DNA flow cytometric analysis of primary operable breast cancer. Relation of ploidy and S-phase fraction to outcome of patients in NSABP B-04. *Cancer* **68,** 1465–1475.
8. Frierson, H. F., Jr. (1991) Ploidy analysis and S-phase fraction determination by flow cytometry of invasive adenocarcinomas of the breast. *Am. J. Surg. Pathol.* **15,** 358–367.
9. Joensuu, H., Toikkanen, S., and Klemi, P. J. (1990) DNA index and S-phase fraction and their combination as prognostic factors in operable ductal breast carcinoma. *Cancer* **66,** 331–340.
10. Ngan, B. Y., Chen-Levy, Z., Weiss, L. M., Warnke, R. A., and Cleary, M. L. (1988) Expression in non-Hodgkin's lymphoma of the bcl-2 protein associated with the t(14;18) chromosomal translocation. *N. Engl. J. Med.* **318,** 1638–1644.

11. Negrini, M., Silini, E., Kozak, C., Tsujimoto, Y., and Croce, C. M. (1987) Molecular analysis of mbcl-2: structure and expression of the murine gene homologous to the human gene involved in follicular lymphoma. *Cell* **49,** 455–463.
12. Tsujimoto, Y. and Croce, C. M. (1986) Analysis of the structure, transcripts, and protein products of bcl-2, the gene involved in human follicular lymphoma. *Proc. Natl. Acad. Sci. USA* **83,** 5214–5218.
13. Hayes, D. F., Zurawski, V. R., Jr., and Kufe, D. W. (1986) Comparison of circulating CA15-3 and carcinoembryonic antigen levels in patients with breast cancer. *J. Clin. Oncol.* **4,** 1542–1550.
14. O'Brien, D. P., Horgan, P. G., Gough, D. B., Skehill, R., Grimes, H., and Given, H. F. (1992) CA15-3: a reliable indicator of metastatic bone disease in breast cancer patients. *Ann. Roy. Coll. Surg. Engl.* **74,** 9.
15. van Dalen, A. (1996) New markers for breast carcinoma-associated antigen in comparison with CA 15-3. *Anticancer Res.* **16,** 2339–2343.
16. Anonymous (1996) *Federal Register,* FDA Docket No. 96M-0216 61: 38206.
17. Fisher, B., Redmond, C., Fisher, E. R., and Caplan, R. (1988) Relative worth of estrogen or progesterone receptor and pathologic characteristics of differentiation as indicators of prognosis in node negative breast cancer patients: findings from National Surgical Adjuvant Breast and Bowel Project Protocol B-06. *J. Clin. Oncol.* **6,** 1076–1087.
18. Bloom, H. J. G. and Richardson, W. W. (1957) Histological grading and prognosis in node-negative breast cancer. *Br. J. Cancer* **11,** 359–377.
19. Sigurdsson, H., Baldetorp, B., Borg, A., Dalberg, M., Ferno, M., Killander, D., and Olsson, H. (1990) Indicators of prognosis in node-negative breast cancer. *N. Engl. J. Med.* **322,** 1045–1053.
20. McGuire, W. L. and Clark, G. M. (1992) Prognostic factors and treatment decisions in axillary-node-negative breast cancer. *N. Engl. J. Med.* **326,** 1756–1761.
21. Gerdes, J., Lemke, H., Baisch, H., Wacker, H. H., Schwab, U., and Stein, H. (1984) Cell cycle analysis of a cell proliferation-associated human nuclear antigen defined by the monoclonal antibody Ki-67. *J. Immunol.* **133,** 1710–1715.
22. Brown, D. C. and Gatter, K. C. (1990) Monoclonal antibody Ki-67: its use in histopathology. *Histopathology* **17,** 489–503.
23. Schwarting, R., Gerdes, J., Niehus, J., Jaeschke, L., and Stein, H. (1986) Determination of the growth fraction in cell suspensions by flow cytometry using the monoclonal antibody Ki-67. *J. Immunol. Methods* **90,** 65–70.
24. Cattoretti, G., Becker, M. H., Key, G., Duchrow, M., Schluter, C., Galle, J., and Gerdes, J. (1992) Monoclonal antibodies against recombinant parts of the Ki-67 antigen (MIB 1 and MIB 3) detect proliferating cells in microwave-processed formalin-fixed paraffin sections. *J. Pathol.* **168,** 357–363.
25. Schwartz, G. F., Finkel, G. C., Garcia, J. C., and Patchefsky, A. S. (1992) Subclinical ductal carcinoma in situ of the breast. Treatment by local excision and surveillance alone. *Cancer* **70,** 2468–2474.
26. Mathews, M. B., Bernstein, R. M., Franza, B. R., Jr., and Garrels, J. I. (1984) Identity of the proliferating cell nuclear antigen and cyclin. *Nature* **309,** 374–376.

27. Hall, P. A., Levison, D. A., Woods, A. L., Yu, C. C., Kellock, D. B., Watkins, J. A., Barnes, D. M., Gillett, C. E., Camplejohn, R., and Dover, R. (1990) Proliferating cell nuclear antigen (PCNA) immunolocalization in paraffin sections: an index of cell proliferation with evidence of deregulated expression in some neoplasms. *J. Pathol.* **162,** 285–294.
28. Leonardi, E., Girlando, S., Serio, G., Mauri, F. A., Perrone, G., Scampini, S., Dalla Palma, P., and Barbareschi, M. (1992) PCNA and Ki67 expression in breast carcinoma: correlations with clinical and biological variables. *J. Clin. Pathol.* **45,** 416–419.
29. Tervahauta, A., Eskelinen, M., Syrjanen, S., Lipponen, P., Pajarinen, P., and Syrjanen, K. (1991) Immunohistochemical demonstration of c-erb B-2 oncoprotein expression in female breast cancer and its prognostic significance. *Anticancer Res.* **11,** 1677–1681.
30. Clark, G. M. and McGuire, W. L. (1991) Follow-up study of HER-2/neu amplification in primary breast cancer. *Cancer Res.* **51,** 944–948.
31. Treurniet, H. F., Rookus, M. A., Peterse, H. L., Hart, A. A., and van Leeuwen, F. E. (1992) Differences in breast cancer risk factors to neu (c-erbB-2) protein overexpression of the breast tumor. *Cancer Res.* **52,** 2344,2345.
32. Kaufman, P. A., Guyre, L. D., and Lewis, F. H. (1996) HER-2/neu targeted immunotherapy: a pilot study of multi-dose MCX-210 in patients with breast or ovarian cancers that overexpress HER-2/neu and a report of an increased incidence of HER-2/neu overexpression in metastatic breast cancer. *Tumor Targeting* **2,** 17–27.
33. Harris, C. C. and Hollstein, M. (1992) p53 tumor suppressor gene, in *PPO Updates* (DeVita, V. T., Hellman, S., and Rosenberg, S. A., eds.), Bristol Myers Oncology Division, p. 1.
34. Thor, A. D., Moore, D. H., Edgerton, S. M., Kawasaki, E. S., Reihsaus, E., Lynch, H. T., Marcus, J. N., Schwartz, L., Chen, L. C., Mayall, B. H., et al. (1992) Accumulation of p53 tumor suppressor gene protein: an independent marker of prognosis in breast cancers. *J. Natl. Cancer Inst.* **84,** 845–855.
35. Mousses, S., Ozcelik, H., Lee, P. D., Malkin, D., Bull, S. B., and Andrulis, I. L. (1995) Two variants of the CIP1/WAF1 gene occur together and are associated with human cancer. *Hum. Mol. Genet.* **4,** 1089–1092.
36. Musgrove, E. A., Lilischkis, R., Cornish, A. I., Lee, C. S., Setlur, V., Seshadri, R., and Sutherland, R. L. (1995) Expression of the cyclin-dependent kinase inhibitors p16INK4, p15INK4B and p21WAF1/CIP1 in human breast cancer. *Int. J. Cancer* **63,** 584–591.
37. Jeoung, D. I., Tang, B., and Sonenberg, M. (1995) Induction of tumor suppressor p21 protein by kinase inhibitors in MCF-7 cells. *Biochem. Biophys. Res. Commun.* **214,** 361–366.
38. Sheikh, M. S., Rochefort, H., and Garcia, M. (1995) Overexpression of p21WAF1/CIP1 induces growth arrest, giant cell formation and apoptosis in human breast carcinoma cell lines. *Oncogene* **11,** 1899–1905.
39. Stoscheck, C. M. and King, L. E., Jr. (1986) Role of epidermal growth factor in carcinogenesis. *Cancer Res.* **46,** 1030–1037.

40. Klijn, J. G., Berns, P. M., Schmitz, P. I., and Foekens, J. A. (1992) The clinical significance of epidermal growth factor receptor (EGF-R) in human breast cancer: a review on 5232 patients. *Endocr. Rev.* **13,** 3–17.
41. Field, J. K. and Spandidos, D. A. (1990) The role of ras and myc oncogenes in human solid tumours and their relevance in diagnosis and prognosis. *Anticancer Res.* **10,** 1–22.
42. Escot, C., Theillet, C., Lidereau, R., Spyratos, F., Champeme, M. H., Gest, J., and Callahan, R. (1986) Genetic alteration of the c-myc protooncogene (MYC) in human primary breast carcinomas. *Proc. Natl. Acad. Sci. USA* **83,** 4834–4838.
43. Joensuu, H., Pylkkanen, L., and Toikkanen, S. (1994) Bcl-2 protein expression and long-term survival in breast cancer. *Am. J. Pathol.* **145,** 1191–1198.
44. Gee, J. M., Robertson, J. F., Ellis, I. O., Willsher, P., McClelland, R. A., Hoyle, H. B., Kyme, S. R., Finlay, P., Blamey, R. W., and Nicholson, R. I. (1994) Immunocytochemical localization of BCL-2 protein in human breast cancers and its relationship to a series of prognostic markers and response to endocrine therapy. *Int. J. Cancer* **59,** 619–628.
45. Sierra, A., Lloveras, B., Castellsague, X., Moreno, L., Garcia-Ramirez, M., and Fabra, A. (1995) Bcl-2 expression is associated with lymph node metastasis in human ductal breast carcinoma. *Int. J. Cancer* **60,** 54–60.
46. Silvestrini, R., Veneroni, S., Daidone, M. G., Benini, E., Boracchi, P., Mezzetti, M., Di Fronzo, G., Rilke, F., and Veronesi, U. (1994) The Bcl-2 protein: a prognostic indicator strongly related to p53 protein in lymph node-negative breast cancer patients. *J. Natl. Cancer Inst.* **86,** 499–504.
47. Leek, R. D., Kaklamanis, L., Pezzella, F., Gatter, K. C., and Harris, A. L. (1994) bcl-2 in normal human breast and carcinoma, association with oestrogen receptor-positive, epidermal growth factor receptor-negative tumours and in situ cancer. *Br. J. Cancer* **69,** 135–139.
48. Spyratos, F., Maudelonde, T., Brouillet, J. P., Brunet, M., Defrenne, A., Andrieu, C., Hacene, K., Desplaces, A., Rouesse, J., and Rochefort, H. (1989) Cathepsin D: an independent prognostic factor for metastasis of breast cancer HER-*2/neu* oncogene protein and prognosis in breast cancer. *Lancet* **7,** 1115–1118.
49. Tandon, A. K., Clark, G. M., Chirgwin, M., and McGuire, W. L. (1990) Cathepsin D and prognosis in breast cancer. *N. Engl. J. Med.* **322,** 297–302.
50. Aranda, F. I. and Laforga, J. B. (1996) Microvessel quantitation in breast ductal invasive carcinoma. Correlation with proliferative activity, hormonal receptors and lymph node metastases. *Pathol. Res. Pract.* **192,** 124–129.
51. Kohlberger, P. D., Obermair, A., Sliutz, G., Heinzl, H., Koelbl, H., Breitenecker, G., Gitsch, G., and Kainz, C. (1996) Quantitative immunohistochemistry of factor VIII-related antigen in breast carcinoma: a comparison of computer-assisted image analysis with established counting methods. *Am. J. Clin Pathol.* **105,** 705–710.
52. Axelsson, K., Ljung, B. M., Moore, D. H., 2nd, Thor, A. D., Chew, K. L., Edgerton, S. M., Smith, H. S., and Mayall, B. H. (1995) Tumor angiogenesis as a prognostic assay for invasive ductal breast carcinoma. *J. Natl. Cancer Inst.* **87,** 997–1008.

53. Fox, S. B., Turner, G. D., Leek, R. D., Whitehouse, R. M., Gatter, K. C., and Harris, A. L. (1995) The prognostic value of quantitative angiogenesis in breast cancer and role of adhesion molecule expression in tumor endothelium. *Breast Cancer Res. Treat.* **36,** 219–226.
54. Bevilacqua, P., Barbareschi, M., Verderio, P., Boracchi, P., Caffo, O., Dalla, P., Palma, S., Meli, N., Weidner, N., and Gasparini, G. (1995) Prognostic value of intratumoral microvessel density, a measure of tumor angiogenesis, in node-negative breast carcinoma—results of a multiparametric study. *Breast Cancer Res. Treat.* **36,** 205–217.
55. Toi, M., Inada, K., Suzuki, H., and Tominaga, T. (1995) Tumor angiogenesis in breast cancer: its importance as a prognostic indicator and the association with vascular endothelial growth factor expression. *Breast Cancer Res. Treat.* **36,** 193–204.
56. Gasparini, G. and Harris, A. L. (1995) Clinical importance of the determination of tumor angiogenesis in breast carcinoma: much more than a new prognostic tool. *J. Clin. Oncol.* **13,** 765–782.
57. Vartanian, R. K. and Weidner, N. (1994) Correlation of intratumoral endothelial cell proliferation with microvessel density (tumor angiogenesis) and tumor cell proliferation in breast carcinoma. *Am. J. Pathol.* **144,** 1188–1194.
58. Sawan, A., Lascu, I., Veron, M., Anderson, J., Wright, C., Horne, C. H., and Angus, B. (1994) NDP-K/nm23 expression in human breast cancer in relation to relapse, survival, and other prognostic factors: an immunohistochemical study. *J. Pathol.* **172,** 27–34.
59. Tokunaga, Y., Urano, T., Furukawa, K., Kondo, H., Kanematsu, T., and Shiku, H. (1993) Reduced expression of nm23-H1, but not of nm23-H2, is concordant with the frequency of lymph-node metastasis of human breast cancer. *Int. J. Cancer* **55,** 66–71.
60. Steeg, P. S., de la Rosa, A., Flatow, U., MacDonald, N. J., Benedict, M., and Leone, A. (1993) Nm23 and breast cancer metastasis. *Breast Cancer Res. Treat.* **25,** 175–187.
61. Kobayashi, S., Iwase, H., Itoh, Y., Fukuoka, H., Yamashita, H., Kuzushima, T., Iwata, H., Masaoka, A., and Kimura, N. (1992) Estrogen receptor, c-*erb*B-2 and nm23/NDP kinase expression in the intraductal and invasive components of human breast cancers. *Jpn. J. Cancer Res.* **83,** 859–865.
62. Bevilacqua, G., Sobel, M. E., Liotta, L. A., and Steeg, P. S. (1989) Association of low nm23 RNA levels in human primary infiltrating ductal breast carcinomas with lymph node involvement and other histopathological indications of high metastatic potenial. *Cancer Res.* **49,** 5185–5190.
63. Nakamura, G., Shikata, N., Shoji, T., Hatano, T., Hioki, K., and Tsubura, A. (1995) Immunohistochemical study of mammary and extramammary Paget's disease. *Anticancer Res.* **15,** 467–470.
64. Simpson, J. F., O'Malley, F., Dupont, W. D., and Page, D. L. (1994) Heterogeneous expression of nm23 gene product in noninvasive breast carcinoma. *Cancer* **73,** 2352–2358.

65. Goodall, R. J., Dawkins, H. J., Robbins, P. D., Hahnel, E., Sarna, M., Hahnel, R., Papadimitriou, J. M., Harvey, J. M., and Sterrett, G. F. (1994) Evaluation of the expression levels of nm23-H1 mRNA in primary breast cancer, benign breast disease, axillary lymph nodes and normal breast tissue. *Pathology* **26,** 423–428.
66. Cox, L. A., Chen, G., and Lee, E. Y. (1994) Tumor suppressor genes and their roles in breast cancer. *Breast Cancer Res. Treat.* **32,** 19–38.
67. Howlett, A. R., Petersen, O. W., Steeg, P. S., and Bissell, M. J. (1994) A novel function for the nm23-H1 gene: overexpression in human breast carcinoma cells leads to the formation of basement membrane and growth arrest. *J. Natl. Cancer Inst.* **86,** 1838–1844.
68. Hahnel, E., Dawkins, H., Robbins, P., and Hahnel, R. (1994) Expression of stromelysin-3 and nm23 in breast carcinoma and related tissues. *Int. J. Cancer* **58,** 157–160.
69. Yamashita, H., Kobayashi, S., Iwase, H., Itoh, Y., Kuzushima, T., Iwata, H., Itoh, K., Naito, A., Yamashita, T., and Masaoka, A. (1993) Analysis of oncogenes and tumor suppressor genes in human breast cancer. *Jpn. J. Cancer Res.* **84,** 871–878.
70. Leone, A., Flatow, U., VanHoutte, K., and Steeg, P. S. (1993) Transfection of human nm23-H1 into the human MDA-MB-435 breast carcinoma cell line: effects on tumor metastatic potential, colonization and enzymatic activity. *Oncogene* **8,** 2325–2333.
71. Royds, J. A., Stephenson, T. J., Rees, R. C., Shorthouse, A. J., and Silcocks, P. B. (1993) Nm23 protein expression in ductal in situ and invasive human breast carcinoma. *J. Natl. Cancer Inst.* **85,** 727–731.
72. Sastre-Garau, X., Ovtracht, L., Lascu, I., Lacombe, M. L., Veron, M., Bourdache, K., and Thiery, J. P. (1992) Ultrastructural immunocytochemical localization of diphosphate kinase/Nm23 in human cancer cells. *Bull. Cancer (Paris)* **79,** 465–470.
73. Sastre-Garau, X., Lacombe, M. L., Jouve, M., Veron, M., and Magdelenat, H. (1992) Nuclcoside diphosphate kinase/NM23 expression in breast cancer: lack of correlation with lymph-node metastasis. *Int. J. Cancer* **50,** 533–538.
74. Okubo, T., Inokuma, S., Takeda, S., Itoyama, S., Kinoshita, K., and Sugawara, I. (1995) Expression of nm23-H1 gene product in thyroid, ovary, and breast cancers. *Cell Biophys.* **26,** 205–213.
75. Beatson, G. T. (1896) On the treatment of inoperable cases of carcinoma of the mamma: suggestions for a new method of treatment with illustrative cases. *Lancet* **2,** 104.
76. Toft, D. and Gorski, J. (1966) A receptor molecule for estrogens: isolation from the rat uterus and preliminary characterization. *Proc. Natl. Acad. Sci. USA* **55,** 1574–1581.
77. Knight, W. A., Livingston, R. B., Gregory, E. J., and McGuire, W. L. (1977) Estrogen receptor as an independent prognostic factor for early recurrence in breast cancer. *Cancer Res.* **37,** 4669–4671.
78. Sunderland, M. C. and McGuire, W. L. (1990) Prognostic indicators in invasive breast cancer. *Surg. Clin. North Am.* **70,** 989–1004.

79. Carter, S. L., Negrini, M., Baffa, R., Gillum, D. R., Rosenberg, A. L., Schwartz, G. F., and Croce, C. M. (1994) Loss of heterozygosity at 11q22-q23 in breast cancer. *Cancer Res.* **54,** 6270–6274.
80. Patel, U., Grundfest-Broniatowski, S., Gupta, M., and Banerjee, S. (1994) Microsatellite instabilities at five chromosomes in primary breast tumors. *Oncogene* **9,** 3695–3700.
81. Gudmundsson, J., Barkardottir, R. B., Eiriksdottir, G., Baldursson, T., Arason, A., Egilsson, V., and Ingvarsson, S. (1995) Loss of heterozygosity at chromosome 11 in breast cancer: association of prognostic factors with genetic alterations. *Br. J. Cancer* **72,** 696–701.
82. Kashiwaba, M., Tamura, G., and Ishida, M. (1995) Frequent loss of heterozygosity at the deleted in colorectal carcinoma gene locus and its association with histologic phenotypes in breast carcinoma. *Virchows Arch.* **426,** 441–446.
83. Radford, D. M., Fair, K. L., Phillips, N. J., Ritter, J. H., Steinbrueck, T., Holt, M. S., and Donis-Keller, H. (1995) Allelotyping of ductal carcinoma in situ of the breast: deletion of loci on 8p, 13q, 16q, 17p and 17q. *Cancer Res.* **55,** 3399–3405.
84. Contegiacomo, A., Pizzi, C., De Marchis, L., Alimandi, M., Delrio, P., Di Palm, E., Petrella, G., Ottini, L., French, D., Frati, L., et al. (1995) High cell kinetics is associated with amplification of the int-2, bcl-1, myc and erbB-2 proto-oncogenes and loss of heterozygosity at the DF3 locus in primary breast cancers. *Int. J. Cancer* **61,** 1–6.
85. Iwase, H., Greenman, J. M., Barnes, D. M., Bobrow, L., Hodgson, S., and Mathew, C. G. (1995) Loss of heterozygosity of the oestrogen receptor gene in breast cancer. *Br. J. Cancer* **71,** 448–450.
86. Zhuang, Z., Merino, M. J., Chuaqui, R., Liotta, L. A., and Emmert-Buck, M. R. (1995) Identical allelic loss on chromosome 11q13 in microdissected in situ and invasive human breast cancer. *Cancer Res.* **55,** 467–471.
87. Chen, L. C., Matsumura, K., Deng, G., Kurisu, W., Ljung, B. M., Lerman, M. I., Waldman, F. M., and Smith, H. S. (1994) Deletion of two separate regions on chromosome 3p in breast cancers. *Cancer Res.* **54,** 3021–3024.
88. Hampton, G. M., Mannermaa, A., Winquist, R., Alavaikko, M., Blanco, G., Taskinen, P. J., Kiviniemi, H., Newsham, H., Cavenee, W. K., and Evans, G. A. (1994) Loss of heterozygosity in sporadic human breast carcinoma: a common region between 11q22 and 11q23. 3. *Cancer Res.* **54,** 4586–4589.
89. Cornelis, R. S., van Vliet, M., Vos, C. B., Cleton-Jansen, A. M., van de Vijver, M. J., Peterse, J. L., Khan, P. M., Borresen, A. L., Cornelisse, C. J., and Devilee, P. (1994) Evidence for a gene on 17p13. 3, distal to TP53, as a target for allele loss in breast tumors without p53 mutations. *Cancer Res.* **54,** 4200–4206.
90. Kirchweger, R., Zeillinger, R., Schneeberger, C., Speiser, P., Louason, G., and Theillet, C. (1994) Patterns of allele losses suggest the existence of five distinct regions of LOH on chromosome 17 in breast cancer. *Int. J. Cancer* **56,** 193–199.
91. Casey, G., Plummer, S., Hoeltge, G., Scanlon, D., Fasching, C., and Stanbridge, E. J. (1993) Functional evidence for a breast cancer growth suppressor gene on chromosome 17. *Hum. Mol. Genet.* **2,** 1921–1927.

92. Eiriksdottir, G., Sigurdsson, A., Jonasson, J. G., Agnarsson, B. A., Sigurdsson, H., Gudmundsson, J., Bergthorsson, J. T., Barkardottir, R. B., Egilsson, V., and Ingvarsson, S. (1995) Loss of heterozygosity on chromosome 9 in human breast cancer: association with clinical variables and genetic changes at other chromosome regions. *Int. J. Cancer* **64,** 378–382.
93. Sheng, Z. M., Marchetti, A., Buttitta, F., Champeme, M. H., Campani, D., Bistocchi, M., Lidereau, R., and Callahan, R. (1996) Multiple regions of chromosome 6q affected by loss of heterozygosity in primary human breast carcinomas. *Br. J. Cancer* **73,** 144–147.
94. Nagai, M. A., Medeiros, A. C., Brentani, M. M., Brentani, R. R., Marques, L. A., Mazoyer, S., and Mulligan, L. M. (1995) Five distinct deleted regions on chromosome 17 defining different subsets of human primary breast tumors. *Oncology* **52,** 448–453.
95. Cleton-Jansen, A. M., Collins, N., Lakhani, S. R., Weissenbach, J., Devilee, P., Cornelisse, C. J., and Stratton, M. R. (1995) Loss of heterozygosity in sporadic breast tumours at the BRCA2 locus on chromosome 13q12-q13. *Br. J. Cancer* **72,** 1241–1244.
96. Yaremko, M. L., Recant, W. M., and Westbrook, C. A. (1995) Loss of heterozygosity from the short arm of chromosome 8 is an early event in breast cancers. *Genes Chromosomes Cancer* **13,** 186–191.
97. Koreth, J., Bethwaite, P. B., and McGee, J. O. (1995) Mutation at chromosome 11q23 in human non-familial breast cancer: a microdissection microsatellite analysis. *J. Pathol.* **176,** 11–18.
98. Winqvist, R., Hampton, G. M., Mannermaa, A., Blanco, G., Alavaikko, M., Kiviniemi, H., Taskinen, P. J., Evans, G. A., Wright, F. A., Newsham, I., et al. (1995) Loss of heterozygosity for chromosome 11 in primary human breast tumors is associated with poor survival after metastasis. *Cancer Res.* **55,** 2660–2664.
99. Kerangueven, F., Essioux, L., Dib, A., Noguchi, T., Allione, F., Geneix, J., Longy, M., Lidereau, R., Eisinger, F., Pebusque, M. J., et al. (1995) Loss of heterozygosity and linkage analysis in breast carcinoma: indication for a putative third susceptibility gene on the short arm of chromosome 8. *Oncogene* **10,** 1023–1026.
100. Nagai, M. A., Yamamoto, L., Salaorni, S., Pacheco, M. M., Brentani, M. M., Barbosa, E. M., Brentani, R. R., Mazoyer, S., Smith, S. A., Ponder, B. A., et al. (1994) Detailed deletion mapping of chromosome segment 17q12-21 in sporadic breast tumours. *Genes Chromosomes Cancer* **11,** 58–62.
101. Hall, J. M., Lee, M. K., Newman, B., Morrow, J. E., Anderson, L. A., Huey, B., and King, M. C. (1990) Linkage of early-onset familial breast cancer to chromosome 17q21. *Science* **250,** 1684–1689.
102. Miki, Y., Swensen, J., Shattuck-Eidens, D., Futreal, P. A., Harshman, K., Tavtigian, S., Liu, Q., Cochran, C., Bennett, L. M., Ding, W., et al. (1994) A strong candidate for the breast and ovarian cancer susceptibility gene BRCA1. *Science* **266,** 66–71.

103. Wooster, R., Neuhausen, S. L., Mangion, J., Quirk, Y., Ford, D., Collins, N., Nguyen, K., Seal, S., Tran, T., and Averill, D. (1994) Localization of a breast cancer susceptibility gene, BRCA2, to chromosome 13q12-13. *Science* **265,** 2088–2090.
104. FitzGerald, M. G., MacDonald, D. J., Krainer, M., Hoover, I., O'Neil, E., Unsal, H., Silva-Arrieto, S., Finkelstein, D. M., Beer-Romero, P., Englert, C., Sgroi, D. C., Smith, B. L., Younger, J. W., Garber, J. E., Duda, R. B., Mayzel, K. A., Isselbacher, K. J., Friend, S. H., and Haber, D. A. (1996) Germ-line BRCA1 mutations in Jewish and non-Jewish women with early-onset breast cancer. *N. Engl. J. Med.* **334,** 143–149.
105. Struewing, J. P., Abeliovich, D., Peretz, T., Avishai, N., Kaback, M. M., Collins, F. S., and Brody, L. C. (1995) The carrier frequency of the BRCA1 185delAG mutation is approximately 1 percent in Ashkenazi Jewish individuals. *Natl. Genet.* **11,** 198–200.
106. Neuhausen, S., Gilewski, T., Norton, T., Tran, L., McGuire, P., Swensen, J., Hampel, H., Borgen, P., Brown, K., Skolnick, M., Shattuck-Eidens, D., Jhanwar, S., Goldgar, D., and Offit, K. (1996) Recurrent BRCA2 6174delT mutations in Ashkenazi Jewish women affected by breast cancer. *Natl. Genet.* **13,** 126–128.
107. National Cancer Institute (1988) *Clinical Alert.* NCT, Bethesda, MD.
108. Glick, J. A. (1990) Adjuvant therapy for node-negative breast cancer: a proactive view, in *Important Advances in Oncology* (DeVita, V. T., Hellman, S., and Rosenberg, S. A., eds.), J. B. Lippincott Co., Philadelphia, p. 183.
109. National Cancer Institute (1990) Treatment of early stage breast cancer. *NIH Consens Dev. Conf. Consens. Statement* **8** (6).
110. Glick, J. A. (1990) in National Cancer Institute: Treatment of early stage breast cancer. *NIH Consens. Dev. Conf. Consens. Statement* **8,** Table 11a-1, p. 186.
111. McGuire, W. L. (1988) Estrogen receptor versus nuclear grade as prognostic factors in axillary node negative breast cancer. *J. Clin. Oncol.* **6,** 1071,1072.

II

Immunodiagnosis of Cancer: Serum and Tissue Markers

5

Serum and Tissue Biomarkers in the Prognosis and Treatment of Breast Cancer

Tina J. Hieken and Rajeshwari R. Mehta

1. Introduction

Several nuclear and membrane proteins are emerging as important markers of breast-cancer behavior. There is increasing evidence that these markers have potential therapeutic as well as prognostic value in the care of breast-cancer patients.

Various methods are available for detecting the expression of these proteins in tumor, blood, and other biologic fluids. In this chapter, we describe a method for detecting the expression of two oncoproteins (HER-2/*neu* and mutant p53) in the tumors and plasma of breast-cancer patients by a quantitative enzyme-linked immunosorbent assay (ELISA) assay. In our experience, this method appears to provide an accurate and useful means of detecting overexpression of HER-2/*neu* and mutant p53 protein in clinical samples. Although immunohistochemistry for the detection of oncoproteins is faster, requires only a small amount of tissue, and can give information on the intratumoral oncoprotein distribution, it has several drawbacks. Tissue processing, fixation procedures, and methods used for antigen retrieval influence immunohistochemical staining and can lead to considerable variation in results. Also, the method is semiquantitative, so interpretation of results may be difficult. Western blot analysis can be used to obtain biochemical information on the protein of interest, but it is time-consuming and ill-suited for studying large numbers of patient samples. It also is only semiquantitative.

In contrast, ELISA assay is not subject to the tissue-processing and antigen-retrieval variability associated with immunohistochemistry, and it may be used

From: *Methods in Molecular Medicine, Vol. 14: Tumor Marker Protocols*
Edited by: M. Hanausek and Z. Walaszek © Humana Press Inc., Totowa, NJ

to assay oncoprotein expression in biologic fluids as well as in tumor tissue. Although it requires greater sample preparation time than does immunohistochemistry, ELISA immunoassay generally produces readily reproducible results. Also, large numbers of specimens can be assayed at the same time. Disadvantages include the requirement for frozen tissue, in greater amounts than for immunohistochemistry, and potential cross-reactivities with endogenous peroxidases (when horseradish peroxidase conjugates are used). Thus, it is important to run appropriate controls to assess crossreactivity or nonspecific binding to other host proteins. Overall, ELISA is a safe and versatile technique, widely used in both clinical and research laboratories ***(1–7)***.

2. Materials

2.1. Preparation of Nuclear and Membrane Extracts from Tumor Specimens

1. 5 *M* NaCl.
2. Ice-cold phosphate-buffered saline (PBS).
3. Glycerol.
4. Tween-20 (or other nonionic agent).
5. Membrane extraction buffer: 10 m*M* Tris-HCl, pH 7.4, 1.5 m*M* ethylenediamene tetra-acetic acid (EDTA), pH 7.4, containing 0.5 m*M* phenylmethylsulfonyl fluoride (PMSF), 1 μg/mL leupeptin, and 1 μg/mL pepstatin A. Add proteinase inhibitors just before using buffer.
6. Nuclear extract swelling buffer: 20 m*M* Tris-HCl, pH 8.0, 5 m*M* EDTA, pH 8.0, containing 1.0 m*M* PMSF, 1 μg/mL leupeptin, and 1 μg/mL pepstatin A. Add proteinase inhibitors just before using buffer.
7. Ice bucket, ice.
8. Liquid nitrogen.
9. Dry ice.
10. Single-channel micropipets (1–200 μL), pipet tips.
11. Pasteur pipets.
12. 1- and 10-mL pipets.
13. Eppendorf tubes.
14. 10-mL (16 × 76 mm) centrifuge tubes (plastic).
15. 50- to 200-mL plastic beakers.
16. Metal spoon or spatula.
17. Balance.
18. Timer.
19. Ultracentrifuge tubes with caps for Ti 70.1-type rotor.
20. Tissue pulverizer.
21. Polytron homogenizer.
22. Table-top centrifuge.
23. Ultracentrifuge with Ti 70.1 rotor.

2.2. ELISA Assays

1. p53 mutant selective quantitative ELISA assay kit, cat. no. Q1A03 and/or human *neu* quantitative ELISA assay kit, cat. no. Q1A10, Oncogene Research Products Division, Calbiochem (Cambridge, MA).
2. Deionized water.
3. Multichannel micropipeter, pipet tips, troughs.
4. Multichannel suction for aspirating buffer from 96-well plates.
5. 500- or 1000-mL graduated cylinder.
6. Paper towels or teri-wipes.
7. Timer.
8. Vortex mixer.
9. Multichannel spectrophotometer (plate reader), able to measure absorption at 405 and 490 nm.

3. Methods

3.1. Preparation of Nuclear and Membrane Extracts from Tumor Specimens

1. Specimen preparation is performed entirely on ice. Precool *all* centrifugation tubes and centrifuges to 4°C. Precool pulverizer and metal spatula in liquid nitrogen (*see* **Notes 1** and **2**).
2. Add proteinase inhibitors to membrane buffer.
3. Aliquot 2.5 mL membrane buffer to each 10-mL (16 × 76 mm) plastic centrifuge tube (one per specimen) on ice.
4. Weigh out 2–5 g of frozen (–80°C) tumor.
5. Pulverize tumor tissue on liquid nitrogen, using a precooled pulverizer.
6. Transfer pulverized tissue powder to 10-mL (16 × 76 mm) polystyrene centrifuge tubes containing 2.5 mL membrane buffer (use precooled metal weighing spoon or spatula).
7. Homogenize the tissue with polytron, in 5- to 10-s bursts, with centrifuge tube kept chilled by immersion in a 50-mL plastic beaker of ice (*see* **Note 3**).
8. Replace tube in ice bucket, and repeat **steps 4–7** for each sample (*see* **Note 4**).
9. Centrifuge at 1000*g* for 10 min at 4°C.
10. Transfer supernatant to another tube (on ice). Supernatant contains membrane extract. Pellet contains nuclei.
    a. Add glycerol to supernatant to 10% (v/v) final concentration.
    b. Aliquot membrane extract into Eppendorf tubes to store at –80°C until assayed.
    c. Reserve aliquot for determination of protein concentration.
11. For nuclear extract, wash pellet once in ice-cold PBS (~2 mL).
12. Vortex gently to mix.
13. Centrifuge at 1000*g* for 10 min at 4°C in table-top centrifuge.
14. Add proteinase inhibitors to nuclear extract swelling buffer.
15. Resuspend pellet in 2.5 mL ice-cold nuclear swelling buffer.
16. Incubate on ice for 30 min. Vortex every 5 min to resuspend.

17. Start cooling ultracentrifuge to 4°C now.
18. Add Tween-20 to 1% (v/v) final concentration.
19. Incubate on ice for 30 min. Vortex every 10 min to resuspend.
20. Add 5*M* NaCl solution to 0.5*M* final concentration.
21. Incubate on ice for 15 min. Vortex every 5 min to resuspend.
22. Transfer to ultracentrifuge tubes with caps.
23. Ultracentrifuge for 45 min at 70,000*g* at 4°C in Ti 70.1 rotor.
24. Transfer and aliquot supernatant into Eppendorf tubes for storage at –80°C. Reserve aliquot for protein analysis.
25. Determine protein concentration of nuclear and membrane preparations spectrophotometrically (*see* **Note 5**).

3.2. Preparation of Other Specimens

1. Tissue culture cells: Pellet harvested cells of interest by centrifugation at 1000*g* for 10 min at 4°C. Wash pellet in 20-pellet volumes of ice-cold PBS, resuspend, and centrifuge at 1000*g* for 10 min at 4°C. Repeat twice. Resuspend pellet in 20-pellet volumes of membrane buffer. Continue with the protocol outline in **Subheading 3.1., step 9**.
2. Other biologic fluids: Assay for mutant p53 oncoprotein may be performed on plasma, serum, and other fluids without further preparation. If specimens are frozen, be sure to thaw them completely and homogenize before using. For the HER-2/*neu* assay, plasma should be diluted 1:50 in sample buffer containing 20% mouse serum. (Both are provided in the Oncogene human *neu* quantitative ELISA assay.)

3.3. ELISA Assays

To efficiently quantify the expression of mutant p53 and HER-2/*neu* protein in human breast cancers and in the plasma of breast-cancer patients, we have used commercially available kits (Oncogene Research Products Division, Cal Biochem) as described in **Subheading 2.2.** These are both indirect "sandwich" ELISA assays, with the capture antibody preadsorbed to a polystyrene 96-well flat-bottomed microtiter plate. Instruction pamphlets accompany each kit. The steps for each assay are outlined below.

3.3.1. Mutant p53 ELISA Assay

1. Reconstitute wash buffer as instructed.
2. Remove the desired number of precoated wells and snap into 96-well plate holder. Remember to allow 12 extra wells for standards. Assay each specimen in duplicate (*see* **Note 6**).
3. Wash wells with 200 µL wash buffer. Invert on stack of teri-wipes or paper towels to empty wells (*see* **Note 7**).
4. Reconstitute p53 protein standards as instructed. Mix thoroughly but gently. Incubate on ice for 15 min (*see* **Note 8**).

5. Pipet 100 µL of standards and samples into wells, in duplicate. Be sure to use a new clean pipet tip for each sample. Pipet precisely into the center of the well without touching the bottom or sides.
6. Cover plate with parafilm or plastic wrap, and incubate overnight (14–18 h) at 4°C.
7. Invert wells on a stack of paper towels to empty.
8. Wash four times with 200 µL wash buffer per well. Use multichannel pipeter. Use multichannel aspirator to empty buffer from wells. An automated plate washer may also be used (*see* **Note 9**).
9. Reconstitute reporter antibody.
10. Use the multichannel pipeter to add 100 µL of reporter antibody to each well.
11. Cover plate with parafilm or plastic wrap. Incubate at room temperature (record room temperature for future reference) for 2 h.
12. Invert wells on a stack of paper towels to decant supernatant.
13. Wash four times with 200 µL of wash buffer per well. Use the multichannel pipeter. Use the multichannel aspirator to empty buffer from wells. An automated plate washer may also be used.
14. Reconstitute peroxidase conjugate.
15. Add 100 µL peroxidase conjugate to each well.
16. Cover plate and incubate at room temperature for 1 h.
17. Invert on stack of paper towels and tap to empty solution from wells.
18. Wash four times with 200 µL of wash buffer per well.
19. Prepare substrate/chromogen by mixing equal volumes of "substrate A" and "substrate B." The resultant solution should be colorless.
20. Pipet 100 µL of substrate solution into each well. Incubate at room temperature for 30 min.
21. Read absorbance at 405 nm with the plate spectrophotometer.
22. Use average absorbance of standards to calculate a standard curve by linear regression (*see* **Note 10**).
23. Calculate mutant p53 protein concentration in each sample using mean absorbance plotted on the standard curve.

*3.3.2. HER-2/*neu *ELISA Assay*

1. For each specimen, add 20 µL of antigen extract agent (included with kit) to 100 µL of membrane extract. Vortex to mix, and incubate on ice for 5 min. Further dilute specimen to 10 µg/mL final concentration with sample diluent.
2. Prepare wash buffer.
3. Reconstitute HER-2/*neu* standards (*see* **Notes 10** and **11**).
4. Set up the desired number of precoated wells in the plate frame.
5. Vortex diluted specimens thoroughly and add 100 µL to each of duplicate wells. Set up four wells with the 0 HER-2/*neu* per mL standard. The upper left-hand corner "A" well is used as the substrate blank well.
6. Cover the microplate with plastic wrap and incubate overnight (12–18 h).
7. Invert on paper towels to empty wells.
8. Wash wells six times with 300 µL of plate wash.

9. Add 100 μL of detector antibody to all wells except the substrate blank well. Incubate at room temperature (15–30°C) for 1 h.
10. Prepare working conjugate by diluting the conjugate concentrate with conjugate diluent in a clean reagent reservoir.
11. Wash wells as in **step 8** above.
12. Add 100 μL of working conjugate to all wells except the substrate blank well. Incubate at room temperature (15–30°C) for 30 min.
13. Prepare working substrate. Vortex well to mix. Cover with foil to minimize exposure to light.
14. Wash wells as in **step 8**.
15. Including the substrate blank well, add 100 μL of working substrate to all wells. Cover with foil to block out light. Incubate the microplate at room temperature for 1 h.
16. Add 100 μL of stop solution to each well.
17. Read the absorbance at 490 nm within 30 min of adding stop solution. Zero the plate reader on the substrate blank well.
18. Calculate standard curve and specimen HER-2/*neu* concentrations as described for the p53 assay.

4. Notes

1. Specimen preparation: Before homogenization, tumor specimens should be kept in their storage containers on dry ice in a covered bucket. Tissue pulverization should be performed in liquid nitrogen.
2. Preparation of membrane and nuclear extracts should be performed on ice.
3. Clean pulverizer between each sample. Clean homogenizer between samples by rinsing with distilled H_2O and drying with a lint-free tissue.
4. Twelve to 24 samples can be processed easily at one time, depending on the capacity of your ultracentrifuge rotor.
5. We use the Lowry method to determine protein concentration of nuclear and membrane preparations.
6. In ELISA assays it is easiest to configure the microtiter plate as shown in **Table 1**.
7. Do not allow wells to become completely dry.
8. When reconstituting standards, tap or swirl vials gently to mix. Check to make sure that all lyophilized protein is completely dissolved. If using previously reconstituted protein standards, make sure that they are completely thawed and well-mixed. In our experience, the low-concentration standards do not store well. We have generated standard curves by serially diluting the 2- or 4-ng standards.
9. When using the multichannel pipeter, check the following to ensure accuracy: Make sure tips are attached securely and aspirated volumes look identical. Also check that expelled volumes are equal by looking at the side of the microtiter plate. Make sure tips are not damaged or blocked.
10. A new standard curve must be generated every time a group of samples is assayed.
11. Standard curves for serum or plasma samples should be generated from standards reconstituted in 10% normal mouse serum.

Table 1
Recommended Configuration of the Microtiter Plate[a]

	1	2	3	4	5	6	7	8	9	10	11	12
A	0	Std 1	Std 2	Std 3	Std 4	Std 5	Spec 1	Spec 2	Spec 3	Spec 4	Spec 5	Spec 6
B	0	Std 1	Std 2	Std 3	Std 4	Std 5	Spec 1	Spec 2	Spec 3	Spec 4	Spec 5	Spec 6
C	Spec 7	Spec 8	Spec 9	Spec 10	Spec 11	Spec 12	Spec 13	Spec 14	Spec 15	Spec 16	Spec 17	Spec 18
D	Spec 7	Spec 8	Spec 9	Spec 10	Spec 11	Spec 12	Spec 13	Spec 14	Spec 15	Spec 16	Spec 17	Spec 18
E	Spec 19	Spec 20	Spec 21	Spec 22	Spec 23	Spec 24	Spec 25	Spec 26	Spec 27	Spec 28	Spec 29	Spec 30
F	Spec 19	Spec 20	Spec 21	Spec 22	Spec 23	Spec 24	Spec 25	Spec 26	Spec 27	Spec 28	Spec 29	Spec 30
G	Spec 31	Spec 32	Spec 33	Spec 34	Spec 35	Spec 36	Spec 37	Spec 38	Spec 39	Spec 40	Spec 41	Spec 42
H	Spec 31	Spec 32	Spec 33	Spec 34	Spec 35	Spec 36	Spec 37	Spec 38	Spec 39	Spec 40	Spec 41	Spec 42

[a]Abbreviations: Std, standard; Spec, sample.

References

1. Borg, A., Lennertsrand, J., Stenmark-Askmalm, M., Ferno, M., Brisfors, A., Ohrvik, A., Stal, O., Killander, D., Lane, D., and Brundell, J. (1995) Prognostic significance of p53 overexpression in primary breast cancer: a novel luminometric immunoassay applicable on steroid receptor cytosols. *Br. J. Cancer* **71,** 1013–1017.
2. Crowther, J. R. (1995) Stages in ELISA. *Meth. Mol. Biol.* **42,** 1–218.
3. Dittadi, R., Catozzi, L., Gion, M., Brazzale, A., Capitanio, G., Gelli, M. C., Menegon, A., Gardini, G., Malagutti, R., and Piffanelli, A. (1993) Comparison between western blotting, immunohistochemical, and ELISA assay for p185*neu* quantitation in breast cancer specimens. *Anticancer Res.* **13,** 1821–1824.
4. El-Gendy, S., Tahin, Q., El-Merzabani, M., El-Aaser, A. A., Barnabas, N. J., and Russo, J. (1995) Co-expression of c-erbB2 and int-2 oncogenes in invasive breast cancer. *Int. J. Oncol.* **6,** 977–984.
5. Engrall, E. and Perlman, P. (1971) Enzyme-linked immunosorbent assay (ELISA): quantitative assay of immunoglobulin G. *Immunochemistry* **8,** 871–879.
6. Gannon, J. V., Greaves, R., Iggo, R., and Lane, D. P. (1990) Activating mutations in p53 produce a common conformational effect. A monoclonal antibody specific for the mutant form. *EMBO J.* **9,** 1595–1602.
7. Kurstak, E. (1986) *Enzyme Immunodiagnosis*. Academic, Orlando, FL.

6

The Oncofetal Protein p65 in Breast Cancer Detection

Margaret Hanausek, Zbigniew Walaszek, Ute Sherman, and Jerzy T. Klijanienko

1. Introduction

It has been known for some time that steroid and thyroid hormones can regulate cell proliferation *(1,2)* by stimulating cell division, as in the case of the thyroid hormone; by preventing proliferation, as in the case of the glucocorticoid hormone; or by inducing differentiation, as in the case of retinoic acid. Recent studies of nuclear hormone receptors have shown that hormone function is mediated by nuclear hormone receptors acting as transcriptional enhancers to stimulate expression of a set of genes *(1)*. The genes for nuclear hormone receptors and for receptors of other regulatory molecules appear to have a common underlying structure *(1,2)*, which in turn suggests that these genes have evolved as duplications of a single ancestral gene. Recent studies of the complex interactions of these receptors with each other and with their DNA targets have led to some understanding of these interactions and their role in tumorigenesis and tumor progression *(3)*.

We have discovered and characterized in our laboratory a 65-kDa oncofetal protein (p65), highly conserved in different species, a potential tumor marker *(4–7)*. Its amino acid composition, peptide map, and N-terminal and internal peptide sequences are very similar if not identical in humans and rodents *(6)*. We have now identified the *p65* gene as a novel member of the superfamily of genes that encode nuclear receptors for various hydrophobic ligands, such as steroids, vitamin D, retinoic acid, and thyroid hormones *(8)*. The p65 protein is highly homologous to estrogen receptor (ER) in its DNA-binding domain but not in other regions of the sequence, indicating that p65 is a new receptor, with an as yet unknown ligand, or transcription factor. In addition, we have identified

From: *Methods in Molecular Medicine, Vol. 14: Tumor Marker Protocols*
Edited by: M. Hanausek and Z. Walaszek © Humana Press Inc., Totowa, NJ

in the cloned *p65* cDNA fragment sequences encoding five peptides, obtained by CNBr cleavage, whose amino acid sequences were previously established ***(6)***.

The intracellular localization of steroid receptors after synthesis has been studied extensively, and a theory proposes that cytoplasmic receptors bind hormones and rapidly translocate to the nucleus ***(1,2)***. This may explain our immunostainings showing the presence of p65 not only in nuclei, but also in the cytoplasm of rat liver preneoplastic foci ***(4)*** as well as rat and human mammary cancer cells ***(7,9)***. Nuclear localization of the rat and human p65 proteins may occur by two different mechanisms. One is the diffusion of proteins through the nuclear membrane; the other is interaction of proteins with the nuclear pores. This process is mediated by a translocation signal in the protein like the one we have found behind the DNA-binding domain (C1) in the C2 region ***(8)***.

The studies described earlier indicate that the *p65* gene may have transcriptional modulatory effects mediated through the receptor molecules themselves ***(8)***, and that antibodies made to its gene product may be successfully used in cancer detection ***(7,9)***. There is, in fact, a strong indication that p65 belongs to a superfamily of receptors or transcription factors that regulate the expression of specific target genes, and have profound effects on many physiological processes as well as the development and growth of malignant tumors, including breast cancer ***(8–10)***.

We have developed several highly specific anti-p65 polyclonal antibodies, made in rabbits, against the human and rat p65 antigens ***(11)***, as well as a line of mouse monoclonal antibodies (MAbs) against the human and rat p65 antigens ***(7,9)***. These antibodies were used in the following assays and procedures:

1. Enzyme-linked immunosorbent assay (ELISA) to test sera from cancer patients for the presence of p65 ***(7,9,12)***;
2. Immunohistochemical staining;
3. Immunofluorescence staining; and
4. Immunoblotting.

The enzyme-linked immunosorbent assay for p65 is described in our earlier publications ***(7,9,12)*** in great detail. In this chapter, we would like to concentrate on the use of polyclonal and monoclonal anti-p65 antibodies for immunohistochemical staining of breast cancer tissue, immunofluorescence staining of breast cancer cell lines and breast cancer tissue, as well as immunoblotting.

2. Materials

2.1. Equipment

1. Microtome suitable for cutting thin sections from paraffin blocks.
2. Light microscope equipped for fluorescence, with appropriate filters for fluorescein and/or rhodamine (Texas red) and objectives suitable for oil immersion.

3. Humidified chamber.
4. Mini slot-blot or dot-blot apparatus.
5. Rocker platform.
6. Microwave oven with the maximum wattage available (i.e., 750 W).
7. Photographic equipment.

2.2. Immunostaining

1. Monoclonal and polyclonal anti-p65 antibodies, obtained as previously described (*see* **refs.** ***7*** and ***11***, respectively). Because of existing patents for the anti-p65 antibodies, their use requires negotiations with The University of Texas, M. D. Anderson Cancer Center Office of Technology Development, Houston, TX.
2. Biotinylated goat antimouse or antirabbit IgGs (Amersham, Arlington Heights, IL) (*see* **Note 1**).
3. Streptavidin–horseradish peroxidase or ABC Vectastain reagent (Vector Laboratories, Burlingame, CA).
4. Streptavidin–fluorescein conjugated goat antirabbit or goat antimouse IgG (Amersham).
5. Phosphate-buffered saline (PBS): In ~1 L of distilled water, dissolve 8.0 g sodium chloride, 1.3 g dibasic sodium phosphate, 0.2 g monobasic sodium phosphate, and adjust pH to 7.4. Add distilled water to 1 L.
6. PBS–BSA: 2% (w/w) bovine serum albumin ([BSA] Sigma, St. Louis, MO) in PBS.
7. Blocking solution: 5–10% normal goat serum in PBS.
8. Solvents: 100, 90, and 70% ethanol; acetone; xylene.
9. Digesting solutions: 0.1% solution of trypsin in 0.05*M* Tris-HCl buffer, with 0.1% $CaCl_2$, pH 7.7; 0.0025% pronase in 0.05*M* Tris-HCl buffer, pH 7.6; 0.1% pepsin in 0.01*N* HCl, pH 2.25; 0.05% saponin in distilled water.
10. Staining solutions: 3,3'-diaminobenzidine tetrahydrochloride (DAB). Dissolve 5 mg DAB (Sigma) in ~100 mL of PBS. Add 0.1 mL of 30% hydrogen peroxide and PBS to 100 mL and filter. Prepare fresh DAB solution daily. *DAB is a suspected carcinogen.* Care should be taken in handling and disposing of all peroxide substrates.
11. DAB/$NiCl_2$: Mix 0.5 mL 0.1% $NiCl_2 \cdot 6\ H_2O$ with 5 mL 0.5 mg/mL DAB.
12. Ponceau S solution: Make a 0.2% Ponceau S solution in 3% trichloroacetic acid or 50 m*M* sodium acetate, pH 5.0.
13. 3-Aminopropyltriethoxysilane (APES): Mix 0.5 mL APES and 25 mL dry acetone.
14. Chrome-alum-gelatin: 2 g of $KCr(SO_4)_2 \cdot 12\ H_2O$ and 2.5 g gelatin in 500 mL of distilled water (dissolve at 40–50°C).
15. 0.01% Solution of poly-L-lysine (molecular weight >300,000) (Sigma).
16. 10 mM Citrate buffer, pH 6.0: Make the following stock solutions and working solution. Stock solution A: 0.1*M* citric acid (21.01 g in 1000 mL); Stock solution B: 0.1*M* solution of sodium citrate (28.41 g of sodium citrate dihydrate in 1000 mL). Working solution: 9 mL of A plus 41 mL of B; dilute to 500 mL.
17. 0.25–0.5% H_2O_2 in absolute methanol.

18. 0.5% Triton X-100 in PBS.
19. Permanent mounting medium: Permount.
20. Mounting solution: 90% glycerol/0.5 *M* carbonate buffer, pH 9.0. Dissolve 1.378 g sodium bicarbonate and 3.108 g sodium carbonate to make 500 mL of 0.5 *M* carbonate buffer, pH 9.0.

2.3. Immunoblotting

1. Nitrocellulose (0.45 µm and 0.2 µm) from Bio-Rad (Hercules, CA) or Schleicher–Schuell (Keene, NH).
2. Tris-buffered saline/Tween-20 (TBST): 10 m*M* Tris-HCl, pH 8.2, 150 m*M* NaCl, 0.05% Tween-20.
3. Monoclonal and/or polyclonal anti-p65 antibodies (*see* **Subheading 2.2.**).
4. Goat antirabbit IgG alkaline phosphatase conjugate (Amersham).
5. Substrate buffer: 100 m*M* Tris-HCl, pH 9.5, 100 m*M* NaCl, 5 m*M* $MgCl_2$.
6. Alkaline phosphatase substrate solution: For each milliliter of alkaline phosphatase substrate solution, combine 1 mL of substrate buffer with 4 µL of substrate component A (nitroblue tetrazolium), mix, and add 4 µL of substrate component B (5-bromo-4-chloro-3-indolyl-phosphate). Mix again and use within 30 min. Alternatively, avidin–biotin peroxidase complex and DAB substrate may be used.
7. Stop solution: 20 m*M* Tris-HCl, pH 8.2, 5 m*M* ethylenediamine tetra-acetic acid (EDTA).

3. Methods

3.1. Immunohistochemical Procedure for Paraffin-Embedded Sections

3.1.1. Glass Slide Pretreatment

Glass slide pretreatment can be achieved by incubation of cleaned microscopy glass slides with either 3-aminopropyltriethoxysilane (APES), chrome-alum-gelatin, or poly-L-lysine.

3.1.1.1. Apes Method

1. Incubate slides in a mixture of 0.5 mL APES and 25 mL dry acetone for 20 s.
2. Wash slides two times in acetone and two times in double-distilled water.
3. Dry slides at room temperature.

3.1.1.2. Chrome-Alum-Gelatin Method

1. Dip slides for 1 s at room temperature in a solution of 2 g of $KCr(SO_4)_2 \cdot 12\ H_2O$ and 2.5 g gelatin in 500 mL of distilled water (*see* **Subheading 2.2.**).
2. Dry slides at room temperature.

3.1.1.3. Poly-L-Lysine Method

1. Dip slides into a 0.01% solution of poly-L-lysine for 5 min.
2. Rinse well in double-distilled water.

3.1.2. Mounting of Sections on Slides

1. Cut sections at 4–6 µm and float on slides in water bath.
2. Dry sections either at 37°C or at room temperature for 48 h before staining.

3.1.3. Removal of Paraffin and Rehydration

1. Remove paraffin in xylene and rehydrate tissue sections through graded concentrations of alcohol and water.
2. Rinse in distilled water for 5 min.

*3.1.4. Pretreatment of Tissue Sections (see **Notes 1–7**)*

The alterations of antigenicity occurring during fixation and embedding may be restored to a certain degree ***(13)*** by incubating sections in one of the following digesting solutions (*see* **Subheading 2.2.**).

3.4.1.1. TRYPSIN (*SEE* **NOTE 3**)

1. Incubate sections in 0.1% trypsin solution at 37°C for 5 to 15 min.
2. Terminate digestion with a soybean trypsin inhibitor applied to slides in PBS buffer.

3.4.1.2. PRONASE

1. Incubate sections in 0.0025% pronase solution at 37°C for 5 to 6 min.
2. Stop the reaction by washing in PBS containing 0.2% glycine.

3.4.1.3. PEPSIN

1. Incubate sections in 0.1% pepsin at 37°C for 15–20 min.
2. Rinse well in distilled water.

3.4.1.4. SAPONIN

1. Incubate sections in 0.05% saponin solution at room temperature for 20 to 30 min.
2. Rinse well in distilled water.

3.1.5. Microwave Method of Slide Processing

Alternatively, the following method of slide processing may be used (*see* **Note 6**).

1. Place slides in a thermoresistant plastic dish filled with 10 m*M* citrate buffer, pH 6.0.
2. Process the slides in a microwave oven (750 W) three to five times for 5 min each (boiling is normal) (*see* **Notes 6** and **7**).

3.1.6. Quenching of Endogenous Peroxidase Activity

1. If quenching of endogenous peroxidase activity is required, prepare 0.25–0.5% H_2O_2 solution in absolute methanol and incubate sections for 30 min.
2. Wash for 20 to 30 min in PBS buffer.

3.1.7. Blocking of Unspecific Tissue Staining

1. Incubate sections in blocking solution for 20 min. Blocking solution contains normal serum (up to 10%) from the species in which secondary antibody was made. In our laboratory we use goat serum.
2. Carefully blot excess of blocking solution from sections.

3.1.8. Immunoperoxidase Staining (see **Notes 8–16**)

1. Apply primary antibody (anti-p65 monoclonal or polyclonal antibody) diluted 1:200 in PBS buffer, or as necessary for the optimum results.
2. Carry out incubations in a humidified chamber at room temperature for 30 min or at 4°C overnight (*see* **Note 8**).
3. Wash slides in PBS buffer, at least twice for 10 min each wash.
4. Incubate sections with biotinylated secondary antibody diluted 1:250 in PBS buffer, for 30 to 60 min. Biotinylated goat antimouse IgG or biotinylated goat anti-rabbit IgG may be used for monoclonal and polyclonal antibodies, respectively.
5. Wash slides in PBS buffer for 10 min.
6. Incubate sections with streptavidin–horseradish peroxidase for 15 min. Concentration of the streptavidin–horseradish peroxidase solution should be determined by titration ***(14)***. The usual range of concentrations is 1:100 to 1:500. Our best results were achieved with the dilution 1:300 in PBS buffer. Alternatively, use avidin–biotin complex (ABC) Vectastain reagent ***(14,15)***.
7. Wash slides in PBS buffer for 10 min.
8. Rinse slides in 0.5% Triton X-100 solution in PBS.
9. Incubate in DAB solution for 5 min. Check intensity of staining under the microscope. If section requires additional staining, prolong incubation with DAB for an additional 1 to 5 min (*see* **Note 10**).
10. Rinse slides in distilled water and counterstain with hematoxylin if desired.
11. Dehydrate through a graded series of ethanol, immerse in xylene, and mount sections using a permanent mounting medium, for example, Permount.

3.1.9. Enhancing of the DAB Staining with Heavy Metal (see **Note 11**)

1. After incubating sections with streptavidin–horsradish peroxidase or alternatively ABC Vectastain reagent, incubate them in nickel-complexed DAB solution (*see* **Subheading 2.2.**) at room temperature for 5 min, and then in the same solution supplemented with 0.01% H_2O_2 for 5 min.
2. Perform rinsing and dehydration steps as described in **Subheading 3.1.8. (steps 10** and **11**).

3.2. Immunofluorescence Staining of Cells in Culture or Frozen Tissue Sections

3.2.1. Slide Preparation for Cells in Culture

1. Grow cultured cells on sterile glass cover slips or slides at 37°C overnight.
2. Rinse cells briefly with PBS buffer.
3. Fix cells in cold acetone for 2 min and air-dry.

3.2.2. Slide Preparation for Tissue Sections

1. Cut 5-8-μm cryostat sections of tissue block embedded in embedding medium for frozen tissue specimens (OCT compound, Fisher Scientific, Pittsburgh, PA) and stored at –70°C.
2. Apply freshly cut frozen sections on clean, uncoated slides and air-dry for 2 h at room temperature or overnight in a refrigerator.
3. Allow sections to warm to room temperature for about 30 min.
4. Fix slides in cold acetone at –20°C for 5 min and then air-dry. Slides may be stored at –20°C until staining.
5. Before staining, rinse slides three times in PBS at room temperature for 5 min.

3.2.3. Immunofluorescence Staining

Incubations should be carried out in a humidified chamber at room temperature. A sufficient reagent volume should be used to cover the specimens adequately; usually 50–100 μL per specimen is satisfactory.

1. Incubate specimens with 10% normal goat serum in PBS for 30 min to suppress unspecific binding of IgG. This step is considered optional; however, we never omit it while staining for p65.
2. Wash slides with PBS.
3. Incubate slides with primary antibody (anti-p65 monoclonal or polyclonal antibodies) at room temperature for 1 h. Always determine the optimal antibody concentration by titration before the staining procedure is carried out. We have found the concentration of 2 μg/mL in PBS–BSA solution to be optimal. The usual range is 2–30 μg/mL in PBS–BSA.
4. Wash slides in PBS buffer at least twice for 10 min each wash.
5. Incubate slides with biotin-conjugated secondary antibody (biotinylated antimouse or antirabbit IgG) for 1 h. The optimal antibody concentration should be verified by titration (the usual range is 2–20 μg/mL in PBS).
6. Rinse slides in PBS buffer three times for 10 min each wash.
7. Incubate with streptavidin–fluorescein or streptavidin–Texas Red (Amersham) in a dark chamber for 15 min. The optimal concentration usually ranges from 1:100 to 1:500; it was 1:200 in our usual procedure as determined by titration.
8. Wash slides with PBS buffer three times for 10 min each wash.
9. Mount in an aqueous mounting medium or 90% glycerol/0.5 *M* carbonate buffer, pH 9.0 (*see* **Subheading 2.2.**).
10. View slides using a fluorescence microscope with appropriate filters.
11. Store slides in the dark at room temperature (semipermanent mounting medium) or in 4°C (glycerol/carbonate).

3.3. Immunoblotting

1. Transfer proteins either from the electrophoretic gels or apply serum or tissue extracts onto nitrocellulose membrane using a slot- or dot-blot apparatus (*see* **Note 17**).

2. Block the nitrocellulose membrane by soaking in Tris-buffered saline, pH 7.2, with 1% Tween-20 (TBST buffer) containing 1% powdered dry milk. Carry out blocking at room temperature using at least 1 mL/cm^2 of membrane (*see* **Note 12**).
3. Wash the membrane in TBST buffer for 10 min, repeating this procedure three times.
4. Incubate the nitrocellulose membrane for 1 h with primary antibody at a concentration of 10–20 μg/mL using TBST as diluent (*see* **Notes 14** and **15**).
5. Wash the nitrocellulose membrane by soaking in TBST solution for 10 min; repeat washing three times.
6. Develop color with either alkaline phosphatase substrate solution or ABC/DAB reagents (*see* **Subheading 2.3.**). As soon as the bands reach the desired intensity, stop the reaction by washing the blot in stop solution, and then air-dry. The blot is ready for photographing or scanning.

4. Notes

1. The condition of tissue sections cut from paraffin blocks is very important. Pores must be created in the membranes of cells to enable passage of antibodies in immunostaining procedures. The pores are created by sectioning or in intact cells by proper fixation, freeze-thawing for at least three cycles, or incubation with detergents such as Triton X-100 or digitonin. Minimizing the sizes of the reagents allows for easier tissue penetration; therefore immunoglobulin fragments are often used instead the whole immunoglobulins. In our laboratory we used, for example, biotinylated F(ab')2 antirabbit IgG fragments, species specific, from Amersham. These F(ab')2 fragments are produced by digestion of the whole antibodies with pepsin; undigested fragments as well as pepsin are removed by gel filtration, and the purity of F(ab')2 is always checked by gel electrophoresis.
2. The fixation method using formaldehyde stabilizes the proteins in the tissue by forming covalent crosslinking (methylene bridges), but compromises the access of the antibody conjugates. Methods that use immunofluorescence require well-preserved tissue, usually obtained by use of alcohol and acetone for dehydration and fixation. In our hands, the best results were obtained by fixing the tissue samples in acid alcohol. Also, digestion of the paraffin-embedded section using trypsin solution gave satisfactory results. Another excellent fixative that is particularly suited for immunostaining is formalin-free Stat-Fix (Stat Path, Riderwood, MD). It replaces formalin and does not contain toxic substances such as formaldehydes, aldehydes, or mercury. It also requires less staining time. Stat-Fix is a blend of buffered alcohols and thermoprotective ingredients, penetrating tissues rapidly and effectively. It reduces fixation, processing, and staining times and preserves excellent tissue quality. In our hands it was the most satisfactory fixative that allowed for crisp nuclear outlines and very-well-defined morphological features. We can recommend this fixative especially for tissues where crosslinking to antigen sites presents a significant problem.

3. Trypsin can be substituted with either pronase or pepsin solution. Pretreatment with saponin is recommended in cases that do not require enzymatic digestion. Saponin is a very mild detergent that causes minimal damage to cell ultrastructure, and we have used this method with great success. Saponin selectively removes cholesterol from membranes and may be included in all buffers to allow permeability of antigens.
4. The incubation time in the protease solutions may need to be increased for tissues fixed for a prolonged time, or for tissue embedded in glycol methacrylate. For example, tissue fixed for 6 wk in formaldehyde may require incubation for up to 2 h.
5. Incubation time may be changed (shortened), if necessary, by increasing concentration of digestion solutions, or by omitting this step if endogenous peroxidase activity does not present a problem.
6. The microwave step is very important in order to allow the fixed or paraffin-embedded tissue antigens to react with monoclonal or polyclonal antibodies. Slides should not dry during the incubation in the microwave. Watch the level of solution in the container and add distilled water to replace the evaporated quantity as necessary.
7. Many times a complication of microwave processing or proteolytic digestion is a loss of adherence of tissue sections to the glass slides. The precoated slides prevent tissue loss.
8. Overnight incubation is recommended for formalin-fixed, paraffin-embedded sections. It is important to establish an optimal concentration of the antibody for a given application by titration. The most commonly used concentration of our anti-p65 antibody was 2 μg/mL (diluted in PBS buffer). The usual range of concentration may vary from 2 to 20 μg/mL.
9. Solutions containing sodium azide or other inhibitor of peroxidase activity should be avoided in diluting the peroxidase substrate or Vectastain ABC reagents (use according to manufacturers' suggestions).
10. Other peroxidase substrates, such as 3-amino-9-ethylcarbazole (0.25 mg/mL in 100 m*M* sodium acetate, pH 5.2) may be substituted for the DAB. Make sure to establish proper conditions for substitute substrate.
11. Sensitivity of antigen detection can be enhanced by several methods. One that was used in our studies enhanced the color intensity of DAB with heavy metal (nickel) counterstaining. Nickel provides the most sensitive enhancement of the metals and was used in our experiments with success.
12. Nonfat powdered milk works well as a blocking agent, but other proteins, such as bovine serum albumin, ovalbumin, or casein, can be substituted for it. Gelatin also works very well as a blocking agent. Blots should be rocked during all blocking and washing steps, as well as during reaction with primary and secondary antibodies.
13. Any washing step can be extended overnight at 4°C if necessary.
14. Make sure to test titer of both primary and secondary antibodies in order to obtain optimal sensitivity and lowest possible background.

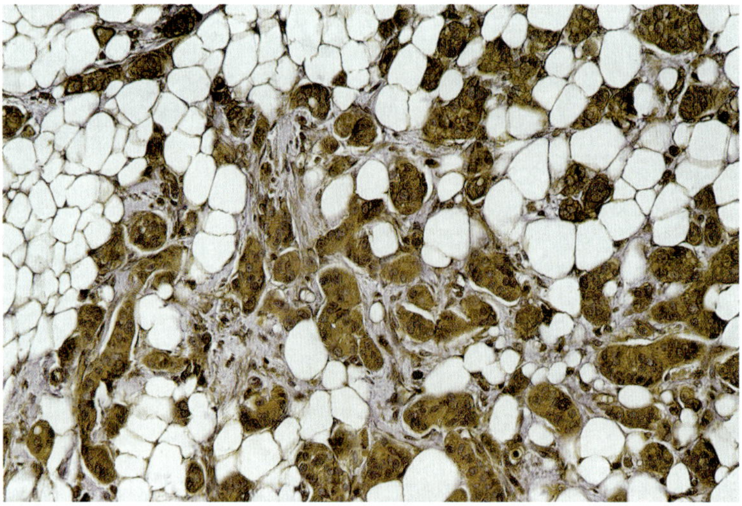

Fig. 1. Immunohistochemical staining of paraffin-embedded breast adenocarcinoma section with monoclonal anti-p65 antibodies. Avidin–biotin–peroxidase/DAB; original magnification ×100. Note strong cytoplasmic and nuclear staining of cancer cells.

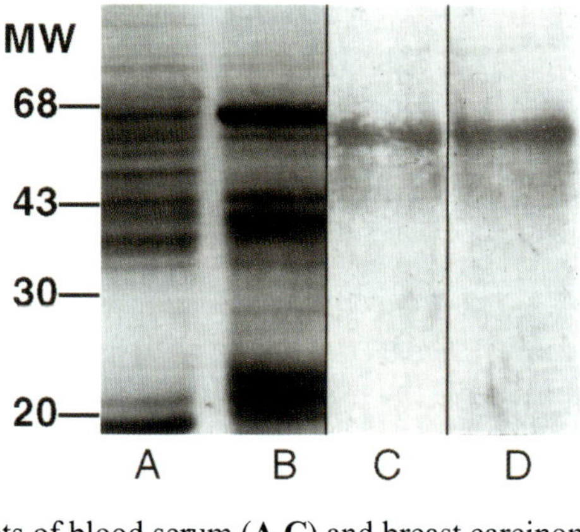

Fig. 2. Western blots of blood serum (**A**,**C**) and breast carcinoma tissue extract (**B**,**D**). Serum and cancer tissue extract (30 µg protein/5 µL) from a breast cancer patient were electrophoresed on a 10% SDS-PAGE gel, transferred onto nitrocellulose membrane, and stained with Ponceau S to visualize the proteins (*see* panels A and B). After destaining blots in double-distilled water, the blots (panels C and D) were immunostained using monoclonal anti-p65 antibodies, biotinylated secondary antibodies, and ABC/DAB reagents.

15. Sometimes substrate solutions may develop precipitates during storage at 4°C or –20°C. To remedy this, warm them to room temperature and mix. A sonicating water bath may be helpful in solubilization of precipitates. If a small amount of

precipitate stays in the solution, the solution can still be used, but expect a slight increase in background.

16. Our immunohistochemical stainings have demonstrated nuclear localization in virtually all p65 positive breast cancer lesions with some cytoplasmic localization (**Fig. 1**). The staining of p65 positive cancer tissues is fairly labile. Optimum antigenicity is retained only if tissue is fixed briefly and preferably not in formalin. If formalin fixative is not avoidable, significant antigenicity may be recovered by either using the microwave method or digestion of formaldehyde-fixed, paraffin-embedded tissue with proteolytic enzymes.
17. It is very important to transfer proteins either from the electrophoretic gels or from serum or tissue extracts (improved performance may be observed when using subcellular fractions such as nuclei for nuclear proteins or membranes for membrane receptors) onto nitrocellulose membrane (**Fig. 2**). The sensitivity of the immunoblotting procedure depends on proper transfer. The efficiency of transfer may be checked by staining proteins with Ponceau S. It is not very sensitive staining, but membranes can be easily and completely destained using neutral pH buffers or double-distilled water.

References

1. O'Malley B. W. (1989) Did eukaryotic steroid receptors evolve from "intracrine" gene regulators? *Endocrinology* **125,** 1119–1127.
2. Evans, R. M. (1988) The steroid and thyroid hormone receptor superfamily. *Science* **240**, 889–895.
3. Beato, M. (1989) Gene regulation by steroid hormones. *Cell* **56,** 335–344.
4. Hanausek-Walaszek, M., Del Rio, M., and Adams, A. K. (1989) Immunohistochemical demonstration of mRNA-transport protein in rat liver putative preneoplastic foci. *Cancer Lett.* **48,** 105–108.
5. Mirowski, M., Sherman, U., and Hanausek, M. (1992) Purification and characterization of a 65-kDa tumor-associated phosphoprotein from rat transplantable hepatocellular carcinoma 1682C cell line. *Protein Expr. Purif.* **3,** 196–203.
6. Mirowski, M., Walaszek, Z., Sherman, U., Adams, A. K., and Hanausek, M. (1993) Comparative structural analysis of human and rat 65 kDa phosphoprotein. *Int. J. Biochem.* **25**, 1865–1871.
7. Wang, S., Mirowski, M., Sherman, U., Walaszek, Z., and Hanausek, M. (1993) Monoclonal antibodies against a 65 kDa tumor-associated phosphoprotein: development and use in cancer detection. *Hybridoma* **12,** 167–176.
8. Hanausek, M., Szemraj J., Adams. A. K, and Walaszek, Z. (1996) The oncofetal protein p65: a new member of the steroid/thyroid receptor superfamily. *Cancer Detect. Prev.* **20,** 94–102.
9. Mirowski, M., Klijanienko, J., Wang, S., Vielh, P., Walaszek, Z., and Hanausek, M. (1994) Serological and immunohistochemical detection of a 65 kDa protein breast cancer. *Eur. J. Cancer* **30A,** 1108–1113.
10. Hanausek, M., Szemraj, J., Adams, A. K., and Walaszek, Z. (1996) Use of RT-PCR to study expression of a novel tumor marker p65 and estrogen receptor in breast cancer patients. *Breast Cancer Res. Treat.* **37(Suppl.),** 40.

11. Coghlan, L. and Hanausek, M. (1990) Subcutaneous immunization of rabbits with nitrocellulose paper strips impregnated with microgram quantities of protein. *J. Immunol. Meth.* **129,** 135–138.
12. Hanausek, M., Wang, S. C., Blonski, J. Z., Polkowska-Kulesza, E., Walaszek, Z., and Mirowski, M. (1996) Expression of an oncofetal 65-kDa phosphoprotein in lymphocytic and granulocytic leukemias. *Int. J. Hematol.* **63,** 193–203.
13. Battifora, H. and Kopinski, M. (1986) The influence of protease digestion and duration of fixation on the immunostaining of keratins. *J. Histochem. Cytochem.* **34,** 1095–1100.
14. Farr, A. G. and Nakane, P. K. (1981) Immunohistochemistry with enzyme labeled antibodies. A brief review. *J. Immunol. Meth.* **47,** 129–144.
15. Hsu, S. M. and Raine L. (1984) The use of avidin–biotin–peroxidase complex (ABC) in diagnostic and research pathology, in *Advances in Immunohistochemistry* (Dellelis, R. A., ed.), Masson, New York, pp. 33–38.

7

Determination of Tumor Ferritin Concentration in Breast Cancer

Jonathan F. Head and Robert L. Elliott

1. Introduction

Ferritin is a cellular-storage protein with the main function of sequestering excess ferric iron and thus preventing high concentrations of soluble iron from becoming toxic to the cells. Dividing cells, both normal and neoplastic, have been shown to increase transferrin receptors in response to increase demand for iron, an essential micronutrient for cell division. However, if too much soluble iron is released into the cytoplasm of the cell, it will become toxic and damage or even kill the cell. Thus, ferritin binds up the excess iron in order to prevent toxicity.

Serum levels of ferritin are often increased in cancer patients *(1,2)*, and therefore serum ferritin was investigated to see if it could be used to screen for breast cancer and for following patients for recurrence and metastatic spread *(2)*. However, it was found that serum levels are not very sensitive and are often not increased until very late in the course of the disease *(3)*. Serum ferritin concentrations, when elevated in advance disease, can be used to follow therapeutic responses *(4)*. Isomers of ferritin have also been investigated but their determination did not increase the sensitivity for screening or for following breast-cancer patients for recurrence.

The concentration of ferritin in the carcinoma cells of tumors from breast-cancer patients has been shown to be of prognostic significance *(5,6)*. High concentrations (≥1000 ng/mg cytosol protein) of ferritin in the tumor, as determined by microparticle enzyme immunoassay (MEIA) of a cytoplasmic preparation, have been associated with poorer outcome. Ferritin concentration in breast tumors is not related to the common prognostic indicators of tumor size, nodal status, and patient age (<50 vs ≥50 yr). Tumor ferritin concentration is inversely related to steroid receptor status and directly related to Ki-67

From: *Methods in Molecular Medicine, Vol. 14: Tumor Marker Protocols*
Edited by: M. Hanausek and Z. Walaszek © Humana Press Inc., Totowa, NJ

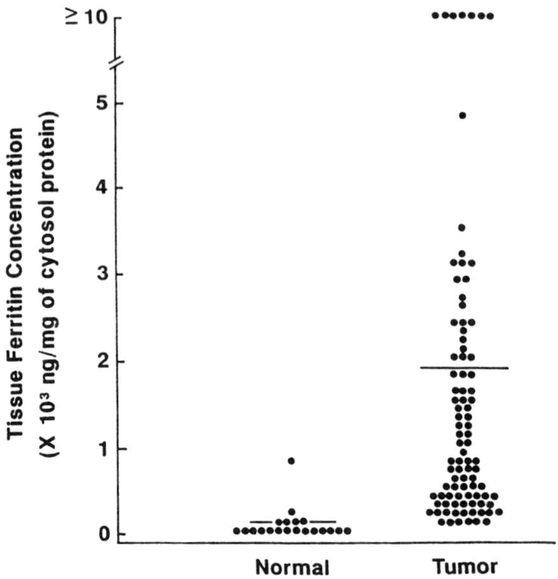

Fig. 1. Concentration of ferritin in normal and tumor tissue from breast-cancer patients.

(proliferation associated nuclear antigen), pathological differentiation of tumor, stage (I–IV) of the disease at presentation, and infrared imaging results (asymmetric heat pattern of breast associated with poorer prognosis). These associations suggest that breast tumors with higher growth rates (therefore carrying a poorer prognosis for the patient) have higher concentrations of ferritin than the less aggressive tumors with their associated better prognosis.

It is becoming more difficult to quantitate prognostic indicators in breast tumors by biochemical methods because of increasing demand for tumor tissue for the ever-increasing number of prognostic indicators. Also, increasing use of screening mammography has resulted in a reduction of the average size of breast tumors and thus has decreased the amount of tumor available to prepare cytoplasmic supernatant for biochemical assays. Therefore, it is desirable to develop methods of quantitating prognostic indicators with smaller pieces of breast-tumor tissue. Immunocytochemical analysis (ICA) and electron microscopy (EM) for determination of cellular ferritin in frozen and fixed sections of breast tumors are attempts to do this.

1.1. Demonstration of Increased Ferritin in Breast Tumors

One hundred twelve samples from patients for which tumor ferritin (FT) concentration had been determined by MEIA were also subjected to ICA and EM in order to see if the three techniques produce the same results. **Figure 1** shows that when ferritin is quantitated by MEIA of a supernatant fraction of homogenized tumor tissue, its concentration is greatly elevated in the breast-tumor tissue (i.e., in normal breast tissue 124 ± 184 [n = 20], and breast tumor tissue 1893 ± 2631 [n = 92] ng/mg of cytosol protein; $p < 0.001$ by Student's t test). Photomicrographs (*see* **Fig. 2**) of tumor tissue stained by ICA for ferritin demonstrate that ferritin is found in the cytoplasm of the tumor cells. The absence of ferritin particles in normal breast tissue and the abundance of ferritin in a breast-carcinoma tissue is clearly demonstrated in **Fig. 3**.

1.2. Comparison of Ferritin Concentration Found by MEIA with Levels Found by ICA

To compare the ferritin concentrations obtained by MEIA with the ferritin levels from ICA, the mean concentration of ferritin by MEIA was determined for each group (low, medium, and high levels of ferritin in the cytoplasm of the tumor cells) resulting from ICA. **Table 1** shows the results of this analysis, and it can be seen that the concentration of ferritin by MEIA is significantly higher (compared to the low group) in the medium ($p = 0.022$) and high ($p = 0.013$) groups of ferritin as determined by Student's t test. **Table 2** presents the distribution of patients with <1000 or ≥1000 ng ferritin/mg cytosol protein by ferritin levels (low, medium, or high) determined by ICA. When a FT level of 1000 ng/mg cytosol protein by MEIA is used as the normal cutoff, 21 (11/52), 46 (18/39), and 76% (16/21) of the breast tumors had high tissue FT in the low, medium, and high FT groups by ICA, respectively. The results from these two methods were equivalent ($p = 0.0001$, χ-square analysis for independence), and therefore FT levels determined by ICA have prognostic value.

1.3. Comparison of Ferritin Concentration Found by MEIA with Levels Found by EM

To compare the ferritin concentrations obtained by MEIA with the ferritin levels from electron microscopy, the mean concentration of ferritin by MEIA was determined for each group (low, medium, and high levels of ferritin in the cytoplasm of the tumor cells) resulting from EM observation. **Table 1** shows the results of this analysis, and it can be seen that the concentration of ferritin by MEIA is significantly higher (compared to the low group) in the medium ($p = 0.031$) group of ferritin by EM as determined by Student's t test but not the high group ($p = 0.060$). **Table 2** presents the distribution of patients with <1000 or ≥1000 ng ferritin/mg cytosol protein by ferritin levels (low, medium, or high) determined by EM. When a FT level of 1000 ng/mg protein by MEIA is used as the normal cutoff, 27 (13/49), 49 (21/43), and 55% (11/20) of the breast tumors had high tissue FT in the low, medium, and high FT groups by EM, respectively. The results from these two methods were similar ($p = 0.0307$, χ-square analysis for independence), and therefore FT levels determined by EM may have prognostic significance.

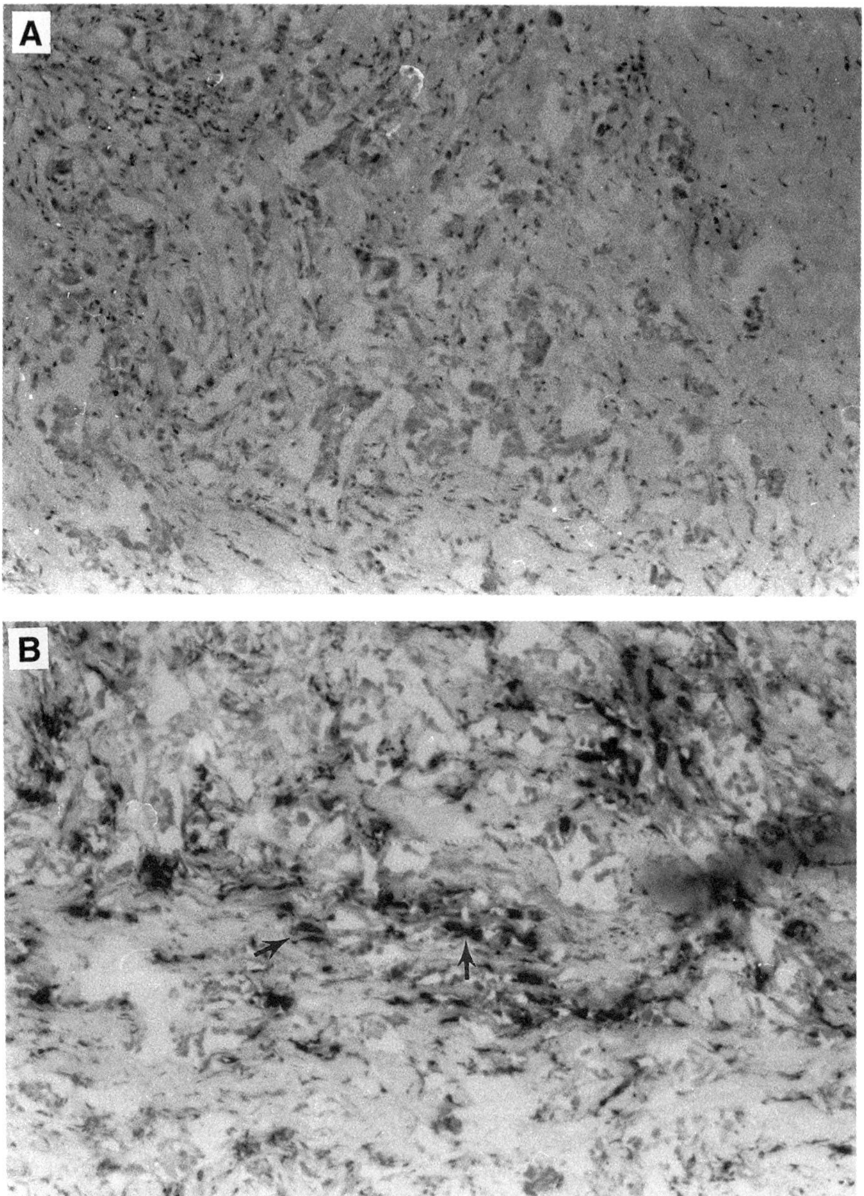

Fig. 2. Photomicrographs of a control slide (**A**, ×1300) without immunoperoxidase staining and a slide (**B**, ×1300) after immunoperoxidase staining for ferritin (note intense cytoplasmic staining at arrows).

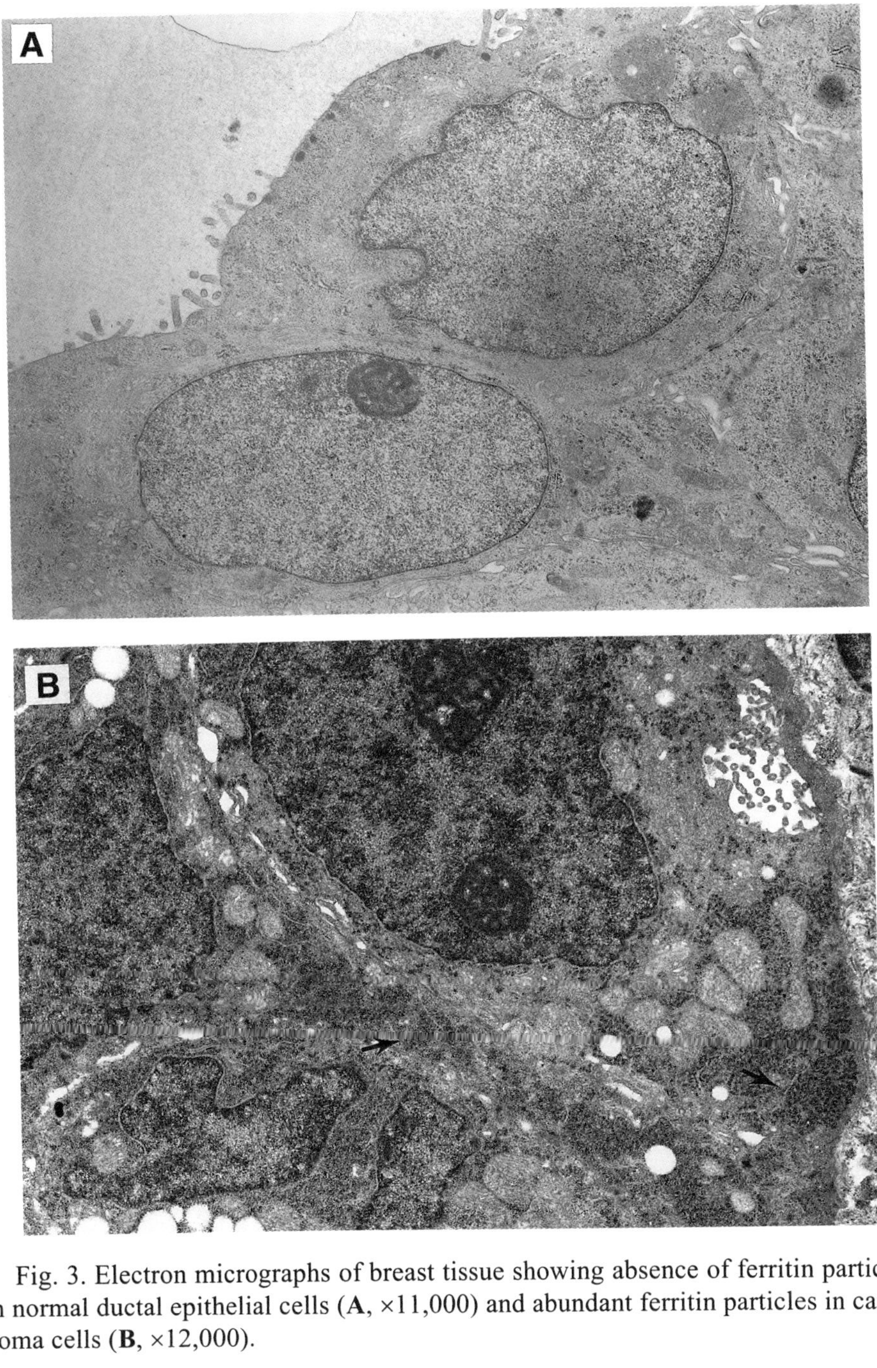

Fig. 3. Electron micrographs of breast tissue showing absence of ferritin particles in normal ductal epithelial cells (**A**, ×11,000) and abundant ferritin particles in carcinoma cells (**B**, ×12,000).

Table 1
Mean MEIA Ferritin Concentration When Grouped by ICA or EM Ferritin Levels

Ferritin results by ICA or EM		Ferritin concentration by MEIA[a] Mean ± SD	Number	Probability
Low	ICA	815 ± 713	52	
	EM	880 ± 780	49	
Medium	ICA	1271 ± 1041	39	0.022[b]
	EM	1279 ± 946	43	0.031[c]
High	ICA	2205 ± 2307	21	0.013[b]
	EM	2008 ± 2488	20	0.060[c]

[a]Values are ng/mg of cytosol protein.
[b]Probability from Student's *t* test compared to low ferritin ICA group.
[c]Probability from Student's *t* test compared to low ferritin EM group.

Table 2
Distribution of Patients Within Groups Derived by Semiquantitation of Ferritin by ICA or EM and Quantitation by MEIA[a]

Ferritin results by ICA or EM		Percent (number/total) of patients, ferritin concentration <1000 ng/mg protein	≥1000 ng/mg protein
Low	ICA	79% (41/52)[b]	21% (11/52)
	EM	73% (36/49)[c]	27% (13/49)
Medium	ICA	54% (21/39)	46% (18/39)
	EM	51% (22/43)	49% (21/43)
High	ICA	24% (5/21)	76% (16/21)
	EM	45% (9/20)	55% (11/20)

[a]Patients having tumors with ferritin concentrations <1000 ng/mg of cytosol protein have a better prognosis.
[b]$p = 0.0001$, χ-square analysis for independence.
[c]$p = 0.0307$, χ-square analysis for independence.

1.4. Overview of Data

The data presented demonstrate that FT concentration is an independent prognostic indicator that predicts disease outcome in breast-cancer patients, and ferritin results should be integrated into the decision process for selecting node-negative breast-cancer patients for adjuvant chemotherapy. The semi-quantitation of tumor ferritin by immunocytochemical analysis of frozen sections of tumor from breast-cancer patients can be achieved with much less tissue than is required for MEIA and produces very similar results. The semiquantitation of tumor ferritin by EM analysis of fixed tumor tissue from breast-cancer patients can also be achieved with much less tissue than is required for MEIA but produces somewhat different results. ICA and EM analysis produce very different results, because they are independent from each other by χ-square analysis ($p = 0.2438$). Therefore, recurrence and survival data for all methods will have to be collected in order to determine which single or combination of results has the greatest prognostic significance.

2. Materials

All reagents should be prepared with sterile distilled water or sterile deionized water and stored at room temperature unless stated otherwise.

2.1. Microparticle Enzyme Immunoassay (MEIA)

1. Buffer for homogenization of normal and tumor tissue from human breast: IMX buffer (MEIA Diluent Buffer, cat. no. 8374-04; Abbott Laboratories, Abbott Park, IL).
2. IMX Ferritin Reagent Pack, 100 tests, cat. no. 2219-20 (Abbott Laboratories).
3. Bio-Data abnormal control, cat. no. 10500 (Bio-Analytics, Palm City, FL).

2.2. Immunocytochemical Analysis (ICA)

1. Phosphate-buffered saline (PBS), pH 7.6: 7.75 g NaCl, 1.50 g K_2HPO_4, 0.20 g KH_2PO_4 in 1 L of deionized water; adjust pH to 7.6 with either 3 *M* NaOH or 0.5% phosphoric acid.
2. 3.7% formaldehyde: Dilute 37% formaldehyde 1:10 with PBS, pH 7.6.
3. HistoGen peroxidase-anti-peroxidase (PAP) immunostaining system (BioGenex, San Ramon, CA).
4. Antibody diluent: PBS, pH 7.6, 1% bovine serum albumin (BSA), 0.1% sodium azide.
5. Primary antibody: Monoclonal antibody (MAb) to human ferritin (cat. no. MU010-UC; BioGenex).
6. 3,3'-Diaminobenzidine tetrahydrochloride (DAB): DAB tablet substrate pack (BioGenex). Dissolve 1 tablet in 5 mL of solution for 15 slides.
7. Hematoxylin (Harris' Alum Hematoxylin): 4.8 g/L hematoxylin, 48.0 g/L aluminum ammonium sulfate, 2.4 g/L mercuric oxide (red).

2.3. Electron Microscopy (EM)

1. EM fixative (Zamboni's Fixative): 20 g of paraformaldehyde in 150 mL of filter-saturated aqueous picric acid; heat to 60°C, add 2.5% NaOH until solution is clear, filter, and bring cooled solution to 1 L with phosphate buffer (3.3 g monobasic sodium phosphate, 17.8 g dibasic sodium phosphate, 1 L distilled water). Stable for 1 yr.
2. 0.2 *M* Cacodylate-HCl buffer: solution A, 42.8 g/500 mL $Na(CH_3)_2AsO_2 \cdot 3H_2O$ or 31.99 g/500 mL $Na(CH_3)_2AsO_2$; solution B, 0.2 *M* HCl (17.2 mL concentrated HCl/L); to make buffer add 50 mL solution A and 8.0 mL solution B, bring up to 100 mL, pH 7.4.

3. Osmium tetroxide: Use clean glassware (glass-stoppered Erlenmeyer flask), prepare 2% OsO_4 (1 g/49 mL deionized water) just prior to use and store at 4°C.
4. Epon-araldite mixture: 25 mL (60 g) epon, 20 mL (60 g) araldite, and 60 mL (126 g) dodecenyl succinic anhydride are mixed thoroughly; mixture can be stored for 1 mo at 4°C. When ready to use add 1 mL of (dimethylaminomethyl) phenol (DMP-30-Tris, Ernest F. Fullman, Inc., Latham, NY).
5. Methylene blue-azure II stain: stock solution A, 1% methylene blue and 1% sodium borate; stock solution B, 1% azure II; to prepare working stain, mix equal parts of stock solution A and stock solution B.
6. Uranyl acetate: 7 g of uranyl acetate are added to a mixture of 50 mL of methanol and 50 mL of deionized water; mix for 30 min; store in dark at 4°C.
7. Lead citrate: 1.33 g lead nitrate [$Pb(NO_3)_2$] and 1.76 g sodium citrate [$Na_3(C_6H_50_7) \cdot 2H_2O$] are dissolved in 39 mL deionized H_2O, shaken for 30 min, and stored in dark at 4°C.

3. Methods

3.1. Ferritin by Microparticle Enzyme Immunoassay (MEIA)

Normal- and tumor-tissue ferritin are quantitated using MEIA technology and the Abbott IMX, as follows:

1. Weigh out approx 300 mg of tissue that was frozen immediately at surgery and stored at –80°C.
2. Add IMX buffer to tissue in a ratio of approx 1:10 (w/v), i.e., 300 mg tissue in 3.0 mL buffer.
3. Homogenized with two 5-s bursts with a polytron.
4. Transfer tissue homogenate to a centrifuge tube and spin at 50,000*g* at 4°C for 1 h.
5. Pipet off supernatant and place in a tube on ice. Make sure to leave any solids in the centrifuge tube (*see* **Note 1**).
6. Quantitate protein in supernatant using the Waddell method *(7)*. Turn on spectrophotometer and let it warm up. The assay contains a blank, standards (1, 2, 4, 6, and 8 mg/mL) of BSA, a 1:25 dilution of Bio-Data abnormal control, and patient unknowns. In all cases, 100 µL of blank, standard, control, or patient cytosol is added to 10 mL of deionized water and vortexed. Read OD at 215 and 225 nm for the blank, standards, control, and patient samples, and determine the difference (OD_{215}–OD_{225}). Plot the concentration (mg/mL) of the standards vs the differences in the two ODs on an *xy* plot. Determine patient-sample values by direct interpolation (*see* **Note 2**).
7. Dilute (1:10) patient samples with IMX MEIA buffer; measure ferritin on IMX as described by the manufacturer (Abbott Laboratories). Multiply the answer by 10.
8. Divide ferritin concentration (ng/mL) by protein concentration (mg/mL cytosolic protein) to yield ng ferritin/mg cytosol protein (*see* **Note 3**).

3.2. Ferritin by Immunocytochemical Analysis (ICA)

The method for staining ferritin using an MAb for ferritin and a DAB reaction are as follows:

1. Cut frozen sections (6 μm) of breast-tumor tissue and fix tissue sections attached to slides by immersing patients' slides, positive control slide, and negative control slide in 3.7% formaldehyde-PBS for 10 min (*see* **Notes 4** and **5**).
2. Rinse slides twice in PBS for 5 min.
3. Incubate sections with 2 drops of peroxide block for 5 min at 37°C (*see* **Note 6**).
4. Rinse sections twice in PBS for 5 min.
5. Incubate section with protein block for 5 min at 37°C.
6. Incubate sections with the ferritin primary antibody (BioGenex, San Ramon, CA, cat. no. MU010-UC) and control antibody for 2 h at 37°C (10 μL antibody in 400 μL antibody diluent).
7. Rinse sections twice in PBS for 5 min.
8. Incubate sections with the linking antibody for 5 min at 37°C.
9. Rinse sections twice in PBS for 5 min.
10. Incubate sections with the labeling antibody for 5 min at 37°C.
11. Rinse sections in PBS for 5 min.
12. Incubate slides with DAB for 6–7 min at room temperature.
13. Rinse slides twice in H_2O for 5 min.
14. Stain slides in hematoxylin for 2 min.
15. Rinse slides twice in H_2O for 5 min.
16. Dehydrate slides twice in 95% EtOH for 5 min.
17. Put slides in xylene and cover slip with permount.
18. Examine slides under light microscope (100×) and grade low, medium, or high level of ferritin.

3.3. Ferritin by Electron Microscopy (EM)

1. Breast-cancer tissue is taken during surgery, minced with razor blades, and fixed in Zamboni's fixative.
2. It is washed overnight in 0.2 *M* cacodylate-HCl buffer and then postfixed in 2% osmium tetroxide for 1 h. The tissue is then washed in 0.2 *M* cacodylate-HCl buffer for 15 min with changes every 5 min (*see* **Note 7**).
3. The tissue is dehydrated using increasing concentrations of alcohol: 30% for 5 min, 50% for 5 min, 70% for 5 min, 80% twice for 5 min, 90% twice for 5 min, 100% three times for 5 min, and propylene oxide three times for 5 min.
4. Infiltration of the tissue with plastic is then started using epon araldite plastic and propylene oxide. Start with a 1:3 ratio of epon araldite to propylene oxide for 1 h, 1:1 ratio overnight, 3:1 ratio for 4 h, and then pure epon araldite overnight.
5. The tissue is then placed into the conical bottom of a Beem capsule and imbedded in epon araldite mixture and polymerized for 2 d in an oven at 60–65°C. Each patient has five blocks made from her biopsy.

6. Sections (800 nm) are cut from each block with a glass knife on the ultra-microtome and stained with methylene blue-azure II. The sections are evaluated under the light microscope, and blocks are chosen to be thin-sectioned under the electron microscope.
7. Thin sections of 100 nm are cut with a diamond knife on the ultramicrotome and stained with 7% uranyl acetate for 15 min followed by lead citrate staining for 15 min.
8. The sections are observed by the pathologist, and photographs are taken of areas of interest. The photographic film is developed and pictures (electronmicrographs), are printed for a permanent record.

4. Notes

1. When separating the cytosol from the pelleted particulate material, it is important not to take the fatty layer above the aqueous cytosol when removing the cytosol from the tube.
2. When plotting the Wadell protein standards concentration against Δ OD to be used for extrapolation of protein concentration, do not force the line through the origin.
3. The cytosol preparation for ferritin determination by MEIA can be stored at 4°C for 24 h but must be stored at –20°C if analysis is delayed any longer than 24 h.
4. Different sources and batches of primary antibody for ferritin determination by ICA can produce differing amounts and intensities of staining. It is therefore recommended that crossover testing be done with varying concentrations of the primary antibody to determine the appropriate dilution of the primary antibody to reduce interassay variability and to provide consistent results over time.
5. Inclusion of a positive and negative control in the ICA is another attempt to increase reliability of the assay.
6. For 37°C incubation, use an oven or incubator.
7. Osmium tetroxide is very hazardous and should be used under a fume hood and handled as instructed by the MSDS.

Acknowledgments

We thank Lindsay Ledford and Marcia Brewer for their technical expertise that was so necessary for developing these assays and the testing of patients' samples.

References

1. Marcus, D. M. and Zinberg, N. (1975) Measurement of serum ferritin by radioimmunoassay: results in normal individuals and patients with breast cancer. *J. Natl. Cancer Inst.* **55,** 791–795.
2. Jacobs, A., Jones, B., Ricketts, C., Bulbrook, R. D., and Wang, D. Y. (1976) Serum ferritin concentration in early breast cancer. *Br. J. Cancer* **34,** 286–290.
3. Worwood, M. (1986) Serum ferritin. *Clin. Sci.* **70,** 215–220.
4. Williams, M. R., Turkes A., Pearson D., Griffiths, K., and Blamey, R. W. (1990) An objective biochemical assessment of therapeutic response in metastatic breast cancer: a study with external review of clinical data. *Br. J. Cancer* **61,** 126–132.

5. Weinstein, R. E., Bond, B. H., and Silberberg, B. K. (1982) Tissue ferritin concentration in carcinoma of the breast. *Cancer* **50,** 2406–2409.
6. Weinstein, R. E., Bond, B. H., Silberberg, B. K., Vaughn, C. B., Subbaiah, P., and Pieper, D. R. (1989) Tissue ferritin concentration and prognosis in carcinoma of the breast. *Breast Cancer Res. Treat.* **14,** 349–353.
7. Waddell, W. J. (1956) Methods of determination of protein concentration. *J. Lab. Clin. Med.* **48,** 311–314.

8

Immunodiagnosis of Childhood Malignancies

David M. Parham and Hallie Holt

1. Introduction

Diagnosis of childhood malignancies, in particular solid tumors, is an enterprise that can be laden with a variety of uncertainties, inconsistencies, and morphologic subtlety, primarily caused by their tendency to recapitulate early organogenesis. This dilemma is manifested by the common sobriquet "small, round, blue-cell tumor," which reflects their frequent composition by primitive, undifferentiated cells with circular nuclei, dense chromatin, and minimal amounts of discernible cytoplasm. Nevertheless, to varying degrees these tumor cells exhibit heterogeneously expressed cytodifferentiation that may be easily observed via light microscopy or require a search via transmission electron microscopy for specific subcellular organelles. Two examples are the neuroblastoma, which is usually composed of a mixture of primitive neuroblasts and partially differentiated ganglion cells, and the rhabdomyosarcoma, which is typically composed of a blend of primitive mesenchyme and incompletely developed rhabdomyoblasts.

Beginning in 1970 *(1)*, a series of techniques were devised that can be used to overcome diagnostic uncertainty by allowing visualization of cell-typic proteins via ordinary light microscopy. These methods are based on the exquisite specificity of the "lock and key" interaction between a protein antigen and its corresponding antibody. Thus, pathologists began the study and routine diagnostic use of cellular markers, which permit one to divine the biologic nature of cell differentiation in normal and neoplastic cells. Because the nosology of tumor diagnosis is based on the normal cell types mimicked by their neoplastic counterparts, one could determine tumor cell differentiation by these techniques and thus infer the tumor type by correlation with normal protein expression. An example of this system of reasoning is the use of intermediate

From: *Methods in Molecular Medicine, Vol. 14: Tumor Marker Protocols*
Edited by: M. Hanausek and Z. Walaszek © Humana Press Inc., Totowa, NJ

Table 1
Categorization of Neoplasms Based on Intermediate Filaments

Filament type	Normal cells	Tumors, general	Tumors, pediatric
Cytokeratin	Epithelium	Carcinoma	Undifferentiated carcinoma (e.g., nasopharyngeal)
Vimentin	Mesenchyme	Sarcoma	Undifferentiated sarcoma
Desmin	Muscle	Myoma	Rhabdomyosarcoma
Glial fibrillary acidic protein	Glia	Glioma	Glioblastoma multiforme, astrocytoma, ependymoma
Neurofilaments	Neurons	Neuroma	Neuroblastoma, medulloblastoma

filaments, which are cytoskeletal elements that appear identical by electron microscopy, but which can be divided into separate biochemical classes that correspond to their presence in epithelial, mesenchymal, glial, muscle, or neural cells *(2)*. Identification of these intermediate filament classes in neoplastic cells can thus place the tumor within the subcategories of carcinoma, sarcoma, glioma, myoma, or neuroma, respectively *(3)*. The implication for childhood neoplasms is that primitive cells that may be difficult to identify with standard stains often exhibit protein expression that corresponds to early differentiation *(4)*. Using the above schema in this manner would allow identification of primitive carcinoma, primitive sarcoma, glioblastoma, rhabdomyosarcoma, or neuroblastoma, respectively (**Table 1**) *(5)*.

Using these principles and a number of cell-typic markers, immunohistochemistry has proven to be invaluable in the diagnosis of primitive pediatric neoplasms and is widely used in general and research hospital laboratories. However, a number of caveats exist, primarily due to the multipotential capacity for differentiation that is displayed by embryonal cells *(6)*.

1.1. Theory of Immunohistochemistry

As noted above, immunohistochemistry is based on the principle that specific proteins can be detected by their interaction with corresponding antibodies. This interaction permits a bonding that resists the dilutional effect of vigorous washing. A second interaction, building a sandwich-like structure, allows attachment of a molecule that will produce a visible reaction product at the end of the procedure. The net result thus causes the affected cell to exhibit a color that is detectable via routine microscopy. An earlier modification of this procedure used attachment of a fluorescent molecule, which requires a fluorescence microscope for observation and does not allow visualization of cell morphology in nonfluorescent areas.

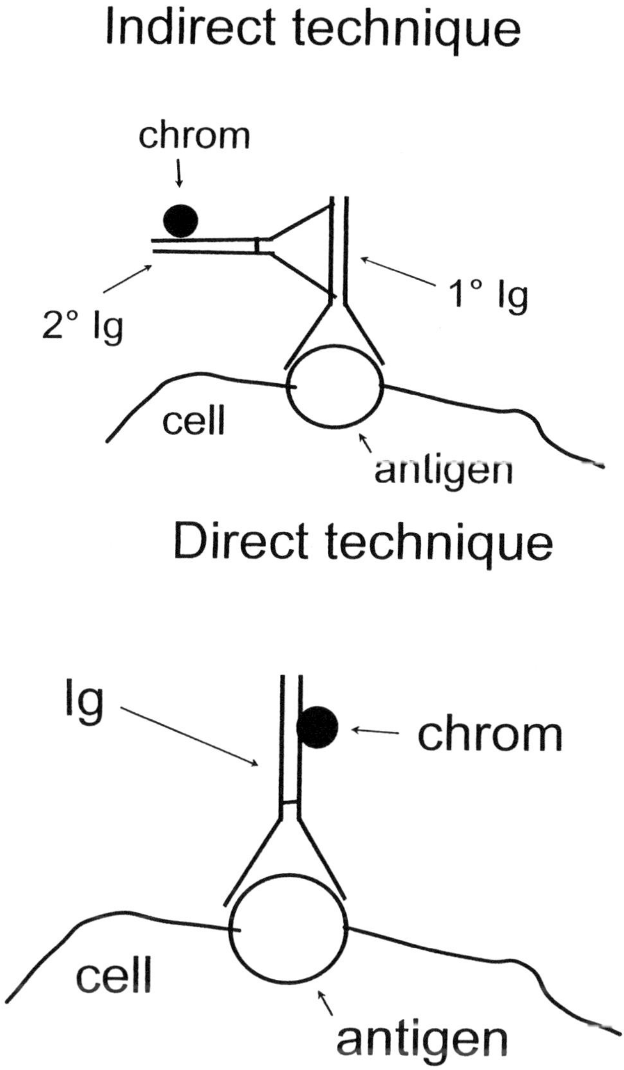

Fig. 1. Schematic of direct **(A)** and indirect **(B)** immunostain procedures. In the direct procedure, the chromogen (chrom) is attached directly to the primary immunoglobulin molecule (Ig), whereas in the indirect procedure, the chromogen is attached to a secondary immunoglobulin (2° Ig), which is in turn attached to the primary immunoglobulin (1° Ig).

1.2. Direct and Indirect Procedures

The simplest and least sensitive of these techniques are the direct and indirect procedures. These are illustrated in **Fig. 1**. In these procedures, only one chromogen is ultimately attached to the antigen in question, so that fewer signals are available per cell. The resultant decrease in sensitivity can be a disadvantage if the protein examined is in low quantity, but this factor can be used to advantage in working with impure or difficult antibodies that yield high background staining. We have particularly enjoyed success in using these techniques with novel antibodies that are not well characterized or that are polyclonal in nature and with which more sensitive procedures yielded an unacceptable level of nonspecific discoloration. Use of the direct procedure requires direct attachment of a chromogen, such as horseradish peroxidase, to the primary antibody. This can be a drawback for novel or obscure antibodies for which commercial products are not readily available. However, the indirect procedure generally can be applied to all antisera, and the reagents are readily available from commercial sources. One comparative drawback is that an extra step is required, which increases the complexity and turnaround time of the procedure.

1.3. Peroxidase–Antiperoxidase (PAP) Method

The PAP method, which was popularized by Sternberger *(1)*, is a more laborious and time-consuming method than the preceding ones, but it offers the advantage of adding more chromogen molecules per tagged antigen (**Fig. 2**). This modification of the former procedures adds a third layer to the "sandwich"—an antiperoxidase molecule that is produced by the same species as the one forming the primary antiserum. Because of this species identity, the molecule is tagged by the secondary immunoglobulin in a manner analogous to the primary reaction. The antiperoxidase complex contains several chromogen molecules, instead of only one, and is therefore more sensitive than the direct or indirect procedures.

This method was popularized in the early 1980s by a number of commercial vendors, who marketed kits containing prediluted reagents that simplified or eliminated the preparatory steps of the procedure. Although less popular today, these kits are still readily available, as is the peroxidase–antiperoxidase reagent. However, in our experience they have suffered from relative insensitivity, lack of sufficient documentation regarding protein concentrations, and incalcitrance to user manipulation to obtain more optimal results. These disadvantages were offset by production of kits for the performance of the avidin–biotin-complex (ABC) procedure, described below.

1.4. Avidin–Biotin-Complex Procedure

The ABC procedure, developed by Hsu in 1981 *(7)*, came into early recognition as a more facile and sensitive technique than the PAP, so that numerous publications using this methodology rapidly appeared. This development to a

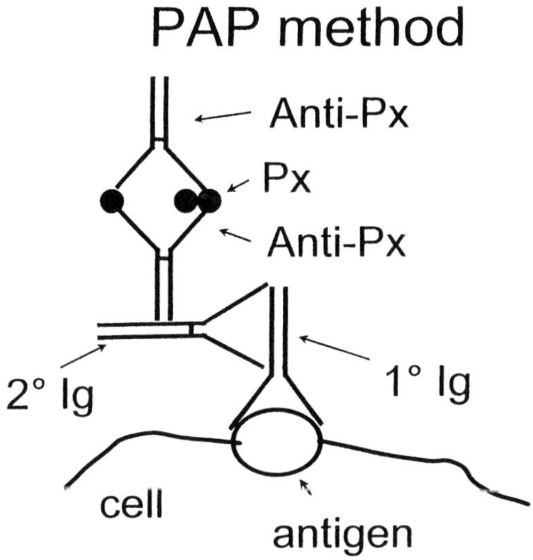

Fig. 2. Schematic of peroxidase–antiperoxidase (PAP) technique. In this method, antiperoxidase molecules (anti-Px) are produced by an animal of the same species as the one making the primary antibody (1° Ig). These molecules attach to the peroxidase chromogen (Px), yielding the PAP reagent. Because of the species specificity, the secondary antibody (2° Ig) binds to the PAP structure and serves as a "bridge" reagent.

large part was fostered by the wide availability and marketing of commercial kits, such as the Vectastain kit sold by Vector Laboratories (Burlingame, CA). The technique is based on the addition of an avidin–biotin bond in the final step of the immunoperoxidase sandwich (**Fig. 3**), instead of an antigen–antibody bond. This biochemical linkage is based on the attraction of the vitamin, biotin, to its natural ligand, avidin, and is a strong bond that resists stringent washing. The relative sensitivity of the technique is caused by both the strength of the avidin–biotin interaction and the fact that additional signal moieties can be added to the sandwich complex (**Fig. 3**).

Further addition of signal molecules was attained by a refinement of the ABC procedure, known as the labeled avidin–biotin procedure *(8,9)*, or the labeled streptavidin–biotin (LSAB) procedure if streptococcal-derived avidin is used as the biotin ligand (**Fig. 4**). Because of this modification, even more sensitivity is attained, as a large number of signals are attached to each antigen. Like ABC kits, LSAB kits are heavily marketed by companies such as Dako (Carpinteria, CA), which offers the "Supersensitive" kit.

Two additional technical modifications, the development of methods to pretreat formalin-fixed, paraffin sections and the invention of instruments to auto-

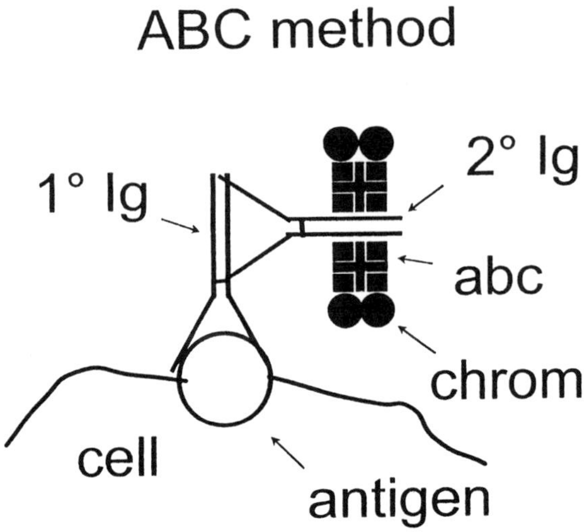

Fig. 3. Schematic of avidin–biotin-complex (ABC) procedure. In this method, the secondary antibody (2° Ig) contains attached biotin moieties, which then form a complex with avidin molecules (abc). The chromogen (chrom) is attached to the avidin particles, so that a color signal results upon peroxidation.

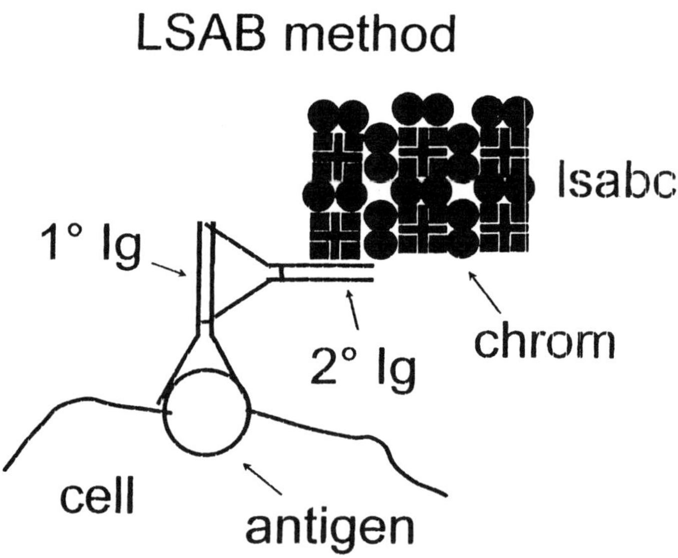

Fig. 4. Schematic of labeled streptavidin–biotin (LSAB) procedure. The labeled streptavidin–biotin complex (lsabc) is a large molecular structure with many chromogen (chrom) moieties, making this procedure very sensitive. Note that there is a large dose of signal attached to each antigenic binding site.

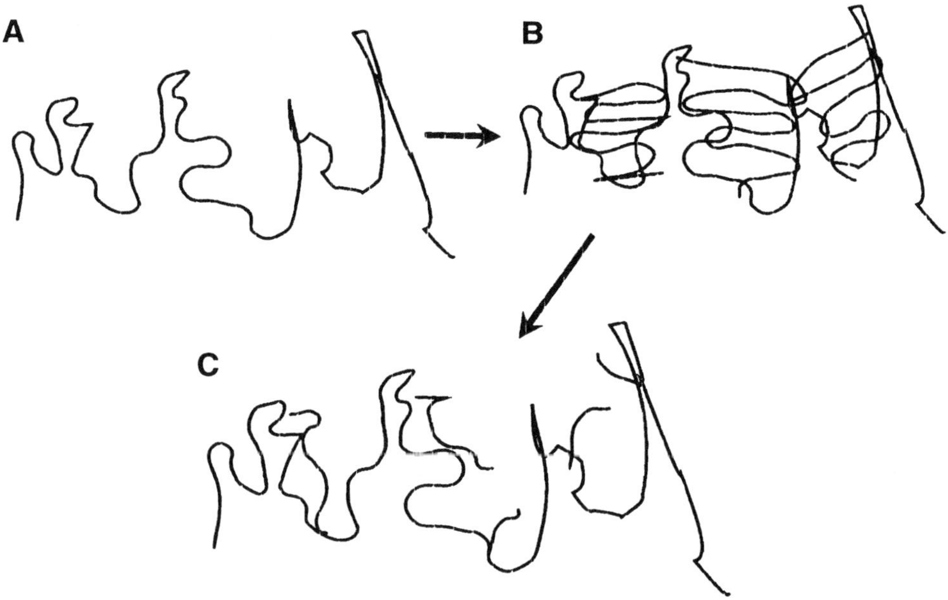

Fig. 5. Conceptual illustration of antigen retrieval. In unfixed tissue **(A)**, there is an irregular macromolecular topography, with free ingress to all potential antibody binding sites. After formalin fixation **(B)**, molecular crosslinks form bridges that prevent ingress of antibodies to all but the most superficial portions of cellular milieu. The application of microwave excitation breaks these linkages **(C)**, again allowing ingress of antibodies into the interior of the macromolecules.

mate the procedure, have greatly influenced this methodology. Pretreatment is necessary for many antigens because of the heavy bondage placed on the cellular milieu by the crosslinking induced by formalin fixation (**Fig. 5**). This fortress of molecular crosslinkages renders many antigenic sites impermeable to the penetration of immunoglobulin molecules, so that the reactivity of tissues is greatly weakened. Inadequate or prolonged exposure to formalin leads to even more problems; we recommend that thin slices of tissue (<5 mm in thickness) be immersed in formalin for 18 h during the initial processing steps. Another way to overcome the deleterious effects of formalin is to try other fixatives, such as 100% ethanol or Carnoy's fixative, or to simply use frozen sections of fresh tissue in lieu of fixed tissue. Each of these options has its drawbacks: alcohol-fixed tissues have peculiar staining qualities because of the severe dehydration, Carnoy's fixative contains chloroform and so must be used with caution, and frozen sections can be difficult to immunostain and interpret. For these reasons, we prefer the use of formalin-fixed material if crosslinking can be overcome.

Early methods of pretreatment of formalin sections utilized protein digestion by enzymes such as ficin, trypsin, or pepsin. In this manner, a weak enzyme solution (e.g., 0.01% [w/v] of trypsin) may be overlaid on the section for 20 min at 37°C prior to beginning the immunostaining procedure and after dewaxing and hydrating the tissue. The digestion step must be carefully controlled, however, for overdigestion can lead to destruction of the integrity of the tissue itself. To further complicate matters, the time of exposure may have to be adjusted according to the time of initial fixation, as the additional crosslinking caused by overfixation requires more vigorous digestion. Finally, a strong adhesive is required to prevent detachment of the histologic section from the slide. We have tried various adhesives, such as albumin or Elmer's glue (Borden, Columbus, OH), which are mixed with water in a 3% (v/v) solution and applied to blank acid-cleaned glass slides, which are then allowed to dry overnight at room temperature before the histologic sections are placed on them. To acid-clean slides, use a solution of 0.12*M* HCl and 70% (v/v) ethanol into which the slides are dipped, followed by rinsing in distilled water and drying at 37°C. More recently, we have used precleaned slides (Superfrost Plus, Fisher Scientific, Pittsburgh, PA); these have given us superior results in preventing section detachment.

A newer and superior method of pretreatment is the use of a microwave oven ***(10,11)*** or pressure cooker ***(12)***, which ameliorates crosslinks by bombarding them with controlled thermal energy. With these methods, the section-bearing slide is placed in a special solution and then subjected to brief heating in the microwave oven for a period of 15 min at a medium-high setting. Original solutions contained heavy metals, such as lead, with potential health and disposal problems, but today we generally use a modified citrate solution. Sections are still easily detached during or following pretreatment, so that we continue to use precleaned slides.

Finally, use of automation has become standard practice in these procedures, and a number of different instruments are being marketed. Automation offers the ability to do more sections per unit time with better consistency than by doing the procedure by hand, so that there are obvious advantages. A major drawback has been the adequacy and timing of the wash cycles, but the instruments yield stunning results if the steps are carefully evaluated and regulated for each antigen. Some instruments use capillary action to spread reagents over the sections, but special slides and/or slide holders may be required. Others use a flat rotating wheel with fixed pipets that apply reagents and washing solutions. These solutions may contain a wetting agent to fully spread the reagents onto the slides. With all instruments, the length and nature of each step is programmed into a computer that regulates the procedure.

2. Materials

2.1. Indirect Immunoperoxidase Procedure

1. Methanolic peroxide: 10 mL 3% (v/v) H_2O_2 in 90 mL methanol.
2. Hydrogen peroxide: 3 and 30% (v/v) solutions. Store at 4°C.
3. Citrate buffer, pH 6.0: 0.1 *M* solution of citric acid. Weigh out 21.01 g citric acid, monohydrate and dissolve in 1 L distilled water. 0.1 *M* solution of sodium citrate: Weigh out 29.41 g trisodium citrate dihydrate, and dissolve in 1 L distilled water.

   To make working solution, add 9 mL of 0.1 *M* citric acid and 41 mL of 0.1 *M* sodium citrate to 450 mL of distilled water. Adjust pH to 6.0 ± 0.1 using 1 *M* NaOH or 1 *M* HCl.
4. Phosphate-buffered saline (PBS), pH 7.4: Dissolve 100 g Difco Fn buffer (Fisher, cat. no. DF-2314-15-0) in 10 L distilled water. Adjust to pH 7.2 using 1 *M* NaOH or 1 *M* HCl.
5. Normal animal serum (*see* **Note 1**). Store at 4°C.
6. Primary antibody (store at 4°C). Expiration dates are listed on commercial reagents, and aliquots may be frozen for longer storage. Refrigerators with automatic defrosting cycles should be avoided to maintain antibody potency. See discussion of positive and negative controls in **Note 2**.
7. Antibody diluent (Dako, cat. no. S3022) (*see* **Note 5**). Store at 4°C.
8. 3,3'-Diaminobenzidine tetrahydrochloride (DAB) (Sigma, St. Louis, MO, cat. no. D-5637). This has been identified as a potential carcinogen, so that handling and disposal precautions should be observed.
9. Peroxidase-tagged secondary antiserum (Sigma). Store at 4°C.
10. Hematoxylin (Richard Allen Medical, Richland, MI, cat. no. 7221).
11. Mounting medium (SurgiPath Medical, Richmond, IL, cat. no. MX004). Keep air-tight to prevent drying.
12. Precleaned glass slides (Superfrost Plus, Fisher Scientific).
13. Ethanol.
14. Xylene: Vapors may be toxic or carcinogenic, so that optimally this reagent should be used with a fume hood. It is also inflammable and potentially explosive.
15. Microwave oven.
16. Humidity chamber (*see* **Note 4**) (Lipshaw-Shandow, Pittsburgh, PA, cat. no. 197).
17. Suction device (Erlenmeyer vacuum flask).
18. Slide oven (Fisher).
19. Tissue or cell block section (*see* **Note 5**).
20. Photographic equipment (*see* **Note 6**).

2.2. LSAB Procedure

1. Xylene.
2. Ethanol.
3. Methanolic peroxide: 10 mL 3% (v/v) H_2O_2 in 90 mL methanol.
4. Hydrogen peroxide: 3 and 30% (v/v) solutions.
5. Citrate buffer (*see* **Subheading 2.1., item 3**).

6. Dako protein block (Dako, cat. no. X0909). Store at 4°C.
7. Primary antibody (as above). Store at 4°C.
8. Dako LSAB + kit (Dako, cat. no. K0690). Store at 4°C.
9. DAB (*see* **Subheading 2.1., item 8**).
10. PBS, pH 7.2 (*see* **Subheading 2.1., item 4**).
11. Hematoxylin.
12. Slide oven.
13. Microwave oven.
14. Precleaned glass slides.
15. Humidity chamber (*see* **Note 4**).
16. Suction device (as above).
17. Tissue section (5 μ in thickness) or cell pellet (*see* **Note 5**).
18. Plastic staining dishes (Hema-Tek, Scientific Products–Baxter, McGaw Park, IL; cat. no. S7626-3).
19. Plastic slide holders (Baxter, cat. no. S7636-1).

3. Methods

3.1. Indirect Immunoperoxidase Procedure (for High Avidity, Impure Reagents such as Polyclonal Antisera)

3.1.1. Antigen Retrieval Method

1. Deparaffinize wax from tissue sections using overnight incubation in a 60°C oven, followed by sequential immersion in three changes of xylene (5 min each) and two changes of 100% ethanol (three min each).
2. Immerse slides in methanolic H_2O_2 (optional) for 30 min (*see* **Note 1**).
3. Begin antigen retrieval (optional).
   a. The sections should be deparaffinized and hydrated in PBS.
   b. Place slides in a plastic slide dish containing citrate buffer working solution, cover loosely with cap, and heat in a microwave oven at medium-high setting (92–98°C) for 10 min. The buffer will come to a boil; this is not a cause for alarm, but be sure that tissue sections are covered adequately by solution.
   c. After heating, remove slides from the microwave and allow to cool at room temperature for 25 min.
   d. Wash in running water for 5 min.
   e. Immerse slides in PBS for 10 min.
   f. Resume immunostaining procedure.
4. Wash in water for 5 min.
5. Immerse in PBS for 10 min.
6. Immerse in 10% (v/v) normal rabbit serum (optional; *see* **Note 1**) in PBS for 30 min.
7. No rinse (*see* **Note 1**).
8. Overlay tissue section with primary antibody and incubate in moisture chamber (*see* **Note 4**) for 1 h. Determine optimal dilution using checkerboard titration (*see* **Note 3**).
9. Rinse in PBS for 10 min.
10. Overlay tissue section with link reagent (secondary antibody) and incubate for 1 h in moisture chamber.

11. Rinse in PBS for 10 min.
12. Incubate in DAB-H_2O_2 solution (add 0.12 g DAB to 200 mL of PBS and then add 0.1 mL 30% v/v H_2O_2) for 10 min.
13. Wash 5 min in running water.
14. Dip in hematoxylin 20 times.
15. Dehydrate by sequential immersion in graded series of alcohols (90% v/v and 100% ethanol; two changes each), for 3 min each.
16. Immerse in four changes of xylene for 3 min each.
17. Mount cover slip on tissue section with mounting medium.
18. Examine under standard brightfield microscopy.
19. Use special photomicroscopy techniques for illustration (*see* **Note 6**).

3.2. LSAB Method (Sensitive Method for Low Avidity Monoclonal Antisera)

1. Deparaffinize slides as above (**step 1**).
2. Perfrom methanolic peroxide block (optional; *see* **Note 1**), as above.
3. Begin antigen retrieval (optional; *see* **Note 1**), as above.
4. Wash in running water and incubate in PBS for 10 min.
5. Overlay with Dako protein block for 30 min at room temperature in moisture chamber (optional; *see* **Note 1**).
6. Suction off excess fluid, but do not rinse. Leave thin coat of serum completely covering section (*see* **Note 1**).
7. Overlay with primary antiserum and incubate in moisture chamber for 1 h at room temperature. For increased sensitivity, slides may be incubated overnight at 4°C during this step.
8. Rinse in PBS for 5 min.
9. Overlay with link (secondary antiserum) from LSAB kit (Dako) and incubate in moisture chamber at room temperature for 20 min.
10. Rinse in PBS for 5 min.
11. Overlay with peroxidase–streptavidin conjugate in moisture chamber and incubate at room temperature for 20 min.
12. Immerse in DAB, as per preceding procedure.
13. Wash in running water.
14. Dip in hematoxylin for 20 s.
15. Wash in running water for 5 min.
16. Dehydrate in graded alcohols and xylene as per preceding procedure.
17. Mount cover slip over section using mounting medium.
18. Examine using standard brightfield microscopy.
19. Use special photomicroscopy techniques for illustration (*see* **Note 6**).

4. Notes

1. Preincubation steps are necessary in many instances to prevent excessive background. There are two main preincubation procedures, involving methanolic peroxide and normal serum. Methanolic peroxide is used to quench the endogeneous

peroxidase activity possessed by a variety of cell types, particularly erythrocytes. This procedure works by depleting the cells of this enzyme by overlaying them with a solution of hydrogen peroxide in methanol (*see* **Subheading 2.**). Normal serum can be used as an additional preincubation step to prevent the excessive nonspecific binding of immunoglobulin molecules to highly charged proteins, as are found in collagen. To perform this procedure, one overlays the tissue section with a 10% (v/v) solution of nonimmune serum taken from the same species as is used to prepare the secondary (or "link") antibody. This coats the charged binding sites with a layer of immunoglobulin, thus preventing attachment by the specific primary antibody. Dako also supplies a protein block for this purpose (*see* **Subheading 2.**). Because these molecules are not tagged with a chromogen, a visible signal is not produced by their attachment to these nonspecific sites. However, it is important not to wash the slides after this step, before overlaying the primary antibody, as the washing will delete the effect. Instead, one uses a suction device with a trap, such as an Erlenmeyer flask, to remove excessive reagent, leaving a coating of attached serum.

With some tissues, it will not be necessary to use these preincubation steps, and the methanolic peroxide can on occasion be detrimental to the antigenicity of some markers.

2. Positive and negative controls *must* be included in each run of immunostaining. Positive controls generally consist of a separate section containing the antigen in question, stained using the appropriate primary antibody at the same dilution as one used for the other sections. The final results are examined to verify that the appropriate staining occurred. Perhaps a more reliable positive control, but one not always available, is the "internal control" furnished by the appropriate staining of the normal cells in the tissue to be tested.

   Negative controls generally consist of sections in which nonimmune serum is substituted for the primary antiserum. If polyclonal antiserum is tested, the negative control should come from the same species, and if monoclonal antiserum is tested, the negative should be the same immunoglobulin isotype. The *protein concentration* used (and *not* necessarily the dilution) should be the same as that of the primary antiserum. Monoclonal antiserum used as a negative control should be selected with the knowledge that it will not stain the tissue to be tested; otherwise aberrant results can be obtained. A similar phenomenon occasionally may happen with polyclonal antiserum, if the animal's serum contains autoantibodies of various types (most commonly against intermediate filaments). These can usually be controlled by decreasing the titer of the stain or the sensitivity of the procedure. Finally, a commonly used but inappropriate negative stain is the substitution of saline or PBS for the primary antiserum.

3. To obtain optimal staining, it is necessary to individualize the concentration of each immunoglobulin reagent by titration experiments. The goal is to obtain strong, clean, specific staining of cells while eliminating excessive background. Also, antibody reagents are expensive, so that lower dilutions are economically attractive. To this end, one obtains multiple identical sections of tissue contain-

Table 2
Example of Checkerboard Titration Procedure

	Secondary antibody 1:40	Secondary antibody 1:100	Secondary antibody 1:300
Primary antibody 1:10	Primary 1:10 Secondary 1:40	Primary 1:10 Secondary 1:100	Primary 1:10 Secondary 1:300
Primary antibody 1:100	Primary 1:100 Secondary 1:40	Primary 1:100 Secondary 1:100	Primary 1:100 Secondary 1:300
Primary antibody 1:1000	Primary 1:1000 Secondary 1:40	Primary 1:1000 Secondary 1:100	Primary 1:1000 Secondary 1:300

ing the marker in question and prepares a series of dilutions to use as the primary reagent. A good starting point for commercially obtained reagents is usually suggested by the package insert. One then uses diluents such as PBS or the Dako premade diluent (*see* **Subheading 2.**) to prepare dilutions containing from 1/100 to 10X the recommended antibody strengths. If information is unavailable concerning an optimal dilution, we would recommend starting at 1:10, 1:100, and 1:1000. Following the titration experiment, the slides are examined, and the optimal dilution with strong, specific staining and the least background is used for further experimentation. Negative controls are not necessary at this point, nor do we recommend using "experimental" tissues, but rather routine positive controls, if possible. "Sausage blocks" containing multiple tissues in one section ***(13)*** have also been advocated for this purpose.

With the indirect procedure, it is necessary to find the optimal titer for the secondary antibody as well as the primary one. For this purpose, the "checkerboard" procedure is used. To perform this maneuver, one constructs a table of tested dilutions using the primary antibody as one parameter and the secondary antibody as the other. An example is found in **Table 2**. Following the experiment, one examines the slides for the optimal combination as above.

4. It is necessary to conduct all immunoglobulin incubations in an airtight chamber to prevent evaporation of solution from the slide, for evaporation will result in loss of reactivity. To prepare a chamber, one may use a flat box constructed of clear plastic. Moistened paper towels are used to coat the bottom of the chamber. Upon the towels one places thin, level supports such as the flat plastic tabs that are supplied with glass slide boxes. These are used to suspend the slides so that they do not come into contact with the moistened towels. The lid of the box is then tightly closed until the next incubation or rinsing step. The clear plastic will allow observation to prevent untoward evaporation and exposure of the tissue sections (*see* **Fig. 6**).
5. Whole cells from cell lines, suspensions, or explants can be used for immunohistochemistry. For the cleanest results, we prepare a cell pellet, which is then treated as a block of tissue, as per the following procedure:

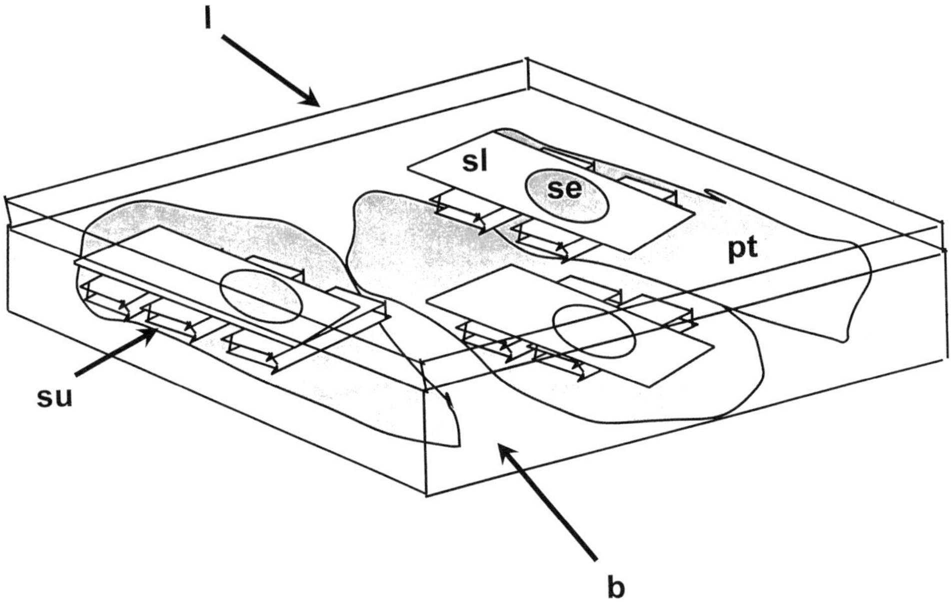

Fig. 6. Illustration of humidity chamber used for immunohistochemistry. A clear plastic box or tray (b) with a tight-fitting lid (l) is partially lined with moistened paper towels (pt). Glass slides (sl) rest upon supports (su) that prevent contact with the towels. Tissue sections (se) are covered with an overlay of immunoglobulin reagent.

a. Suspend approximately 2.5×10^7 cells in culture media such as RPMI-1640.
b. Spin down in table top centrifuge at 200*g* for 10 min.
c. Resuspend in 0.5 mL of human plasma (can be obtained from blood bank).
d. Add 25 λ of thrombin (Parke–Davis, Morris Plains, NJ, cat. no. N0071-4173-35).
e. When clot forms, put immediately in 10% buffered formalin (CMS, Houston, TX, cat. no. 245-685).
f. Fix for 3 h.
g. Process overnight in tissue processor and embed in paraffin as single fragment.

6. Photography of immunohistochemical stains is not a simple manner. The best and simplest way to document the results of staining is to use color photography. However, the publishing of color figures is a very expensive proposition, and journal publishers generally expect the author to pay at least $1000 (U.S.) per page. This cost is circumvented by the use of black and white photography, but one cannot obtain a suitable distinction of colors on the "gray scale" without special maneuvers. Thus, we typically employ the same color dyes, i.e., DAB stain (brown or black) and hematoxylin counterstain (blue), so that a standard procedure can be used. The major problem is the elimination of the deep blue hematoxylin stain, which merges with the brown specific stain on black and white film. Thus, for best results we use panchromatic film (Kodak T-Max ASA 100

[Kodak, Rochester NY]) and a Kodak gelatin 47B blue filter (Kodak), as the filter accentuates the brown stain and neutralizes the blue on black and white photographs. The filter can simply be placed over the lower condenser of the photomicroscope.

Acknowledgments

Supported in part by Cancer Center Support CORE grant P30 CA 21765 and by American Lebanese Syrian Associated Charities (ALSAC). The authors also thank John Zacher, RBP, for his contribution to the manuscript.

References

1. Sternberger, L. A., Hardy, P. H., Cuculis, J. J., and Meyer, H. G. (1970) The unlabeled antibody enzyme method for immunohistochemistry. Preparation and properties of soluble antigen-antibody complex (horseradish peroxidase–anti-horseradish peroxidase) and its use in the identification of spirochetes. *J. Histochem. Cytochem.* **18,** 315–333.
2. Lazarides, E. (1980) Intermediate filaments as mechanical integrators of cellular space. *Nature* **283,** 249–256.
3. Miettinen, M., Lehto, V. P., and Virtanen, I. (1984) Antibodies to intermediate filament proteins in the diagnosis and classification of human tumors. *Ultrastruct. Pathol.* **7,** 83–107.
4. Triche, T. J., Askin, F. B., and Kissane, J. (1986) Neuroblastoma, Ewing's sarcoma, and the differential diagnosis of small-, round-, blue-cell tumors, in *Pathology of Neoplasia in Children and Adolescents* (Finegold, M., ed.), W. B. Saunders, Philadelphia, PA, pp. 145–195.
5. Parham, D. M. (1993) Immunohistochemistry of childhood sarcomas: old and new markers. *Mod. Pathol.* **6,** 133–138.
6. Parham, D. M., Weeks, D. A., and Beckwith, J. B. (1994) The clinicopathologic spectrum of putative extrarenal rhabdoid tumors: an analysis of 42 cases studies with immunohistochemistry and/or electron microscopy. *Am. J. Surg. Pathol.* **18,** 1010–1029.
7. Hsu, S., Raine, L., and Fanger, H. (1981) A comparative study of the peroxidase-antiperoxidase method and an avidin-biotin complex method for studying polypeptide hormones with radioimmunoassay antibodies. *Am. J. Clin. Pathol.* **75,** 734–738.
8. Elias, J. M., Margiotta, M., and Gaborc, D. (1989) Sensitivity and detection efficiency of the peroxidase antiperoxidase (PAP), avidin–biotin peroxidase complex (ABC), and peroxidase-labeled avidin–biotin (LAB) methods. *Am. J. Clin. Pathol.* **92,** 62–67.
9. Giorno, R. (1984) A comparison of two immunoperoxidase staining methods based on the avidin-biotin interaction. *Diagn. Immunol.* **2,** 161–166.
10. Dookhan, D. B., Kovatich, A. J., and Miettinen, M. (1993) Nonenzymatic antigen retrieval in immunohistochemistry: comparison between different antigen retrieval modalities and proteolytic digestion. *App. Immunohistochem.* **1,** 149–155.

11. Leong, A. S.-Y. and Milios, J. (1990) Accelerated immunohistochemical staining by microwaves. *J. Pathol.* **161,** 327–334.
12. Chan, J. K. C. (1995) Kitchen ideas in the immunohistochemistry laboratory? *Adv. Anat. Pathol.* **2,** 1.
13. Battifiora, H. (1986) The multitumor (sausage) tissue block: novel method for immunohistochemical antibody testing. *Lab. Invest.* **55,** 244–248.

9

Immunohistochemical Evaluation of Biomarkers in Prostatic and Colorectal Neoplasia

Principles and Guidelines

William E. Grizzle, Russell B. Myers, Upender Manne, and Sudhir Srivastava

1. Introduction

In Chapter 10 of this volume and in prior publications *(1,2)*, the materials and methods used in immunohistochemical techniques are described, and factors that affect the immunohistochemical detection of biomarker expression are discussed. In this chapter, our semiquantitative method of evaluation of biomarker expression is explained. Also, the use of immunohistochemistry (IHC) to characterize biomarker expression in neoplastic lesions of the prostate and colorectum is discussed. The use of biomarkers in the study of neoplasia has expanded in recent years as new uses for biomarkers have been identified. For example, our laboratory has studied biomarkers to characterize tumorigenesis, support novel therapies, aid in diagnosis, predict aggressive tumor subtypes, and monitor the response of oral dysplasia to chemoprevention with retinoids. In studying oral neoplasia, we found that the expression of transforming growth factor α (TGFα) was increased in leukoplakia and that treatment with 13-cis-retinoic acid caused complete or partial resolution of leukoplakia and concomitant decreased expression of TGFα *(3)*. Thus, TGFα could be considered a candidate surrogate intermediate endpoint biomarker for chemoprevention studies of oral dysplasia. Similarly, we have used the evaluation of biomarker expression in tissues to qualify patients with tumors of the prostate, colorectum, breast, ovary, and lung for specific types of immunotherapy and for immunization with tumor vaccines *(4)*. We also have studied biomarker expression in order to characterize the putative preinvasive neoplas-

From: *Methods in Molecular Medicine, Vol. 14: Tumor Marker Protocols*
Edited by: M. Hanausek and Z. Walaszek © Humana Press Inc., Totowa, NJ

tic lesion of the prostate, prostatic intraepithelial neoplasia (PIN), and to relate PIN to invasive prostatic adenocarcinomas ***(5–8)***. Determination of biomarker expression in colorectal tumors has been useful in identifying racial differences in these tumors as well as differences in histopathological tumor types ***(9)***. In this chapter, we demonstrate the advantages of semiquantitatively evaluating biomarker expression in neoplastic lesions of the prostate and colorectum.

2. Materials and Methods

Our immunohistochemical methods are described in detail in Chapter 10 and are not reviewed in this chapter. The evaluation of the expression of biomarkers in neoplasia using IHC depends on the availability of specific and sensitive antibodies to the biomarker of interest. We prefer monoclonal antibodies (MAbs) for the evaluation of biomarker expression in tumors because of their consistency and specificity.

3. Evaluation of Biomarker Expression Using Immunohistochemistry

The interpretation of immunohistochemical results are confused by the need to incorporate both the pattern and the extent of staining. For some biomarkers, such as p185^{erbB-2}, the pattern of expression (e.g., membrane vs cytoplasmic vs membrane plus cytoplasmic), has been postulated to be very important when p185^{erbB-2} is used as a biomarker of prognosis in breast adenocarcinoma ***(10)***. Similarly, most evaluations of p53 expression ignore cytoplasmic staining of p53. Unfortunately, for most antigens, the significance of the pattern of expression has not been determined.

The extent of staining is also an important consideration. Many reports describe the intensity of staining of positive cells, and some describe the proportion of positive cells staining at any intensity. Few reports have evaluated the proportion of cells staining at different intensities. We have tried to consider the proportion of cells staining at specific intensities and have developed a semiquantitative score to incorporate these two parameters.

3.1. Evaluation of Patterns of Expression of Biomarkers

Several "rules" have developed for the classification of breast tumors as being either "positive" or "negative" for antigens. For example, the "poor" prognostic feature in breast cancer of overexpression of p185^{erbB-2} relies on p185 expression on the cellular membranes of a proportion of breast-cancer cells ***(10)***. However, in some cases of breast cancer, p185^{erbB-2} expression has primarily a strong cytoplasmic pattern ***(2)***, and cytoplasmic expression of p185^{erbB-2} has not been associated with a bad prognosis. The study of De Potter et al. ***(10)*** suggests that cytoplasmic immunoreactivity may result from the crossreaction of certain MAbs with a 155-kDa protein localized to the mitochondria. We believe that not all cytoplasmic expression of $p185^{erbB-2}$ is crossreactive with mitochondrial proteins ***(1,5)***, and that the cytoplasmic pattern of expression of $p185^{erbB-2}$ in tissues other than breast, such as prostate, may indicate a poor prognosis ***(11,12)***. Similarly, the proportion of breast-cancer cells that must express $p185^{erbB-2}$ on their membranes in order for a tumor to be classified as "positive" for erbB-2 has not been determined.

The interpretation of whether cases of breast cancer are positive for p53 (e.g., exhibit nuclear accumulation of the p53 protein) is another area of confusion. Although some studies ***(13)*** indicate that p53 nuclear accumulation is associated in breast cancer with mutations in p53, this may depend on how positive p53 cases are defined, and which antibodies are used to identify the p53 protein. IHC using an antibody like PAb240, which does not detect wild-type p53 as well as some mutated forms of p53, correlates with mutations in p53. However, PAb240 is not as sensitive in detecting p53 mutations as the antibodies CM1, PAb1801, and BP53-12-1, which detect both wild-type and mutated p53. Since native p53 is thought to have a half-life too short to be detected by IHC, such antibodies as PAb1801, CM1, and BP53-12-1 are considered very sensitive at detecting mutations. Thor et al. ***(13)***, using PAb1801 in a study of breast cancers, reported that p53 nuclear accumulation was associated with 15 identifiable p53 mutations, whereas nuclear accumulation was not detected in those cases that did not exhibit p53 mutations. However, since better antibodies have been developed and immunohistochemical detection techniques have become more sensitive, detection of p53 nuclear accumulation may not always correlate with p53 mutations ***(14)***. Specifically, p53 nuclear accumulation may indicate cellular damage and/or dysregulation of p53. In addition, we have observed that certain antigen-retrieval techniques, especially those that use urea solutions, cause positive staining in the nuclei of some normal adjacent cells (e.g., stromal cells) so care must be used in the classifying cases as p53 positive when antigen-retrieval techniques are involved ***(2)***.

A major problem in using p53 as a biomarker is that the cut-off points for defining p53 positivity have not been established. In some cases of breast cancer, the majority of nuclei have accumulation of p53, but in other cases, <1% of nuclei may be stained for p53. Barnes et al. ***(15)***, whose sensitive immunohistochemical methods detect some p53 accumulation in up to 80% of breast cancers, reported that cases of breast cancer with >75% of nuclei positive for p53 correlate with a poor prognosis; Thor et al. ***(13)***, who found about 25% of cases of breast cancer positive, demonstrated correlations with a poor prognosis if >1% of breast-cancer cells were positive for p53.

There are several major issues that should be clarified in future studies that evaluate p53 mutations in human cancers by IHC. These include:

1. What proportion of cells should demonstrate p53 accumulation in order for a specimen to be defined as p53-positive?
2. How intense must staining be for a nucleus to be considered positive?
3. Should cytoplasmic staining be evaluated in addition to nuclear staining?
4. How are specimens classified if there are focal areas of p53 accumulation?
5. How do you classify tumors in which one histologic variant of a tumor demonstrates p53 nuclear accumulation whereas another histologic variant in the same tumor does not?
6. Should antigen-retrieval techniques be used in evaluation of p53 positivity?

We have found that the antibody BP53-12-1 (Biogenex, San Ramon, CA) is similar to PAb1801 in detecting p53 nuclear accumulation in prostate cancer and colorectal tumors ***(9,16)***. Using BP53-12-1, most breast cancers have rare cells in which p53 nuclear accumulation can be detected, as has been described by Barnes et al. ***(15)***, and about 30% of breast cancers have 10% or more cells with nuclear p53 accumulation, as is discussed subsequently. In addition, use of antigen-retrieval techniques in breast adenocarcinomas markedly increases the proportion of cells with p53 nuclear accumulation compared to standard immunohistochemical techniques. It should be emphasized that the rules for understanding the pattern and extent of antigen expression in one tissue do not necessarily translate to other tissues. For example, the rules for interpreting positive expression of $p185^{erbB\text{-}2}$ and $p160^{erbB\text{-}3}$ in prostatic adenocarcinoma are likely different from those in breast cancer ***(1)***. Also, although the majority of adenocarcinomas of the breast have p53 nuclear accumulation, p53 "positivity" is less obvious in prostate cancers, in which most investigators report <20% of nonmetastatic prostate cancers to be positive for p53 by IHC ***(1,16–18)***. Use of semiquantitative analysis of immunohistochemical results may aid in determining the standards for classifying cases as positive or negative.

The wide variations in the literature concerning antigen expression in various tumor types are indicative of varying primary antibodies, different secondary detection systems, multiple criteria for antigen identification, ill-defined cut-off points, numerous types of tissue preparations, varying tumor stages, and ultimately, different approaches to interpretation of results. As is discussed in Chapter 10 in this volume, in evaluating information based on immunohistochemical methods, all of the above "problems" must be considered by the informed clinical scientist. Similarly, scientists considering the use of immunohistochemical techniques to evaluate clinical outcomes/therapies/diagnoses must realize that extensive expertise is necessary for proper performance/interpretation of immunohistochemical studies.

Table 1
Numerical Analysis of Tumor Staining: Example of Weak Staining

Intensity	0	1	2	3	4	
% Cell staining	40	20	30	10	0	Total score
Score	0	0.2	0.6	0.3	0	1.1

Table 2
Numerical Analysis of Tumor Staining: Example of Strong Staining

Intensity	0	1	2	3	4	
% Cell staining	0	10	20	40	30	Total score
Score	0	0.1	0.4	1.2	1.2	2.9

3.2. Semiquantitation of Biomarker Expression

Although IHC is not a linear technique, a reproducible semiquantitative index of immunohistochemical reactions, incorporating both the intensity of staining of individual cells and the portion of cells staining at each intensity, can be developed and is needed to evaluate whether biomarkers can be used as surrogate endpoint biomarkers.

The tumor cells selected for evaluation can be selected randomly by a point-counting technique. Specifically, a grid can be used to select tumor cells for evaluation (those cells that fall on the intersection of grid lines are selected). This is a standard technique of stereological analysis and is statistically valid for randomly identifying cells/organelles/areas of tissue. The tumor cells are classified with respect to immunostaining for each antigen with the percentage of cells estimated at each staining intensity from 0 to +4. Alternately, an experienced pathologist/microscopist can estimate the proportion of cells that fall in each staining category to yield an equivalent numerical index.

To permit numerical analysis, the proportion of cells at each intensity (decimal equivalent of percentage of cells) can be multiplied by the numerical value of that intensity. A score can be developed that ranges from 0 to 4 (maximum). For example, consider the tumor staining for p185$^{erbB\text{-}2}$ (*see* **Table 1**). This tumor would be classified as staining relatively weakly, with a total score of 1.1. For comparison, consider a tumor staining for p185$^{erbB\text{-}2}$ (*see* **Table 2**). This tumor would be classified as staining strongly, with a total score of 2.9. As can be seen from the technique, a score of 1 would be obtained if 100% of cells stained at +1. Similarly, a score of 1 would be obtained if 50% of cells stained at +2, or alternatively if 20% of cells stained at +4 and 10% of cells stained at +2. Usually only a few patterns of staining occur in tumors. The most common pattern is one that occurs when the majority of cells stain to some extent. Typically there is a modal value (e.g., +3) and an approximate bell-shaped distribution of cells staining around this modal value (e.g., 40% at +3, 30% at +2, 30% at +4). Another typical pattern is where most cells are negative, but a small subpopulation of cells stain. This would be the case when 80% of cells do not stain and 20% of cells stain strongly (e.g., +3).

These examples demonstrate why the semiquantitation of IHC by this method cannot be considered a linear assay. Another reason is that the method of secondary detection is not linear; most methods saturate the signal at the higher intensities of staining (e.g., +3 and +4) and demonstrate accentuated staining at the lowest intensities of staining (e.g., +1). Thus, our approach to quantitative IHC differs from that typically used in determining whether or not an antigen is present, because we do not saturate staining in the immunohistochemical method.

The stoichiometry of the reaction between an antigen and the ultimate colored precipitate deposited at the antigen location by the primary antibody detection system is not one-to-one and hence not linear. Specifically, one molecule of substrate does not interact with one molecule of antigen identified by the primary antibody. This is in contrast to some fluorescent immunochemical techniques in which one antibody molecule with a single fluorescent probe reacts with one molecule of antigen. However, the method of IHC that uses peroxidase probes and biotin-avidin intensification techniques has an advantage over fluorescent techniques in tissue sections because it can be interpreted more easily in tissue without the extensive background that sometimes occurs when fluorescence is used in tissue sections. Also, this technique can be used to more easily localize and identify the specific cells/tissues stained for the antigen.

Because the stoichiometry of the IHC detection system is unknown but not linear, an intensity of +3 is unlikely to indicate three times the expression of antigen than is identified by an intensity of +1. However, one can demonstrate using sequential antibody dilutions that an intensity of +3 indicates more antigen is present than an intensity of +1 or less. Nevertheless, we require assay results to be as consistent as possible so that the standard error of repeat assays can be reduced, so statistical changes in response to chemopreventive agents can be identified more easily.

In small biopsies, field-selection bias is a potential problem. The main bias would be introduced by the biological variability of biomarker expression in tumors and therefore the sampling inherent in small samples of an unhomogeneous whole. To demonstrate reproducibility of immunohistochemical assay and semiquantitative evaluations, we stained tissues of seven surgical specimens with three dilutions of several antibodies. The staining was repeated four times each on a different day, and the results were analyzed by one observer without knowledge of the stain, concentration, or repeat status of the stain. Typical results are shown in **Table 3**.

Table 3
Reproducibility of Staining with the Antibody CC-49 (TAG-72)

Tissue	Antibody CC-49 concentration (μg/mL)	Repeat measurements[a] R_1	R_2	R_3	R_4	Average	SD
Spleen white pulp	0.1	0	0	0	0	0	0
Spleen white pulp	1.0	0	0	0	0	0	0
Spleen white pulp	10.0	0	0	0	0	0	0
Colon tumor 1	0.1	1.8	1.7	1.55	1.7	1.69	0.10
Colon tumor 1	1.0	2.1	1.4	1.7	1.6	1.70	0.29
Colon tumor 1	10.0	2.5	2.0	2.4	2.6	2.37	0.26
Smooth muscle in colonic wall 1	0.1	0	0	0	0.3	0.07	0.15
Smooth muscle in colonic wall 1	1.0	0	0	0	0	0	0
Smooth muscle in colonic wall 1	10.0	0.1	0	0.3	0	0.10	0.14
Colon tumor 2	0.1	3.2	1.6	1.55	2.9	2.31	0.86
Colon tumor 2	1.0	3.2	2.3	2.6	2.9	2.67	0.51
Colon tumor 2	10.0	3.2	2.0	2.4	3.0	2.65	0.55
Smooth muscle in colonic wall 2	0.1	0	0	0	0	0	0
Smooth muscle in colonic wall 2	1.0	0	0.5	0	0	0.12	0.25
Smooth muscle in colonic wall 2	10.0	0	0.5	0.7	0	0.30	0.36

[a]Each repeat measurement (e.g., R_1, R_2, R_3, and R_4) represents a different staining run and separate grading (blinded) by the same observer.

4. Examples of the Application of Immunohistochemistry to Understanding the Development of Prostatic Adenocarcinoma

Our laboratory has characterized biomarker expression in prostatic adenocarcinoma (PCa) and the putative preinvasive lesion PIN. Because PIN lesions are frequently focally distributed, IHC and *in situ* hybridization are well-suited for analysis of biomarker expression within these lesions. The following is a summary of our studies of two biomarkers in normal prostatic epithelium, PIN and PCa.

4.1. $p185^{erbB\text{-}2}$

The product of the c-erbB-2 proto-oncogene is a 185-kDa protein ($p185^{erbB\text{-}2}$), which is structurally similar to the EGF receptor ***(19)***. Overexpression of $p185^{erbB\text{-}2}$ has been detected in several malignancies, including those of the breast ***(20)*** and ovary ***(21)***. Studies from our laboratory demonstrate strong

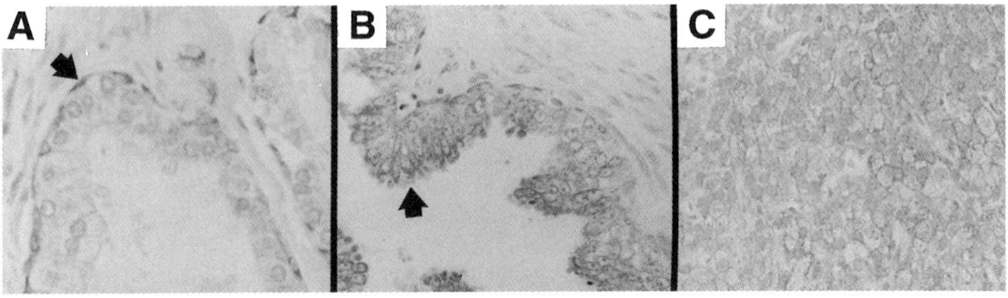

Fig. 1. Immunostaining of prostatic tissue for p185^{erbB-2}. **(A)** Normal epithelium. **(B)** Prostatic intraepithelial neoplasia. **(C)** Prostatic adenocarcinoma.

immunoreactivity for p185^{erbB-2} among the basal-cell layer of the normal epithelium (**Fig. 1A**). Immunostaining for p185^{erbB-2} is typically weak and focal among the luminal cells of the normal epithelium (**Fig. 1A**). In contrast, strong immunostaining is detected among the dysplastic luminal cells of PIN lesions (**Fig. 1B**). Likewise, the cells of invasive PCa frequently demonstrate p185^{erbB-2} immunoreactivity (**Fig. 1C**).

4.2. nm23-H1

The product of the nm23-H1 gene is a nucleoside diphosphate kinase that was initially shown to inhibit metastasis in breast malignancies and melanoma *(22,23)*. Subsequent studies, however, have demonstrated that the nm23 gene product is not an inhibitor of metastasis in all tumors, and in fact elevated nm23 expression may be an indicator of poor prognosis in some tumors *(24,25)*. Within the benign prostatic epithelium, strong expression of the nm23-H1 protein is restricted to the basal-cell layer. The dysplastic luminal cells of PIN lesions, however, frequently demonstrate strong immunostaining for the nm23-H1 protein. In addition, the cells of the invasive PCa consistently demonstrate strong nm23-H1 protein immunostaining. **Figure 2** illustrates the use of the semiquantitative analysis of immunostaining described in this chapter to compare the expression of the nm23-H1 protein in the normal and dysplastic luminal cells as well as the cells of invasive PCa.

The examples described above demonstrate the use of IHC to localize and compare biomarker expression in normal, dysplastic, and malignant prostatic tissues. The expression of p185^{erbB-2} and the nm23-H1 protein in the dysplastic luminal cells of PIN lesions and invasive PCa supports the concept that PIN represents a preinvasive lesion of the prostate. The value of this technique is further illustrated by the ability to precisely detect biomarker expression in specific cell types, such as basal cells of the normal epithelium or dysplastic luminal cells. IHC as well as *in situ* hybridization are likely to be of great value in understanding the molecular and cellular events involved in the development of PCa.

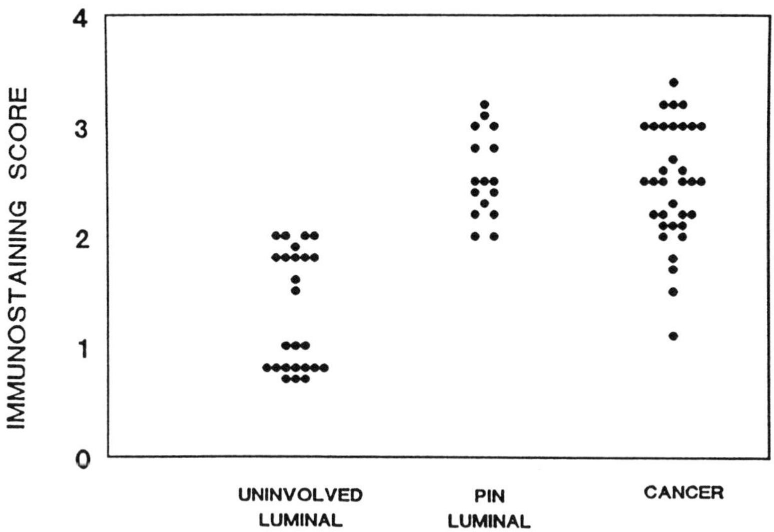

Fig. 2. Immunohistochemical analysis of nm23-H1 expression in prostatic tissue. The immunostaining score was derived as described in this chapter.

5. The Application of Immunohistochemistry to Visualize Oncoprotein p53 and *bcl-2* in Colorectal Adenocarcinoma

Our laboratory is interested in the expression of malignant-tumor factors that control cellular proliferation and apoptosis. Two examples of genes involved in the pathways are *bcl-2* and *p53*. Study of the expression of the protein products of these genes demonstrates general technical issues.

IHC is the most widely used technique to detect the accumulation of abnormal p53 protein and expression of bcl-2. Studies have emphasized the importance of the role of p53 and bcl-2 proteins in the process of programmed cell death (apoptosis) ***(26,27)***, cell proliferation ***(28)***, and tumor development ***(29,30)***. In addition, these biomarkers have been implicated in resistance to chemotherapy and radiation therapy ***(31,32)*** in different malignancies.

The expression of *bcl-2* and the mutated forms of p53 protein have been demonstrated by IHC in several malignancies, including breast ***(33,34)***, ovarian ***(35,36)***, prostate ***(16,37)***, colorectal ***(38–40)***, nonsmall cell lung carcinomas ***(41)***, medullary carcinoma of the thyroid ***(42)***, and non-Hodgkin's lymphoma ***(43)***.

Several earlier studies ***(44–46)*** and our study examined the sensitivity of IHC to detect mutant p53 protein by comparing IHC with single-strand conformation polymorphism (SSCP), a molecular technique often used for detection of genomic mutations ***(47,48)***, and demonstrated strong concordance between

Table 4
Evaluation of p53 Nuclear Accumulation in Colorectal Adenocarcinoma Before and After Antigen Retrieval

IHC	SSCP Negative ($n = 53$) N (%)	SSCP Point mutations ($n = 37$) N (%)	SSCP Nonpoint mutations ($n = 17$) N (%)
Nonantigen retrieval			
Positive ≥10% ($n = 48$)	8 (15)	35 (95)	5 (29)
Negative <10% ($n = 59$)	45 (85)	2 (5)	12 (71)
	79.4% concordance (κ coefficient = 0.589; $p < 0.0001$)		
Antigen retrieval[b]			
Cut-off 10%[a]			
Positive ≥10% ($n = 75$)	30 (57)	35 (95)	10 (59)
Negative <10% ($n = 32$)	23 (43)	2 (5)	7 (41)
	63% concordance (κ coefficient = 0.29; $p < 0.001$)		
Antigen retrieval			
Cut-off 50%[a]			
Positive ≥50% ($n = 51$)	9 (17)	35 (95)	7 (41)
Negative <50% ($n = 56$)	44 (83)	2 (5)	10 (59)
	80% concordance (κ coefficient = 0.61; $p < 0.001$)		

[a]Percent of malignant cells with p53 nuclear accumulation after staining with BP53-12-1 MAb.
[b]Deparaffinized and rehydrated tissue sections treated with 0.01 *M* citrate buffer and heated in a microwave oven at high power for two 5-min intervals before probing with primary antibody.

the two methods. However, our study in colorectal adenocarcinoma indicated that IHC is a more sensitive method for detecting SSCP-identified missense point mutations (about 95%) than the nonpoint mutations in p53 gene (**Table 4**). Although in the majority of malignancies the results of IHC are synonymous with genomic mutations, there are some discrepancies between these two techniques. For instance, the cytoplasmic accumulation of p53 protein in breast carcinoma ***(49)***, p53 overexpression in hepatic tumors of childhood ***(50)***, head and neck squamous carcinoma ***(51)***, and testicular carcinoma ***(52,53)*** have been reported to occur without apparent genetic alteration in *p53* genes.

Table 5
Effect of Antigen Retrieval (AR) on Immunohistochemical Detection of *bcl-2* in Colorectal Adenocarcinomas

	Without AR (n = 173)		With AR[b] (n = 133)	
Immunostaining score[a]	*N*	%	*N*	%
Negative (0.0–<0.5)	156	90	67	50
Positive (≥0.5–4)	17	10	66	50

[a]Immunostaining score, estimated absolute percentage of cells at each intensity on a scale of 0 (no staining) to +4 (strongest intensity), multiplied by the appropriate intensity score to obtain a weighted average score. The scores of the four authors were combined to obtain an average score.

[b]Deparaffinized and rehydrated tissue sections treated with 0.01 *M* citrate buffer and heated in a microwave oven at high power for two 5-min intervals, before probing with primary antibody.

The application of IHC to formalin-fixed, paraffin-embedded tissue is made difficult because of loss of antigenicity following formalin fixation and tissue processing. To overcome this problem, several antigen-retrieval (AR) techniques are suggested for different antigens and for different tissues ***(2,54–60)***. These methods are helpful in rapid revelation of various antigenic epitopes of many proteins and improve immunochemical detectability in paraffin sections. For example, only about 10% of tumors express detectable levels of bcl-2 in colorectal adenocarcinomas. However, when the tissue sections are treated with citrate buffer and microwave-heated, the intensity of bcl-2 immunostaining is enhanced remarkably (*see* **Table 4**) and the incidence of *bcl-2* expressing tumors increases to about 50% ***(40)***.

As discussed in Chapter 10, AR improves the detection of specific antigens in some tissues but its use also may introduce problems in interpretation. For example, specific antigen retrieval methods do not improve the IHC detection of specific antigens using some antibodies to $p185^{erbB-2}$ and EGF receptor in breast ***(2)***; similarly, the increased detection of p53 nuclear accumulation in prostate adenocarcinoma following the use of AR ***(16)*** may not affect interpretation when specimens are evaluated, because the cut-off points for designating a case as positive change following the use of AR. In colorectal adenocarcinomas, the immunochemical detection of mutant forms of p53 protein was more intense and demonstrated an increase in the number of cells with nuclear accumulation, but showed no significant change in the incidence of positive cases when a microwave heating in citrate buffer AR method was employed (*see* **Table 5**). In addition, the significant association between p53 nuclear accumulation and prognosis was lost when AR was used to detect p53 nuclear accumulation.

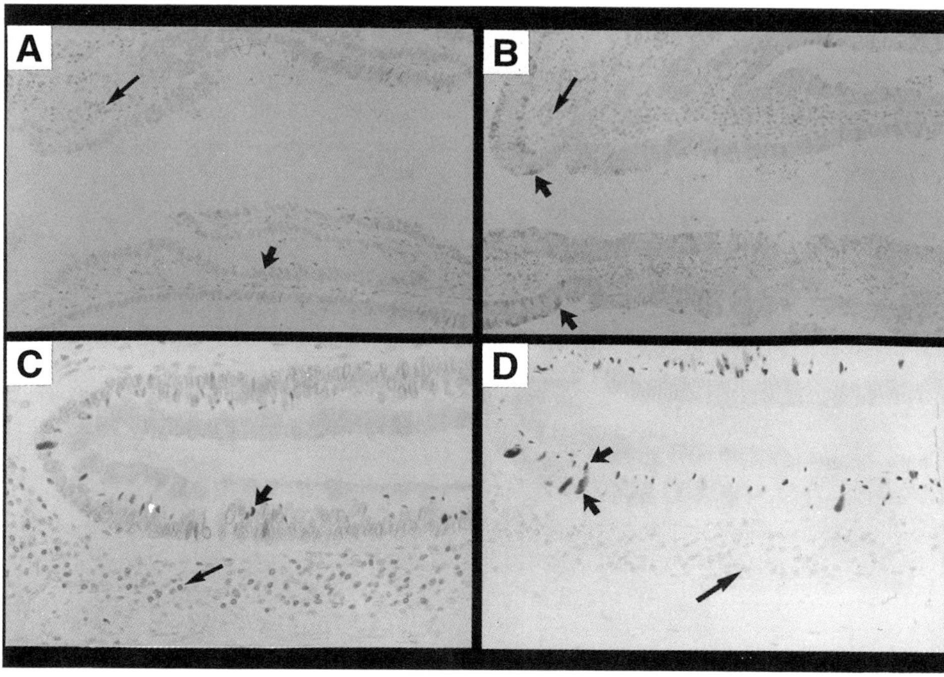

Fig. 3. Nonspecific immunostaining in control (delete) tissue sections of colorectal adenocarcinoma. **(A)** Comparison of nonspecific cytoplasmic immunostaining in control (delete) tissue sections that were not exposed to primary antibody. No immunostaining was observed in a tissue section subjected to microwave heating with citrate buffer (antigen retrieval, AR) and exposed to DAB alone (short arrow in A). **(B)** Strong to moderate cytoplasmic staining pattern was observed in a serial tissue section that was subjected to AR and exposed to avidin-peroxidase complex and DAB (short arrows in B). **(C)** Comparison of nonspecific nuclear immunostaining in control (delete) tissue sections that were not exposed to primary antibody. Weak to moderate nuclear and perinuclear staining pattern was observed in scattered tumor cells in a tissue section routinely stained and exposed to avidin-peroxidase complex and DAB (short arrow in C). **(D)** Strong to moderate nuclear and perinuclear staining was observed in high proportion of malignant cells in a serial tissue section subjected to citrate buffer AR and exposed to avidin-peroxidase complex and DAB (short arrows in D). Note the lack of staining in the lymphocytes (long arrows in A–D).

Even though in our study AR improved detection of some of the nonpoint mutations the detection of missense point mutations remained the same (**Table 4**) *(40)*.

Following an AR method (microwave heating with citrate buffer), some tissue sections (about 5–10%) from malignant colorectal tumors demonstrate moderate-to-strong membrane, cytoplasmic or nuclear patterns of staining, even though the sections are not exposed to primary antibody (**Fig. 3**). This pattern was observed when the cells were exposed to the avidin-horseradish-peroxidase complex and 3,3'-diaminobenzidine tetrahydrochloride (DAB) (**Fig. 3B–D**). In contrast, no immunostaining was evident when cells were exposed to DAB alone (**Fig. 3**). These results suggest that the nonspecific pattern of staining that appears specific is actually caused by endogenous biotin. Earlier studies have reported that endogenous biotin is abundant in several tissues, and about one-half of intracellular biotin can be stored in nuclei. Furthermore, biotin may interfere with the avidin-biotin-peroxidase complex method on formalin-fixed, paraffin-embedded tissue sections ***(62,63)***. We propose that AR may emphasize nonspecific patterns of staining that are caused by endogenous biotin in some of the colorectal adenocarcinoma archival tissues as well as in other tissues.

The interpretation of immunostaining is another important issue in IHC. Several different criteria are followed in the analysis and interpretation of different biomarkers. In addition, the method of evaluation of a specific biomarker may differ among various investigators. There are some good examples that examine the perplexity of the immunostaining evaluation issue. Noguchi et al. ***(46)***, in lung carcinomas, has demonstrated about 90% concordance between IHC and PCR-SSCP techniques in detecting nuclear accumulation of p53 when any immunoreactivity in malignant cell nuclei was considered as positive, irrespective of the percentage of positive cells. However, in breast carcinoma, Allred et al. ***(45)***, using a similar evaluation method, could only find 62% concordance between these two techniques. We have found (**Table 4**) that selection of a cut-off value for percent p53 immunopositive cells that would consider a tumor as positive for p53 nuclear accumulation would affect the comparison between SSCP and IHC methods. Thus, it is justifiable to follow an appropriate immunostaining evaluation method that is suitable to certain IHC studies.

In conclusion control (delete) tissue sections must be examined for nonspecific cytoplasmic and/or nuclear staining, particularly in malignant colorectal malignant tissue sections processed with AR. Both cytoplasmic and/or nuclear immunostaining patterns that appear specific can be observed in tissue sections exposed only to the biotinylated secondary detection systems. These patterns are nonspecific, i.e., independent of the primary antibody used. These technical issues should be considered for studies using archival tissues exposed to various antigen retrieval methods in order to evaluate the expression of biomarkers.

References

1. Grizzle, W. E., Myers, R. B., Arnold, M. M., and Srivastava, S. (1994) Evaluation of biomarkers in breast and prostate cancer. *J. Cell. Biochem.* **19(Suppl.)**, 259–266.
2. Grizzle, W. E., Myers, R. B., and Oelschlager, D. K. (1995) Prognostic biomarkers in breast cancer: factors affecting immunohistochemical evaluation. *Breast* **1**, 243–250.

3. Beenken, S., Huang, P., Sellers, M., Listinsky, C., Stockard, C., Hubbard, W., Wheeler, R., and Grizzle, W. E. (1994) Retinoid modulation of biomarkers in oral leukoplakia/dysplasia. *J. Cell. Biochem.* **19(Suppl.),** 270–277.
4. Myers, R. B., Meredith, R. F., Schlom, J., LoBuglio, A. F., Bueschen, A. L., Wheeler, R. H., Stockard, C. R., and Grizzle, W. E. (1994) Tumor associated glycoprotein-72 is highly expressed in prostatic adenocarcinomas. *J. Urol.* **152,** 243–246.
5. Myers, R. B., Srivastava, S., Oelschlager, D. K., and Grizzle, W. E. (1994) Expression of $p160^{erbB-3}$ and $p185^{erbB-2}$ in prostatic intraepithelial neoplasia and prostatic adenocarcinoma. *J. Natl. Cancer Inst.* **86,** 1140–1145.
6. Myers, R. B., Schlom, J., Srivastava, S., and Grizzle, W. E. (1995) Expression of tumor associated glycoprotein-72 in prostatic intraepithelial neoplasia and prostatic adenocarcinoma. *Modern Pathol.* **8,** 260–265.
7. Myers, R. B., Srivastava, S., and Grizzle, W. E. (1995) Lewis Y antigen as detected by the monoclonal antibody BR96 is expressed strongly in prostatic adenocarcinoma. *J. Urol.* **153,** 1572–1574.
8. Myers, R. B., Srivastava, S., Oelschlager, D. K., Brown, D., and Grizzle, W. E. (1996) Expression of nm23-H1 in prostatic intraepithelial neoplasia and prostatic adenocarcinoma. *Hum. Pathol.* **27,** 1021–1024.
9. Manne, U., Myers, R. B., Moron, C., Srivastava, S., and Grizzle, W. E. (1996) Racial distribution of p53 mutations in colorectal adenocarcinoma. *Modern Pathol.* **1,** 62A.
10. De Potter, C. R., Quatacker, J., Maertens, G., Van Daele, S., Pauvels, C., Verhofstede, C., Eechaute, W., and Roels, H. (1989) The subcellular localization of the *neu* protein in human normal and neoplastic cells. *Int. J. Cancer* **44,** 969–974.
11. Sadasivan, R., Morgan, R., Jennings, S., Austenfeld, M., Van Veldhuizen, P., Stephens, R., and Noble, M. (1993) Overexpression of HER-2/neu may be an indicator of poor prognosis in prostatic cancer. *J. Urol.* **150,** 126–131.
12. Veltri, R. W., Parfin, A. W., Epstein, J. P., Marley, G. M., Miller, G. M., Singer, D. S., Patton, K. P., Criley, S. R., and Coffey, D. S. (1994) Quantitative nuclear morphometry, Markovian texture descriptors and DNA content captured on a CAS-200 image analysis system, combined with PCNA and HER-2-NEU immunohistochemistry for prediction of prostate cancer progression. *J. Cell. Biochem.* **19(Suppl.),** 249–258.
13. Thor, A. D., Moore, D. H., Edgerton, S. M., Kawasaki, E. S., Reinhouse, E., Lynch, H. T., Marcus, J. N., Schwartz, L., Chen, L. C., and Mayall, B. H. (1992) Accumulation of p53 tumor suppressor gene protein: an independent marker of prognosis in breast cancers. *J. Natl. Cancer Inst.* **84,** 845–855.
14. Lohman, D. L., Ruhri, Ch., Schmitt, M., Graeff, H., and Hofler, H. (1993) Accumulation of p53 protein as an indicator for p53 gene mutation in breast cancer. Occurrence of false-positives and false negatives. *Diagn. Mol. Pathol.* **2,** 36–41.
15. Barnes, D. M., Dublin, E. A., Fisher, C. J., Levison, D. A., and Millis, R. R. (1993) Immunohistochemical detection of p53 protein in mammary carcinoma: an important new independent indicator of prognosis? *Hum. Pathol.* **24,** 469–476.

16. Myers, R. B., Oelschlager, D., Srivastava, S., and Grizzle, W. E. (1994) Accumulation of the p53 protein occurs more frequently in metastatic than in localized prostatic adenocarcinomas. *Prostate* **25,** 243–248.
17. Visakorpi, T., Kallioniemi, O. P., Heikkinen, A., Koivula, T., and Isola, J. (1992) Small subgroup of aggressive highly proliferative prostatic carcinomas defined by p53 accumulation. *J. Natl. Cancer Inst.* **84,** 883–887.
18. Navone, N. M., Troncoso, P., Pisters, L. L., Goodrow, T. L., Palmer, J. L., Nichols, W. W., Von Eschenbach, A. C., and Conti, C. J. (1993) p53 protein accumulation and gene mutation in progression of human prostate carcinoma. *J. Natl. Cancer Inst.* **85,** 1657–1669.
19. Yamamoto, T. M., Ikawa, S., Akiyama, T., Semba, K., Nomura, N., Miyajima, N., Saito, T., and Toyoshima, K. (1986) Similarity of protein encoded by the human c-erbB-2 gene to epidermal growth factor receptor. *Nature* **319,** 230–234.
20. Slamon, D. J., Clark, G. M., Wong, S. G., Levin, W. J., Ullrich, A., and McGuire, W. L. (1987) Human breast cancer: correlation of relapse and survival with amplification of the HER2/neu oncogene. *Science* **235,** 177–182.
21. Slamon, D. J., Godolphin, W., Jones, L. A., Holt, J. A., Wong, S. G., Keith, D. E., Levin, W. J., Stuart, S. G., Udove, J., Ullrich, A., and Press, M. F. (1989) Studies of the HER2/neu proto-oncogene in human breast and ovarian cancer. *Science* **244,** 707–712.
22. Hennessy, C., Henry, J. A., May, F. E., Wesfly, B. R., Angus, B., and Lennard, T. W. (1991) Expression of the antimetastatic gene in human breast cancer: an association with good prognosis. *J. Natl. Cancer Inst.* **83,** 281–285.
23. Florenes, V. A., Aamdal, S., Myklebost, O., Maelandsmo, G. M., Bruland, O. S., and Fodstad, O. (1992) Levels of nm23 messenger RNA in metastatic malignant melanomas. *Cancer Res.* **52,** 6088–6091.
24. Leone, A., Seeger, R. C., Hong, C. M., Hu, Y. Y., Arboleda, M. J., Brodeur, G. M., Stram, D., Slamon, D. J., and Steeg, P. S. (1993) Evidence for nm23 RNA overexpression, DNA amplification and mutation in aggressive childhood neuroblastomas. *Oncogene* **8,** 855–865.
25. Engel, M., Theisinger, B., Seib, T., Seitz, G., Huwer, Ii., Zang, K. D., Welter, C., and Dooley, S. (1993) High levels of nm23-H1 and nm23-H2 messenger RNA in human squamous-cell lung carcinoma are associated with poor differentiation and advanced tumor stages. *Int. J. Cancer* **55,** 375–379.
26. Diller, L., Kassel, J., Nelson, C. E., Gryka, M. A., Litwak, G., Gebhardt, M., Ozturk, M., Baker, S. J., and Vogelstein, B. (1990) p53 functions as a cell cycle control protein in osteosarcomas. *Mol. Cell. Biol.* **10,** 5772–5781.
27. Yonisch-Rovach, E., Resnitzky, D., Lotem, J., Sachs, L., Kimchi, A., and Oren, M. (1991) Wild-type p53 induces apoptosis of myeloid leukaemic cells that is inhibited by interleukin-6. *Nature* **352,** 345–347.
28. Kobayashi, M., Watanabe, H., Ajioka, Y., Yoshida, M., Hitomi, J., and Asakura, H. (1995) Correlation of p53 protein expression with apoptotic incidence in colorectal neoplasia. *Virchows Archiv.* **427,** 27–32.

29. Vaux, D. L., Cory, S., and Adams, T. M. (1988) bcl-2 promotes the survival of haemopoietic cells and cooperates with c-myc to immortalize pre-b cells. *Nature* **335,** 440–442.
30. Clarke, A. R., Purdie, C. A., Harrison, D. J., Morris, R. G., Bird, C. C., Hooper, M. L., and Wyllie, A. H. (1993) Thymocytic apoptosis induced by p53-dependent and independent pathways. *Nature* **362,** 849–852.
31. Lotera, J. and Sachs, L. (1993) Regulation by bcl-2, c-myc, and p53 of susceptibility to induction of apoptosis by heat shock and cancer chemotherapy compounds in differentiation-competent and defective myeloid leukemic cells. *Cell Growth Differ.* **4,** 41–47.
32. Chang, E. H., Pirollo, K. F., Zou, Z. Q., Cheung, H. Y., Lawler, E. L., Garner, R., White, E., Bernstein, W. B., Fraumeni, J. F., and Blattner, W. A. (1987) Oncogenes in radioresistant noncancerous skin fibroblasts from a cancer-prone family. *Science* **237,** 1036–1039.
33. Silvestrini, R., Veneroni, S., Daidone, M. G., Benini, E., Boracchi, P., Mazzetti, M., Di Fronzo, G., Rilke, F., and Veronesi, U. (1994) The bcl-2 protein: a prognostic indicator strongly related to p53 protein in lymph node-negative breast cancer patients. *J. Natl. Cancer Inst.* **86,** 499–504.
34. Gorczyca, W., Markiewski, M., Kram, A., Tuziak, T., and Domagala, W. Immunohistochemical analysis of bcl-2 and p53 expression in breast carcinomas: their correlation with Ki-67 growth fraction. *Virchows Archiv.* **426,** 229–233.
35. Henriksen, R., Wilander, E., and Oberg, K. (1995) Expression and prognostic significance of bcl-2 in ovarian tumors. *Br. J. Cancer* **72,** 1324–1329.
36. Herold, J. J. O., Eliopoulos, A. G., Warwick, J., Niedobitek, G., Young, L. S., and Kerr, D. J. (1996) The prognostic significance and p53 expression in ovarian carcinoma. *Cancer Res.* **56,** 2178–2184.
37. McDonnell, T. J., Troncoso, P., Brisbay, S. M., Lagothetis, C., Chung, L. W. K., Hsieh, J.-T., Tu, S.-M., and Campbell, M. L. (1992) Expression of the proto-oncogene bcl-2 in the prostate and its association with emergence of androgen independent prostate cancer. *Cancer Res.* **52,** 6940–6944.
38. Hague, A., Moorghen, M., Hicks, D., Chapman, M., and Paraskeva, C. (1994) bcl-2 expression in human colorectal adenocarcinomas and carcinomas. *Oncogene* **9,** 3367–3370.
39. Bosari, S., Moneghini, L., Graziani, D., Lee, A. K., Murray, J. J., Coggi, G., and Viale, G. (1995) bcl-2 oncoprotein in colorectal hyperplastic polyps, adenomas, and adenocarcinomas. *Hum. Pathol.* **26,** 534–540.
40. Manne, U., Myers, R. B., Moron, C., Poczantek, R. B., Dillard, S., Weiss, H., Brown, D., Srivastava, S., and Grizzle, W. E. (1996) Prognostic significance of Bcl-2 expression and p53 nuclear accumulation in colorectal adenocarcinoma. *Int. J. Cancer* **74,** 346–358.
41. Fontanini, G., Vignati, S., Bigini, D., Mussi, A., Lucchi, M., Angeletti, C. A., Basolo, F., and Bevilacqua, G. (1995) bcl-2 protein: a prognostic factor inversely correlated to p53 in non-small-cell lung cancer. *Br. J. Cancer* **71,** 1003–1007.

42. Roncalli, V., Grimelius, L., Graziani, D., Wilander, E., Johansson, H., Bergholm, U., and Coggi, G. (1995) Prognostic value of bcl-2 immunoreactivity in medullary thyroid carcinoma. *Hum. Pathol.* **26,** 945–950.
43. Pezella, F., Morrison, H., Jones, M., Gatter, K. C., Lane, D., Harris, A. L., and Mason, D. Y. (1993) Immunohistochemical detection of p53 and bcl-2 proteins in non-Hodgkin's lymphoma. *Histopathology* **22,** 39–44.
44. Wynford-Thomas, D. (1992) p53 in tumor pathology—can we trust immunocytochemistry? *J. Pathol.* **166,** 329,330.
45. Allred, C. D., Clark, M. G., Elledge, R., Fuqua, S. A., Brown, R. W., Chamness, G. C., Osborne, C. K., and McGuire, W. L. (1993) Association of p53 protein expression with tumor cell proliferation rate and clinical outcome in node-negative breast cancer. *J. Natl. Cancer Inst.* **85,** 200–206.
46. Noguchi, M., Maezawa, N., Nakawishi, Y., Matsuno, Y., Shimosata, Y., and Hirohashi, S. (1993) Application of the p53 gene mutation pattern for differential diagnosis of primary versus metastatic lung carcinomas. *Diagn. Mol. Pathol.* **2,** 29–35.
47. Orita, M., Iwahana, H., Kanazawa, H., Hayashi, K., and Sekiya, T. (1989) Detection of polymorphisms of human DNA by gel electrophoresis of single-strand conformation polymorphisms. *Proc. Natl. Acad. Sci. USA* **86,** 2776–2780.
48. Murakami, Y., Hayashi, K., Hirohashi, S., and Sekiya, T. (1991) Aberrations of the tumor suppressor-p53 and retinoblastoma genes in human hepatocellular carcinoma. *Cancer Res.* **51,** 5520–5525.
49. Moll, U. M., Riou, G., and Levine, A. J. (1992) Two distinct mechanisms alter p53 in breast cancer mutation and nuclear exclusion. *Proc. Natl. Acad. Sci. USA* **89,** 7262–7266.
50. Kennedy, S. M., Macgeogh, C., Jaffe, R., and Spurr, N. K. (1994) Overexpression of the oncoprotein p53 in primary hepatic tumors of childhood does not correlate with gene mutations. *Hum. Pathol.* **25,** 438.
51. Xu, L., Chen, Y.-T., Huvos, A. G., Zlotolow, I. M., Rettig, W., Old, L. J., and Ganin Chess, P. (1994) Overexpression of p53 in squamous cell carcinomas of head and neck without apparent gene mutations. *Diagn. Mol. Pathol.* **3,** 83.
52. Heimdal, K., Lothe, R. A., Lystad, S., Holm, R., Fossa, S. D., and Borresen, A. L. (1993) No germline Tp53 mutations detected in familial and bilateral testicular cancer. *Genes Chrom. Cancer (Philadelphia)* **6,** 92–97.
53. Peng, H. Q., Hogg, D., Malkin, D., Bailey, D., Gallie, B. L. Bulbul, M., Jewell, M., Buchanan, J., and Goss, P. E. (1993) Mutations of p53 gene do not occur in testis cancer. *Cancer Res.* **53,** 3574–3578.
54. Battifora, H. and Kopinski, M. (1986) The influence of protease digestion and duration of fixation on the immunostaining of keratins: a comparison of formalin and ethanol fixation. *J. Histochem. Cytochem.* **34,** 1095–1100.
55. Ordonez, M. A., Manning, J. T., and Brooks, T. E. (1988) Effect of trypsinization on the immunostaining of formalin-fixed paraffin embedded tissues. *Am. J. Surg. Pathol.* **12,** 121–129.
56. Andrade, R. E., Hagen, K. A., and Swanson, P. E. (1988) The use of proteolysis with ficin for immunostaining of paraffin sections. *Am. J. Clin. Pathol.* **90,** 33–39.

57. Shi, S.-R., Key, M. E., and Kalra, K. L. (1991) Antigen retrieval in formalin-fixed, paraffin-embedded tissues: an enhancement method for immunohistochemical staining based upon microwave heating of tissue sections. *J. Histochem. Cytochem.* **39,** 741–748.
58. Cattoretti, G., Becker, H. M. G., and Key, G. (1992) Monoclonal antibodies against recombinant parts of the Ki-67 antigen (MIB 1 and MIB 3) detect proliferating cells in microwave-processed formalin-fixed paraffin sections. *J. Pathol.* **168,** 357–363.
59. Gown, A. M., de Wever, N., and Battifora, H. (1993) Microwave-based antigenic unmasking: a revolutionary new technique for routine immunohistochemistry. *Appl. Immunohistochem.* **1,** 256–266.
60. Brown, R. W. and Chirala, R. (1995) Utility of microwave-citrate antigen retrieval in diagnostic immunohistochemistry. *Modern Pathol.* **8,** 515–520.
61. Wood, G. S. and Warnke, R. (1981) Suppression of endogenous avidin-binding activity in tissues and its relevance to biotin-avidin detection systems. *J. Histochem. Cytochem.* **29,** 1196–1204.
62. Hsu, S. M., Raine, L., and Fanger, H. Use of avidin-biotin peroxidase complex (ABC) in immunoperoxidase technique: a comparison between ABC and unlabeled antibody (PAP) procedures. *J. Histochem. Cytochem.* **29,** 577–580.
63. Mazur, M. T., Hendrickson, M. R., and Kempson, R. L. (1983) Optically clear nuclei: an alteration of endometrial epithelium in the presence of trophoblast. *Am. J. Surg. Pathol.* **7,** 415–423.

10

Factors Affecting Immunohistochemical Evaluation of Biomarker Expression in Neoplasia

William E. Grizzle, Russell B. Myers, Upender Manne, Cecil R. Stockard, Lualhati E. Harkins, and Sudhir Srivastava

1. Introduction

Archival paraffin blocks frequently are studied to evaluate the expression of biomarkers in tumors and disease states. The identification of biomarkers is becoming increasingly important to identify and characterize early preinvasive neoplastic lesions *(1–3)*, and to correlate the expression of specific biomarkers with diagnosis and prognosis of invasive tumors *(4–6)*. Also, the expression of biomarkers in archival tissues is sometimes used to identify patients who may be eligible for novel therapies, including gene therapy and immunotherapy, as well as to monitor the effectiveness of conventional and novel therapies using changes in the expression of biomarkers as surrogate endpoints *(7–12)*. In addition, the response of some tumors to specific therapies has been correlated with specific biomarker expression *(13)*.

Many factors may influence the detection of biomarkers using immunohistochemistry. These include the tissue, the tissue collection procedure, the method of fixation-tissue processing, the antigen, the antibody, and the secondary detection system. Also, some of these factors may interact with one another; for example, for a specific antigen one antibody may work in formalin-fixed paraffin tissues, whereas another antibody that binds to this same antigen but a different epitope will work only in frozen sections of the same tissue.

This chapter focuses primarily on performing immunohistochemistry and on factors that affect this method. Chapter 9 of this volume discusses biomarker expression in prostate and colorectal neoplasia as well as the methods used to evaluate biomarker expression in neoplasia using immunohistochemistry.

From: *Methods in Molecular Medicine, Vol. 14: Tumor Marker Protocols*
Edited by: M. Hanausek and Z. Walaszek © Humana Press Inc., Totowa, NJ

1.1. Antibodies Used in Immunohistochemistry

1.1.1. Primary Antibodies

The primary antibody used in immunohistochemistry binds to the antigen of interest. A secondary detection system deposits a colored precipitate at the sites where the primary antibody binds in order to detect and localize the antigen in the tissue. The three main factors that affect the performance of a primary antibody in immunohistochemical techniques are specificity, affinity, and concentration. To obtain the best results, strong positive staining with low background, a high signal to low noise ratio, optimize the concentration of the primary antibody as well as the detection system. In actual practice, the detection system is kept constant and the concentration of the primary antibody is varied. Either ready-to-use antibodies and detection systems that have been optimized by the vendor, or concentrated primary or secondary antibodies that are optimized by the user are utilized. Even for optimized antibodies, all users should validate concentrations above and below the vendor-recommended concentrations (*see* **Notes 1** and **2**).

The main problems associated with the primary antibody are crossreactivity, nonspecific binding, and background staining. These problems are exacerbated by weak or low affinity binding by the primary antibody or via contaminating antibodies and proteins. The sensitivity and the specificity of an antibody are affected by the type of antibody, i.e., whether the antibody is monoclonal or polyclonal, and by the number and types of epitopes that the antibody recognizes.

A "polyclonal antibody" that detects a single antigen is actually a heterogeneous population of antibodies directed against several antigenic determinants (sites) on this single antigen. The specific site (i.e., 5–15 amino acids not necessarily in sequence but in very close proximity) on a peptide antigen with which one antibody from a polyclonal population of antibodies reacts is the "epitope" of that antibody. Polyclonal antibodies are produced by collecting serum from such animals as rabbits, horses, sheep, and goats previously immunized with antigens or in some cases by impure preparations of antigens (e.g., extract of tumor cells).

Because a "polyclonal antibody" is composed of a mixture of multiple antibodies that react with different epitopes on the antigen of interest, there is both increased sensitivity as well as increased probability that at least one component of the mixture of antibodies will bind to a similar epitope on a different antigen and result in crossreactivity, i.e., binding with two (or more) different antigens. Decreased specificity of polyclonal antibodies also results when the serum of the source animal contains antibodies to other antigens that either were present prior to immunization or were produced by immunization with "impure" antigens. Multiple contaminating antibodies or antibodies reacting with a ubiquitous interstitial protein (albumin or immunoglobulin) would cause crossreactivity and an apparent increase in background staining (*see* **Note 3**).

A monoclonal antibody (MAb) reacts with a single epitope (antigenic site) on a molecule. An "immortal" cancer cell line is fused with a lymphocyte that is secreting multiple copies of a single antibody to produce an immortal hybridoma cell that maintains the ability to synthesize and secrete the single antibody. Antibody is obtained from the spent culture medium after growing the cells in culture or from ascitic fluid after growing cells in the peritoneal cavities of mice or rats. Because MAbs bind to a single epitope, they provide increased specificity but reduced sensitivity when compared to polyclonal antibodies (*see* **Notes 4** and **5**).

1.1.2. Secondary Detection System

The unbound primary antibody is washed away and the secondary detection system is used to identify the binding sites of the primary antibody, i.e., the localization in the tissue of the antigen of interest. The "direct" method of immunostaining uses a probe attached directly to the primary antibody. Indirect methods, in which the primary antibody is detected by a labeled secondary antibody or by a bridging antibody, solved problems with reagent preparation and stability as well as increased sensitivity and reduced background by increasing the number of labeling sites per primary antibody molecule.

The peroxidase–antiperoxidase (PAP) immunodetection system utilizes a linking antibody (secondary antibody) that binds to both the primary antibody and an antibody against peroxidase (antiperoxidase antibody) because both the primary antibody and the antibody to peroxidase are produced by the same animal species (*see* **Note 6**).

The alkaline phosphatase–antialkaline phosphatase method (APAAP) uses the same techniques as the PAP method, except that alkaline phosphatase and an antibody to it rather than peroxidase are used as the labeling-enzyme–antibody complex. The APAAP method may be combined with peroxidase in double-labeling techniques.

The PAP, APAAP, and avidin–biotin complex (ABC) methods have been replaced largely by the biotin–streptavidin (BSA) method, which relies on the high binding affinity of streptavidin for four molecules of biotin. The labeling reagent (an enzyme-labeled streptavidin) binds to the biotin residues on the link antibody (a biotinylated immunoglobulin, such as antimouse Ig). In the enhanced biotin–streptavidin (EBSA) system, the link antibody is modified to allow the attachment of more biotin molecules without adversely affecting the binding affinity of the link antibody, and the streptavidin has been optimized for maximum labeling of the streptavidin molecule. These improvements increase the amount of signal generated per antigen–antibody binding event (*see* **Fig. 1** and **Note 7**).

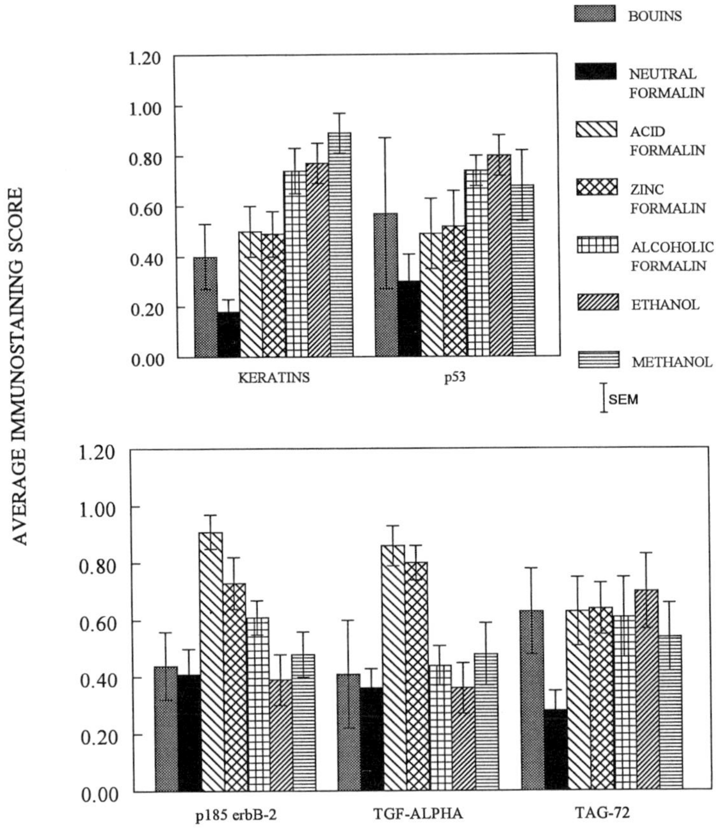

Fig. 1. Immunostaining scores. This graph demonstrates the effects of various methods of fixation on immunohistochemistry for five different antigens performed on paraffin-processed–embedded tissues. Each bar represents an average based on 7–11 tumor specimens. Error bars represent standard errors of the mean. This figure demonstrates that for each antigen, fixation in neutral buffered formalin is the least optimal fixative.

Once a detection system is chosen, a substrate, such as 3,3'-diaminobenzidine tetrahydrochloride (DAB) or 3-amino-9-ethylcarbazole (AEC), is selected for peroxidase-labeled systems. AEC is alcohol-soluble; therefore aqueous mounting medium must be used (*see* **Note 8**). Substrates including New Fuchsin and Fast Red can be used with alkaline phosphatase and x-gal with β-galactosidase. Users should understand the uses and limitations of these substrates.

The most common counterstain for immunostaining is hematoxylin; however, standard hematoxylin staining procedures produce nuclei that are stained too darkly for some evaluations, especially with nuclear antigens, such as PCNA, p53, Ki67, or Rb. For weak hematoxylin staining, the hematoxylin can be diluted 1:1 and/or the staining time shortened. For nuclear antigens we prefer a methyl green counterstain ***(14)***.

The choice of dilution buffers is important because they can affect the intensity of immunohistochemical staining and the shelf-life of diluted antibodies and detection systems. Sometimes a carrier protein is useful in a buffer solution to prevent the antibody from sticking to the container.

1.2. Fixation and Tissue Processing

Variable and inconsistent immunostaining can be caused by overfixation, underfixation, high temperature (>60°C for slides or processing), delays in tissue processing, or reagents in tissue processing. Alternative reagents to replace xylene, formalin, and so on should be tested regarding their effects on immunohistochemistry prior to being placed in use. Since biomarker expression in neoplasia is usually evaluated in diagnostic pathology material, fixation and tissue processing are variables over which usually there is little control. Unfortunately, tissue processing and fixation have as much effect on the quality of the immunostains as do the antibody and the secondary detection systems.

Tissue is fixed to stop degradation by endogenous enzymes (autolysis) as well as by bacteria and fungi. Fixation immobilizes antigens and provides added rigidity to make tissue easier to cut. Fixation also can affect the results of immunostaining. Tissue should be put into fixative as soon as possible after removal from the body, and it should be removed from the fixative as soon as fixation is adequate to preserve the morphology and to immobilize antigens. The longer the tissue is in the fixative, the more the immunodetection of susceptible epitopes may be affected.

The three main categories of fixatives are additive (primarily crosslinking), nonadditive (usually coagulant or dehydrant), and combination (crosslinking plus dehydrant). All types of fixation change the molecular shape of proteins and modify other cellular structures. Depending on the tissue structure and the type of fixative, this process may be reversible or irreversible.

Formalin is the protypical additive fixative, which crosslinks proteins by inserting a methylene bridge between reactive amine groups that are in close proximity. Fixation by neutral buffered formalin (NBF) changes the structure of many protein antigens and hence greatly affects immunorecognition. It is the fixative of choice for diagnostic pathology based on routine staining, and hence, most archival tissue collections contain paraffin blocks previously fixed in NBF.

In nonadditive fixation, the fixative (i.e., ethanol, methanol, or acetone) does not combine with the tissue, although some crosslinking may occur as reactive groups come in closer contact with one another. Excessive or prolonged exposure to a nonadditive fixative results in excessive shrinkage and dry and hardened tissue blocks, owing to the removal of more and more free water and hence tighter and tighter folding of proteins. Nonadditive fixatives seldom are used in routine diagnostic pathology because they do not preserve nuclear chromatin patterns in the formats that are important in diagnostic pathology. There are several fixatives that are hybrids of crosslinking and coagulative fixatives designed to combine the best aspects of both types. For example, alcoholic formalin is a fixative combining dehydrant and additive crosslinking activities.

The effect of fixation on antigen recognition varies primarily with the antigen, the fixative, and the tissue processing following fixation. The same fixative that decreases immunorecognition of an antigen if the tissue is processed to paraffin blocks may increase immunorecognition of this same antigen if frozen sections are fixed under identical conditions. From the standpoint of immunohistochemistry, there is no "perfect" fixative; however, standard neutral buffered formalin is usually the worst fixative for paraffin-processed tissues. **Figure 1** shows that for paraffin sections, certain antigens (i.e., TGFα and $p^{185erbB\text{-}2}$) are better recognized by immunohistochemistry if the initial fixative is acid formalin or unbuffered zinc formalin (both weaker crosslinkers than NBF), whereas other antigens, such as p53 and keratin, have better immunorecognition if the initial fixation is in a fixative with dehydrant properties (ethanol, methanol, or alcoholic formalin) ***(15)*** (*see* **Notes 9** and **10**).

1.3. Overfixation of Tissue

1.3.1. Vimentin Control to Monitor Overfixation

For most archival tumors, the quality and extent of fixation are unknown but can be evaluated by using vimentin as an internal control. Vimentin is a ubiquitous molecule that displays a partial sensitivity to formalin fixation. The longer a tissue is exposed to formalin, the more vimentin epitopes are altered or destroyed. By staining with a monoclonal vimentin antibody that recognizes a sensitive epitope, usually one that is expressed in endothelial cells, the quality of fixation can be determined.

1.3.2. Recovery from Overfixation

Enzyme predigestion recovers masked antigens by enzymatically cleaving the crosslinking produced by fixation. Its effectiveness depends on the fixative used, the time of fixation, and the antigen to be identified. Although enzyme digestion has been proven useful in formalin-fixed tissues for some antibodies, it has several limitations, including difficulty of performance, only partial recovery of very overfixed tissue, and potential tissue damage. A recommended antigen retrieval method is discussed in **Subheading 3.3.**

New procedures for recovery from formalin fixation are based on observations that suggest boiling formalin-fixed tissue can "break" the methylene cross-bridges and hence reverse crosslinking of proteins caused by additive fixatives. Newer modifications are based on microwave boiling of tissue sections in the presence of various solutions ***(16,17)***.

A critical factor in all antigen-retrieval techniques is that the tissue sections be exposed to a boiling solution for a minimum of 10 min. Another critical aspect of this procedure is that the sections do not dry during the antigen-retrieval procedure. These two factors are controlled when using a microwave by boiling the sections for two 5-min intervals, with the time measured from when boiling begins. It may be necessary to add additional solutions if too much boils away during the first interval. Boiling frequently causes sections to detach from slides, especially if the tissue contains extensive fat, such as breast sections. Recent methods to avoid this utilize steam for antigen retrieval. Such antigen-retrieval techniques can be accomplished in a pressure cooker autoclave or in a microwave tenderizer.

Some of the solutions reported to be most effective for antigen retrieval include the following: urea, citrate, lead isothiocyanate, and zinc chloride. The intensity of staining of specific antigens as well as the background activity varies with the solution used in antigen retrieval as well as the specific tissue to which antigen retrieval is applied.

Immunostaining following antigen-retrieval techniques cannot be interpreted unless a control using an equivalent section from the same tissue is performed concomitantly in the same antigen-retrieval solution. Antigen retrieval sometimes causes nonspecific nuclear staining that is tissue-specific (i.e., is reproducible for a specific tissue specimen). Similarly, colorectal tumors may have tissue-reproducible but nonspecific cytoplasmic staining induced by antigen-retrieval techniques. Again, the cytoplasmic staining is specific for individual tumors that are stained following antigen retrieval.

The use of antigen retrieval on alcohol-fixed tissues produces no enhancement of reactivity, suggesting that the antigen retrieval works by reversing the crosslinking fixation.

Antigen retrieval techniques have been shown to permit positive staining in tissues that were overfixed for over 2 yr in formalin and that stained negatively even when enzyme predigestion techniques were performed. The antigen-retrieval procedure increases the staining intensity with many important biomarkers, including cytokeratin, vimentin, NSE, κ, λ, LCA, bcl-2, p53, and Ki67 (MIB-1). In many cases, the background staining is reduced following antigen retrieval. The immunodetection of some antigens by specific antibodies is not improved by antigen retrieval, i.e., $p185^{erbB\text{-}2}$ ***(4,5)***.

1.4. Quality Assurance

Quality assurance is monitoring and evaluating all elements, including reagents, tissue, and technique, as well as the final product in order to ensure consistent and accurate immunohistochemical results. It is important to demonstrate reproducibility of day-to-day immunohistochemistry techniques as well as continued consistency when new lots of reagents are introduced into standard techniques. Sections from a tissue block stained at each run with the same primary antibodies aid in this effort. In addition, 10% of cases should be selected randomly, and both staining and evaluation should be repeated on these cases. As part of a quality assurance program the quality of reagents is monitored, including attention to expiration.

1.5. Controls

The use of adequate positive and negative control slides is one of the most important and often misunderstood aspects of immunohistochemistry. When used correctly, control slides help reduce the possibilities of false negative and false positive interpretations.

Negative control slides are used to confirm that a positive test reaction is the result of specific antigen–antibody binding, rather than nonspecific staining. The negative control slide is run using the same procedure and the same tissue as the actual test, excluding the primary antibody. This negative control is sometimes referred to as the "delete." There is, however, some disagreement regarding what should be substituted for the primary antibody.

Buffer alone is not optimal for negative control slides. Also, for mouse monoclonals, mouse serum is frequently not a good negative control because it may contain immunoglobulins that bind nonspecifically to human tissues and cause high background. A better control is a similar dilution of an isotypic immunoglobulin from the same species but directed against clearly different antigens, preferably localized at different cellular sites (i.e., nuclear vs membrane). When running a panel of antibodies, different primary antibodies can be used as controls for each other. Although this should be adequate if the antibodies are of the same species and at similar dilutions, this approach is not yet widely utilized.

Positive control slides confirm that the primary antibody and the detection system reagents are working. The positive control slide is run using the same procedure as the test, and on a tissue that is known to contain the antigen. Because many antigens are adversely affected by fixation and processing, it is important to use a positive control slide that has been fixed in the same manner as the tissue that is being tested. Thus, manufacturers' positive control slides may not be adequate. Normal tissue often contains more antigen than tumor tissue. If normal tissues are used as a positive control, the positive control slides will often stain much more intensely than the tumor tissues. To keep the possibility of false negatives and false positives as low as possible, tumor tissue should be used as positive control tissue for tumors, and normal tissues should be used for normal tissue.

Many tissues, including tumors, contain "built-in" positive and negative controls. In evaluating immunohistochemistry, it should become a habit to check all the tissues on each slide and evaluate tissues that should stain positive or negative. Multitissue slides can be used when evaluating new antibodies and screening for crossreactivity and affinity. Multitissue control slides are available from several sources and may contain many different tissues on one slide. With one slide, the performance of a new antibody can be evaluated and the fear of false positives dramatically reduced.

General multitissue control blocks can be prepared from common tumors and tissues. For biomarkers associated with neoplasia, useful tissues include colonic adenocarcinomas, spleen with lymphoma (B-type), squamous cell carcinomas (lung or oral cavity), neuroendocrine tumors (i.e., carcinoid, islet cell tumor, or pheochromocytoma), and normal tissues, including cerebellum, tail of pancreas, and adrenal medulla. Using multiple pieces from the same tissues, multiple control blocks can be prepared that are approximately identical and will be useful as positive and negative controls for both immunohistochemistry and histochemical staining (*see* **Note 11**).

2. Materials

2.1. Buffers

1. 4 L Tris-HCl buffer: 50 m*M* Tris-HCl base (24.23 g), 150 m*M* NaCl (35.06 g), 0.01% Triton X-100 (16 drops). Bring solution to 4 L with deionized H_2O and adjust pH to 7.6 with concentrated HCl.
2. 1 L Phosphate-buffered saline (PBS): 137 m*M* NaCl (8.0 g), 2.7 m*M* KCl (0.2 g), 8.1 m*M* Na_2HPO_4 (1.5 g), 1.5 m*M* KH_2PO_4 (0.2 g). Bring volume to 1 L with deionized H_2O.
3. 100 mL of PBS with 1% bovine serum albumin (BSA) fraction V (1 g), 15 m*M* NaN_3 (9.75 mg), 1 m*M* ethylenediamine tetra-acetic acid (EDTA) (29.2 mg). This buffer is referred to as PBE.

2.2. Antigen-Retrieval Solution

1. 0.01*M* Citric acid: 1.92 g anhydrous citric acid in 1 L deionized H_2O; adjust pH to 6.0 with NaOH.

2.3. Enzymes for Digestion

1. Several companies supply the nonspecific protease from *Streptomyces griseus* for this procedure:

a. Protease Type XIV (Sigma [St. Louis, MO], cat. no. P5147);
b. Pronase (Boehringer Mannheim [Mannheim, Germany], cat. no. 165 921). Dilute the enzyme to a 1 mg/mL stock solution with deionized H_2O. The unused portion of the stock solution can be aliquoted and frozen at –20°C. Thawed enzyme must be used within 1.5 h after thawing. The protease is further diluted to a final concentration of 0.15 mg/mL.

2.4. Reagents for Quenching Endogenous Peroxidase Activity and for Blocking Nonspecific Binding of the Antibodies

1. 3% H_2O_2 in deionized H_2O.
2. Casein Biogenex Universal Blocking Reagent (10X) from Biogenex (San Ramon, CA). Dilute 1:10 with deionized H_2O and add 9.75 mg NaN_3 per 100 mL (*Caution*: Casein also may suppress specific antigen-antibody reactions).

2.5. Negative Control Serum

1. Rabbit, swine, or goat serum diluted to 1% with PBE buffer (*see* **Note 12**). Filter with a 0.22-µm syringe filter.

2.6. Primary Antibody

1. Antibodies are obtained from various vendors or other sources (*see* Chapter 9 of this volume).

   Appropriate dilutions are determined for each antibody and diluted in PBE buffer.

2.7. Detection System

1. Biogenex StrAvigen Supersensitive Detection System. Dilute the secondary or link antibody (biotinylated goat antimouse or goat antirabbit) and the streptavidin–peroxidase 40:1 with PBE. This should be diluted just prior to immunostaining. Unused solutions degrade within several days.

2.8. Chromogen

1. 3,3'-Diaminobenzidinetetrahydrochloride (DAB) kit from Biogenex. Add 0.5 mL of substrate buffer to 4.5 mL of deionized H_2O. Add four drops of DAB solution and two drops of H_2O from kit.

2.9. Other Materials

1. Superfrost/Plus Slides (Fisher Scientific, Pittsburgh, PA).
2. Xylene.
3. 70, 95, and 100% ethanol.
4. PAP pen (Biogenex).
5. Plastic microwaveable container.
6. Plastic Coplin jars.
7. Racks for holding slides.

8. Humidity chambers.
9. Magnetic Immunostaining Tray (MIST) (Marketing International, Topeka, KS).
10. Glass staining dishes with glass slide holders.
11. 700–800-W microwave oven with carousel.
12. Counterstain: hematoxylin or methyl green.

3. Methods

3.1. Slide Preparation

1. Cut 5-µm sections from formalin-fixed paraffin-embedded specimen and mount on plus slides.
2. Heat the slides for 1 h at 58°C.

3.2. Preparation for Immunostaining

1. Remove the paraffin with three changes of xylene for 5 min each. Rehydrate tissue by placing slides in 100, 95, 70% ethanol (3 min each).
2. Place slides in Tris-HCl buffer for 3–30 min.
3. Dry the area around the tissue with a laboratory wipe and make a hydrophobic ring around the tissue with a PAP pen (*see* **Note 13**).
4. Add 50–200 µL of Tris-HCl buffer to cover the specimen (*see* **Note 14**). The specimens are stable for several hours at this point.

3.3. Antigen Retrieval (Optional)

1. Drain Tris-HCl buffer from the slides and place them in plastic Coplin jars filled with deionized H_2O for 1–3 min.
2. Preheat an 8 × 8-in. microwave-oven-safe dish, filled with 1-in. of H_2O, in a microwave oven for 8 min.
3. Remove H_2O from a Coplin jar and completely fill with antigen-retrieval solution.
4. Place Coplin jar in the 8 × 8-in. previously preheated dish (*see* **step 2**), which now has 1-in. of hot H_2O in bottom.
5. Heat the slides in the microwave oven on "high" and continue heating for 5 min after the solution begins to boil. Check Coplin jars to make sure that the antigen-retrieval solution has not evaporated to the point that the tissue sections dry. If needed, add more antigen-retrieval solution to the plastic Coplin jar (*see* **Note 15**). Heat the slides for an additional 5 min of boiling.
6. Remove the plastic Coplin jar from the microwave oven and allow the slides to cool for 10–15 min.
7. Rinse the slides twice with deionized H_2O and place the slides in Tris-HCl buffer.
8. Mark the slides with a PAP pen.

3.4. Quenching Endogenous Peroxidase

1. Drain the Tris-HCl buffer from the slides and add 50–200 µL of 3% H_2O_2 for 3 min to quench the endogenous peroxidase activity. Rinse in Tris-HCl buffer for 3 min.

3.5. Enzymatic Digestion (Optional)

1. Add 50–200 µL of protease (0.15 mg/mL) to the slides and incubate for 8 min at 37°C.
2. Rinse the slides with Tris-HCl buffer for 5 min.

3.6. Immunostaining Procedure

1. Drain the Tris-HCl buffer from the slides and place them in a humidity chamber. Add 50–200 µL of casein to the tissue for 10 min (*see* **Note 16**) to reduce nonspecific staining.
2. Drain the blocking agent from the slides and add the appropriate primary antibody diluted in PBE buffer. The appropriate negative control serum should be added to the "delete" slides. Place the slides in a humidity chamber for 1 h at room temperature.
3. Rinse the slides in Tris-HCl buffer for 10 min (two 5-min rinses). Drain the excess Tris-HCl buffer from the slides and place them on the slide racks.
4. Add the appropriate secondary antibody diluted in PBE buffer to the slides for 10 min.
5. Rinse the slides for 10 min (two 5-min rinses) in Tris-HCl buffer and drain the excess buffer from the slides.
6. Add the streptavidin–peroxidase diluted in PBE buffer to the slides for 5 min.
7. Rinse the slides as in **step 5**.
8. Add the DAB mixture to the slides for 7 min or until the desired intensity of staining is achieved. This may take 2–15 min. Rinse the slides with deionized water.

3.7. Counterstain Protocol (see Note 17)

3.7.1. Hematoxylin Counterstain

1. Place slides in Mayers hematoxylin for 1 min. One-minute staining in hematoxylin will provide a light nuclear counterstain that will not interfere with the detection of immunostaining.
2. Rinse slides with tap water for 1 min.
3. Dehydrate slides through graded alcohols (3 min each in 70, 95, and 100% ethanol) and three changes of xylene (3 min each).
4. Mount cover slips with permount.

3.7.2. Methyl Green Counterstain

1. Rinse slides in deionized H_2O (three 1-min rinses and one 5-min rinse).
2. Place slides in methyl green for 2–3 min.
3. Rinse slides in deionized H_2O for 2 min.
4. Dehydrate the slides as follows: 70% ethanol (3 min), 95% ethanol (3 min), 1-butanol (3 min), 95% ethanol (3 min), and 100% ethanol (3 min). The slides are then placed in xylene (three changes of 3 min each).
5. Mount cover slips with permount.

3.8. Troubleshooting

3.8.1. Weak or No Staining in Control and Specimen

1. Procedure: Failure to follow the manufacturers' immunostaining procedures, including interchanging the link and labeling steps or omitting the primary antibody, as well as use of shorter incubation times or lower temperatures than recommended may result in little or no staining. Ensure that the secondary detection system is designed to detect the exact primary antibody, i.e., correct species, immunoglobulin type, and substrate. Use of old or poorly prepared solutions of substrates, i.e., New Fuchsin or DAB, is one of the most common errors in immunostaining.
2. Buffers: Immunostaining is very dependent on buffers. Care should be taken to ensure that a buffer containing azide is not used when preparing secondary labels or chromogens; however, azide is washed away in early steps, so it is frequently used as a preservative to prevent bacterial/fungal attack on primary antibodies. The buffer's pH suggested by the immunostaining reagent's manufacturer should be adhered to.
3. Washing: Excess buffer on the slide after washing acts as a diluent diminishing the reagent's potency.
4. Deparaffinization: Residual paraffin or xylene acts as a barrier to aqueous reagents.
5. Titer: The antibody and/or the detection system components are too dilute to detect the antigens.
6. Counterstain, dehydration, mounting media: Aqueous chromogens (i.e., AEC) may be removed by alcohols and xylene. Counterstain should be as light as practicable to prevent masking of light immunostaining and should be chosen to contrast with the color of the chromogen.
7. Reagents: Reagents should be used prior to their expiration date.

3.8.2. No Staining or Weak Staining in Specimen Only

1. Fixation: Overfixation with aldehyde crosslink may mask or destroy the antigenic sites. Underfixation or delayed fixation may cause antigen loss owing to autolysis or loss of antigen into solutions.
2. Pretreatment: Recommended pretreatments will be found on antibody specification sheets supplied by manufacturers. Be sure the epitope can withstand the pretreatment.
3. Temperature: Tissues should not be exposed to temperatures above 60°C, which causes some epitopes to be destroyed.
4. Antigen: The antigen may not be present at levels that can be detected by immunohistochemistry.

3.8.3. Background Staining

1. Procedure: Tissue drying or overincubation during any step of the procedure causes increased background staining.

2. Washing: Slides should be washed thoroughly between each incubation step. Any residual reagent that is not bound specifically will result in increased background.
3. Titer: Too concentrated reagents will bind nonspecifically to the tissue, increasing background staining.
4. Nonspecific binding: A protein block reduces nonspecific staining in most procedures.
5. Endogenous staining activity: Endogenous peroxidase, alkaline phosphatase, or biotin may result in noninformative and background staining. Red blood cells contain endogenous peroxidase activity as do white blood cells. Therefore, there is increased endogenous peroxidase activity in "bloody" tissue specimens such as liver, kidney, and muscle or in areas of inflammation. Endogenous peroxidase is partially blocked with hydrogen peroxide as an aqueous solution or in methanol.

   Endogenous alkaline phosphatase can cause background with the APAPP detection system. An acid alcohol block or addition of levamisole (all tissue except intestine) to the substrate will alleviate this problem. With the avidin–biotin methodology endogenous biotin can cause nonspecific staining in certain tissues (mainly liver and kidney). This is limited primarily to frozen tissue since biotin is destroyed by usual tissue processing protocols; however, it is also seen after paraffin-processed tissue has been treated with antigen-retrieval techniques.

   Endogenous brown (black) pigments, such as melanin, hemosiderin, or carbon, also can cause a problem with interpretation of immunohistochemistry using DAB or chromogens with blue-black coloration. These problems can be reduced by shifting to AEC as a chromogen or to other systems that use chromogens that produce colors distinct from the endogenous pigment.
6. Tissue/slide preparation: Underfixation may cause nonspecific binding. Thick sections of tissues tend to overstain, as do irregularly cut areas of a tissue section. Similarly, tissue folds and areas of a tissue section not uniformly attached to the slide will overstain. Excess adhesive causes background staining owing to nonspecific binding by the adhesive as well as background staining by the counterstain. Egg white adhesive contains avidin and will bind with biotinylated immunoglobulin, causing background.
7. Moisture loss: "Ring around the tissue" may be caused by the outer edges of the tissue drying out during background staining. Similarly, drying of part of the specimen or the whole tissue section during any step will greatly increase background staining. Ample reagent on the slide at each step will reduce evaporation. Utilization of a humidity chamber also reduces evaporation. A PAP pen, which produces a fluid barrier, can be used to prevent solutions from running off tissue sections and hence drying.
8. Chromogen preparation: Speckled or streaked patterns secondary to insufficiently mixed or dissolved substrate adding chromogen (DAB or Fast red tablets) to the substrate solution. Centrifugation of DAB solutions can be used to decrease precipitation/insoluble components. In automated systems, increasing the wash time or number of washes should reduce such background deposits.

Table 1
Background Source Investigation: Guide to Procedure

Protocol	Slide 1	Slide 2	Slide 3	Slide 4
Antibody	Antibody	PBS	PBS	PBS
Link	Link	Link	PBS	PBS
Label	Label	Label	Label	PBS
Chromogen	Chromogen	Chromogen	Chromogen	Chromogen

Table 2
Background Source Investigation: Evaluation of Results

Protocol	Slide 1	Slide 2	Slide 3	Slide 4
Antibody	Background	Negative	Negative	Negative
Link	Background	Background	Negative	Negative
Label	Background	Background	Background	Negative
Chromogen	Background	Background	Background	Background

3.8.4. Staining Limited to the Negative Control

1. Titer: The negative control is too concentrated. The immunoglobulin concentration should match that of the primary antibody. Control serum containing antibodies that react with the specimen was used.
2. Preservative: A preservative (sodium azide) was not added to the control immunoglobulin to prevent bacterial growth.
3. Freeze–thawing: Freezing and thawing of the antibodies or detection systems causes protein aggregation and deposition on the tissue section.
4. Crossreaction: The primary antibody may be crossreactive or may contain crossreactive contaminant antibodies to ubiquitous proteins (i.e., albumin). This can be evaluated by Western blotting.

3.8.5. Investigating Background Source

1. The following approach helps to determine which immunostaining reagent is causing background staining: Vary the procedure on four specimen slides by substituting PBS in the procedure using the guide below (*see* **Table 1**). (Example: Slide 4 would only be using the chromogen in the staining procedure.)
2. Evaluate the results of the above test as follows (*see* **Table 2**): If slides 1 and 2 have background staining and slides 3 and 4 do not, the link (secondary antibody) may be causing the nonspecific staining. If all slides (1–4) have equivalent background staining, the chromogen is probably causing the problem.

Table 3
A Hypothetical Checkerboard Titration Pattern

Secondary detection system	Primary antibody		
	Dilution 1	Dilution 2	Dilution 3
Dilution 1	0	0	+1
Dilution 2	+1	+2	+3
Dilution 3	+2	+3	+4

4. Notes

1. Prediluted ready-to-use reagents are typically more standardized and usually undergo testing by the vendor to ensure lot-to-lot consistency and stability so there is less chance of error owing to inaccurate titers. In contrast, concentrated antibodies require more technician attention because the reagent testing is performed in the laboratory, introducing a greater chance of error. It is important that correct dilutions be determined since a too concentrated dilution of antibody or detection system fields false positive results; too dilute concentrations give false negative results. The specificity and sensitivity of the titered reagent are tested on a known positive tissue or multitissue slides. Since each new lot of a reagent may differ from the last, each should be retitered as well as checked for sensitivity and specificity. It is of great advantage to use a composite control block that can monitor the quality of staining between various lots of immunochemicals.
2. To optimize immunostaining results with concentrated reagents, perform a checkerboard titration on reagents (primary antibody, secondary detection system) to find the concentration that delivers consistent results. **Table 3** demonstrates a hypothetical checkerboard pattern with staining intensities at the various conditions graded from 0 to +4. For evaluating and comparing the expression of biomarkers in neoplastic processes, it is best to select conditions in which the average immunostaining score is +2 to +3.
3. Contaminating antibodies can be detected using Western blotting, and removed by affinity purification.
4. Preparations of monoclonal antibodies (MAbs) may be contaminated by immunoglobulins or proteins from ascitic fluid or from calf serum used in cell culture. It is important to know how antibodies have been purified, what the immunoglobulin concentration is, and with which proteins the antibody reacts in Western blotting. Adequate immunostaining with respect to signal-to-noise (background) ratio is obtained at final dilutions of the serum that are more dilute than 1:75. Purified polyclonal immunoglobulin preparations can be used at final concentrations of up to 10 μg/mL, and affinity purified polyclonals and monoclonals can be used up to 20 μg/mL. More concentrated solutions of antibodies may produce high background as well as nonspecific staining.
5. Dilute solutions (<1 mg/mL) of antibodies should not be frozen or subjected to freeze–thaw cycles to prevent molecular damage or failure of the antibody to

return completely into solution. Storage for up to 2 yr can be used with sodium azide (15 m*M*) to prevent bacterial/fungal growth. Storage in aliquots minimizes contamination when the vials are opened. If longer storage is necessary, the solution can bc frozen before dilution or carrier protein can be added before freezing and storage at –70°C or less. However, storage in too concentrated solutions may induce oligomer formation.

6. If PAP is to identify a mouse IgG primary antibody, the linking antibody is directed against mouse IgG (one binding site) as well as to the PAP complex mouse antibody directed against peroxidase (second binding site).
7. Unlike avidin, streptavidin contains no carbohydrate molecules that can bind nonspecifically to lectin-like substances found in normal tissues, such as kidney, liver, brain, and mast cells. In addition, neutral streptavidin conjugates do not bind nonspecifically, as do positively charged avidin conjugates. Also, the enzyme is directly conjugated to streptavidin, resulting in a highly stable reagent, unlike the avidin–biotin complex, which should be prepared immediately prior to use.
8. AEC staining is semipermanent and is susceptible to oxidation; however, clear nail polish can seal the edges of the cover slip, or newer aqueous mounting media that do not require the use of nail polish can be used. Another disadvantage of AEC is that photographs are frequently bluffed and the slides deteriorate and fade on long-term (>6 mo) storage. DAB is permanent and resistant to alcohol and xylene. Although DAB is a potential carcinogen, new delivery systems for staining aid in minimizing exposure. There is great variation in the quality of DAB depending on the delivery and stabilization systems. Also, the inclusion of peroxide in the newly prepared DAB solutions is critical.
9. Use of unbuffered zinc formalin or alcoholic formalin as the primary fixative for one or two small pieces of each tumor can produce optimal immunohistochemical staining. Following overnight fixation of small pieces of tumor in zinc formalin or alcoholic formalin, the tissue can be processed routinely and still produce results in most immunohistochemical assays that are much better than routinely neutral-buffered-formalin fixed and processed tissues from tumors.
10. For prospective studies in which the antigens of interest are known, fixation using multiple fixatives matched to the antigen of interest is recommended.
11. To prepare multiple tissue control blocks, various tissucs can be cut into multiple small pieces, fixed, and processed but not embedded. When enough tissues are collected, embed one piece of each tissue type in the same block to prepare multiple identical blocks.
12. The negative control is commonly referred to as the "delete" because of the deletion of the primary antibody. A dilution of the serum from the species in which the secondary antibody was developed can be used for this purpose. Under these conditions, nonspecific staining likely reflects the binding of the secondary or link antibody to the specimen. In addition, an isotypic immunoglobulin from the same species as the primary antibody, which does not recognize an antigen in specimen, can be used. This may provide insight into the potential "stickiness" of a tissue to the immunoglobulin isotype of the primary antibody.

13. Boiling the slides in the antigen-retrieval solution frequently reduces the effectiveness of the PAP pen lines. For this reason, slides should be marked with the PAP pen following antigen retrieval.
14. The volume of reagents required to cover the tissue specimen is dependent on the size of the specimen. By surrounding the specimen closely with a PAP pen line, the amount of reagents required can be reduced. It should be pointed out, however, that if drying of the reagents occurs during the immunostaining procedure, high nonspecific staining can occur.
15. The level of the antigen retrieval solution must be checked during this procedure. If the tissue sections dry owing to evaporation, high nonspecific staining can result.
16. Several solutions can be used as blocking reagents. A 1–3% solution of serum from the species from which the secondary antibody was developed is appropriate. The milk protein casein also can be used.
17. If the antigen of interest is localized to the cytoplasm or cell membrane, the nuclear stain hematoxylin can be used as a counterstain. If the slides are left in hematoxylin for an extended period of time (2–5 min), cytoplasmic staining may occur. This may interfere with the detection of light immunostaining. If the antigen is localized to the nucleus, a lighter nuclear counterstain, such as methyl green ***(14)***, is recommended. Sections that are subjected to antigen retrieval may require more than 3–5 min of counterstaining with methyl green. Antigen retrieval significantly reduces the ability to stain with methyl green.

Acknowledgments

Supported, in part, by the Early Detection Research Network Contract NO1-CN-15340-02, funded by the National Cancer Institute. Thanks to Libby Chambers for typing this manuscript.

References

1. Myers, R. B., Srivastava, S., Oelschlager, D. K., and Grizzle, W. E. (1994) Expression of p160^{erbB-3} and p185^{erbB-2} in prostatic intraepithelial neoplasia and prostatic adenocarcinoma. *J. Natl. Cancer Inst.* **86,** 1140–1145.
2. Myers, R. B., Schlom, J., Srivastava, S., and Grizzle, W. E. (1995) Expression of tumor associated glycoprotein-72 in prostatic intraepithelial neoplasia and prostatic adenocarcinoma. *Modern Pathol.* **8,** 260–265.
3. Myers, R. B., Srivastava, S., and Grizzle, W. E. (1995) Lewis Y antigen as detected by the monoclonal antibody BR96 is expressed strongly in prostatic adenocarcinoma. *J. Urol.* **153,** 1572–1574.
4. Grizzle, W. E., Myers, R. B., and Oelschlager, D. K. (1995) Prognostic biomarkers in breast cancer: factors affecting immunohistochemical evaluation. *Breast* **1,** 243–250.
5. Grizzle, W. E., Myers, R. B., Arnold, M. M., and Srivastava, S. (1994) Evaluation of biomarkers in breast and prostate cancer. *J. Cell. Biochem.* **19(Suppl.),** 259–266.
6. Myers, R. B., Oelschlager, D., Srivastava, S., and Grizzle, W. E. (1994) Accumulation of the p53 protein occurs more frequently in metastatic than in localized prostatic adenocarcinomas. *Prostate* **25,** 243–248.

7. Kelloff, G. J., Boone, C. W., Crowell, J. A., Steele, V. E., Lubet, R., and Doody, L. A. (1994) Surrogate endpoint biomarkers for phase II cancer chemoprevention trials. *J. Cell. Biochem.* **19(Suppl.),** 1–9.
8. Boone, C. W. and Kelloff, G. J. (1994) Development of surrogate endpoint biomarkers for clinical trials of cancer chemopreventive agents: relationships to fundamental properties of preinvasive (intraepithelial) neoplasia. *J. Cell. Biochem.* **19(Suppl.),** 10–22.
9. Myers, R. B., Kudlow, L. E., and Grizzle, W. E. (1993) Expression of transforming growth factor alpha, epidermal growth factor and the epidermal growth factor receptor in benign prostatic hyperplasia and adenocarcinoma of the prostate. *Modern Pathol.* **6,** 733–737.
10. Myers, R. B., Meredith, R. F., Schlom, J., LoBuglio, A. F., Bueschen, A. L., Wheeler, R. H., Stockard, C. R., and Grizzle, W. E. (1994) Tumor associated glycoprotein-72 is highly expressed in prostatic adenocarcinomas. *J. Urol.* **152,** 243–246.
11. Deshane, L., Leochel, F., Conry, R. M., Siegal, G. P., King, C. R., and Curiel, D. T. (1994) Intracellular single-chain antibody directed against erbB2 down-regulates cell surface erbB2 and exhibits a selective anti-proliferative effect in erbB2 overexpressing cancer cell lines. *Gene Therapy* **1,** 332–337.
12. Conry, R. M., LoBuglio, A. F., Kantor, J., Schlom, J., Loochel, F., Moore, S. E., Sumerd, L. A., Barlow, D. L., Abrams, S., and Cureil, D. T. (1994) Immune response to a carcinoembryonic antigen polynucleotide vaccine. *Cancer Res.* **54,** 1164–1168.
13. Muss, H. B., Thor, A. D., Berry, D. A., Kute, T., Liu, E. T., Koemer, F., Cirrincione, C. T., Budman, D. R., Wood, W. C., Barcos, M., and Henderson, I. C. (1994) c-erbB-2 expression and response to adjuvant therapy in women with node positive early breast cancer. *N. Engl. J. Med.* **330,** 1260–1266.
14. Staples, T. C. and Grizzle, W. E. (1986) A methyl green nuclear stain for argyrophil procedures. *Lab. Med.* **17,** 532–534.
15. Arnold, M. M., Srivastava, S., Fredenburgh, L., Stockard, C. R., Myers, R. B., and Grizzle, W. E. (1996) Effect of fixation-tissue processing on immunohistochemical demonstration of specific antigens. *Biotech. Histochem.* **71,** 224–230.
16. Shi, S.-R., Key, M. E., and Kalra, K. L. (1991) Antigen retrieval in formalin-fixed, paraffin-embedded tissues: an enhancement method for immunohistochemical staining based on microwave oven heating of tissue sections. *J. Histochem. Cytochem.* **39,** 741–748.
17. Taylor, C. R., Shi, S.-R., Chaiwun, B., Young, L., Imam, S. A., and Cote, R. J. (1994) Strategies for improving the immunohistochemical staining of various intranuclear prognostic markers in formalin-paraffin sections: androgen receptor, estrogen receptor, progesterone receptor, p53 protein, proliferating cell nuclear antigen, and Ki-67 antigen revealed by antigen retrieval techniques. *Hum. Pathol.* **25,** 263–270.

11

Instrumentation, Accuracy, and Quality Control Issues in Development of Quantitative Fluorescence-Image Analysis (QFIA)

Rebecca B. Bonner, Robert E. Hurst, Jian Yu Rao, and George P. Hemstreet

1. Introduction

Many fundamental questions of biology reduce to a common technical problem: how to assess the phenotype of cells in relation to other cells or in relation to morphology. Traditionally, this problem has been approached by classical biochemical analysis, in which heterogeneous tissue is disrupted and proteins or nucleic acid are determined on the extracts. The hope is that if one particular cell type manifests the interesting biology, the overall change in the presence of other cells will be large enough to detect. Microdissection and application of techniques such as polymerase chain reaction (PCR) and reverse transcription-polymerase chain reaction (RT-PCR) have extended the biochemical approach for DNA and RNA to a few cells *(1)*, although quantitation is not precise and the techniques are technically demanding. This approach is not useful for determining phenotypic properties relating to the amount of particular proteins in cells, and ultimately, the behavior of cells is established by their phenotype. Flow cytometry has been used to quantify protein biomarkers in individual cells, but the method is limited by requiring large numbers of single cells. Further, it lacks the ability to correlate biochemical analysis with morphology *(2)*.

The technique of quantitative fluorescence-image analysis (QFIA) provides a unique solution to many fundamental problems of biology by providing an ability to quantify proteins or nucleic acids on individual cells in relation to their morphology or relation to other cells. Problems such as understanding tumor heterogeneity, cancer detection and individual risk classification, carcinogenesis, normal development, cell-signaling, and many diseases can be

From: *Methods in Molecular Medicine, Vol. 14: Tumor Marker Protocols*
Edited by: M. Hanausek and Z. Walaszek © Humana Press Inc., Totowa, NJ

usefully addressed with the ability to assess the concentrations of individual proteins within single cells ***(3)***. The technique embodies two concepts: quantitative fluorescence and image analysis. Quantitative fluorescence means that the intensity of a fluorescent signal emitted from cells labeled with fluorescent molecular probes is proportional to the concentration of target biomolecule within cells. The process of quantitatively labeling cells is often referred to as "biophysical cytochemistry" ***(4)***. Image analysis is used to isolate the signal from specific cells within a heterogeneous cell mixture or subcellular locales within cells and processes the signal in order to extract quantitative information from the microscopic images of cells quantitatively labeled with fluorochromes. The technique differs from flow cytometry in that the image of the target object is available for examination ***(4)*** and from immunocytochemistry in offering true quantitation ***(5)***.

This chapter reviews: collection of cellular samples and preservation of target biomolecules; preparation and quantitative labeling of cells; instrumentation needed for image analysis; and quality control of biomarker analysis to yield consistent analyses over time.

2. Materials

2.1. QFIA Reagents (see Note 1)

1. 10X Modified 3-[Morpholino]-2 hydroxypropanesulfonic acid (Sigma, St. Louis, MO) (MOPSO) buffer: 298.2 g KCl, 450.8 g MOPSO, 3630-mL double-distilled water. Dissolve thoroughly. Freeze in 400 mL aliquots (measured accurately).
2. Buffer (enough for 8 L of QFIA Fixit): Thaw 400 mL aliquot of 10X modified MOPSO buffer. Combine with 3600-mL deionized double-distilled water, 14.74 g dipotassium ethylenediamine tetra-acetic acid (EDTA), and 0.8 g sodium azide. Dissolve thoroughly, and adjust pH to 6.5 with KOH.
3. QFIA Fixit: Filter 1896 mL Buffer from **item 2** above (*see* **Notes 2** and **3**). Add 2104 mL filtered 95% ethanol EtOH (use a 0.45 μm filter). Stir till thoroughly mixed. May store at room temperature but 4°C is preferred. Mark each storage vessel with expiration date (expires in 1 yr).
4. Cell wash solution: Filter 1053 mL 95% EtOH. Combine and stir with 2947 mL double-distilled water and 7.37 g dipotassium EDTA; adjust pH to 5.5 with KOH. Filter into the filtered EtOH and stir.
5. Cell adherent fluid: 7.36 g Dipotassium EDTA, 4000 mL deionized distilled water, and 0.8 g sodium azide. Stir until dissolved. Adjust pH to 5.5. Filter and store at –20°C. When needed, thaw and adjust pH prior to use.
6. Buffered-filtered saline (BFS, 1 L): 9.0 g Sodium chloride, 980 mL double-distilled water, and 20 mL Buffer B (add after warming until crystals dissolve). Adjust pH to 7.0 with Buffer A, filter, and refrigerate. Discard after 1 wk.
   a. Buffer A (0.1 *M* citric acid): 5.25 g citric acid and 250 mL double-distilled water. Refrigerate.

   b. Buffer B (0.2 *M* dibasic sodium phosphate): 7.1 g anhydrous sodium phosphate (dibasic) and 250 mL double-distilled water. Refrigerate.
7. Immunofix (modified Saccomanno fixative): 20 mL Polyethylene glycol 1540 (Union Carbide), 516 mL 95% EtOH, and 464 mL BFS. Melt polyethylene glycol (PEG) at 60°C. Prepare 50% EtOH solution. Add stirring bar and begin stirring. Slowly add 20 mL of the melted PEG to the stirring ethanol solution. Let stir 1 h. Store at room temperature.
8. 10X MOPSO/NaCl: Combine and stir thoroughly 233.76 g NaCl, 45.08 g MOPSO, and 4000 mL double-distilled water. Freeze in 400-mL aliquots (measured accurately). Use for Hoechst dye only (Hoechst working solution).
9. MOPSO/EDTA: Thaw 10X MOPSO/NaCl. Combine 400 mL 10X MOPSO/NaCl with 3600 mL double-distilled water, and add 14.88 g sodium EDTA. Dissolve thoroughly, adjust pH to 6.8, and filter. Freeze in approx 500 mL aliquots. Use for Hoechst dye only.
10. Hoechst 33258 (Molecular Probes, Eugene, OR) working solution (10 μ*M* alcoholic Hoechst, 40.0 mL): 0.2 mL Hoechst stock solution (0.2 m*M*), 29.3 mL MOPSO/EDTA (pH 6.8), and 10.5 mL 95% ETOH.

2.2. Immunofluorescence Reagents

1. 10X Autobuffer (Fisher Scientific, Plano, TX) with Brij: Add 25 mL of Brij 35 to 1 L of 10X autobuffer and mix well. Filter solution into a clean container.
2. 1X Autobuffer with Brij: Combine 100 mL 10X autobuffer with Brij and 900 mL filtered, double-distilled water into a clean container that has been used only for autobuffer. Mix well.
3. Primary diluent: Combine 0.1 g sodium azide, 0.5 mL bovine serum albumin (BSA, Sigma, cat. no. B2518), 500 mL 1X autobuffer, and 1.25 mL Brij. Filter into a clean container.
4. 0.09 *M* n-Propyl gallate (NPG) mounting media:
   a. Trisma buffer: Combine 0.709 g preset Trisma crystals (pH 8.0) with 100 mL filtered, double-distilled water. Stir until dissolved and filter. Cover bottle with aluminum foil. Store up to 1 mo at room temperature. It is not necessary to adjust pH.
   b. Combine 1.91 g NPG (Sigma, cat. no. P-3130) and 90 mL (112.5g) glycerol (spectranalyzed grade). Stir overnight, then slowly add 10 mL Trisma buffer while stirring.

3. Methods

3.1. Overview of Sample Collection and Fixation

Sample collection methods for QFIA utilize methods developed for cytopathology and histopathology with usually only minor, but crucial, modifications to preserve the stoichiometry of biophysical cytochemical probes in single cells or tissues. Cell suspensions obtained from exfoliated cells or fine needle aspirates allow the laboratory to prepare a monolayer of cells at the optimal cell density on slides specially treated to enhance cell adherence and minimize background fluorescence. Long-term storage of aliquots of precious patient samples is also feasible with suspensions. Rapid fixation is the single most important step needed to obtain reproducible, accurate data, requiring a high degree of awareness on the part of clinicians, scientists, technicians, and support personnel of the need to fix samples or put them on ice within 15 min of collection.

Proper sample fixation is critical to quantitation of biomarkers *(6)*. Fixation requires optimization of two opposing processes: immobilization of target biomolecules to prevent their loss, and permeabilization, which permits ingress and egress of reagents. Many biomarkers will be lost from cells by fixation in alcohol alone, which is the typical fixative of choice for cytopathology. Excessive crosslinking destroys antigenic epitopes, increases background fluorescence, and often increases the coefficient of variation (CV) of measurements. Insufficient permeabilization blocks reagent access but excessive permeabilization facilitates antigen loss unless the proteins are crosslinked. A "universal fixative" that seems to be effective for most biomarkers has been developed in our laboratories. This fixation method consists of two basic steps: crosslinking proteins and permeabilization. Certain grades of reagent alcohol may destroy target biomolecules or produce undesirable background noise. Substitution of reagents, even what is nominally the same reagent from different vendors, or even impurities in the water can cause errors, and all substitutions should be tested prior to implementation. Finally, the antigen of interest should be tested with controls to validate the fixation methods employed. For illustrative purposes, details of urine sample collection are provided. This basic approach has been successfully applied to a wide variety of sample types, including clinical samples such as esophageal, pulmonary lavage, cervical, and fine needle aspirates as well as basic research samples such as cultured cells.

3.2. Voided Urine Samples

1. Collect voided urine prior to catheterization and cystoscopy to prevent instrumentation artifact. The first morning void should **NOT** be collected because morning voids contain a high prevalence of degenerated cells. Instruct the patient to arrive for the procedure with a full bladder.
2. The patient collects 100 mL of nonfirst void urine specimen directly in a container marked "QFIA." The cup is labeled and the specimen type recorded on the collection cup.
3. Add the contents of the vial attached to the collection container labeled "10% formaldehyde" to the specimen as described on the instructions attached to the vial.
4. Mix the specimen and formaldehyde and allow to sit at room temperature for 15 min. This results in a final concentration of 0.5% formaldehyde.
5. Add an equal volume of "QFIA Fixit" (a solution containing buffered 50% ethanol with an inhibitor of crystal formation) to the urine specimen.
6. Place the specimen in the refrigerator to await shipment.

3.3. Bladder Wash for QFIA

Bladder wash specimens should be collected prior to surgical resection (TURBT).

1. Lubricate the catheter (22 French Red Robinson), and insert (the procedure is performed by the physician or nurse) and perform bladder barbotage with sterile saline. Saline (150 mL) is instilled into the bladder. Perform barbotage by withdrawing the fluid with the syringe and forcibly expelling the fluid back into the bladder five times, rotating the catheter, prior to removal of fluid from the bladder.
2. Place 100 mL of the fluid in the provided QFIA container and label "Bladder Wash." If additional sample is required, repeat the above procedure. **DO NOT** increase the volume of saline introduced because this diminishes the turbulence and, hence, the efficiency of washing.
3. Add the contents of the vial attached to the collection container labeled "10% formaldehyde" to the specimen as described on the instructions attached to the vial.
4. Mix the specimen and formaldehyde and allow to sit at room temperature for 15 min. This results in a final concentration of 0.5% formaldehyde.
5. Add an equal volume of "QFIA Fixit" to the urine specimen.
6. Place the specimen in the refrigerator to await shipment.

3.4. Biopsy Specimens

The ideal battery consist of three samples: tumor; a site within 1 cm of resected tumor (adjacent to tumor site); and a site 5 cm or more from the resected tumor site (distant from tumor site) *(7)*.

1. Collect biopsy specimens and label them "tumor," "near," or "far."
2. Embed the tissue in OCT and snap freeze in an isopentane bath available in pathology.
3. Transfer the frozen specimen to the pre-labeled freezing container and store at –70°C until shipping. Frozen tissue specimens must be shipped on dry-ice on Monday–Wednesday, avoiding weekends and holidays.

3.5. Cultured Cells

1. Pour off nutrient solution and wash the cells once or twice with saline.
2. Remove cells from the plastic or other matrix by scraping, EDTA treatment, or trypsinization. However, with trypsinization, the investigator should be certain the target biomolecule is either not trypsin-sensitive or is not a surface protein.
3. Wash cells once or twice by centrifugation and take up in buffered, filtered saline (about 5 mL per flask of cells). If a protease is used to release the cells, the cells should be washed three times.
4. Fix cells with formaldehyde and "QFIA-Fixit" as described above (*see* **Subheadings 3.2.** and **3.3.**).

3.6. Shipping

Stability experiments should be performed on each biomarker before determination of acceptable shipping conditions can be established. Most biomarkers are stable under refrigeration temperatures but may degrade rapidly at room temperature. Other biomarkers such as p300, a tumor-associated glycoprotein detected by the M344 antibody *(8,9)*, are stable under a wide range of conditions for extended periods of time. Shipping conditions in the summer months are drastically different than those of other times of the year, and many of the transport delivery trucks and warehouses are not air conditioned. These conditions can be simulated by placing control samples in a 60°C oven and removing aliquots at various time intervals to evaluate the effect of high temperatures on fixed cells.

3.7. Sample Processing and Storage

Each biomarker has its own unique characteristics and should be tested for stability under various conditions. Samples fixed by the methods described generally are stable at 4°C for most biomarkers for two weeks. This allows adequate time to ship samples from remote collection sites for multi-institutional or international trials and reference laboratories.

1. Allow samples to fix overnight at 4°C prior to freezing. Allow samples to warm to room temperature prior to counting.
2. Pipet 20 mL of Isoton into the Accuvette II (Coulter Corp.) container used to count cells.
3. Shake vigorously and pipet 1 mL of urine into Accuvette II container with Isoton.
4. Count the number of cells in specimen using the Coulter ZM Particle Counter (*see* **Note 4**) to determine the number of vials you can freeze from that particular specimen. Four vials containing at least 90,000 cells per vial are typically stored.
5. Divide specimen evenly among 2–4 polypropylene 5-mL centrifuge tubes.
6. Centrifuge specimen at 600*g* for 10 min.
7. Decant enough of the supernatant so that 4–4.5 mL of the specimen remains in centrifuge tube.
8. Aspirate cells gently in centrifuge tube and pipet into 5-mL Accu-Nunc cryotube, making sure each tube is labeled with appropriate specimen lab number.

3.8. Slide Preparation

The goal of slide preparation for QFIA is to prepare a nonoverlapping monolayer of cells with good cytomorphology and biomarker preservation with minimal loss of cells during processing. In contrast to many applications, air-drying of slides must be avoided because it induces artifacts. At present, slide preparation represents some art, and laboratory personnel are routinely tested for consistency of measurements with control cells to ensure accuracy of their technique.

3.9. Filter Imprint Method of Slide Preparation

1. Thoroughly mix the sample and count so that not more than 45,000 cells are placed in the filtration funnel. The number of cells on a slide required for proper

coverage will vary with the cell type and must be determined empirically for samples other than urothelial cells.

2. Place a Nuclepore filter into the filtration column and clamp into place. An 8-μm filter is used for urine samples, a 5-μm filter for cultured cells. Other applications will require empirical determination of the maximum size that will not pass the cells of interest. The column is obtained from Millipore and holds 15 mL of fluid.
3. Pour the calculated amount of specimen into the filtration funnel and gently vacuum, being careful not to let filter air dry, until 2 mm of liquid remains.
4. Rinse funnel with approximately 15 mL of cell adherent fluid.
5. Pour approx 5 mL of modified Saccamanno fixative into funnel and let sit for two min.
6. Label two probe-on slides: one "+" and one "–" along with lab number, the date, and the technician's initials.
7. Lift filter off with clean forceps and place on probe-on slide with cell side facing down. Uncharged slides are used because charged slides tend to bind antibody reagents, thereby leading to unacceptable background fluorescence.
8. Gently press down on the positive slide with a *moistened* Kimwipe for 7 s.
9. Place two drops of cell adherent fluid on the negative slide.
10. Lift filter off of positive slide; spray 2–3 times with Carbofix-E (Statlabs, Inc.).
11. Place filter on negative slide and gently press with moistened Kimwipe for 7 s.
12. Lift filter off slide and spray with Carbofix-E.
13. Let slides dry for at least 15 min prior to freezing. Slides may be stored frozen at –20°C for a maximum of 2 wk.

3.10. CytoRich™ Urine Sample Preparation

The CytoRich™ was developed by Roche® Image Analysis Systems to automate cervical Papanicolaou cytology examination. This instrument can be readily adapted to cytopreparation of almost any body fluid with the final result of a nice monolayer of unstained cells suitable for quantitation. The principle of the device relies on the fact that cells in suspension will settle onto a microscope slide. Inflammatory and red blood cells can even be removed using differential centrifugation through a gradient solution.

1. Concentrate the fixed urine sample by centrifugation. If sample is frozen at –70°C, thaw samples.
2. Combine pellets into one 50-mL conical polypropylene centrifuge tube rinsing each tube after pellet transfer with 5–10 mL of cell wash fluid. If frozen, agitate cryovial to resuspend cells and pour sample into a labeled 50-mL centrifuge tube; rinse cryotube with double-distilled water.
3. Aspirate cells in centrifuge tube to disperse cell clumps and discard transfer pipet.
4. Fill centrifuge tube up to 50 mL mark with cell wash fluid.
5. Let set for 20 min.
6. Note that counting is not necessary because the CytoRich™ will only create monolayers and unused sample is retrieved by the instrument for other studies.

7. Enter count into database; this calculates the volume of sample to use to prepare one slide.
8. Pour calculated amount of specimen into a labeled 15-mL centrifuge tube and concentrate by centrifugation in CytoRich™ centrifuge.
9. Decant supernatant and place in CytoRich™ tube rack position.
10. Place water tubing #1 in cell adherent fluid. Place HEMAT tubing #2 and 95% EtOH tubing #4 in reagent vessel containing modified Saccamanno fluid. Leave EA-OG tubing #3 in double-distilled water.
11. Turn on vacuum pump and start the CytoRich™ program.
12. Prime tubing at least twice, watching for air bubbles (gaps) in the reagent lines.
13. When program is completed, decant rack.
14. Remove the tophat (Roche® Image Analysis Systems, Inc.) and spray slide with Carbofix-E.
15. Let slides dry for 15 min.

3.11. Overview of Immunofluorescence Assays and Quantitative Labeling (see Notes 5–14)

In order to achieve quantitative labeling of biomolecules, the cell must be treated as a chemical system ***(4)***. The objective becomes chemical analysis rather than to provide a visually pleasing appearance though morphology is generally well maintained. Because chromogenic absorption measurements are, in general, of poorly defined stoichiometry and often of complex chemistry, the relationship between absorbance and quantity is nonlinear and may be entirely nonstoichiometric ***(4,5)***. Further, when multiple markers are labeled simultaneously, the cells of interest may be impossible to accurately isolate when there are overlapping spectra. In contrast, when approached from a biophysical paradigm, fluorescent-labeled probes can have a defined stoichiometry wherein the intensity of emitted light is directly proportional to the quantity or concentration of the target biomolecule, even in complex systems ***(4,6–8,10–22)***. Complementary fluorochromes can be selected that allow discrete, accurate measurements of multiple target biomolecules simultaneously ***(7,8)***. As described below, stoichiometry is established by titration of reagents and serves as the basis for standardization of fluorescence measurements in absolute terms.

The major drawbacks with fluorochrome techniques have been the lack of permanency of the preparation and experience with fluorescent microscopy. The emergence of antifading agents ***(23)*** has revolutionized fluorochrome technology providing slides that are now stable for several years. The main problem with fluorescence assays is their unsuitability for use on paraffin sections because of the high level of background fluorescence.

Immunofluorescence assays have traditionally been performed by manual application of each reagent. This requires careful constant attention throughout the entire assay. The technician must be extremely methodical so that each slide is incubated for exactly the same amount of time with precisely the same amount of reagent. In our hands, we were unable to produce the accuracy and volume of assays by any technician. The Fisher Code-On™ Immunostainer was developed by Dr. David Brigati and uses a unique capillary action to apply and remove reagents to cells and tissues placed on microscope slides *(24)*. This computer-controlled device is capable of performing unattended immunoassays and DNA hybridization. The advent of this technology made it possible to accurately quantify biomarkers in cells. BioTek Solutions Inc. (recently merged with Vantana, Inc., Santa Barbara, CA) improved and further automated the device, in addition to developing specially designed microscope slides that increase the capillary gap that drastically improved performance of immunoassays of tissue sections and plain slides. To prevent drying, the cabinet should be kept humidified. Filtration of all reagents to remove lint and dust are imperative. Immunoreagents must be aliquoted and stored frozen. Other automated immunoassay devices are available today but have not been tested by our laboratory. A typical program for labeling with three reagents is shown in **Table 1**.

3.12. Image Analysis Instrumentation (see *Notes 5–14*)

Until recently, image analysis hardware with the power and sophistication required for automated scanning required special-purpose (and expensive) computer boards in order to achieve the computational power needed for sophisticated applications. Recent developments of Pentium and other powerful computing platforms capable of handling the computations required in image analysis without add-on boards should bring the price of such instrumentation to within the reach of many laboratories.

3.13. Summary of a Basic Approach to Biomarker Development

1. Establish a positive control using the manufacturer's recommended dilution of the primary reagent (note that secondary and tertiary reagents should be titrated previously).
2. Titrate to determine saturation point.
3. Determine slide to slide reproducibility using control cells. Modify techniques, measurements or control cells to achieve less than at least 10% CV's. Control cells will be assayed in every batch as a calibration standard.
4. Assay 14 cases and controls to determine the potential value of the biomarker using control cells to achieve absolute units. For example, the assay can be standardized against an arbitrary standard of a cultured cell line (indicated with subscript "s" in the following equations) expressing the DD23 antigen (UM-UC-13 cells). The concentration of DD23 antigen/cell, in "DD23 Units," D_i, is calculated using the following equations:

Table 1
Typical Program for Triple Label Immunofluorescence Assay

Step	Position	Minutes	Special activity	Position description
1	100	0		Oven at 40°C
2	0	0.01	Hold	Hold for oven warmup
3	11	1	Mix	Pad
4	14	1	Mix	Block
5	7	15		Incubator
6	8	0.5		Pad
7	10	0.1		1X Autobuffer (BM-M30)
8	9	0.3		Pad
9	10	0.1		1X Autobuffer (BM-M30)
10	11	0.3		Pad
11	10	0.5		1X Autobuffer (BM-M30)
12	11	1	Mix	Pad
13	15	1	Mix	Primary antibody
14	7	30		Incubator
15	8	0.5		Pad
16	10	0.1		1X Autobuffer (BM-M30)
17	9	0.3		Pad
18	10	0.1		1X Autobuffer (BM-M30)
19	11	0.3		Pad
20	10	0.5		1X Autobuffer (BM-M30)
21	12	1	Mix	Pad
22	16	1	Mix	Biotinylated secondary Ab
23	7	30		Incubator
24	8	0.5		Pad
25	10	0.1		1X Autobuffer (BM-M30)
26	9	0.5		Pad
27	10	0.1		1X Autobuffer (BM-M30)
28	11	0.5		Pad
29	10	0.1		1X Autobuffer (BM-M30)
30	12	1	Mix	Pad
31	17	1	Mix	Texas red
32	7	30		Incubator
33	8	0.5		Pad
34	10	0.5		1X Autobuffer (BM-M30)
35	9	0.5		Pad
36	10	0.5		1X Autobuffer (BM-M30)
37	11	0.5		Pad
38	10	0.5		1X Autobuffer (BM-M30)
39	12	0.5		Pad
40	10	0.5		1X Autobuffer (BM-M30)

(continued)

Table 1 *(continued)*

Step	Position	Minutes	Special activity	Position description
41	8	1		Pad
42	10	0.6		1X Autobuffer (BM-M30)
43	9	1	Mix	Pad
44	6	2		Hoechst
45	11	0.6		Pad
46	6	0.5		Hoechst
47	12	0.6		Pad
48	6	0.5		Hoechst
49	8	0.6		Pad
50	6	0.5		Hoechst
51	9	0.6		Pad
52	6	0.5		Hoechst
53	11	0.5		Pad
54	6	2		Hoechst

$$\text{DD23 Units/cell} = D_i = G_i/\overline{G}_S \times 100 \quad (1)$$

$$\overline{G}_S = 1/N_s \times \Sigma G_{s,i} \quad (2)$$

where $\overline{G}_S$ is the population mean of all the standard cells and N_s is the number of cells measured. The proportionality constant between immunofluorescence and biochemical content can be determined, in principle, by measuring the mean immunofluorescence of a population of cells and the mean content per cell by biochemical analysis.

5. Determine stability of biomarker in fixed control cells stored at 4°C by refrigerating a batch of control cells and freezing aliquots at various time intervals. Assay these aliquots on the same batch same day. We usually carry this experiment out to at least 6 wk to evaluate how well the biomarker might perform in internationally conducted studies that are often thwart with delays in shipping samples.
6. Determine stability of prepared slides stored at –20°C. This step is to designed to solve logistical issues in the laboratory to determine if controls cell standard slides can be made on a single day and used on subsequent assays on different days. It will also indicate acceptable limits to manage weekends, holidays, instrument repairs and other downtimes.
7. Determine assay batch-to-batch variability.
8. Determine control cell line batch-to-batch variability.
9. Assay 50 cases and controls to refine receiver operator curve (ROC) plots.

4. Notes

1. Many of the reagents published herein are protected by patent and require negotiations with The University of Oklahoma Health Science Center prior to commercial use.

2. All reference to filtering of reagents indicate that a 0.22-µm Magna nylon filter be used with vacuum filtration techniques to remove dust, debris and bacteria.
3. Filtering the Buffer is the rate limiting step, therefore we recommend you filter items separately to save time.
4. We use the Coulter ZM particle counter fitted with a 100-µm aperture tube. This size prevents obstruction of the aperture by large cells and tissue fragments and increases accuracy of the epithelial cell counts. Use of a hemacytometer may result in a gross underestimation of the cell count when large epithelial cells are present because this apparatus was design for counting blood cells which are much smaller.
5. Image Analysis Instrument Requirements:
   a. The basic components of an image analysis system include a microscope, an image detector, an image frame grabber, and a computer for image analysis. The microscope may be significantly modified even to the point of omitting eyepieces for viewing. It can be adapted with motors to control movement of the slide in X, Y, and Z planes. Shutters are used to control presence and absence of light, while motorized filter wheels and sliders are used to control excitation and emission wavelengths and signal intensity. Image detectors most commonly are some form of camera such as black and white charge-coupled device (CCD) or silicon-intensifier target (SIT) tube; intensified versions of each, a color charge-coupled device (CCD), or even a photomultiplier tube can be used. Multiple detectors can be placed on the same system, but then require careful registration with each other. Computers interfacing with all of this hardware can handle all image processing or have an image processor interface to increase speed. Data and images can be stored on local hard drives or removable drives such as optical disks and Bernoulli cartridges. Video printers and video cameras are interfaced with image analysis systems.
   b. The highest quality optical system is required to acquire images for biomarker analysis. A poor microscope will result in blurring of the image and degrade the accuracy of the data. Apochromatic lenses are essential for quantifying objects of different colors. The portion of the microscope field that will be digitized must be flat so that the entire digitized image is in focus. The lower the numerical aperture the more reproducible the focus. The lowest magnification possible for the degree of accuracy should be determined because fluorochrome fading is minimized at lower magnifications. The magnification used is a major rate-limiting factor in calculating machine time that determines the number of fields required for quantitation. In some instances, a triage at a lower magnification to identify objects of interest followed by quantitation at a higher magnification may solve the problem of speed for markers requiring a high degree of accuracy. Excitation and emission filter sets should be carefully selected with quantitation in mind. Microscope vendors in the past have been mostly interested in providing an easy to see bright signal using long-pass filters usually at the expense of allowing nonspecific fluorescence to pass through the emission filters. AR-coated band-pass excitation

and emission filters optimized for the fluorochrome should be used in the biomarker assay. The *camera* can detect and measure a specific signal even though this signal may be difficult for the human eye to see.

c. The detector in an image analysis system is the most critical part of the instrument. The device is usually a camera but can be a photomultiplier tube in which the face of the tube is bitmapped. SIT cameras and intensified SIT cameras are the most sensitive to low light levels. The dynamic range of SIT cameras must be addressed particularly in multiple marker assay analysis because the absolute range of signal expression is marker- and fluorochrome-dependent. Excessive signal can be overcome by using neutral density interference filters mounted on a computer-controlled filter wheel. CCD cameras have a greater dynamic range, but the usual digital image point is an 8-bit number with 256 gray value bins. Increasing dynamic range could result in wider bins (hence, less precision) unless a sliding scale or 16-bit image system is used.
d. CCD cameras have recently approached the speed and sensitivity necessary to acquire fluorescent images. These are still slow for practical consideration in multiple marker rare event detection. Image acquisition requires 3–4 s compared to 30 ms (i.e., 0.03 s) for a SIT camera. Cooled CCDs rapidly return to the zero state. Their advantage is in increased dynamic range, convenience, lack of geometric distortion and low noise. Similarly, color CCDs have advanced rapidly but still are not at the level required for quantitative measurements as defined in this communication. They do provide excellent images for visual morphology and are acceptable for nonquantitative biomarkers.
e. Sources of error in the image detector include nonlinearity at all emission wavelengths. Neutral density filters may be used to test the linearity of the camera at various emission wavelengths, a very important quality control and preventive maintenance procedure for any quantitative device. There may be nonuniform response across the image plane or geometric distortion may need to be corrected by software. One must also determine how much time is required for the camera to return to the zero state to prevent imprint traces of the previous acquired image.
f. Quantitative image analysis is still in its infancy. Virtually no software packages are available commercially to accomplish the tasks of fluorescence image analysis and multiple marker quantitative analysis. A few instruments are available with software packages for Feulgen DNA quantitation and percent area positive for a few biomarker probes in immunocytochemistry. Roche image analysis systems (RIASs) and SAMBA have versatile software packages available that can be used for fluorescence or chromogen assays in tissue sections. New biomarkers are being developed daily that need special image analysis software for research purposes that later can be modified for reduction to clinical practice. The computer software for image analysis must be flexible, allowing the user to write software routines for quantitation, image processing, selection of features for measurement, and develop calibration

methods for assays. Serious consideration of speed of image processing, scene segmentation, measurement algorithms, image storage, and communication with various hardware devices attached to the microscope is required in addition to technical support, programming support, hardware reliability, and service. User groups and electronic bulletin board systems, where users can share experience and help solve each others problems, are also convenient. The Zeiss/Roche IBAS-VIDAS-Videoplan users group consists of several hundred users using these devices in many disciplines. NIH-Image has the largest known group of image analysis users for a wide variety of applications. NIH-Image analysis software, originally developed for the Macintosh platform, was released for Microsoft WindowsNT in 1996. Both versions are available on the World Wide Web for free download of executables and source code.

g. In selecting instrumentation to perform quantitative fluorescence-image analysis, one must first consider all variables in order to find a device that will be flexible enough to meet current and future needs. If multiple markers will be evaluated, then an excitation and emission filter changer is required that has an interface with the image analysis software. A motorized computer controlled magnification device will be required if different magnifications within the same analysis are desired. If low-expressed biomarkers with low-light emission are to be measured, then a SIT camera may be required. The image analysis software must be capable of processing the format of the image produced by the detector (PAL, NTSC, digital, etc.). The focus mechanism must be proven to operate efficiently with fluorescent assays and be a rapid, reliable focus with a feedback return to the software for testing for focus failures. Some focus mechanisms may not function if only one cell is present in the field. Others focus on the brightest area that may be dirt or lint above the plane of the cells. An automatic gain device to enhance the focus input signal may be used to improve autofocus of sparse cell preparations. Automatic gain may not be used to acquire an image for quantitation nor should the image that will be measured be processed with enhancements etc. other than flattening the field and correcting geometric distortion produced by the optical/camera system. This means that several images should be allowed in memory simultaneously to allow for gray-level image acquisition, background correction, flat field reference, image enhancement, and binary scene segmentation algorithms. The speed of image processing, measurement, image storage, motor speeds of the stage focus, and filter wheels should be calculated before purchase to decide time limitations and feasibility of the instrument. We have calculated that a dual marker quantitative test should require no more than 15 min of instrument time to be cost effective in the clinical environment. This becomes a difficult goal if every field on the microscope slide must be evaluated. Some time restraints may be overcome by allowing the machine to run unassisted overnight. As faster processors and computer technologies continue to improve, the speed of image analysis systems will approach that of flow cytometry systems.

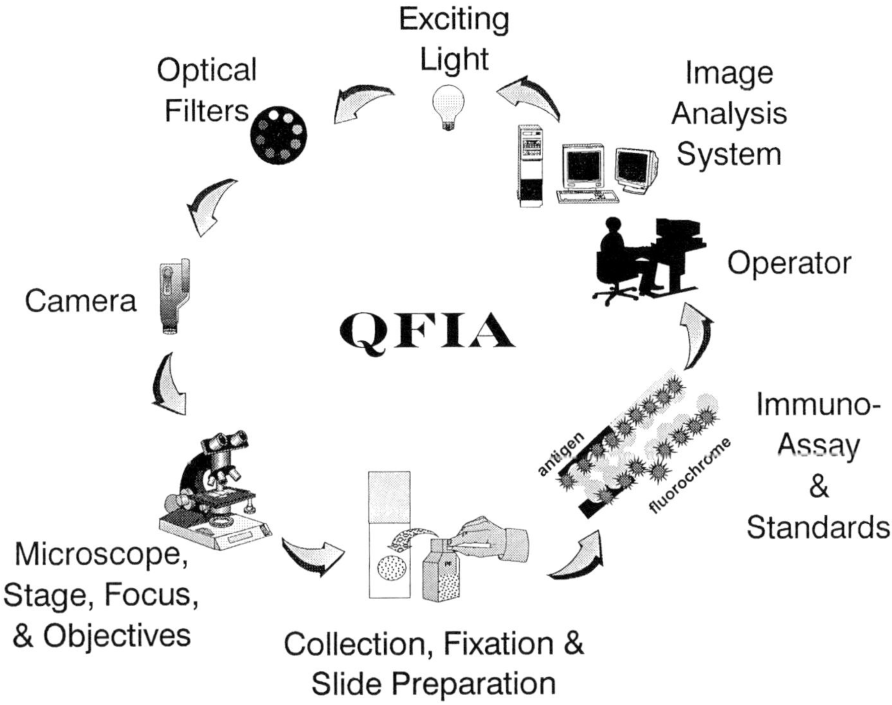

Fig. 1. The components of the entire system where sources of variability and error occur in an immunofluorescence quantitative assay.

6. Calibration Variability and Error:
   a. **Figure 1** depicts the components of the entire system where sources of variability and error occur in an immunofluorescence quantitative assay. These can be divided into three main components: the microscope, including the excitation source, objectives and filters; the image digitizing equipment, including the image detection device and AD converter, the immunofluorescent labeling system, including slide preparation, antibodies, assay techniques, fluorochrome selections and coverslipping. Each area must be addressed by the image system to achieve accurate absolute quantitative measurements.
   b. The mercury lamp is the greatest source of error in the system. Flickering of the lamp causes variability in emission intensity. The mercury arc must be carefully aligned and defocused to prevent hot spots in the rectangular image captured. Approximately 200 h of operation per lamp is usual before flickering is observed in a new lamp. The two most important keys to good lamp stability are a stable power supply (a better grade than the one normally sold with microscopes) and secure oxidation-free contacts. Mercury-xenon lamps

with 2000 h of stable life may solve many of these problems and are easier to align owing to the larger arc. Optiquip manufactures a universal lamphouse that can be fitted on most microscopes regardless of vendor. This product and accompanying power supply provide the most versatile choice on the market and allows xenon, mercury or mercury-xenon lamps to be used with the same hardware. A stable fluorescent object such as a phosphor particle can be continuously measured over a period of a few minutes using the CV and distribution of the gray value measurements as a guide to lamp stability. This object may also be used to determine the uniformity of the field of illumination and establish a shading reference for field flatness correction when measuring cells. The shading reference includes transmission artifacts of the entire optical path and may be different for each filter set and magnification, therefore the best shading references are obtained all under conditions in which the cells would be measured. Fluorescent beads such as Fluorospheres® (Molecular Probes) or DNA Check Beads (Coulter Corp.) can be measured to calibrate the exciting light and establish the accuracy of measurements.

c. Quality assurance of the instrumentation should also include linearity checks of the camera, digitizer and AD converter at various emission wavelengths, and should be an integral component of preventive maintenance. This can be done with a stable fluorescent object such as a phosphor particle and a set of calibrated neutral density interference filters ***(15)***. Image registration at each of the excitation wavelengths and dichroic filter alignment must be determined as well as reproducibility when changing filters through motor control. Reproducibility of the microscope hardware is essential as consistent image shifts and different focal planes can be corrected through software.

d. Established cell lines provide a system for testing assay reproducibility and a convenient means of standardization for future assays ***(18)***. An assay baseline calibration point must be established for quantitative assays. At least one and preferably two points of high and low expression arc established for each assay in which the slope and other statistics are used for quality assurance of the assay. Many laboratories merely use a case that was previously positive for the marker similar to methods that have been used for special stains in histotechnology for many years. This is not satisfactory for quantitation standards since the amount of expression is unknown and the supply of this control is limited. The amount of biomarker present in a given cell line can be measured by other methods such as ELISA or RIA and regression coefficients can be used to determine reliability of image data ***(18,22,25)***. **Figure 2** shows a regression curve of transglutaminase activity with ELISA and our quantitative image analysis ($R = 0.99$). In this instance, the same antibody and cells from a single harvest were used to determine comparability of quantitative methods. Different antibodies, cells at different stages of the cell cycle, number of passes and degree of confluence can affect biomarker expression.

e. **Figure 3** shows a comparison of EGFR quantitation using antibodies from two different vendors demonstrating that the sensitivity and specificity of anti-

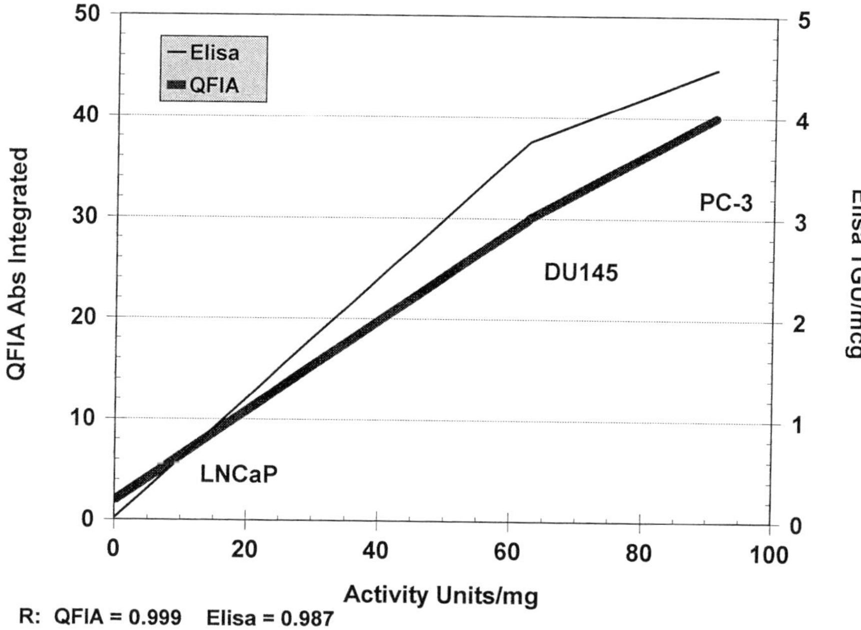

Fig. 2. Regression curves of transglutaminase activity measured by QFIA on the Y1 axis ($R = 0.999$) and ELISA ($R = 0.987$) on the Y2 axis against activity measurements of the enzyme. Results are expressed as arbitrary, absolute units of transglutaminase immunofluorescence per cell and are reported as the mean of at least 100 cells. This value is directly proportional to the biochemical assay of activity/mg of cell protein.

bodies for a quantitative biomarker assay must be evaluated. Titration of reagents is crucial to maintaining accuracy because it ensures saturation of binding sites without providing excessive nonspecific binding. The antibody from Vendor B shows no evidence of binding specifically to the target protein. In contrast, the titration curve of the antibody from Vendor A shows virtually ideal properties with a clearly defined equivalence point as indicated. In addition, the antibody shows no evidence of nonspecific binding, which is manifested by a continuous positive slope with increasing antibody concentration. If the system is not saturated (e.g., the antibody concentration is below that required to flatten the curve), then the intensity will vary with cell number. Excess antibody can lead to nonspecific binding. In using multiple markers, the lack of interference between antibodies and the spectral properties of probes also must be established. **Figure 4** shows an example establishing independence of M344 labeled with Neutralite-Texas Red (Molecular Probes, Inc.) and DNA labeled with Hoechst dye (Polysciences, Inc.).

f. Optimum fixation for all markers in the assay must be established prior to collection of patient samples. Cell lines of the same batch varying concentra-

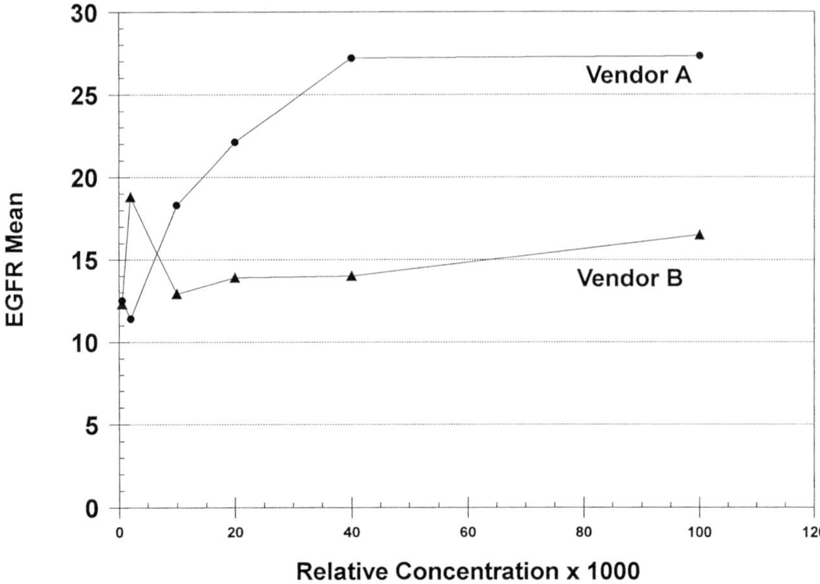

Fig. 3. Titration of EGFR antibody from two different vendors. Note that Vendor B antibody appears to label nonspecifically. Each point represents the mean immunofluorescence of approximately 100 cells as a function of the antibody concentration. The optimum concentration of the antibody from Vendor A is indicated by the arrow.

tions and combinations of various fixation techniques can be compared to fresh unfixed cells. An example of marker degradation with increasing concentrations of alcohol and paraformaldehyde is apparent in assays for F-actin and DNA *(6)*. In this instance, the higher concentrations of paraformaldehyde produced high DNA CV's. This is an important experiment that can be quickly performed. Once optimum fixation has been determined, several control slides are assayed on different days and among slides on the same day, as shown in **Fig. 5**. It is important to establish base line variability of the assay in order to interpret variability among patient samples. Variability of the assay may be caused by many factors including unstable image analysis instrumentation, poor reproducibility of the slide preparation, batch-to-batch variability of assay standards, and variability in the labeling techniques. The causes of significant variability must be decided and corrected as best as possible.

g. Factors that may degrade the quality of the assay include: the number of cells on the slide; degree of cell overlap; quality of the cells deposited; artifacts introduced by slide preparation; individual variation in preparation of slides; and the mix of cell types in the preparation. Slide preparation is an extremely important issue for quantitative assays. For instance, in our laboratory we have spent several years developing techniques to prepare reproducible slides

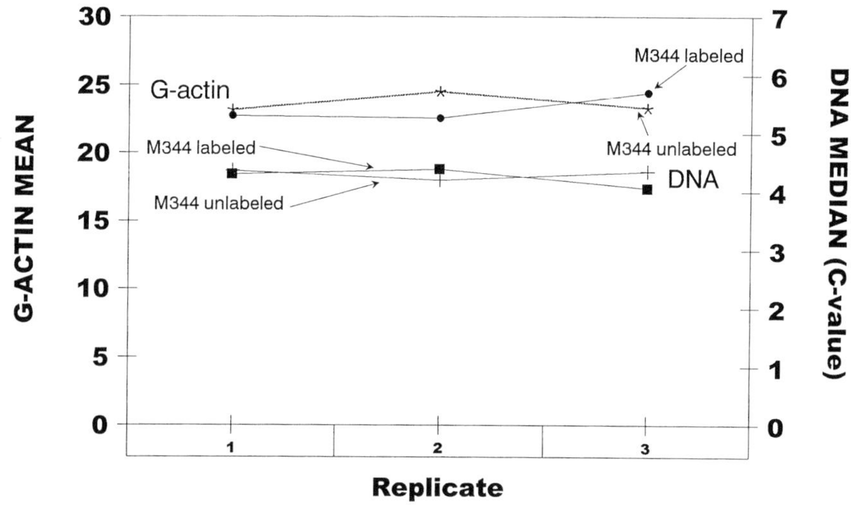

Fig. 4. Independence of M344 antibody labeled with biotinylated goat antimouse IgG-neutralite avidin labeled with Texas Red and Hoechst 33258 dye labeling DNA. The intensities of each fluorophore is independent of the presence of the other fluorophore.

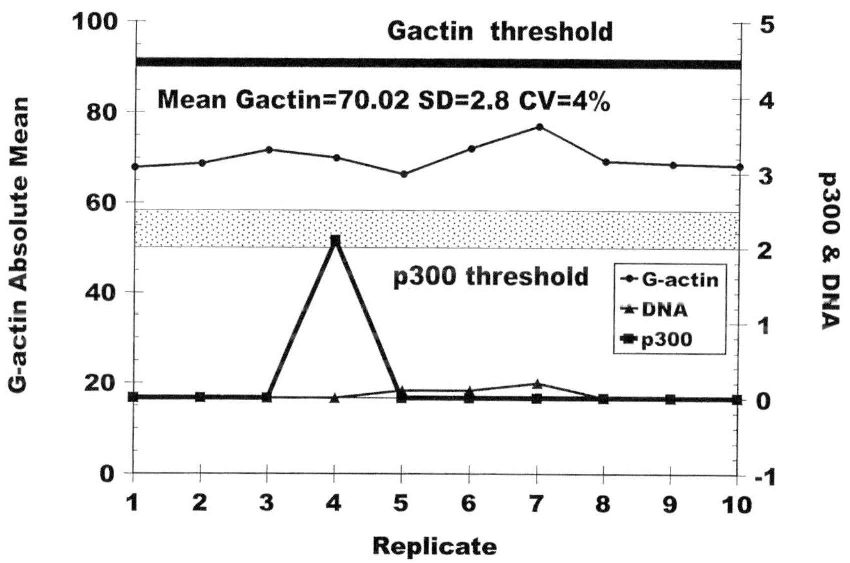

Fig. 5. Replication of G-actin, DNA, and M344 assays. Samples were all prepared and analyzed on the same day.

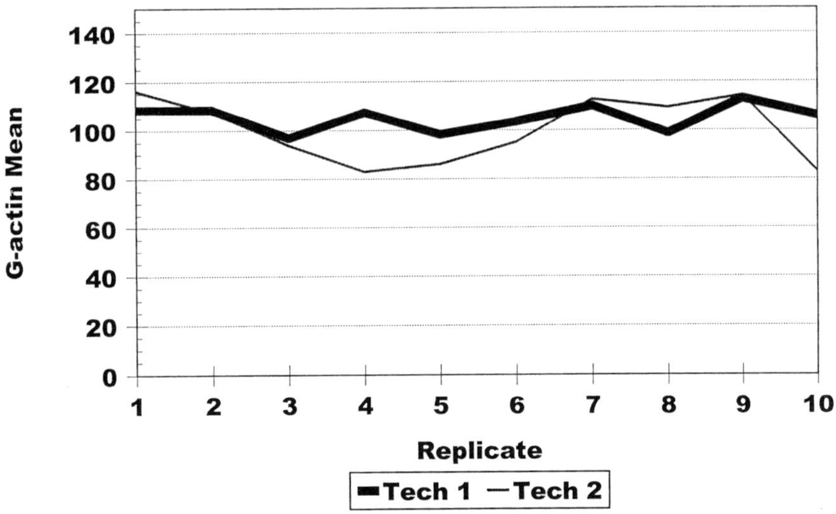

Fig. 6. Demonstration that without specific training and testing, technicians do not all prepare identical slides.

using manual methods and the same preparatory technician for each experiment. We then trained several other technicians and compared results. **Figure 6** shows that some individuals have much greater variability in preparing the same cells on slides. We now use this technique to certify slide preparation technicians in our laboratory and continue to train them until they can achieve an acceptable degree of variability. Instruments are now being developed to prepare slides for automated cytology such as the CytoRich (Roche) and the ThinPrep (Cytec) that may solve the slide preparation problem.

The integrity of all cell types in all states of degeneration may not be preserved by the slide preparation method or in reality in the sample collected for quantitation. As cells degenerate, they become more fragile and release enzymes that degrade biomarkers of interest. Acceptable methods of sample collection must be established for each biomarker evaluated. The biology of the biomarker provides clues about which sample types may not be acceptable. For instance, markers that were abnormal when expressed at levels lower than normal, would probably not be suitable for voided urine since this type of sample contains such a wide range of degenerated cells that may lead to false positive diagnoses. Shipping samples may also degrade quantitative markers depending on the method of shipment. Some biomarkers have "abnormal" expression in immature cells from normal individuals but not in mature normally exfoliated cells. An example is p300 where test positive thresholds are adjusted based on sample type ***(8)***. Here urine samples yield higher sensitivity and specificity than bladder irrigation. Therefore, we expect that tissue touch preps would also have a high rate of p300 "false positive"

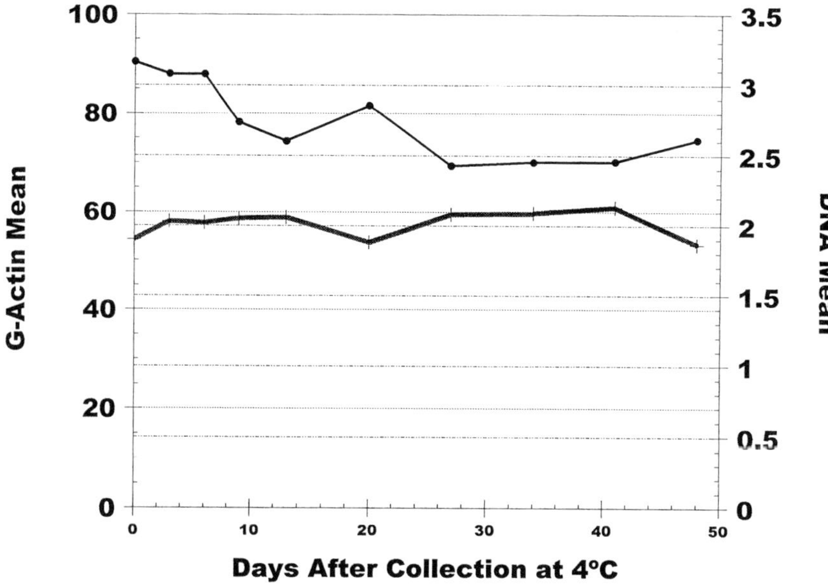

Fig. 7. Stability of G-actin in voided urine samples stored at 4°C over time.

results as was demonstrated by Rao et al. *(7)*. Other methods of sample collection include fine needle aspiration (FNA) in vivo or of the excised tissue. This is an important method of sample collection for prostate and kidney due to anatomic position of these organs. Most FNA samples do not yield sufficient cells for flow cytometry and are mixed with tumor infiltrating lymphocytes, normal connective tissue, tissue fragments and tissue debris that complicate the results.

h. Stability of the biomarker with each type of sample collection must be evaluated to establish storage and sample handling requirements. For example, **Fig. 7** shows the stability of G-actin in voided urine samples as a function of storage time at 4°C. Similar experiments should be performed on frozen samples to determine how long an archived sample is stable at –70° and –20°C. Stability of prepared slides under various storage conditions should also be evaluated. Ambient temperature can have a major impact on shipped samples if no refrigeration is supplied. Shipped samples may be exposed to extreme heat during the summer months of up to 110°C (e.g., 230°F, typical of delivery trucks parked in the sun) requiring that the stability of biomarkers be tested in the worse case scenario to determine acceptable shipping requirements. This also establishes reliability of data collected in multi-institutional trials and after reduction to clinical practice.
i. Once all aspects of biomarker assay variability have been examined, one can then begin to determine overall accuracy of the method. Each aspect of assay

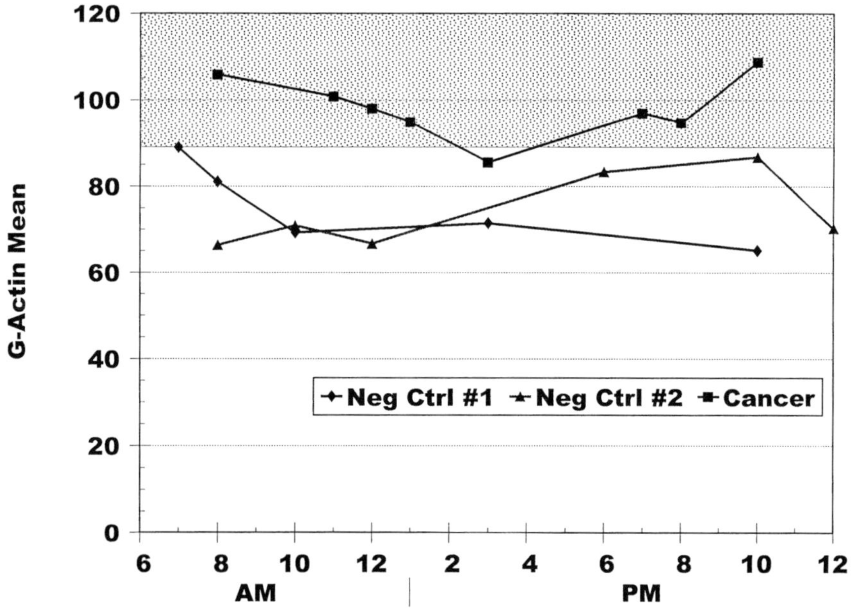

Fig. 8. Lack of diurnal variation in G-actin concentrations in exfoliated urinary tract cells in two controls and one cancer patient.

variability contributes to threshold values that will be set by clinical trials and to the overall confidence in a positive or negative test result. We continue to improve our technical assay by reducing the controllable variability. Remaining variability in patient samples must be due to individual variability. **Figure 8** shows no significant difference in G-actin expression in bladder cancer patient urine samples collected consecutively over a 24-h period and illustrates lack of diurnal variability for this biomarker. Intraindividual variability can be estimated by collecting urine samples from the same patient each day for a week (*see* **Fig. 9**). The impact of tumor heterogeneity has little effect on our test result in a given patient but there is a wide range of expression of G-actin among tumors from different individuals. Interindividual variability is the most difficult to measure due to the wide range of population genetic variability, medications, diet, medical conditions, age, smoking history and other xenobiotic effects that may influence biomarker expression.

j. Experiments should be designed to determine the important factors causing variability and how much each contributes to the quantitative result. The primary goal is to determine how much accuracy is required to determine a positive from negative test. If there is a substantial difference between positive and negative ranges, the quantitative measurements can incorporate more variability without compromising diagnostic accuracy. The ROC plots are a primary tool for determining the accuracy and reproducibility needed in any given assay (*see* Chapter 3).

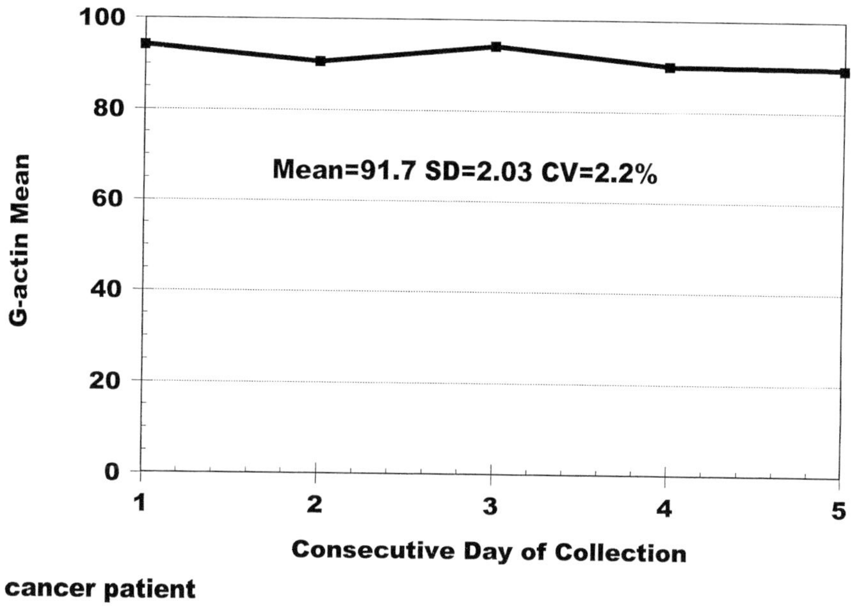

Fig. 9. Long-term intraindividual variability can be estimated by collecting urine samples from the same patient each day for a week.

References

1. Vogelstein, B., Fearon, E., Hamilton, S., Kern, S., Preisinger, A. C., and Leppert, M. (1988) Genetic alterations during colorectal tumor development. *N. Engl. J. Med.* **319,** 525–532.
2. Koss, L. G., Bogdan, C., Herz, F., and Wersto, R. (1989) Flow cytometric measurements of DNA and other cell components in human tumors: a critical appraisal. *Hum. Pathol.* **20,** 528–548.
3. National Research Council (U.S. Subcommittee on Biological Markers in Urinary Toxicology) (1995) in *Biological Markers in Urinary Toxicology*. National Academy Press, Washington, DC, pp. 1–309.
4. West, S. S. (1970) Biophysical cytochemistry, in *Introduction to Quantitative Cytochemistry* (anonymous, ed.), Academic, New York, p. 451.
5. Nibbering, P. H., Leijh, P. C. J., and van Furth, R. (1985) Quantitation of monoclonal antibody binding to individual cells by cytophotometry, in *Techniques in Immunocytochemistry*, Academic, New York, pp. 97–114.
6. Rao, J. Y., Hurst, R. E., Bales, W. D., Jones, P. L., Bass, R. A., Archer, L. T., and Hemstreet, G. P. (1990) Cellular f-actin levels as a marker for cellular transformation: relationship to cell division and differentiation. *Cancer Res.* **50,** 2215–2220.
7. Rao, J. Y., Hemstreet, G. P., Hurst, R. E., Bonner, R. B., Jones, P. L., Min, K. W., and Fradet, Y. (1993) Alterations in phenotypic biochemical markers in bladder epithelium during tumorigenesis. *Proc. Natl. Acad. Sci. USA* **90,** 8287–8291.

8. Bonner, R. B., Hemstreet, G. P., Fradet, Y., Rao, J. Y., Min, K. W., and Hurst, R. E. (1993) Bladder cancer risk assessment with quantitative fluorescence image analysis of tumor markers in exfoliated bladder cells. *Cancer* **72,** 2461–2469.
9. Fradet, Y., Islam, N., Boucher, L., Parent-Vaugeois, C., and Tardif, M. (1987) Polymorphic expression of a human superficial bladder tumor antigen defined by mouse monoclonal antibodies. *Proc. Natl. Acad. Sci. USA* **84,** 7227–7231.
10. Nakamura, N., Hurst, R. E., West, S. S., Menter, J. M., Golden, J. F., Corliss, D. A., and Jones, D. D. (1980) Biophysical cytochemical investigations of intracellular heparin in neoplastic mast cells. *J. Histochem. Cytochem.* **28,** 223–230.
11. Hurst, R. E., ParmLey, R. T., Nakamura, N., West, S. S., and Denys, F. R. (1981) Heparan sulfate of AH-130 ascites hepatoma cells: a cell-surface glycosaminoglycan not displaced by heparin. *J. Histochem. Cytochem.* **29,** 731–737.
12. Hemstreet, G. P., West, S. S., Weems, W., Echols, C. K., McFarland, S., Lewin, J., and Lindseth, G. (1983) Quantitative fluorescence measurements of AO-stained normal and malignant bladder cells. *Int. J. Cancer* **31,** 577–585.
13. ParmLey, R. T., Hurst, R. E., Takagi, M., Spicer, S. S., and Austin, R. L. (1983) Glycosaminoglycans in human neutrophils and leukemic myeloblasts: ultrastructural, cytochemical, immunologic, and biochemical characterization. *Blood* **61,** 257–266.
14. Bass, R. A., Hemstreet, G. P., Honker, N. A., Hurst, R. E., and Doggett, R. S. (1987) DNA cytometry and cytology by quantitative fluorescence image analysis in symptomatic bladder cancer patients. *Int. J. Cancer* **40,** 698–705.
15. West, S. S., Hemstreet, G. P., Hurst, R. E., Bass, R. A., Doggett, R. S., and Schulte, P. A. (1987) Detection of DNA aneuploidy by quantitative fluorescence image analysis: potential in screening for occupational bladder cancer, in *Biological Monitoring of Exposure to Chemicals* (Dillon, K. and Ho, M., eds.), Wiley, New York, pp. 327–341.
16. McGowan, P., Hurst, R. E., Bass, R. E., Hemstreet, G. P., and Postier, R. (1988) Equilibrium binding of Hoechst 33258 and Hoechst 33342 fluorochromes with rat colorectal cells. *J. Histochem. Cytochem.* **36,** 757–762.
17. Rao, J. Y., Hemstreet, G. P., Hurst, R. E., Bonner, R. B., Min, K. W., and Jones, P. L. (1991) Cellular F-actin levels as a marker for cellular transformation: correlation with bladder cancer risk. *Cancer Res.* **51,** 2762–2767.
18. Hemstreet, G. P., 3d, Rao, J. Y., Hurst, R. E., Bonner, R. B., Jones, P. L., Vaidya, A. M., Fradet, Y., Moon, R. C., and Kelloff, G. J. (1992) Intermediate endpoint biomarkers for chemoprevention. *J. Cell. Biochem.* **16I(Suppl.),** 93–110.
19. Bonner, R. B., Liebert, M., Hurst, R. E., Grossman, H. B., Bane, B. L., and Hemstreet, G. P. (1996) Marker network for bladder cancer characterization of the DD23 tumor-associated antigen for bladder cancer detection and recurrence monitoring. *Cancer Epidemiol. Biomarkers Prev.* **5,** 971–978.
20. Hemstreet, G. P., Bonner, R. B., Hurst, R. E., and O'Dowd, G. A. (1996) Cytology of Bladder Cancer, in *Comprehensive Textbook of Genitourinary Oncology* (Vogelzang, N. J., Scardino, P. T., Shipley, W. U., and Coffey, D. S., eds.), Williams and Wilkins, Baltimore, MD, pp. 338–350.

21. Hemstreet, G., Hurst, R., and Bonner, R. (1998) Selection and development of biomarkers for bladder cancer, in *Methods in Molecular Medicine*, Humana, Totowa, NJ, pp. 37–60.
22. Rao, J. Y., Bonner, R. B., Hurst, R. E., Qiu, W. R., Reznikoff, C. A., and Hemstreet, G. P. (1997) Quantitative changes in cytoskeletal and nuclear actin levels during cellular transformation. *Int. J. Cancer* **70,** 423–429.
23. Krenik, K. D., Kephart, G. M., Offord, K. P., Dunnette, S. L., and Gleich, G. J. (1989) Comparison of antifading agents used in immunofluorescence. *J. Immunol. Meth.* **117,** 91–97.
24. Brigati, D. J., Budgeon, L. R., Unger, E. R., Koebler, D., Cuomo, C., Kennedy, T., and Perdoma, J. M. (1988) Immunocytochemistry is automated: development of a robotic workstation based upon the capillary action principle. *J. Histotechnol.* **11,** 165–183.
25. Jones, P. L., O'Hare, C., Bass, R. A., Rao, J. Y., Hemstreet, G. P., and Hurst, R. E. (1990) Quantitative immunofluorescence, anti-ras p21 antibody specificity and cellular oncoprotein levels. *Biochem. Biophys. Res. Commun.* **167,** 464–470.

III

Cytogenetic Markers

12

Cytogenetics as a Diagnostic Aid for Childhood Hematologic Disorders

Conventional Cytogenetic Techniques, Fluorescence In Situ *Hybridization, and Comparative Genomic Hybridization*

Susana C. Raimondi, Susan Mathew, and Ching-Hon Pui

1. Introduction

Cytogenetic analysis is an important aid in the classification of hematological disorders. Most types of leukemia display either numerical chromosomal abnormalities or structural rearrangements, mainly translocations. Increasingly recognized, nonrandom chromosomal abnormalities are especially useful in diagnosing the leukemic subtype and predicting treatment outcome. Molecular analysis of the genes adjacent to the breakpoints of specific translocations and study of the functions of their gene products have helped to clarify the complex interactions that promote leukemogenesis and perpetuate the leukemic phenotype *(1)*.

Acute lymphoblastic leukemia (ALL) is the most common childhood malignancy, comprising about 30% of all cases of pediatric neoplasia. ALLs can be classified into five subtypes based on the modal number of chromosomes: hyperdiploid (>50), hyperdiploid (47–50), pseudodiploid (46 chromosomes with structural or numerical abnormalities), diploid (2n), and hypodiploid (2n-). Recognition of ploidy as a distinctive cytogenetic feature in ALL has greatly enhanced our ability to predict treatment outcome *(2)*.

Defining ALL by the types of structural abnormalities found in the chromosomes of leukemic clones has led to impressive advances in understanding the biology of the disease and suggests opportunities for risk-specific therapies *(3)*. **Table 1** describes the most common structural abnormalities found in

From: *Methods in Molecular Medicine, Vol. 14: Tumor Marker Protocols*
Edited by: M. Hanausek and Z. Walaszek © Humana Press Inc., Totowa, NJ

Table 1
Recurrent Structural Chromosome Abnormalities in Childhood Acute Lymphoblastic Leukemia (ALL)

	Approximate incidence (%)		Chromosome band/gene involved	
Abnormality	ALL overall	Specific immunophenotype		
B-lineage				
t(8;14)(q24;q32)	3	B-cell, 90	8q24/*MYC*	14q32/*IGH*
t(2;8)(p12;q24)	<1	B-cell, 4–5	2p12/*IGK*	8q24/*MYC*
t(8;22)(q24;q11)	<1	B-cell, 6–10	8q24/*MYC*	22q11/*IGL*
t(1;19)(q23;p13)	5–6	Pre-B, 90	1q23/*PBX1*	19p13/*E2A*
t(9;22)(q34;q11)	2–5	Early pre-B, 75	9q34/*ABL*	22q11/*BCR*
t(4;11)(q21;q23)	2	Early pre-B, 80	4q21/*AF4*	11q23/*MLL (ALL1)*
t(11;19)(q23;p13.3)	1	Early pre-B	11q23/*MLL(ALL1)*	19p13.3/*ENL*
t(5;14)(q31;q32)	<1	Eosinophilia	5q31/*IL3*	14q32/*IGH*
t(17;19)(q22;p13)	<1	Early pre-B	17q22/*HLF*	19p13/*E2A*
t(12;21)(p13;q22)[a]	<1	Early pre-B, Pre-B	12p13/*TEL(ETV6)*	21q22/*AML1 (CBFA2)*
T-lineage				
t(11;14)(p13;q11)	1	T-cell, 7	11p13/*RHOMB2(TTG2)*	14q11/*TCRAD*
t(11;14)(p15;q11)	<1	T-cell, 1	11p15/*RHOMB1(TTG1)*	14q11/*TCRAD*
t(10;14)(q24;q11)	1	T-cell, 5–10	10q24/*HOX11*	14q11/*TCRAD*
t(8;14)(q24;q11)	<1	T-cell, 2	8q24/*MYC*	14q11/*TCRAD*
t(1;14)(p32;q11)	<1	T-cell, 3	1p32/*TAL1(TCL5/SCL)*[b]	14q11/*TCRAD*

inv(14)(q11q32)	<1	14q11/*TCRAD*	14q32/*IGH*
t(1;7)(p32;q35)	<1	1p32/*TAL1(TCL5/SCL)*	7q35/*TCRB*
t(1;7)(p34;q35)	<1	1p34/*LCK*	7q35/*TCRB*
t(7;9)(q35;q34)	<1	7q35/*TCRB*	9q34/*TAN1*
t(7;9)(q35;q32)	<1	7q35/*TCRB*	9q32/*TAL2*
t(7;10)(q35;q24)	<1	7q35/*TCRB*	10q24/*HOX11*
t(7;11)(q35;p13)	<1	7q35/*TCRB*	11p13/*RHOMB2 (TTG2)*
inv(7)(p15q35)	<1	7p15/*TCRG*	7q35/*TCRB*
t(7;19)(q35;p13)	<1	7q35/*TCRB*	19p13/*LYL1*
Nonspecific lineage			
del(6q)	4–13		
t/del(9p)	7–12	9p22/*p16(MTS1)*	
t/del(11q)	3–5	11q23/*MLL(ALL1)*	
t/del(12p)	10–12	12p13/*TEL(ETV6)*	

[a]*TEL(ETV6)* gene rearrangements by Southern blotting/RT-PCR positive for cryptic t(12;21) in 25% of B-lineage immunophenotype.
[b]*TAL1* submicroscopic deletion in 15–26% of T-cells.

ALL blasts and the proto-oncogenes associated with these abnormalities. The protein products of these proto-oncogenes may contribute to the disease process.

Most of the structural chromosome rearrangements found in ALL correlate with the leukemic cell immunophenotype. The t(8;14), t(8;22), and t(2;8) rearrangements, for example, were initially described in Burkitt-type B-cell ALL with the French-American-British (FAB)-L3 morphology. In childhood ALL, illustrations of the close association between chromosomal translocation and phenotype of the leukemic cell include the t(1;19) in pre-B ALL, the t(4;11) and t(9;22) in B-cell precursor ALL, and the 7q35 and 14q11 rearrangements in T-cell ALL.

In children, acute myeloid leukemia (AML) occurs less frequently than ALL, accounting for only one of every five cases of childhood leukemia. In addition, about 75% of childhood AML cases have a modal number of 46 chromosomes; in contrast to its use in childhood ALL, ploidy does not differentiate subtypes of AML. As with ALL, however, specific chromosome abnormalities frequently are associated with particular subtypes of AML (**Table 2**) ***(4)***. For example, the t(15;17) is found primarily in cases of acute promyelocytic leukemia (APL), t(8;21) characterizes AML with differentiation, inv(16) is associated with myelomonocytic leukemia with bone marrow eosinophilia, and t(9;11) with cases of monocytic leukemia.

Chronic myelogenous leukemia (CML) accounts for fewer than 5% of childhood leukemias. The Philadelphia chromosome, t(9;22) (q34;q11), is the hallmark of this hematologic malignancy and is observed in 90% of the cases studied cytogenetically. Light microscopy fails to identify this translocation in the remaining 10%, which frequently are positive for this abnormality by molecular methods ***(5)***.

Myelodysplastic syndromes (MDS) in children are rare. Among the most common cytogenetic alterations in this subgroup of patients are monosomy 7 and trisomy 8 ***(6)***.

Most nonrandom translocations associated with leukemias involve proto-oncogenes. The availability of molecular probes for sequences containing translocation breakpoints has greatly facilitated the development of molecular cytogenetics, including fluorescence *in situ* hybridization (FISH). Comparative genomic hybridization (CGH) gives a comprehensive analysis of aberrations (especially gains and losses of partial or whole chromosomes) in a single hybridization experiment ***(7)***. These tests are especially useful as complements to conventional cytogenetics for monitoring patients with known chromosomal abnormalities and resolving complex chromosome rearrangements (*see* **Notes 1–3**). Clearly, analysis of leukemic cell karyotypes has deepened our understanding of the pathobiologic changes underlying hematologic disorders ***(8–21)***. With the development of additional probes, we should be able to explain

Table 2
Recurrent Chromosome Abnormalities in *De Novo* Childhood Acute Myeloid Leukemia (AML)

	Approximate incidence (%)			
Abnormality	AML overall	Specific FAB	Chromosome band/gene involved	
t(8;21)(q22;q22)	14	M2, M1	8q22/*ETO*	21q22/*AML1 (CBFA2)*
inv(16)(p13q22)/del(16q)	11	M4	16p13/*MYH11*	16q22/*CBFB*
t(9;11)(p22;q23)	7	M4, M5	9p22/*AF9*	11q23/*MLL*(*ALL1*)
t(15;17)(q22;q11–12)	8	M3	15q22/*PML*	17q11–12/*RARA*
t(5;17)(q35;q11–12)	<1	M3, variant	5q35/*NPM*	17q11–12/*RARA*
t(11;17)(q23;q11–12)	<1	M3, variant	11q23/*PLZF*	17q11–12/*RARA*
–7/del(7q)	4	M2, M4		
t(1;22)(p13;q13)	1	M7		
t(3;5)(q25.1;q35)	1	Nonspecific	3q25.1/*MLF1*	5q35/*NPM*
inv(3)(q21q26)	1	Nonspecific	3q21/*RIBOPHORIN1*	3q26/*EVI1*
t(6;9)(p23;q34)	1	M2, M4	6p23/*DEK*	9q34/*CAN*
t(8;16)(p11;p13)	1	M4, M5	8p11/*MOZ*	16p13/*CBP*
t(1;11)(p32;q23)	<1	M4, M5	1p32/*AF1*	11q23/*MLL (ALL1)*
t(6;11)(q27;q23)	<1	M4, M5	6q27/*AF6*	11q23/*MLL (ALL1)*
t(10;11)(p14;q21)	1	M5		
t(10;11)(p12;q23)	1	M4, M5	10p12/*AF10*	11q23/*MLL (ALL1)*
t(11;17)(q23;q21)	<1	M4, M5	11q23/*MLL (ALL1)*	17q21/*AF17*
t(11;19)(q23;p13.3)	1	M4, M5	11q23/*MLL (ALL1)*	19p13.3/*ENL*
t(11;19)(q23;p13.1)	<1	M4, M5	11q23/*MLL (ALL1)*	19p13.1/*ELL*
t(16;21)(p11;q22)	<1	Nonspecific	16p11/*TLS(FUS)*	21q22*(ERG)*
t(12;22)(p13;q11)	<1	Nonspecific	12p13/*TEL(ETV6)*	22q11/*MN1*

hematological disorders in molecular terms and devise treatment accordingly. This chapter describes the methodology for conventional chromosome analysis, FISH, and CGH, with emphasis on techniques used in studying hematologic disorders. A variety of probes, each having a different cytogenetic application, are available for use in FISH (*see* **Notes 4–6**).

2. Materials

2.1. Cytogenetics

1. Centrifuge.
2. Incubator: 37°C.
3. Light microscope.
4. Media: RPMI-1640 with L-Glutamine (JRH Biosciences, Lenexa, KS).
5. Fetal bovine serum (FBS) (Gemini-Bioproducts, Calabasas, CA).
6. Heparin: preservative-free, 1000 U/mL (Fujisawa USA, Deerfield, IL).
7. Colcemid: Karyomax 10 μg/mL (Life Technologies, Grand Island, NY).
8. Stainless-steel wire: diameter 0.29 (Small Parts, Inc. Miami, FL), cat. no. SWX-029.
9. Wright's stain: powder (Sigma, St. Louis, MO).
10. pHydrion buffers in capsules, pH 7.00 ± 0.02 at 25°C (Micro Essential, Brooklyn, NY).
11. Trypsin: 0.25% (Gibco), stored as 5-mL aliquots in the freezer.
12. Hypotonic solution: Add 0.54 g potassium chloride to 100 mL deionized water. Make fresh solution before each experiment.
13. 3:1 Carnoy's fixative: Combine 75 mL ACS-certified methanol and 25 mL ACS-certified glacial acetic acid in a 100-mL glass bottle. Make fresh solution just prior to use.
14. Normal saline solution: Add 27 g sodium chloride to 3 L deionized water.
15. Stock buffer for Wright's stain: Add 1 pHydrion capsule to 100 mL deionized water. Mix well and store at 4°C.
16. Working buffer for Wright's stain: Add 5 mL stock buffer to 95 mL deionized water.
17. Wright's stain stock solution: Place 100 mL methanol in a beaker on a stir-plate. While stirring at medium-high speed, gradually add 0.3 g powdered Wright's stain. Cover beaker to prevent splashing and stir for at least 30 min. Filter solution over double Whatman no. 40 paper and store at 4°C in a brown bottle.

2.2. *Fluorescence* In Situ *Hybridization (FISH) and Comparative Genomic Hybridization (CGH)*

1. Incubator: Set at 37°C.
2. Microcentrifuge.
3. Humidified chamber.
4. Water baths: 37°C, 43–45°C, 70°C.
5. Rubber cement.

6. Precleaned microscopic slides and cover slips.
7. Coplin jars.
8. Polypropylene microcentrifuge tubes: 0.5 mL.
9. Microscope: with epifluorescence, 100-W mercury lamp, and filters described below.
10. Filters (CHROMA Technology, Brattleboro, VT).
    a. Single bandpass filters no. 150191 (spectrum orange and propidium iodine) and no. 150291 (spectrum green and propidium iodide);
    b. Dual bandpass filters no. 151501 (spectrum orange, DAPI) and no. 151533 (spectrum green, DAPI);
    c. Triple bandpass filter no. 152517 (spectrum orange, spectrum green, DAPI, propidium iodide).
11. Charge coupled device (CCD) camera (Photometrics, Tucson, AZ).
12. Computer with software appropriate for FISH and CGH analysis (Vysis, Downers Grove, IL; Applied Imaging, Pittsburgh, PA; and Perceptive Scientific Instruments, League City, TX).
13. Formamide (Fisher Scientific, Fair Lawn, NJ), cat. no. 227-500.
14. Ethanol: 100%.
15. Dextran sulfate (Pharmacia, Piscataway, NJ), cat. no. 17-0340-02.
16. Salmon sperm DNA-sonicated (Pharmacia), cat. no. 27-4565-01.
17. RNase A (Sigma), cat. no. 6513.
18. Oncor *in situ* kits (Oncor, Gaithersburg, MD).
19. Cot-1™ DNA (BRL/Life Technologies, Gaithersburg, MD), cat. no. 15279-011.
20. Proteinase K (Boehringer Mannheim, Indianapolis, IN), cat. no. 745723.
21. NP-40 (Sigma), cat. no. I-3021.
22. Nick translation kit (Life Technologies, cat. no. 18160-010).
23. BioNick labeling system (Life Technologies, cat. no. 18247-015).
24. 20X SSC: 0.3 *M* NaCl, 0.3 *M* Na-citrate, pH 7.4. Combine 175.2 g sodium chloride, 88.2 g sodium citrate, and distilled water to a final volume of 1 L.
25. 2X SSC: Add 50 mL 20X SSC to 450 mL distilled water.
26. Denaturation solution: 70% formamide/2X SSC. Make fresh solution before each experiment. Combine 4 mL 20X SSC, 8 mL distilled water, and 28 mL formamide. Adjust to pH 7.0 with 1 *N* hydrochloric acid.
27. RNase solution: 100 μg/mL in 2X SSC.
28. Phosphate buffered detergent (PBD) (pH 8.0): Add 4 g sodium bicarbonate and 20 mL NP-40 to 4 L distilled water.
29. Proteinase K: 0.1 μL/mL in 20 m*M* Tris-HCl, 2 m*M* $CaCl_2$, pH 7.5.
30. Hybridization buffer:
    a. For centromeric probes: 65% formamide/2X SSC (Oncor, cat. no. S1370-30);
    b. For unique sequence probes and CGH: Combine 5 mL 50% formamide, 1 mL 20X SSC and 2 mL 10% dextran sulfate, 1 mL salmon sperm DNA (10 mg/mL). Bring to 10 mL with distilled water. Store at 4°C.
31. Ethanol: 70, 80, 95, and 100% stored at room temperature and at –20°C.

3. Methods

3.1. Cytogenetics

3.1.1. Collecting the Specimen

Use aseptic technique.

3.1.1.1. Bone Marrow

1. Put 15–20 mL RPMI-1640 with 15% fetal bovine serum (FBS) and 0.05 mL heparin solution into a 50-mL centrifuge tube.
2. Collect the sample into the prepared tube; each chromosome analysis requires 1–2 mL sample of bone marrow aspirate. To prevent clotting, immediately cap the tube and mix by inverting several times.
3. Preparing the sample: In the laboratory, split the collected marrow sample among two or more 15-mL centrifuge tubes. Each tube should contain no more than 0.5 mL of bone marrow aspirate. Therefore, the number of tubes prepared should be adjusted according to the amount of bone marrow collected in the 50-mL centrifuge tube. One of the tubes is processed immediately; the other is incubated at 37°C overnight. Both tubes are processed as described in **Subheading 3.1.2.**

3.1.1.2. Peripheral Blood

Peripheral blood cultures should be performed when circulating blasts are present (preferably >25%).

1. Draw 5–10 mL blood with a syringe coated with preservative-free heparin and transfer to a 15-mL centrifuge tube. Centrifuge at 200*g* for 6 min.
2. Remove buffy coat/plasma and transfer to two to four 15-mL centrifuge tube containing 9–10 mL RPMI-1640 with 15% FBS and 0.05 mL heparin solution. Adjust amount of buffy coat/plasma according to the white blood cell count of the patient: for <30,000/mm^3 add 1 mL of buffy coat/plasma, for 30,000 to 70,000 add 0.5 mL, for 70,000 to 150,000 add 0.2 mL, and for >150,000 add 0.1 mL.
3. Incubate overnight at 37°C and process as described below.

3.1.2. Processing the Specimen

1. Add 0.05–0.1 mL of Colcemid to each centrifuge tube. Recap the centrifuge tube and invert several times to mix. Let it stand 25 min at room temperature.
2. Centrifuge at 200*g* for 10 min.
3. Remove the supernatant, leaving approx 0.25 mL above the cell pellet. Resuspend the cell pellet with a stirring wire or a Pasteur pipet *(22)*.
4. While stirring, add 5 drops of hypotonic solution. Slowly add an additional 10 mL of hypotonic solution while stirring. Recap the tube and let it stand at room temperature for 25–30 min.
5. Centrifuge tube at 200*g* for 6 min. After centrifugation, a layer of white cells above the red cells will be observed.

6. Remove most of the supernatant, leaving approx 0.25 mL above cell pellet.
7. Resuspend cell pellet. While stirring, add 5 drops of 3:1 Carnoy's fixative. This step is important to prevent clumping of cells. Add an additional 10 mL of Carnoy's fixative and resuspend. Recap the tube and let stand for 15 min at room temperature.
8. Centrifuge tube at 200*g* for 6 min.
9. Remove the supernatant and repeat **step 7**. Let the tube stand at room temperature for 10 min.
10. Centrifuge tube at 200*g* for 6 min.
11. Remove supernatant. Repeat the Carnoy's fixative until the cell pellet is white. Prior to slide making, resuspend the cell pellet. Add 3:1 Carnoy's fixative a few drops at a time until the final suspension is slightly cloudy.

3.1.3. Making Slides

For best results in making the slides, adjust the humidity of the area to 75%.

3.1.3.1. Hot-Plate Method

1. Slides are precleaned in 75% alcohol and refrigerated in deionized water.
2. Shake off excess water.
3. Aspirate the final suspension into a siliconized disposable glass Pasteur pipet with a rubber bulb.
4. Hold the filled pipet 6–12 in. above the slide.
5. Tilt the slide at a 45° angle to the floor.
6. Release one to two drops of the cell suspension onto the slide. The drop should land near the frosted end of the slide. Very gently, blow once or twice on the slide.
7. Wipe the bottom of the slide with gauze and place on a warm (>30°C) or hot (60–75°C) plate until the slide is dry.
8. Etch the accession number onto each slide.
9. Examine a test slide under a phase-contrast microscope to ensure that the sample has metaphase chromosomes.

3.1.3.2. Flame Method

Use the flame method only when the metaphases are poorly spread ***(22)***.

1. Follow **steps 1–5** in hot-plate method (**Subheading 3.1.3.1.**).
2. With the slide tilted, release one drop of the cell suspension onto the slide. The drop should land on the bottom half of the slide.
3. Flame the slide, using an alcohol burner, holding the slide at a 45° angle. Move the slide in the flame for approx 2 s. Place slides in drying racks and let them dry completely.
4. Etch each slide with the accession number.
5. Examine a test slide under a phase-contrast microscope to ensure that chromosomes are adequately spread.

3.1.4. Aging the Slide

Optimal aging is an important step in successful chromosome banding for hematologic disorders and may be attained by natural or rapid methods.

3.1.4.1. Natural Aging

The slides are aged by leaving them at room temperature. The time varies (3–10 d) and, on rare occasions, good G-banding may be obtained immediately after harvest.

3.1.4.2. Rapid Aging

Rapid aging can be achieved either by placing the slides in a conventional oven at 90–95°C for 10–30 min or by microwaving the slides at high-setting for 2 to 3 min. Rapid aging leads to adequately banded metaphases for analysis.

3.1.5. G-Banding Technique (using Wright's stain)

1. Line up five Coplin jars and fill one with 5 mL of Trypsin plus 45 mL normal saline solution, one with 50 mL normal saline, one with 10 mL Wright's stain stock solution and 40 mL working buffer for Wright's stain, and two with 50 mL deionized water.
2. Treat slides one at a time through the banding set-up until results are optimized. For fresh slides, start with 10–15 s in the trypsin solution, rinse in saline by dipping the slide two or three times, stain for 1–2 min, and rinse twice in deionized water. Adjust times as needed. Let slides air-dry.

3.2. FISH

An *in situ* hybridization protocol adheres to the following general outline:

1. Preparing slides.
2. Pretreating material on the slides.
3. Denaturating *in situ* target (chromosomal) DNA.
4. Preparing probe.
5. *In situ* hybridization.
6. Posthybridization washes.
7. Probe detection (immunocytochemistry).
8. Microscopy and image analysis.

3.2.1. Preparing Slides

According to standard procedures, prepare metaphase chromosome spreads or interphase nuclei on glass microscope slides. To achieve optimal results, use prepared slides within 2 wk. **Do not bake slides**.

3.2.2. Pretreating Slides

3.2.2.1. RNase Treatment

RNase removes endogenous RNA and minimizes the background.

1. Incubate the slides in DNase-free RNase solution (100 μg/mL in 2X SSC) for 1 h in a 37°C water bath.
2. Wash the slides in 2X SSC twice to remove excess RNase.
3. Dehydrate the slides in 70, 80, and 100% ethanol, for 2 min in each at room temperature.

3.2.2.2. Proteinase K Treatment (Optional)

Proteinase K increases the accessibility of the probe by digesting the chromosomal protein that surrounds the target nucleic acid. Incubate the slides in proteinase K for 7 min at 37°C.

3.2.3. Denaturing In Situ *Target DNA*

Target DNA can be denatured by alkaline (high pH) conditions or by heat.

1. Prewarm the denaturation solution (70% formamide/2X SSC) to 70°C in a water bath.
2. Denature slides for 2 min. Time and temperature are important in order to maintain chromosome morphology. For every slide, there will be a decrease of 1°C.
3. Immediately transfer slides to a Coplin jar containing 40 mL ice-cold 70% ethanol. Rinse slides for 2 min. Rinse for 2 min each in cold 80% ethanol, then 100% ethanol.
4. Allow slides to air-dry or dry under an air jet.

3.2.4. Preparation of Probe

A variety of methods can be used for labeling probes. Nick translation is the most widely used method for introducing labels *(23)*. This section gives general guidelines; manufacturer's recommendations should be followed whenever applicable (*see* **Notes 4–7**).

3.2.4.1. Specific Centromeric Probes (α, β, Classical), and Minisatellite DNA Probes

1. Prewarm the tube containing probe at 37°C for 5 min, vortex, and centrifuge 2–3 s to collect the contents in the bottom of the tube.
2. Combine 1.5 μL biotin- or digoxygenin-labeled probe with 30 μL of hybridization buffer (65% formamide, 2X SSC, Oncor, cat. no. S1370-30) in a 0.5-mL microcentrifuge tube.
3. Denature in 70°C water bath for 5 min. Chill quickly on ice. Centrifuge for 2–3 s.

3.2.4.2. All Human Centromeric, Telomeric, and Total Genomic Painting Probes

1. These probes are provided in 50% formamide/2X SSC. Put 30 μL probe into a microcentrifuge tube.
2. Denature in a 70°C water bath for 5 min.
3. Chill quickly on ice. Centrifuge for 2–3 s.

3.2.4.3. Painting (Coatasome®) Total Chromosome Probes

1. Prewarm probe at 37°C for 5 min, vortex, and centrifuge for 2–3 s to collect the contents in the bottom of the tube.
2. Put 15 μL probe into a microcentrifuge tube.
3. Denature the probe at 70°C for 10 min and centrifuge 2–3 s.
4. Incubate in a 37°C water bath for 2–2.5 h to preanneal the repetitive sequences. Centrifuge 2–3 s.

   Direct labeled probes do not require preannealing.

3.2.4.4. Unique Sequence Probes

1. Mix 100–200 ng labeled probe with the hybridization buffer and 1 μL Cot-1 DNA, which limits nonspecific binding to the chromosomal preparation.
2. Denature the probe at 70°C for 7 min.

3.2.5. In Situ *Hybridization*

1. Place the denatured probe mix on the denatured chromosomal (target) DNA and cover with a glass cover slip.
2. Seal by applying rubber cement along the perimeter of the cover slip.
3. Incubate at 37°C in a humidified chamber for 4–16 h.

3.2.6. Posthybridization Washes

Unhybridized and nonspecifically bound probe is removed by washes of various stringencies. The stringency of these washes can be manipulated by varying temperature and the concentrations of formamide and salt.

1. Prewarm the 50% formamide/2X SSC wash solution at 43–45°C for 30 min.
2. Remove the rubber cement and place the slides in a Coplin jar containing the prewarmed 50% formamide/2X SSC. Incubate for 10 min. The cover slips will fall off.
3. Wash the slides in 50% formamide at 43°C, four times for 5 min each time.
4. Wash the slides four times for 5 min each in 2X SSC at 43°C.
5. Place the slides in PBD until ready for detection. Slides can be stored at 4°C for up to 2 wk.

3.2.7. Probe Detection

Remove the slides from PBD and blot excess fluid. **Do not allow the slides to dry.**

3.2.7.1. Biotin

Biotin-labeled probes can be detected with the Biotin-FITC kit (Oncor, cat. no. S-1333-BF).

1. Apply 50 μL FITC–avidin, cover with a plastic cover slip, and incubate 30 min at 37°C. Remove cover slip and wash slides three times in PBD at room temperature, 2 min each wash.
2. Apply 20 μL propidium iodide to the slide and cover with a glass cover slip. View under a fluorescence microscope.
   If the signal is weak, perform the following steps to amplify the signal.
3. Remove the cover slip and perform three 2-min washes in PBD. Apply 50 μL antiavidin antibody and incubate for 15 min at 37°C. Repeat the washes in PBD.
4. Apply 50 μL FITC–avidin conjugate and incubate for 15 min at 37°C. Repeat the washes in PBD.

3.2.7.2. Digoxigenin

Digoxigenin-labeled probes can be detected with the Rhodamine-Digoxigenin kit (Oncor, cat. no. S-1332DR)

1. Apply 50 μL rhodamine-labeled antidigoxigenin to slide, cover with a cover slip, and incubate at 37°C for 15 min. Remove cover slip and perform three 2-min washes in PBD.
2. Add 20 μL 4,6-diamino-2-phenylindole (DAPI, 0.5 μg/mL in antifade solution) and view under a fluorescence microscope.
   If the signal is weak, perform the following steps to amplify the signal.
3. Add 50 μL rabbit antisheep antibody and incubate for 15 min at 37°C. Repeat the washes in PBD.
4. Apply 50 μL rhodamine-labeled antirabbit antibody and incubate for 15 min at 37°C. Repeat the washes in PBD.

3.2.8. Microscopy

Kodacolor 400 and Fujichrome 400 films are superior for photographing the red of propidium iodide and yellow of fluorescein.

3.3. CGH

3.3.1. CGH Metaphase Spreads

Prepare metaphase spreads of phytohemagglutinin (PHA)-stimulated lymphocytes from healthy individuals by using standard cytogenetic procedures with hypotonic treatment and methanol/acetic acid fixation.

3.3.2. Labeling of Tumor and Normal DNAs

3.3.2.1. Tumor DNA

High-molecular-weight tumor DNA for analysis is labeled by nick translation as follows:

1. In a microcentrifuge tube, combine 1 µg tumor DNA, 5 µL 10X A4 mixture (0.2 m*M* dATP, dCTP, dGTP in 500 m*M* Tris-HCl, pH 7.8, 50 m*M* $MgCl_2$, 100 m*M* β-mercaptoethanol, and 100 µL/mL bovine serum albumin (BSA), 1 µL biotin-14-dATP, 5 µL enzyme mixture (contains DNA polymerase I [2 U] and DNase I [200 pg]; and 1 µL (10 U) DNA polymerase I (Promega, Madison, WI) and make up to 50 µL with distilled water.
2. Incubate 45–60 min at 15°C in a water bath.
3. Stop the reaction by incubating for 10 min at 70°C.

3.3.2.2. Normal DNA

The reference (normal) DNA is labeled as described in **Subheading 3.3.2.1.**, except digoxigenin-11-dUTP is used in place of biotin-14-dATP.

Alternatively, the reference and tumor DNAs can be labeled with direct fluorochromes (e.g., spectrum orange and spectrum green).

3.3.2.3. Determining the Size of DNA

Determine the sizes of both tumor and normal DNA by electrophoresis through a 1% agarose gel. The size range should be about 500–2000 bp. The fragment length should be modified by adjusting the ratio of DNase to DNA polymerase in the nick translation reaction or the incubation time.

3.3.3. Denaturating the Probes

1. In a microcentrifuge mix the following and store at –20°C for 2–12 h:
   a. 200 ng each of the tumor and normal DNAs;
   b. 10–15 µg of unlabeled Cot-1 DNA (to block the binding of repetitive sequences);
   c. 3 µL 2 *M* sodium acetate;
   d. Ethanol to make up to 100 mL.
2. Precipitate by centrifuging for 30 min in a microcentrifuge at 11,000*g*.
3. Remove the supernatant and air or vacuum dry.
4. Dissolve the dried pellet in 10 µL of hybridization buffer.
5. Denature at 70°C for 5 min and preanneal for 60 min at 37°C.

3.3.4. Denaturating and Preparing the Slides

1. Denature metaphase chromosome spreads at 70°C in 70% formamide/2X SSC for 2.5 min.
2. Dehydrate slides through a series of 70, 80, 95, and 100% cold ethanol solutions. Incubate for 2 min at each concentration. Air-dry.
3. Incubate the slides in proteinase K solution (0.1 µg/mL) for 5.5 min at room temperature.
4. Wash in 2X SSC for 2 min. Repeat washes two times.
5. Dehydrate in 70, 80, 100% ethanol series. Air-dry.

3.3.5. Hybridization

Add the denatured probe to the slides and hybridize for 3–4 d at 37°C.

3.3.6. Posthybridization Washes

1. Remove the cover slip and wash the slides (to remove the unbound DNAs) three times in 50% formamide/2X SSC (pH 7.0) for 10 min each wash.
2. Wash twice in 2X SSC, then once in 0.1X SSC at 45°C for 10 min each wash.
3. Wash the slides in PBD three times for 2 min each wash.
   For direct fluorochromes, go directly to **step 5**.
4. Apply 50 µL of rhodamine antidigoxigenin/FITC avidin (Oncor, cat. no. S-1374-1), cover with a cover slip, and incubate at 37°C for 15 min. Remove the cover slip and wash slides three times in PBD for 2 min each wash.
5. Counterstain with DAPI.

3.3.7. Digital Image Acquisition and Processing

1. Acquire blue, red, and green images using a quantitative image processing system with a fluorescence microscope equipped with a cooled charge coupled device (CCD) camera and appropriate filter sets. The software program integrates the green and red fluorescence intensities in stripes orthogonal to the chromosomal axis, subtracts local background, generates the intensity profiles for red and green colors, and calculates the ratio profiles for both colors from pter to qter of each chromosome.
2. Acquire 5–10 metaphases from each case. Average the ratio profiles from a chromosome type. "Copy number karyotype" of the leukemia/tumor can be generated using the ratio profiles for all chromosomes.
3. To obtain the normal thresholds for the analysis of tumor/normal hybridization, a control hybridization with normal/normal DNAs should be performed. Ratios above 1.25 and below 0.85 are generally considered as gain or loss of chromosomal region, respectively.

4. Notes

1. Conventional cytogenetic techniques performed on bone marrow aspirates or peripheral blood (if high percentage of circulating blasts are present) yield karyotypes of the leukemic cells at the light microscopy level. At diagnosis, such cytogenetic findings are useful to identify unique leukemic subtypes, and correlate with prognosis and direct treatment. Although not commonly used to monitor minimal residual leukemia owing to a lack of sensitivity (~5%), karyotypic analysis at relapse is crucial to evaluate the clonal evolution and occasional clonal switch (i.e., development of a new malignancy).
2. FISH techniques enable the detection of specific nucleic acid sequences (DNA or RNA) in metaphase chromosomes, interphase cells, or frozen tissue sections. In combination with immunocytochemistry, *in situ* hybridization relates topographic information to gene activity at the DNA and mRNA levels. Because of the low

proliferative rate of some leukemias (i.e., lack of metaphases), the subtlety of some chromosomal rearrangements and some seemingly identical lesions differs at the molecular level. FISH has numerous applications in the diagnosis and management of patients with neoplastic disorders, particularly those with hematologic malignancies.

3. CGH is a global approach that rapidly (i.e., within a single *in situ* hybridization experiment) screens the entire genome for the presence of imbalances in the genetic material ***(7)***. No specific probes for or prior information of the loci involved are required. CGH also maps the amplified and deleted sequences on normal chromosomes. CGH is based on the competitive *in situ* hybridization of differentially labeled tumor/leukemic cells and normal DNA to a spread of normal metaphase chromosomes. Alterations in the ratio of the two fluorochromes correlate with gains (amplification or differences in copy number) or losses (deletion) of DNA segments. In addition, CGH results provide a copy number of individual chromosomes and preliminary data for isolation of the target genes. This technique has several limitations. Balanced rearrangements cannot be detected with CGH. In addition, CGH results do not provide the investigator any information on the identity of the amplified or deleted segments. Further, these imbalances are apparent only when they are present in most of the cells in the specimen. Thus, CGH is not appropriate for studying the clonal heterogeneity of a neoplastic sample.
4. α Satellite probes (DNA sequences isolated from the centromeric region) are used to detect trisomies/monosomies at much higher frequencies than does routine chromosome analysis. In addition, results from FISH are useful in predicting prognosis, following clonal evolution, and monitoring therapy. For example, FISH techniques aid identification of trisomy 8 and monosomy 7 in cases of MDS and AML ***(6)***, trisomies 3, 12, and 18 in cases of lymphoma ***(8,9)***, and various trisomies in cases of ALL ***(10)***. Identification of X and Y chromosomes make possible the monitoring of bone marrow transplantation (BMT) procedure using opposite-sex donor ***(11)***.
5. Unique sequence probes detect microdeletions, inversions, amplifications, and translocations in interphase nuclei. The presence of these abnormalities is difficult to establish by conventional cytogenetics. The high specificity and efficiency of these probes lead to diagnosis of structural chromosome aberrations at the one-cell level. Even small structural aberrations are easily detected by using the appropriate DNA fragments as probes. For example, detection of *BCR/ABL* genes in CML enables the early detection of minimal residual disease (MRD) and relapse following chemotherapy or BMT ***(12,13)***. A number of studies have shown the importance of FISH with unique DNA sequence probes, and include those of *p16* and *TEL-AML1* (ETV6-CBFA2) in ALL ***(14,15)***, *MLL* in ALL and AML ***(16,17)***, and *MYH11-CBFB* and *PML-RARA* in AML ***(18,19)***.
6. Whole chromosome painting probes contain a complete set of DNA sequences from one chromosome. These probes are useful when the banding pattern suggests rearrangement but is insufficient for identification of the specific chromo-

some involved. Painting probes can be used to define marker chromosomes and to identify deletions and translocations ***(20,21)***.

7. Labeling of probes: Probes for *in situ* hybridization procedures can be labeled through a variety of methods. The most widely used method is nick translation ***(23)***. Two types of fluorochromes for labeling, direct and indirect, currently are in use. In the direct method, the detectable molecule (e.g., spectrum orange, Texas red, spectrum green, fluorescein isothiocyanate [FITC]) is bound to the nucleic acid probe so that the probe:target complexes can be visualized immediately after their hybridization. Probes for indirect labeling methods have been chemically or enzymatically modified to carry a reporter molecule; the probe:target complexes are visible only after affinity cytochemical treatment. Biotin and digoxigenin are widely used as reporter molecules in probes that are indirectly labeled. Directly and indirectly labeled probes for centromeric, whole chromosome, and a few unique DNA sequences are commercially available (Oncor and Vysis are the major distributors of probes.

References

1. Rabbitts, T. H. (1991) Translocations, master genes, and differences between the origins of acute and chronic leukemias. *Cell* **67,** 641–644.
2. Raimondi, S. C. (1993) Current status of cytogenetic research in childhood acute lymphoblastic leukemia. *Blood* **81,** 2237–2251.
3. Pui, C.-H. (1995) Childhood leukemias. *N. Engl. J. Med.* **332,** 1618–1630.
4. Raimondi, S. C., Kalwinsky, D. K., Hayashi, Y., Behm, F. G., Mirro, J., and Williams, D. L. (1989) Cytogenetics of childhood acute nonlymphocytic leukemia. *Cancer Genet. Cytogenet.* **40,** 13–27.
5. Tkachuk, D. C., Westbrook, C. A., Andreeff, M., Donlon, T. A., Cleary, M. L., Suryanarayan, K., Homge, M., Redner, A., Gray, J., and Pinkel, D. (1990). Detection of bcr-abl fusion in chronic myelogenous leukemia by *in situ* hybridization. *Science* **250,** 559–562.
6. Han, K., Lee, W., Harris, C. P., Kim, W., Shim, S., and Meisner, L. F. (1994) Quantifying chromosome changes and lineage involvement in myelodysplastic syndrome (MDS) using fluorescent *in situ* hybridization (FISH). *Leukemia* **8,** 81–86.
7. Kallioniemi, A., Kallioniemi, O.-P., Sudar, D., Rutovitz, D., Gray, J. W., Waldman, F., and Pinkel, D. (1992) Comparative genomic hybridization for molecular cytogenetic analysis of solid tumors. *Science* **258,** 818–821.
8. Nylund, S. J., Ruutu, T., Saarinen, U., Larramendy, M. L., and Knuutila, S. (1994) Detection of minimal residual disease using fluorescence DNA *in situ* hybridization: a follow-up study in leukemia and lymphoma patients. *Leukemia* **8,** 587–594.
9. Whang-Peng, J., Knutsen, T., Jaffe, E. S., Steinberg, S. M., Raffeld, M., Zhao, W. P., Duffey, P., Condron, K., Yano, T., and Longo, D. L. (1995) Sequential analysis of 43 patients with non-Hodgkin's lymphoma: clinical correlation with cytogenetic, histologic, immunophenotyping, and molecular studies. *Blood* **85,** 203–216.
10. Anastasi, J., Vardiman, J. W., Rudinsky, R., Patel, M., Nachman, J., Rubin, C. M., and Le Beau, M. M. (1991) Direct correlation of cytogenetic findings with cell

morphology using *in situ* hybridization: an analysis of suspicious cells in bone marrow specimens of two patients completing therapy for acute lymphoblastic leukemia. *Blood* **77,** 2456–2462.

11. Durnam, D. M., Anders, K. R., Fisher, L., O'Quigley, J., Bryant, E. M., and Thomas, E. D. (1989) Analysis of the origin of marrow cells in bone marrow transplant recipients using a Y-chromosome-specific *in situ* hybridization assay. *Blood* **74,** 2220–2226.
12. Dewald, G. W., Schad, C. R., Christensen, E. R., Tiede, A. L., Zinsmeister, A. R., Spurbeck, J. L., Thibodeau, S. N., and Jalal, S. M. (1993) The application of fluorescence in situ hybridization to detect Mbcr/abl fusion in variant Ph chromosomes in CML and ALL. *Cancer Genet. Cytogenet.* **71,** 7–14.
13. Amiel, A., Yarkonim S., Slavin, S., Or, R., Lorberboum-Galsky, H., Fejgin, M., and Nagler, A. (1994) Detection of minimal residual disease state in chronic myelogenous leukemia patients using fluorescence *in situ* hybridization. *Cancer Genet. Cytogenet.* **76,** 59–64.
14. Okuda, T., Shurtleff, S. A., Valentine, M. B., Raimondi, S. C., Head, D. R., Behm, F., Curcio-Brint, A. M., Liu, Q., Pui, C.-H., Sherr, C. J., Beach, D., Look, A. T., and Downing, J. R. (1995) Frequent deletion of $p16^{ink4a}$*/MTS1* and $p15^{INK4b}$*/MTS2* in pediatric acute lymphoblastic leukemia. *Blood* **85,** 2321–2330.
15. Romana, S. P., Poirel, H., Leconiat, M., Flexor, M.-A., Mauchauffe, M., Jonveaux, P., Macintyre, E. A., Berger, R., and Bernard, O. A. (1995) High frequency of t(12;21) in childhood B-lineage acute lymphoblastic leukemia. *Blood* **86,** 4263–4269.
16. Rowley, J. D., Diaz, M. O., Espinosa, R., III, Patel, Y. D., Van Melle, E., Ziemin, S., Taillon-Miller, P., Lichter, P., Evans, G. A., Kersey, J. H., Ward, D. C., Domer, P. H., and Le Beau, M. M. (1990) Mapping chromosome band 11q23 in human acute leukemia with biotinylated probes: identification of 11q23 translocation breakpoints with a yeast artificial chromosome. *Proc. Natl. Acad. Sci. USA* **87,** 9358–9362.
17. Kearney, L., Bower, M., Gibbons, B., Das, S., Chaplin, T., Nacheva, E., Chessells, J. M., Reeves, B., Riley, J. H., Lister, T. A., and Young, B. D. (1992) Chromosome 11q23 translocations in both infant and adult acute leukemias are detected by *in situ* hybridization with a yeast artificial chromosome. *Blood* **80,** 1659–1665.
18. Dauwerse, J. G., Wessels, J. W., Giles, R. H., Wiegant, J., van der Reijden, B. A., Fugazza, G., Jumelet , E. A., Smit, E., Baas, F., Raap, A. K., Hagemeijer, A., Beverstock, G. C., van Ommen, G. J. B., and Breuning, M. H. (1993) Cloning the breakpoint cluster region of the inv(16) in acute non-lymphocytic leukemia M_4Eo. *Hum. Mol. Genet.* **2,** 1527–1534.
19. Schad, C. R., Hanson, C. A., Paietta, E., Casper, J., Jalal, S. M., and Dewald, G. W. (1994) Efficacy of fluorescence *in situ* hybridization for detecting *PML/RARA* gene fusion in treated and untreated acute promyelocytic leukemia. *Mayo Clin. Proc.* **64,** 1047–1053.
20. Pinkel, D., Landegent, J., Collins, C., Fuscoe, J., Segraves, R., Lucas, J., and Gray, J. (1988) Fluorescence *in situ* hybridization with human chromosome-specific libraries: detection of trisomy 21 and translocations of chromosome 4. *Proc. Natl. Acad. Sci. USA* **85,** 9138–9142.

21. Filatov, L. V., Behm, F. G., Pui, C.-H., Head, D. R., Downing, J. R., and Raimondi, S. C. (1995) Childhood acute lymphoblastic leukemia with equivocal chromosome markers of the t(1;19) translocation. *Genes Chrom. Can.* **13,** 99–103.
22. Rigby, P. W., Dieckmann, M., Rhodes, C., and Berg, P. (1977) Labeling of deoxyribonucleic acid to high specific activity in vitro by nick translation with DNA polymerase I. *J. Mol. Biol.* **113,** 237–251.
23. Williams, D. L., Harris, A., Williams, K. J., Brosius, M. J., and Lemonds, W. (1984) A direct bone marrow chromosome technique for acute lymphoblastic leukemia. *Cancer Genet. Cytogenet.* **13,** 239–257.

13

Preparation of Metaphase Chromosomes for Cytogenetic Analysis

Kevin A. Hahn and Penny K. Riggs

1. Introduction

The in vitro cultivation of cells and tissues forms an integral part of the work of every diagnostic cytogenetics laboratory, since it is from cells undergoing mitosis that metaphase chromosome preparations are obtained. Spontaneously dividing cells suitable for direct chromosome preparations are found only in the rapidly proliferating tissues of the body such as the gonads and bone marrow, or in tissues with malignancies. In order to obtain metaphase spreads from other cells or tissues, it is necessary to induce cell division artificially.

This chapter provides the methods necessary to collect, transport, and cultivate tissues for the harvest and preservation of metaphase chromosomes from lymphocytes, bone marrow elements, solid tissues, and effusions.

2. Materials

2.1. Culture Medium

1. Culture medium: Mix 115.5 mL RPMI-1640 medium (1X) with 30.0 mL calf serum (1X) and 1.5 mL 200 m*M* L-glutamine (100X). Add 1.5 mL of penicillin/streptomycin (10,000 IU/mL + 10,000 μg/mL), and 1.5 mL 7.5% sodium bicarbonate. Add double-distilled (dd) water to a final volume of 150 mL. Stable at 4°C for 4 wk; maintain at pH 7.4 with sodium bicarbonate *(1)*. All reagents available from Sigma, St. Louis, MO.

2.2. Specimen Collection, Transport, and Preparation

1. Heparin, 1 vial, 1000 IU/mL, store at 15–30°C.
2. Sterile 3–8-mL nonadditive blood collection vials.
3. Sterile 15–50-mL conical centrifuge tubes.

From: *Methods in Molecular Medicine, Vol. 14: Tumor Marker Protocols*
Edited by: M. Hanausek and Z. Walaszek © Humana Press Inc., Totowa, NJ

2.3. Rapid Preparation of Metaphase Cells

1. Sterile and nonsterile 2-, 4-, and/or 10-mL pipets.
2. Sterile and nonsterile 15-mL conical centrifuge tubes.
3. Sterile 25-cm^3 tissue culture flasks.
4. Colcemid (Grand Island Biologic Company, Inc., Grand Island, NY), 1 vial. Store at 15–30°C.

2.4. Lymphocyte Culture Materials

1. Sterile and nonsterile 2-, 4-, and/or 10-mL Pasteur pipets.
2. Sterile and nonsterile 15-mL conical centrifuge tubes.
3. Sterile 25-cm^3 tissue culture flasks.
4. Phytohemagglutinin (Grand Island Biologic Company, Inc.), M form, 1 vial, lyophilized. Store at 2–8°C before rehydration, and at –5 to –20°C after rehydration.
5. Pokeweed (Grand Island Biologic Company, Inc.), 1 vial, lyophilized. Store at 2–8°C before rehydration, and at –5 to –20°C after rehydration.
6. Methotrexate (Amethopterin) (Sigma). Store at 15–30°C.
7. Thymidine (Sigma). (1-[2-deoxy-α-D-ribofuranosyl]-5-methyl-uracil). Store at 15–30°C.
8. Colcemid (Grand Island Biologic Company, Inc.), 1 vial. Store at 15–30°C.

2.5. Solid Tissue Culture Materials (see Note 1)

1. Sterile and nonsterile 15–50-mL conical centrifuge tubes.
2. Sterile 2-, 4-, and/or 10-mL Pasteur pipets.
3. Sterile 4.5-cm diameter disposable Petri dishes.
4. Scapular blade and handle.
5. Tissue forceps.
6. 70% Ethanol. Make fresh as needed.
7. Sterile 25-cm^3 tissue culture flasks.
8. Phosphate-buffered saline (PBS) (pH 6.8): 0.025 *M* KH_2PO_4 (3.4 g/L) titrated to pH 6.8 with 50% NaOH.
9. 0.025% Trypsin (Grand Island Biologic Company, Inc.): 1 vial, 2.5% 10X liquid (25 g [1:250] in 8 g NaCl). Store –5 to –20°C after rehydration.
10. Methotrexate (Amethopterin) (Sigma). Store at 15–30°C.
11. Thymidine (1-[2-deoxy-α-D-ribofuranosyl]-5-methyl-uracil) (Sigma). Store at 15–30°C.
12. Colcemid (Grand Island Biologic Company, Inc.), 1 vial. Store at 15–30°C.

2.6. Harvesting of Metaphase Chromosomes

1. Nonsterile 2-, 4-, and/or 10-mL Pasteur pipets.
2. Nonsterile 15-mL conical centrifuge tubes.
3. 0.075 *M* KCl, make fresh as required, maintained at 37°C.
4. Carnoy's fixative: 3:1 methanol-acetic acid. Make fresh as required, maintain at 4°C.

2.7. Cell Count Determination

1. Nonsterile 15-mL conical centrifuge tube.
2. Graduated pipet.
3. Hemocytometer slide and cover slip.

3. Methods

3.1. Lymphocyte Culture

3.1.1. Specimen Collection, Transport, and Preparation

1. Collect 10–20 mL of peripheral blood by venipuncture or marrow by core aspiration (these procedures must be performed by a medically qualified person) (*see* **Note 2**).
2. Dispense the sample immediately into a sterile nonadditive vial containing heparin at a concentration of approx 10–20 IU/mL of blood and shake well to mix.
3. If enough blood is drawn, half of the sample may be dispensed into a nonadditive tube without heparin and centrifuged (100*g* for 15 min) to obtain autologous serum.
4. Transport the sample at no less than room temperature to the laboratory as quickly as possible. For optimum culture results, the sample must be received by the laboratory within 24 h.
5. Centrifugation and sedimentation:
   a. Centrifugation: Leukocytes can be separated by centrifugation (100*g*) and will settle out with the platelets as a pale layer on top of the heavier erythrocytes, and below the plasma supernatant. The cells can be carefully removed with a wide bore syringe needle or a sterile Pasteur pipet.
   b. Sedimentation: Allow the vial of whole blood to stand undisturbed at 37°C for 1–2 h. After this time, the blood will have separated into two layers, the bottom layer comprising mainly red blood cells (RBCs) and the upper layer containing an enrichment of lymphocytes. Sedimentation of the RBCs varies with temperature and the surface tension of the container. Less than 50% of the white blood cells from whole blood may be obtained by this method.
6. Resuspend the leukocyte layer in 10 mL of freshly prepared medium and determine the concentration of cells within the suspension (*see* **Subheading 3.5.**).

3.1.2. Rapid Preparation of Metaphase Chromosomes from Bone Marrow

1. Use methods as described in **Subheading 3.3.**

3.1.3. Nonsynchronized Cell Culture

1. Dispense approx 4.0 × 10^6 cells aseptically into a sterile 25-cm^3 tissue culture flask containing 10.0 mL prepared culture media.
2. Add 200 µL of either phytohemagglutinin or pokeweed mitogen to the culture flask (*see* **Note 3**).

3. Incubate at 37°C in a 5% CO_2 and 97% humidified atmosphere.
4. Tap flask firmly against the hand or lab bench daily to resuspend the contents and reduce the agglutination of any erythrocytes caused by the action of phytohemagglutinin.
5. Seventy hours following culture initiation, add colcemid (0.2 µg/mL final concentration) (*see* **Notes 4** and **5**) and swirl the contents gently ***(1,2)***. There is no need to filter sterilize the colcemid and the aseptic technique need no longer be employed.
6. Ninety minutes later, swirl the contents gently to resuspend the cells and decant the contents of the culture flask into a 15-mL conical centrifuge tube.
7. Centrifuge the tube for 10 min at 100*g*.
8. Discard all but 1 mL of the supernatant.
9. Gently resuspend the cell pellet into a fine cell suspension in the remaining supernatant using the tip of a Pasteur pipet.
10. The sample is now prepared for the harvesting of metaphase chromosomes (*see* **Subheading 3.4.**).

3.1.4. Synchronized Cell Culture

1. Dispense approx 4.0×10^6 cells aseptically into a sterile 25-cm^3 tissue culture flask containing 10.0 mL prepared culture media.
2. Add 200 µL of either phytohemagglutinin or pokeweed mitogen to the culture flask.
3. Incubate at 37°C in a 5% CO_2 atmosphere (*see* **Notes 4** and **5**).
4. Tap flask firmly against the hand or lab bench daily to resuspend the contents.
5. Forty-eight hours following culture initiation, add methotrexate (0.45 µg/mL final concentration) ***(1,3)***.
6. To release the methotrexate block, 17 h later, decant the contents of the tissue culture flask into a 15-mL conical centrifuge tube.
7. Centrifuge the tube at 100*g* for 10 min.
8. Discard the supernatant and gently resuspend the cell pellet in 10 mL of freshly prepared medium.
9. Centrifuge the tube at 100*g* for 10 min.
10. Discard the supernatant and gently resuspend the cell pellet in 10 mL of freshly prepared medium.
11. Add thymidine (0.24 µg/mL final concentration) ***(1,3)***.
12. Incubate the tube at 37°C and 5 h later, add dilute colcemid (0.05 µg/mL final concentration) ***(1,3)***.
13. After 20 min, centrifuge the tube for 10 min at 100*g*.
14. Discard all but 1 mL of the supernatant.
15. Gently resuspend the cell pellet into a fine cell suspension in the remaining supernatant using the tip of a Pasteur pipet.
16. The sample is now prepared for the harvesting of metaphase chromosomes (*see* **Subheading 3.4.**).

3.2. Solid Tissue Culture (see Note 1)

3.2.1. Specimen Collection, Transport, and Preparation

1. Immediately following biopsy/surgery or collection of a serous effusion, the tissue specimen should be placed in a conical centrifuge tube containing 5 mL of transport media for every 1 mL of specimen (*see* **Note 1**).
2. Transport the sample to the laboratory as quickly as possible. If the sample can be received by the laboratory within 24 h, maintain the sample at room temperature; otherwise, the sample may be stored in the transport media (*see* **Subheading 2.1.**) for up to 3 d at 37°C.
3. Add 2 mL of PBS to a 4.5-cm diameter disposable Petri dish.
4. Centrifuge the specimen tube at 100*g* for 10 min, discard the supernatant, and transfer the tissue pellet to the Petri dish (for serous effusions, proceed to **step 8**).
5. Using sterile instruments, cut the tissue using a scalpel blade until all fragments are ≤1 mm in diameter. Attempt to obtain precise smooth cuts to avoid tearing the tissue. All surgical utensils should be kept in a beaker of 70% ethanol and flamed before each use. Carefully wash hands and cleanse the area well. Gloves and a mask will help cut down on contamination. Keep talking to a minimum during culture preparation. Preparation should be carried out under a sterile hood.
6. Collect the cells in suspension into a sterile 15-mL conical centrifuge tube containing on RPMI-1640 medium. Fill the tube with additional media to a final volume of 10 mL.
7. Centrifuge at 100*g* for 10 min, discard the supernatant, and resuspend in 10 mL of fresh complete medium.
8. Determine the concentration of cells within the suspension (*see* **Subheading 3.5.**).

3.2.2. Rapid Preparation of Metaphase Chromosomes from Solid Tissues and Effusions

1. Use methods as described in **Subheading 3.3.**

3.2.3. Nonsynchronized Tissue Culture

1. Dispense approximately 4.0 × 10^6 cells aseptically into a sterile 25 cm^3 tissue culture flask containing 10.0 mL prepared culture media.
2. Incubate at 37°C in a 5% CO_2 and 97% humidified atmosphere and check daily for growth. Replenish the media at least twice weekly.
3. When the surface of the tissue culture flask is 60–70% confluent with cells, add colcemid (0.2 μg/mL final concentration) ***(1)***.
4. Twenty-four hours later, tap flask firmly against the hand or lab bench to dislodge cells. Decant all media into a 15-mL conical centrifuge tube.
5. Rinse the flask with 2 mL of PBS to remove remaining media (serum in the media contains α_1-antitrypsin). Decant the PBS wash into the centrifuge tube.
6. Place 2 mL of 0.025% trypsin into the flask (just so it covers the bottom of the flask). Place the flask into a 37°C incubator for 1–5 min. Check the flask to see if cells have been dislodged (flask may be tapped firmly against the hand or lab

bench to help dislodge cells). If cells are not dislodging from the bottom of the flask, rinse again in PBS and add more trypsin.

7. Decant the trypsin cell suspension into the conical centrifuge tube.
8. Centrifuge the tube for 10 min at 100*g*.
9. Discard all but 1 mL of the supernatant.
10. Gently resuspend the cell pellet into a fine cell suspension in the remaining supernatant using the tip of a Pasteur pipet.
11. The sample is now prepared for the harvesting of metaphase chromosomes (*see* **Subheading 3.4.**).

3.2.4. Synchronized Tissue Culture

1. Dispense approximately 4.0×10^6 cells aseptically into a sterile 25-cm^3 tissue culture flask containing 10.0 mL prepared culture media.
2. Incubate at 37°C in a 5% CO_2 and 97% humidified atmosphere and check daily for growth. Replenish the media at least twice weekly.
3. When the surface of the tissue culture flask is 60–70% confluent with cells, add methotrexate (0.45 µg/mL final concentration) ***(1,3)***.
4. To release the methotrexate block, 16 h later, tap flask firmly against the hand or lab bench to dislodge cells. Decant all media into a 15-mL conical centrifuge tube.
5. Rinse the flask with 2 mL of PBS to remove remaining media (serum in the media contains α_1-antitrypsin). Decant the PBS wash into the centrifuge tube.
6. Place 2 mL of 0.025% trypsin into the flask (just so it covers the bottom of the flask). Place the flask into a 37°C incubator for 1–5 min. Check the flask to see if cells have been dislodged (flask may be tapped firmly against the hand or lab bench to help dislodge cells). If cells are not dislodging from the bottom of the flask, rinse again in PBS and add more trypsin.
7. Decant the trypsin cell suspension into the centrifuge tube.
8. Centrifuge the tube for 10 min at 100*g*.
9. Discard the supernatant and gently resuspend the cell pellet in 10 mL of freshly prepared medium.
10. Centrifuge the tube at 100*g* for 10 min.
11. Discard the supernatant and gently resuspend the cell pellet in 10 mL of freshly prepared medium.
12. Add thymidine (0.24 µg/mL final concentration) ***(1,3)***.
13. Incubate the tube at 37°C and 5 h later, add dilute colcemid (0.05 µg/mL final concentration) ***(1,3)***.
14. After 20 min, centrifuge the tube for 10 min at 100*g*.
15. Discard all but 1 mL of the supernatant.
16. Gently resuspend the cell pellet into a fine cell suspension in the remaining supernatant using the tip of a Pasteur pipet.
17. The sample is now prepared for the harvesting of metaphase chromosomes (*see* **Subheading 3.4.**).

3.3. Rapid Preparation of Metaphase Chromosomes

1. Dispense approximately 4.0×10^6 cells aseptically into a sterile 25-cm^3 tissue culture flask containing 10.0 mL prepared culture media.
2. Add colcemid (0.2 µg/mL final concentration) ***(4,5)***.
3. Incubate at 37°C in a 5% CO_2 and 97% humidified for 30 min and subsequently at 10–12°C for 2–3 h. During this period cells will be able to enter mitosis, but chromosome contraction will be delayed or partly inhibited by the low temperature ***(4,5)***.
4. Decant the cells into a 15-mL conical centrifuge tube.
5. Centrifuge the tube for 10 min at 100*g*.
6. Discard all but 1 mL of the supernatant.
7. Gently resuspend the cell pellet into a fine cell suspension in the remaining supernatant using the tip of a Pasteur pipet.
8. The sample is now prepared for the harvesting of metaphase chromosomes (*see* **Subheading 3.4.**).

3.4. Harvesting of Metaphase Chromosomes

1. Slowly add the fine cell suspension, drop by drop, to a new 15-mL conical centrifuge tube containing 10 mL of hypotonic and incubate at 37°C for 20 min (*see* **Note 6**) ***(1)***.
2. Gently layer 1 mL of Carnoy's fixative on top of the hypotonic mixture, slowly invert the tube to arrest the hypotonic process, and let stand an additional 5 min at room temperature.
3. Centrifuge the tube at 100*g* for 10 min.
4. Discard all but 1 mL of the supernatant.
5. Gently resuspend the cell pellet into a fine cell suspension in the remaining supernatant using the tip of a Pasteur pipet (*see* **Note 7**).
6. Slowly add the cell suspension, drop by drop, to a new 15-mL conical centrifuge tube containing 10 mL of fixative and let stand for 10 min at 4°C ***(1,4,6)***.
7. Centrifuge the tube at 100*g* for 10 min.
8. Discard all but 1 mL of the supernatant.
9. Gently resuspend the cell pellet into a fine cell suspension in the remaining supernatant using the tip of a Pasteur pipet.
10. Slowly add down the side of the centrifuge tube, drop by drop, 10 mL of fixative and let stand for 10 min at 4°C.
11. Repeat **steps 7–10** until the cell pellet appears "snowy" or "fluffy" and is white in color ***(1,4,6)***.
12. Slides should be made as quickly as possible. If slides are to be made at a later time, the fixed cell pellet may be stored in a 10 mL suspension of fresh fixative in a sealed 15-mL conical centrifuge tube at 4°C for a few weeks ***(1,4,6)***.

3.5. Determining Cell Concentration in a Suspension

1. Take 0.1 mL of a cell suspension and dilute to 2.0 mL using a graduated pipet.
2. With the cover slip firmly adhered to a hemocytometer slide, add one drop of the cell suspension to each side of the slide and count the cells in the four outer corner squares.

3. Divide the total cell number by four to obtain the average number of cells in each square. This number is the number of cells × 10^4. Multiply this by 20 to correct for the initial dilution. This figure gives the number of cells per mL of the original suspension.
4. If there are too few cells on the hemocytometer grids to provide a reliable cell count determination, dilute the 0.1 mL aliquot of the original suspension with less medium (e.g., 0.5 or 1.0 mL). If there are still too few cells to evaluate, centrifuge the original suspension, discard the supernatant, and resuspend the cells in less medium.

4. Notes

1. The most common reason for solid tissue culture failure is the tissue sample did not contain a sufficient number of viable cells (samples sizes <0.2 mm^2 are very difficult to grow). Meticulous care in the collection, transport, and preparation of the sample is required. Other common reasons include the selection of an unsuitable cell type that is composed entirely of cells which do not undergo active division (i.e., terminally differentiated nonneoplastic normal tissue) or cells which have had prior in vivo exposure to drugs or disease. Removing the cells from the influence of their own plasma by washing them with medium or a balanced salt solution may improve culture efficiency.
2. Whole blood microculture is the simplest, most widely used method of cell culture for routine cytogenetic purposes ***(1,2)***. There appears to be little advantage in separating white blood cells except, perhaps, with difficult cord or stillbirth blood, or in certain disease situations (i.e., leukemias). The erythrocytes and other blood cell elements seldom interfere with division of lymphocytes in culture (i.e., granulocytes degenerate after 48 h). Occasionally excess RBCs can be detrimental to the culture of lymphocytes because they metabolize vital nutrients in the medium. This problem has been overcome by using a large volume of medium and a relatively small volume of blood.
3. The exact mechanism of phytohemagglutinin- and pokeweed-stimulated lymphocytic proliferation is unknown; however, the effect is only necessary for the first 24 h of the induction process. Both B- and T-lymphocytes respond to the stimulation. The difference between the two mitogens lies entirely in the fact that they initially stimulate different subpopulations or clones of T-lymphocytes ***(7)***. During the next 24 h, the nucleus enlarges and DNA synthesis begins. The first mitoses are seen around 48 h, with waves at 24 h intervals. Thus cultures are normally harvested at 48, 72, or 96 h although divisions will be seen any time after 48 h. In the initial 24 h, the lymphocytes produce interleukin-2 (IL-2), which perpetuates the mitotic process in a chain reaction manner ***(7)***.
4. There have been several recent modifications to the standard cytogenetic culture method. These are aimed at allowing analysis of chromosomes at earlier stages of cell division in order to gain additional information from more elongated chromosomes which permit high resolution banding ***(8)***. The limiting factor in standard nonsynchronized culture is the fact that the proportion of early metaphase

and prometaphase cells is low compared with mid- and late-metaphase cells, because of the nature of the colcemid arrest *(8)*.

Arresting agents which specifically stop prophase and prometaphase have not been identified, but it is possible to introduce a chemical block at an earlier stage of the cell cycle (i.e., prior to DNA synthesis), so that when cultures are subsequently released from the block, the cells proceed in synchrony to complete division (i.e., mitosis) *(3)*. Methotrexate, a commonly used cell synchronizing agent, is an analog of folic acid with a higher affinity for dihydrofolate reductase than folic acid, so that the synthesis of folinic acid is potentially inhibited *(3)*. Folinic acid is required for the production of thymidine which in turn is required for DNA synthesis *(3)*. The cells are therefore blocked prior to DNA synthesis at the G1/S interface. The low thymidine content makes RPMI-1640 a suitable medium for methotrexate-block cell synchronization *(3)*. Cells are released from the methotrexate block by washing and adding thymidine. Alternatively, the thymidine analog bromodeoxyuridine (BrdU) can be effectively substituted for thymidine to release the methotrexate block. BrdU has the added advantage of mildly inhibiting chromosome condensation. The optimum period between release of the block and harvesting must be precisely determined. Although reports vary, the optimum period appears to be 5 h *(8)*. Careful attention to the use of synchronizing agents and their timing allows the harvest of cultures with a high proportion of prophase, prometaphase or early metaphase cells.

5. The recommended use of colcemid with regard to exposure time and concentration varies in the literature. This variation may reflect species and cell specific differences in threshold values. To obtain the long, thin prophase or prometaphase chromosomes, short exposure time and low concentration are normally recommended. Because the threshold of colcemid action is rather sharp, small variations in concentration above the threshold seem to have no real significance, but exposure below results in incomplete spindle inhibition, which may be interpreted as an increase of prophase cells in the preparations *(7)*.
6. Hypotonic treatment prior to fixation is necessary to ensure proper visualization of chromosomes with minimal overlapping of material. Any hypotonic solution induces the swelling of animal cells; however, chromosome contraction and morphology are influenced by the type and concentration of the salt. The standard KCl hypotonic solution has proven sufficient for this purpose.
7. Metaphase chromosomes are mainly composed of DNA, nonhistone and histone proteins. Methanol in the 3:1 methanol-acetic acid fixative denatures and precipitates most of the histone and some of the nonhistone proteins by dehydration *(6)*. Furthermore, the acetic acid coagulates nucleoproteins and causes swelling of the cells, thus counteracting the shrinking caused by the methanol. The fixative penetrates the cells rapidly, preserves the chromosome structure, and to a large extent, strips cytoplasmic proteins from cells *(6)*. The methanol-acetic acid fixation does enhance Giemsa staining and prolonged exposure times appear to induce conformation changes that result in longer and more segmented chromosomes *(6)*.

In whole blood microcultures, hemolysis of the remaining red cells occurs at the first fixation, when red hemoglobin is visibly converted to dark brown hematin. Lack of adequate and careful mixing results in solid brown lumps in the culture and ultimately gives dirty preparations with a low mitotic index. Should any lumps appear, they should be removed with a pipet and up to six fixations may be necessary to clear the metaphase cells of coating substances *(6)*.

Timing between changes in fixative is not critical after the first fixation, providing cells are in freshly prepared fixation for slide preparation. Old fixative will contain large amounts of methyl acetate which is an inadequate fixative *(6)*. Insufficient fixation interferes with the flattening of the preparations on the slides, and coating substances interfere with banding.

References

1. Hahn, K. A., Richardson, R. C., Hahn, E. A., and Chrisman, C. L. (1994) The diagnostic and prognostic importance of cytogenetic aberrations identified in spontaneously occurring canine malignant lymphoma. *Vet. Path.* **31,** 528–540.
2. Morehead, P. S., Nowell, P. C., Mellman, W. J., Battips, D. M., and Hungerford, D. A. (1960) Chromosome preparations of leukocytes cultured from human peripheral blood. *Exp. Cell Res.* **20,** 613–120.
3. Ronne, M., Anderson, O., and Hansen, S. O. (1984) Methotrexate-leucovorin synchronization of human lymphocyte cultures: induction of high resolution R- and G-banding. *Anticancer Res.* **4,** 357–360.
4. Ronne, M. (1984) An easy method for instant preparation of chromosome slides from solid tumors. *Anticancer Res.* **4,** 45,46.
5. Macera, M. J., Szabo, P., and Verma, R. S. (1989) A simple method for short term culturing bone marrow and unstimulated blood from acute leukemias. *Leukemia Res.* **13,** 729–734.
6. Islam, M. Q. and Levan, G. (1987) A new fixation procedure for improved quality G-bands in routine cytogenetic work. *Hereditas* **107,** 127–130.
7. Ronne, M. (1989) Chromosome preparation and high resolution banding techniques: a review. *J. Dairy Sci.* **72,** 1363–1377.
8. Drouin, R., Lemieux, N., and Richer, C. L. (1988) High-resolution R-banding at the 1250-band level: technical considerations on cell synchronization and R-banding (RHG and RBG). *Cytobios* **56,** 107–125.

14

Chromosome Staining and Banding Techniques

Kevin A. Hahn and Penny K. Riggs

1. Introduction

Each chromosome in the somatic-cell complement can be uniquely identified by following a number of different banding procedures. The banding patterns are highly characteristic. The International System for Cytogenetic Nomenclature (ISCN) provides schematic representations, or ideograms, of human chromosomes corresponding to approx 400, 550, and 850 bands per haploid set *(1)*. Although under constant revision, its principles rest on a numbering system based on major bands as they appear from the centromere outward along each chromosome arm. Similar standards have been established for other mammalian species, and recent literature should be reviewed for appropriate standards and revisions before attempting to karyotype a metaphase specimen.

To the cytogeneticist, the appearance of well-prepared, clearly banded chromosomes has an aesthetic appeal that is often difficult for the noncytogeneticist to comprehend. In part, this may be attributable to steps in some procedures that have no obvious scientific explanation but that nevertheless do materially affect results. Many published staining methods devised in one laboratory require modification in another laboratory. Despite these somewhat mystical aspects of the craft, rigid adherence to times, concentrations, temperatures, and pH can result in methods that are highly reproducible and reliable *(2)*.

The finding of a chromosome abnormality does not always imply multiple defects in morphogenesis, growth disturbances, and mental retardation in a patient. Some anomalies cause no harm. Their effects on the phenotype obviously depend on both the quality and quantity of the genetic material involved.

The methods described in this chapter represent procedures that the authors have found useful in their laboratories. Optimal use of the microscope and good photographic procedures are essential.

From: *Methods in Molecular Medicine, Vol. 14: Tumor Marker Protocols*
Edited by: M. Hanausek and Z. Walaszek © Humana Press Inc., Totowa, NJ

2. Materials

2.1. Slide Preparation

1. Absolute methanol.
2. Deionized or distilled water.
3. Microscope slides.
4. Nonsterile 2–4-mL Pasteur pipets.

2.2. Solid Staining

1. 0.025*M* Phosphate buffer (pH 6.8): 0.025*M* KH_2PO_4 (3.4 g/L) titrated to pH 6.8 with 50% NaOH. Make fresh as required.
2. 10% Giemsa stain: 5 mL of Giemsa (Gurr's) plus 45 mL of 0.025*M* phosphate buffer (pH 6.8). Make fresh as required.

2.3. Giemsa Banding (G-Bands)

1. 0.025*M* Phosphate buffer (pH 6.8): 0.025*M* KH_2PO_4 (3.4 g/L) titrated to pH 6.8 with 50% NaOH. Make fresh as required.
2. Deionized or distilled water.
3. 10% Hydrogen peroxide: 33 mL 30% H_2O_2 with 67 mL distilled or deionized water. Maintained at 4°C. Make fresh as required.
4. 0.025% Trypsin (Grand Island Biologic Company, Grand Island, NY): 5 mL of 0.25% trypsin to 45 mL of 0.025*M* phosphate buffer, pH 6.8. Maintain at 4°C. This solution must be used immediately or replaced after 30–60 min of use.
5. 0.02% Fetal bovine serum (FBS): 1 mL serum added to 50 mL phosphate buffer (pH 6.8), maintained at 4°C. Make fresh as required.
6. 10% Giemsa stain: 5 mL of Giemsa (Gurr's) plus 45 mL of 0.025*M* phosphate buffer (pH 6.8). Make fresh as required.

2.4. Reverse Banding (R-Bands)

1. Sörensen's buffer, solution A: 0.5*M* KH_2PO_4 (6.8 g/100 mL deionized or distilled water). Stable at room temperature for 1 mo.
2. Sörensen's buffer, solution B: 0.5*M* Na_2HPO_4 (7.1 g/100 mL deionized or distilled water). Stable at room temperature for 1 mo.
3. Sörensen's buffer (pH 6.8): 31.4 mL of Sörensen's buffer solution A, 22.8 mL of Sörensen's buffer solution B, 945.8 mL deionized or distilled water. Stable at room temperature for 1 mo.
4. Sörensen's buffer (pH 8.0): 2.8 mL of Sörensen's buffer solution A, 32.4 mL of Sörensen's buffer solution B, 964.8 mL deionized or distilled water. Stable at room temperature for 1 mo.
5. Hoechst 33258 (Sigma, St. Louis, MO): 1 mg Hoechst in 1 L Sörensen's buffer (pH 6.8). Make fresh as required.
6. 2X SSC: 0.3*M* NaCl, 0.03*M* trisodium citrate. Make fresh as required.
7. 3% Giemsa stain: 3 mL Gurr's Geimsa in 97 mL Sörensen's, pH 8.0. Make fresh as required.

3. Methods

3.1. Slide Preparation (see Note 1)

1. Soak new microscope slides in absolute methanol overnight.
2. Rinse slides three times in deionized water.
3. Slides can be stored in water and used wet or dry depending on preference.
4. Centrifuge the cell suspension containing metaphase chromosomes (*see* Chapter 13 in this volume) at 100*g* for 10 min.
5. Discard all but 1–2 mL of the supernatant.
6. Gently resuspend the cell pellet into a fine cell suspension in the remaining supernatant using the tip of a Pasteur pipet.
7. Aspirate a small amount of cell suspension into a Pasteur pipet and expel about three drops carefully in three different positions on each slide.
8. Place the slide at a 45° angle and let the slide air-dry. Spreading is achieved by the movement of the periphery of the drop outward until air-dried.

3.2. Solid Staining (see Note 2)

1. Place air-dried slides in the Giemsa stain for 8 min.
2. Rinse the slides twice in deionized or distilled water.
3. Air-dry.
4. Mount, if necessary, with a cover slip.

3.3. Giemsa Banding (G-Bands) (see Note 3)

1. Dry the slides on a 60°C warming tray or incubator for at least 4 h prior to staining.
2. Immerse the slide into a 10% hydrogen peroxide solution for 15 s, rinse in deionized or distilled water and drain slide well (shake off excess water). Cytoplasm that may cover metaphase chromosomes will be removed by this procedure and permit better exposure of the chromatin to the trypsin treatment *(2)*. This will result in more consistent staining of the slides prepared from different samples.
3. Immerse the slide into the trypsin solution for about 10–15 s. This time will vary considerably depending on the quantity of sample on the slide and the activity of the trypsin. Therefore, use test slides to determine optimal time of trypsin exposure and concentration *(2)*.
4. Immerse the slide 5–7 times in FBS solution (serum in the media contains α_1-antitrypsin to arrest the digestion process). Longer treatment at this step may adversely affect banding *(2)*.
5. Rinse the slide with phosphate buffer.
6. Place the slide in Giemsa stain for about 8–10 min. Time may vary.
7. Rinse the slide with phosphate buffer.
8. Rinse the slide with deionized or distilled water.
9. Allow slide to air-dry in a vertical position.
10. Mount, if necessary, with a cover slip.

3.4. Reverse Banding (R-Bands) (see Note 4)

1. Dry slides for at least 1 wk at room temperature or dry overnight on a 60°C slide warmer.
2. Immerse the slides in Hoechst solution for 30 min at room temperature *(3,4)*.
3. Add fresh Hoechst solution to slide and cover with cover slip.
4. Illuminate the slides under uv light for 30 min. The uv lamp should be 2.5 cm from the slide *(3,4)*.
5. Rinse the slides in 2X SSC.
6. Incubate for 60–90 min in 2X SSC at 65°C. Tap occasionally to dislodge bubbles *(3,4)*.
7. Rinse the slides in Sörensen's phosphate buffer, pH 8.0.
8. Stain with 3% Giemsa stain for 10 min.
9. Rinse the slides three times in Sörensen's buffer, pH 8.0, and twice in distilled water.
10. Air-dry slides at room temperature for 30 min and then on a 50°C slide warmer for 1 h.
11. Mount, if necessary, with a cover slip.

4. Notes

1. Laboratories vary in their preparation of microscope slides. Some use slides straight from the manufacturer's box, whereas others soak slides in alcohol, fixative, ether, or chromic acid, and dry and polish slides prior to use. Some use a detergent to remove all traces of grease; however, the detergent may also leave a "coating layer" on the slide. Whether pretreated for extra cleanliness or not, slides should be clean and grease-free to ensure good spreading of chromosomes. There are many variations of the spreading method described in **Subheading 3.1.** The quality of spreading may be influenced by temperature; high temperatures may cause overspreading of chromosomes and cell breakage, whereas low temperatures may inhibit spreading. This is caused, in part, by the different rates of evaporation of the fixative *(3)*. Additionally, chromosome spreading quality may be improved by varying the height from which the cell suspension is dropped onto the slide. Solid-stain a representative slide (**Subheading 3.2.**) and observe for metaphase cells. If protein-stained debris obscures the visualization of chromosomes, recentrifuge the cell suspension, discard all but 1 mL of the supernatant, resuspend the cells in fresh fixative, let stand for 10 min at room temperature, centrifuge, discard all but 1 mL of the supernatant, and make another slide. Once conditions are appropriate (i.e., metaphase chromosomes with minimal overlap and crisp solid-stained chromosomes), make a minimum of 10 nonstained slides for chromosome banding. The cell pellet can then be maintained for 4–6 wk in a sealed centrifuge tube kept under refrigeration.
2. Staining procedures that provide a uniform unbanded appearance to chromosomes are referred to as solid or conventional staining. Although banded chromosome studies are far more informative, solid-stained preparations can be useful for studies on chromosome breakage since scoring gaps and breaks can be diffi-

cult in lightly stained chromosomes. Slides can be destained by soaking in Carnoy's fixative (three parts absolute methanol and one part glacial acetic acid) and subsequently stained by another technique.

3. Giemsa banding (G-banding) has become the most widely used technique for the routine staining of mammalian chromosomes. The most usual methods to obtain this staining are to treat the slides with a protease, such as trypsin, or incubate the slides in hot saline–citrate, although a variety of other methods have been used. The quality of banding is greatly influenced by the trypsinization procedure ***(2)***. Slides should be monitored as they are prepared since it may be necessary to vary the length of trypsin exposure or Giemsa staining time.
4. Bands that are negative, which appear pale by G-banding, stain darkly by R-banding. Conversely, dark positive G-bands appear pale using R-banding techniques. R-banding can be achieved by incubation in hot saline solution followed by Giemsa staining. Although the pattern of staining appears to reflect the structural and functional composition of chromosomes, the chemical basis for the staining reactions remains obscure ***(3,4)***.

References

1. Harnden, D. G. and Klinger, H. P. (1985) ISCN *International System for Human Cytogenetic Nomenclature.* Karger & Basel, New York, pp. 1–117.
2. Hahn, K. A., Richardson, R. C., Hahn, E. A., and Chrisman, C. L. (1994) The diagnostic and prognostic importance of cytogenetic aberrations identified in spontaneously occurring canine malignant lymphoma. *Vet. Path.* **31,** 528–540.
3. Ronne, M. (1989) Chromosome preparation and high resolution banding techniques: a review. *J. Dairy Sci.* **72,** 1363–1377.
4. Drouin, R., Lemieux, N., and Richer, C. L. (1988) High-resolution R-banding at the 1250-band level: technical considerations on cell synchronization and R-banding (RHG and RBG). *Cytobios* **56,** 107–125.

15

Fluorescence *In Situ* Hybridization to Chromosomes

Penny K. Riggs and Kevin A. Hahn

1. Introduction

Development of the technique of in situ hybridization (ISH) propelled the merger of the fields of molecular genetics and cytogenetics. Refinement of the technique through the use of fluorescence (FISH) produced a remarkably powerful research and diagnostic tool.

As early as 1969, Gall and Pardue hybridized tritiated RNA probes to amplified rRNA genes in nuclei of *Xenopus* oocytes ***(1)***. They subsequently reported hybridization of a centromeric repeat to mouse metaphase chromosomes in 1970 ***(2)*** These two papers marked the beginning of ISH, but another decade passed before single copy genes were localized on metaphase chromosomes, and those procedures required isotopically labeled probes ***(3,4)***.

Langer-Safer et al. ***(5)*** developed a method of tagging nucleic acids with biotin and devised a protocol for visualizing a probe hybridized to chromosomes. They produced rabbit anti-biotin antibodies which bound to the biotin-labeled probe hybridization site, then added a layer of fluorescein-labeled antirabbit antibodies. This immunological sandwich resulted in a fluorescent signal that allowed the site of hybridization to be seen directly on *Drosophila* polytene chromosomes. Variations on Langer Safer's work were soon published ***(6)***, and in 1988, FISH was used to demonstrate the integration sites of Epstein-Barr viral DNA in human chromosomes ***(7)***.

For localization of individual genes, however, FISH lacked the sensitivity of isotopic methods, until chromosomal *in situ* suppression (CISS) was introduced ***(8)***. This method enabled cytogeneticists to obtain bright signals by hybridizing large, genomic DNA fragments to unique sequences on chromosomes and suppress hybridization to repetitive elements scattered throughout

From: *Methods in Molecular Medicine, Vol. 14: Tumor Marker Protocols*
Edited by: M. Hanausek and Z. Walaszek © Humana Press Inc., Totowa, NJ

the genome. Since then, innumerable variations and applications of FISH have been described, including hybridization to interphase DNA ***(9)***, development of chromosomal painting probe ***(10–12)***, and polymerase chain reaction (PCR)-based methods ***(13–17)***. Additional reviews on *in situ* hybridization can be found in **refs.** ***18–22***.

2. Materials

2.1. Probe Labeling with Digoxigenin (see Note 1)

1. 10X Nick translation buffer: 500 m*M* Tris-HCl, pH to 7.8, 50 m*M* magnesium chloride, 0.5 mg/mL nuclease-free bovine serum albumin (BSA) (all reagents from Sigma, St. Louis, MO). Filter sterilize, and store in 1 mL aliquots at –20°C.
2. 100 m*M* Dithiothreitol (DTT) (Sigma): rehydrate in sterile, distilled water, store in 1 mL aliquots at –20°C.
3. Dig-nucleotide mix: 0.2 m*M* dATP, 0.2 m*M* dGTP, 0.2 m*M* dCTP, 0.08 m*M* dTTP, 0.2 m*M* digoxigenin-11-dUTP. For best results, purchase nucleotides as 100 m*M* lithium salt solutions (Boehringer-Mannheim Biochemicals or Pharmacia), and alkali-stable dig-11-dUTP in 1 m*M* solution, dilute in sterile, distilled water, and store in small (100 μL) aliquots at –20°C. Avoid freeze–thaw cycles, and keep on ice when in use.
4. DNase I lyophilized (Boehringer-Mannheim Biochemicals): rehydrate in sterile, distilled water to 1 mg/mL (~2000 U/mL). Dilute working stock to 5 U/mL, store at –20°C.
5. DNA polymerase I (Boehringer-Mannheim Biochemicals): 10 U/μL. Store at –20°C
6. 500 m*M* ethylenediamine tetra-acetic acid (EDTA) (Sigma), pH 8.0; filter sterilize and store at 15–30°C.
7. Genius system labeling and detection kit (optional) (Boehringer-Mannheim Biochemicals).

2.2. Purification of Probe by Ethanol Precipitation

1. Glycogen solution (Boehringer-Mannheim Biochemicals): 20 mg/mL in sterile, distilled water. Store at –20°C.
2. 4 *M* Lithium chloride (Sigma): filter sterilize, store at 15–30°C.
3. Absolute ethanol, store at –20°C.
4. TE/SDS buffer: 10 m*M* Tris-HCl, pH 7.5, 1 m*M* EDTA, 0.1% sodium dodecyl sulfate (SDS). All reagents may be from Sigma.

2.3. Preparation of Chromosomes

1. Microscope slides and glass cover slips, clean cover slips with ethanol and lint-free wipes.
2. Soft plastic cover slips (can cut to size from parafilm).
3. Nucleic acid grade deionized formamide (Oncor, Gaithersburg, MD). Store in dark at 4°C. Wear gloves when handling.

4. 20X SSC: 3 *M* NaCl, 0.3 *M* sodium citrate, pH 7.0. Autoclave, store at 15–30°C; dilute as needed to 2X or 4X with sterile water.
5. Absolute ethanol, and ethanol diluted with water to 70% and 85%. Store at –20°C.

2.4. Preparation of Probe

1. Sheared herring sperm DNA (Oncor): prepare 10 mg/mL. Store at –20°C.
2. 50% dextran sulfate (Oncor).
3. 2X SSX, pH 7.0 (make as in **Subheading 2.3.4.**).
4. Nucleic acid grade, deionized formamide.
5. Total genomic DNA or CoT1 fraction of genomic DNA.
6. Absolute ethanol. Store at –20°C.

2.5. Hybridization

1. Rubber cement.
2. 3 mL plastic syringe.
3. Humidified chamber.

2.6. Posthybridization Wash and Signal Detection

1. 50% Formamide/2X SSC, pH 7.0: use 1 part formamide, 1 part 4X SSC, filter sterilize, store at 4°C for up to 1 wk.
2. 2X SSC, pH 7.0.
3. 4X SSC/0.1% Tween-20 wash solution or PBD (Oncor).
4. Fluorescein-labeled antidigoxigenin antibody (Oncor) ***(24,25)***.
5. Rabbit antisheep antibody (Oncor).
6. Fluorescein-labeled antirabbit antibody (Oncor).
7. Antifade mounting medium ***(26,27)***: titrate 0.5 *M* sodium bicarbonate buffer to pH 9.0 with NaOH. Titrate 10 mL phosphate-buffered saline (PBS) to pH 8.0 with 0.5 *M* sodium bicarbonate solution. Add 100 mg p-phenylenediamine to 10 mL PBS and 90 mL glycerol. Aliquot to 1 mL tubes and store in dark at –20°C for up to 1 yr. Solution may darken, but is still usable.
8. Hoescht dye/antifade solution: prepare 50 µg/mL Hoechst 33258 in 2X SSC, pH 7.0, filter sterilize, and store in dark at 4°C. Mix 500 µL antifade mounting medium with 500 µL 4X SSC and 20 µL Hoescht 33258 solution. Store in dark at –20°C for 1–2 d. **Warning: carcinogen.**
9. Propidium iodide/antifade (molecular probes): dissolve 10 mg propidium iodide (PI) in 10 mL sterile distilled water. Dilute 10 µL PI in 10 mL antifade solution described above. Store in 1 mL aliquots in dark at –20°C for up to 1 yr. **Warning: carcinogen.**

Mention of commercial products does not imply any type of endorsement of said products. Specific products used in our labs are mentioned as examples of items used successfully in these procedures in our labs.

3. Methods

3.1. Preparation of Probe (Modified from Refs. 15–17)

3.1.1. Choice of Probe (see **Note 1**)

The type of probe used for FISH may vary depending upon the goal of the diagnostic test. This section will use the general example of a unique-sequence probe, while alternatives are discussed in the notes section (*see* **Notes 1–4**). Obtain microgram quantities of high quality, purified cosmids containing genomic DNA inserts for preparation of FISH probes.

3.1.2. Choice of Label (see **Note 2**)

Many options are available for labeling probes. At present the two most commonly used are biotin and digoxigenin *(24,25)*. These compounds are incorporated into probe DNA and hybridization is detected by posthybridization application of antibodies conjugated to fluorochromes such as fluorescein or rhodamine. This section will describe preparation of a digoxigenin-labeled probe by Nick translation.

3.1.3. Incorporation of Digoxigenin-11-dUTP into FISH Probes by Nick Translation (see **Note 3**)

1. Add the following reagents to a sterile microfuge tube on ice:
   10X Nick translation buffer 5 μL
   100 m*M* DTT, 5 μL;
   Dig-nucleotide mix, 4 μL;
   Cosmid DNA containing probe insert, n μL (1 μg) (*see* **Note 1**);
   DNase I (~5 mU/μL), 2 μL;
   DNA polymerase I (10 U/μL), 1 μL.
   Make up to 50 μL with sterile distilled water.
2. Incubate at 15°C for 1 h.
3. Place reaction on ice and remove 7.5 μL (150 ng). Electrophorese aliquot on 1% agarose gel with size standards from 100 bp to 1 kb. Optimal probe size is 200–400 kb *(7)*. If most of the DNA (>75%) is between 100 and 500 bp, stop the reaction by adding 5 μL 500 m*M* EDTA. If much of the DNA is larger than 500 bp, add 1 μL each DNase I and DNA polymerase, and allow the reaction to progress at 15°C for an additional 30–60 min. Repeat above steps until probe is optimal size.
4. Optional: Estimate the yield of Dig-labeled nucleic acids. This is most easily done by using a commercial kit (e.g., Boehringer-Mannheim Genius System labeling and detection kit or, BRL BluGene Nonradioactive Nucleic acid Detection System) and following manufacturer's suggested protocols.
5. The probe can now be purified by ethanol precipitation. Add 1 μL glycogen (20 mg/mL) to the reaction tube. Glycogen is a carrier to improve the recovery of nucleic acids.

6. Precipitate labeled DNA with 0.1 vol LiCl and 2.5 vol cold, absolute ethanol. Mix and incubate at –70°C for 30 min.
7. Thaw at room temperature and microcentrifuge (13,000*g*) for 15 min.
8. Decant supernatant, wash with 100 µL 70% ethanol, and centrifuge for 5 min.
9. Dry pellet briefly, then resuspend in 85 µL TE/SDS buffer at 37°C for 10 min (assume final concentration is now ~10 ng/µL). Use immediately or store at –20°C for up to 1 yr.

3.2. In Situ *Hybridization of Labeled DNA Probes to Chromosomes*

The basic protocol for FISH can be adapted to meet various goals. The following method (modified from **ref. *15***) outlines steps for hybridizing cosmid DNA containing a genomic insert to metaphase chromosomes. Variations for slightly different approaches are also described.

3.2.1. Preparation of Metaphase Chromosomes (see ***Note 4***)

1. Harvest chromosomes, preferably after BrDU incorporation, and prepare slides containing high-quality metaphase chromosomes and good mitotic index. Chromosome preps should be 1–5 d old.
2. Incubate, or "age," slides overnight at 37°C.
3. Denature chromosomes by immersing in freshly made 70% formamide/2X SSC (4 mL 20X SSC, 8 mL distilled water, 28 mL formamide) for 2 min at 70°C, then transfer immediately to 70% EtOH at –20°C. It is convenient to heat the denaturation solution to 72°C in a coplin jar in a water bath. The temperature will drop when the microscope slides are placed into the solution. Denature chromosomes on only one or two microscope slides at a time, and monitor the temperature carefully.
4. Dehydrate slides in EtOH series: 2 min each 70%, 80%, 95%, then prewarm slides to 37°C. Failure to warm the slides sometimes results in precipitation of probe DNA.

3.2.2. Preparation of Probe (see ***Note 5***)

Calculate amount of probe DNA and volume of hybridization solution. This amount must be determined for each probe. Typical amounts are 1–30 ng/µL hybridization solution; 30 µL per slide. For testing new probes, try 5 ng/µL and 10 ng/µL first.

1. Unique sequence genomic DNA probe or genomic painting probes:
   a. Ethanol-precipitate required amount of probe DNA plus sheared herring sperm DNA (use 500X probe concentration) plus sonicated (0.5–2.5 kb) genomic DNA or CoT1 DNA (same species as chromosomes; 100X probe conc.).
   b. Prepare hybridization solution. For 100 µL:
      50% Dextran sulfate, 20 µL;
      2X SSC, 10 µL;

ddH_2O, 20 µL;

Formamide, 50 µL.

(For best results, dissolve DNA in formamide before adding to the rest of the solution.)

c. Mix well, then denature probe for 5 min at 70°C. Transfer probe to 37°C water bath for 15–30 min, then place on ice. This step allows repetitive sequences to bind "cold" DNA, thus suppressing background signals.

2. cDNA probes: For probes which do not require suppression of genomic repeat sequences, omit genomic or CoT1 DNA. After denaturation of probe, immediately place on ice.
3. Repetitive sequence probes: Omit genomic or CoT1 DNA as above, and increase formamide concentration to 65% in hybridization solution.

3.2.3. *Hybridization* (see **Note 5**)

1. Place 10 µL probe solution under 22 × 22 cm coverglass on slide or 30 µL under 24 × 50 cm µL coverglass. Seal edges with rubber cement.
2. Incubate slides in humidified chamber overnight at 37°C.
3. Place slide carrier in Wheaton jar containing 50% formamide/2X SSC.
4. Peel rubber cement off slides and drop into wash solution. Coverglass should fall off in first wash step, or it can be carefully removed before washing.
5. Wash slides 3 times for 4 min in 2X SSC at 40°C. May need to adjust this temperature, depending upon probe. For highly repetitive probes, increase temperature and decrease wash times to 2 min.
6. Wash slides 3 times for 4 min in 50% formamide/2X SSC.
7. Place slides in PBS/0.1% Tween-20 at room temperature.

3.3.4. *Signal Detection* (see ***Notes 6** and **7***)

This section requires the Oncor Dig-FITC detection kit S1331-DF or similar products.

1. Remove slides, one at a time, from PBS and drain briefly but do not allow slide surface to dry.
2. Apply 30 µL fluorescein-labeled antidigoxigenin antibody and cover with plastic cover slip.
3. Incubate in humidified chamber 20 min at 37°C.
4. Wash slides in PBS 3 times for 2 min at room temperature.
5. Apply 30 µL rabbit anitsheep antibody. Repeat **steps 3** and **4**.
6. Apply 30 µL fluorescein-labeled antirabbit. Repeat **steps 3** and **4**.

3.3.5. *Staining*

1. Stain slides with propidium iodide/antifade solution. Apply 15 µL stain and coverglass. Carefully press out all excess stain between two paper towels (wear gloves). Observe under fluorescence microscope using appropriate filter sets.

2. To observe banding on chromosomes containing incorporated BrDU, apply 15 μL Hoechst dye/antifade and proceed as above.
3. Observe fluorescein isothiocyanate (FITC) signal on chromosomes under fluorescence microscope equipped with appropriate filter sets.

4. Notes

1. Numerous applications for chromosomal FISH exist, and the basic protocols can be adapted for specific uses.
   a. Chromosome painting: Labeled genomic DNA can be utilized as a probe for applications such as painting chromosome preparations from somatic cell ***(17)*** or radiation hybrid cell lines. This type of probe allows the chromosomes of a target species to be differentiated from those of the host. Painting probes can also be constructed from flow-sorted chromosomes ***(12)*** or purified DNA from chromosome libraries ***(11)***. Many painting probes are also commercially available.
   b. To hybridize to single-copy genes or other unique DNA sequences, obtain purified cosmids or phagemids containing large (10–40 kb) genomic DNA inserts. Artificial chromosome inserts (BACs or YACs) containing still larger DNA fragments are also useful. Excision of the insert from the vector is unnecessary in most cases. Detection and visualization of hybridization signals is easier when the probe and/or hybridization target is large.
   c. Plasmids containing cDNA or smaller genomic inserts of repetitive sequences or multicopy genes can also be used successfully. Generally, larger signals are observed from larger targets, but hybridization and observation of small unique sequence probes has been reported ***(23)***.
2. Choice of probe label: Biotin (biotin-11-dUTP) ***(25)*** and digoxigenin (digoxigenin-11-dUTP) are most commonly used. In our hands, digoxigenin produces cleaner hybridization with less nonspecific sticking than biotin. For two-color applications, one may choose to utilize two probes: one labeled with biotin, and the other with digoxigenin. Additionally, many PCR-based applications rely on direct incorporation of various fluorochromes ***(13,14)***. Commercial suppliers are rapidly developing novel fluorochromes in various kits for FISH applications. With slight modifications, the single-probe hybridization technique can be adapted for use with multiple probes and fluorochromes ***(24)***.
3. Incorporation of digoxigenin-11-dUTP into FISH probes by nick translation. Best results will be achieved if probe fragment size is monitored. Numerous other types of fluorochromes and labeling methods are available. Nick translation is a common method and gives good results. Methods can be modified to meet specific needs.
4. Chromosome preparation:
   a. Aging: Good quality chromosome preps are a necessity. We prefer to use fresh slides, but slides containing metaphase chromosomes can be stored at –20°C for several months. If chromosomes are not aged enough, they will be "soft" and denaturation will cause excessive damage and poor morphology. Too much aging will result in chromosomes resistant to denaturation.

In both cases, hybridization will be poor, so a balance must be reached experimentally. Aging can be accomplished by storage of slides at 37°C for up to 2 d, incubation at 55°C for 1–2 h, or incubation in 2X SSC for 30–60 min at 37°C.

b. RNase treatment: In most cases, this treatment is unnecessary. However, when hybridizing certain probes likely to bind to RNA transcripts on the slide (e.g., 28S rRNA probes), RNase treatment may help reduce background hybridization. Add 40 µL of 1 mg/mL RNase A to 40 mL 2X SSC and incubate for 1 h at 37°C. Follow with dehydration in ethanol series before hybridization.

c. Denaturation: The duration of denaturation is critical and optimal time must be determined experimentally. Postdenaturation chromosomes that appear "ghost-like" were denatured too long. If chromosomes retain excellent banding morphology, denaturation time should be increased.

5. Probe hybridization:
   a. Application of rubber cement can be done easily and fairly neatly with a 3-mL syringe. Remove plunger and cover small opening. Pour in ~2-mL rubber cement. Carefully reinsert plunger, remove cover, and squeeze cement around edges of slide.
   b. A hybridization chamber can be constructed from a plastic box with tight-sealing lid. Place wet paper towels on the bottom of the box, and place plastic supports or grids on top of towels. Slides should be placed flat on the supports. Seal lid and incubate in 37°C incubator. Wet towels should be replaced after each experiment.
6. Signal detection: We have experienced good results using the Oncor detection kit. Antibodies can also be purchased separately from other vendors. Additionally, several other detection schemes, such as digoxigenin-rhodamine can be used.
7. Other applications: The methods described for FISH to chromosomes can be easily modified for application to tissue sections ***(13,14)***, although specific details are beyond the scope of this chapter. Briefly, tissue should be fixed in 10% formalin prior to sectioning. Removed paraffin by treating with xylene for 5 min, followed by ethanol dehydration. Samples should then be digested with proteinase K, and FISH can proceed essentially as described above.

References

1. Gall, J. G. and Pardue, M. L. (1969) Formation and detection of RNA-DNA hybrid molecules in cytological preparations. *Proc. Natl. Acad. Sci. USA* **63,** 378–383.
2. Pardue, M. L. and Gall, J. G. (1970) *Science* **168,** 1356.
3. Gerhard, D. S., Kawasaki, E. S., Bancroft, F. C., and Szabo, P. (1981) Localization of a unique gene by direct hybridization in situ. *Proc. Natl. Acad. Sci. USA* **78,** 3755–3759.
4. Harper, M. E. and Saunders, G. F. (1981) Localization of single copy DNA sequences on G-banded human chromosomes by in situ hybridization. *Chromosoma* **83,** 431–439.

5. Langer-Safer, P. R., Levine, M., and Ward, D. C. (1982) Immunological method for mapping genes on *Drosophila* polytene chromosomes. *Proc. Natl. Acad. Sci. USA* **79,** 4381–4385.
6. Bhatt, B., Burns, J., Flannery, D., and McGee, J. O. (1988) Direct visualization of single copy genes on banded metaphase chromosomes by nonisotopic in situ hybridization. *Nucl. Acids Res.* **16,** 3951–3961.
7. Lawrence, J. B., Villnave, C. A., and Singer, R. H. (1988) Sensitive, high-resolution chromatin and chromosome mapping in situ: presence and orientation of two closely integrated copies of EBV in a lymphoma line. *Cell* **52,** 51–61.
8. Lichter, P. (1990) High resolution mapping of human chromosome 11 by in situ hybridization with cosmid clones. *Science* **247,** 61.
9. Lawrence, J. B., Singer, R. H., and McNeil, J. A. (1990) Interphase and metaphase resolution of different distances within the human dystrophin gene. *Science* **249,** 928.
10. Koch, J., Hindkjaer, J., Mogensen, J., Kolvraa, S., and Bolund, L. (1991). An improved method for chromosome-specific labeling of alpha satellite DNA in situ by using denatured double-stranded DNA probes as primers in a primed in situ labeling procedure. *GATA* **8,** 171–178.
11. Lengauer, C., Eckelt, A., Weith, A., Endlich, N., Ponelies, N., Lichter, P., Gruelich, K. O., and Cremer, T. (1991). Painting of defined chromosomal regions by in situ suppression hybridization of libraries from laser-microdissected chromosomes. *Cytogenet. Cell Genet.* **56,** 27–30.
12. Telenius, H., Pelmear, A. H. P., Tunnacliffe, A., Carter, N. P., Behmel, A., Ferguson-Smith, M. A., Nordenskjold, M., Prfagner, R., and Ponder, B. (1992). Cytogenetic analysis by chromosome painting using DOP-PCR amplified flow-sorted chromosomes. *Genes Chrom. Cancer* **4,** 257–263.
13. Komminoth, P. and Long A. A. (1995) Review: In situ polymerase chain reaction—methodology, applications, and nonspecific pathways, in *PCR Applications Manual,* Boehringer Mannheim GmbH, Biochemica, Germany, pp. 97–112.
14. Nuovo, G. J. (1994) *PCR In Situ Hybridization: Protocols and Applications,* 2nd ed., Raven Press, Ltd., New York.
15. Briley, G. P., Riggs, P. K., Womack, J. E., Hancock, D. L., and Bidwell, C. A. (1996). Chromosomal localization of the porcine skeletal muscle calpain gene. *Mamm. Genome* **7,** 226–228.
16. Neibergs, H. L., Gallagher, D. S., Georges, M., Sargeant, L. S., Dietz, A. B., and Womack, J. E. (1993). Physical mapping of inhibin-beta-A in domestic cattle. *Mamm. Genome* **4,** 328–332.
17. Pinkel, D., Straume, T., and Gray, J. W. (1986) Cytogenetic analysis using quantitative, high-sensitivity fluorescence hybridization. *Proc. Natl. Acad. Sci. USA* **83,** 2934–2938.
18. Brandiff, B. F., Gordon, L. A., and Trask, B. J. (1991) DNA sequence mapping by fluorescence in situ hybridization. *Env. Mol. Mutagen.* **18,** 259–262.
19. Chrisman, C. L., Briley, G. P., and Waldbieser, G. C. (1991) In situ hybridization and high-resolution banding of chromosomes, in *Gene Mapping Techniques and*

Applications (Schook, L. B., Lewin, H. A., and McLaren, D. G., eds.), Marcel Dekker, Inc., New York.

20. Dyer, K. and Meyne, J. (1991) Molecular cytogenetics: Use of DNA probes as an adjunct to classical clinical cytogenetics, in *The ACT Cytogenetics Laboratory Manual,* 2nd ed. (Barch, M. J., ed.), Raven, New York, pp. 525–537.
21. Lichter, P., Boyle, A., Cremer, T., and Ward, D. C. (1991). Analysis of genes and chromosomes by nonisotopic in situ hybridization. *GATA* **8,** 24–35.
22. Trask, B. J. (1991) Fluorescence in situ hybridization: applications in cytogenetics and gene mapping. *Trends Genet.* **7,** 149–154.
23. Lemieux, N., Dutrillaux, B., and Viegas-Pequignot, E. (1992) A simple method for simultaneous R- or G-banding and fluorescence in situ hybridization of small single-copy genes. *Cytogen. Cell Genet.* **59,** 311,312.
24. Weigant, J., Wiesmeijer, C. C., Hoovers, J. M. N., Schuuring, E., d'Azzo, A., Vrolijk, J., Tanke, H. J., and Raap, A. K. (1993) Multiple and sensitive fluorescence in situ hybridization with rhodamine-, fluorescein- and coumarin-labeled DNAs. *Cytogenet. Cell Genet.* **63,** 73–76.
25. Mechanic, M. (1988) Biotinylated DNA probes: powerful and practical tools for visualization of complementary nucleic acids on Southern blots or in situ. *Immunochemica* **2,** 12–14.
26. Florijn, R. J., Slats, J., Tanke, H. J., and Raap, A. K. (1995) Analysis of antifading reagents for fluorescence microscopy. *Cytometry* **19,** 177–182.
27. Johnson, G. D. and Nogueira-Araujo, G. M. (1981) A simple method of reducing the fading of immunofluorescence during microscopy. *J. Immunol. Methods* **43,** 349,350.

16

Detection of Rearrangements in the *bcl-2* Gene Using the Polymerase Chain Reaction

Ruth W. Craig, Dorothy R. Belloni,
Ernest S. Kawasaki, and Norman B. Levy

1. Introduction

Described here is a simple polymerase chain reaction (PCR) method for detecting rearrangements involving the *bcl-2* gene. Rearrangements in *bcl-2* are characteristic of follicular B-cell lymphoma (present in >80% of cases) and can occur in other malignancies (**refs. *1–5***; *see* review of *bcl-2* in **ref. *6***). Tumors with these rearrangements usually have a chromosome translocation [t(14;18)(q32;21)] that places *bcl-2* (normally on chromosome 18) adjacent to the joining region (J_H) of the immunoglobulin heavy chain locus (normally on chromosome 14). The translocation results in a high level of expression of *bcl-2*, presumably due to the proximity of enhancers within the immunoglobulin locus *(7)*. The translocation breakpoint in *bcl-2* usually lies downstream of the protein coding region in one of two regions termed the major breakpoint region (MBR) and the minor cluster region (MCR). The MBR is in the 3'-untranslated region of *bcl-2*; approx 50–60% of breakpoints occur within this ~150 bp region ***(8,9)***. The MCR is in 3'-flanking genomic DNA about 20 kb 3' of the MBR ***(8)***; approx 10–25% of breakpoints occur within this ~500 bp region ***(8,9)***.

The method presented here for detecting these rearrangements depends upon the fact that the majority of breakpoints in *bcl-2* lie within these two regions, the MBR and the MCR. The method employs primer sets designed to yield PCR products that span these chromosome translocation breakpoints, one primer set specific for the MBR and one for the MCR. The downstream primer in both these sets represents the immunoglobulin J_H region (a consensus sequence common to many J_H regions since different J_H regions are involved in different patients). Two different upstream primers are used along with this

From: *Methods in Molecular Medicine, Vol. 14: Tumor Marker Protocols*
Edited by: M. Hanausek and Z. Walaszek © Humana Press Inc., Totowa, NJ

downstream primer, one representing a segment of *bcl-2* just upstream of the MBR and the other representing a segment just upstream of the MCR. The two primers in both these sets derive from loci that normally reside on different chromosomes; therefore, no PCR product is generated in the absence of *bcl-2* rearrangement. However, in cells containing an appropriate rearrangement, a PCR product can be generated due to the juxtaposition of sequences represented by the primers.

This method is useful for confirming a diagnosis of follicular lymphoma obtained by pathologic and cytogenetic criteria, particularly in cases where the pathology is not clear-cut or the disease has an abnormal presentation (e.g., **ref. *10***). It can also be used to detect a rearrangement in large cell lymphoma, where *bcl-2* is involved much less frequently than in follicular lymphoma. Importantly, rearrangements can sometimes be detected in non-malignant tissue, such as normal or hyperplastic tissue, or in tumors that do not contain the t(14;18) translocation ***(11–14)***. The method can be used with fresh as well as archival tissues ***(15,16)***. It has the potential to be used for sequencing the breakpoints in individual patients ***(8)***. It also has the potential to be used for the detection of residual disease ***(17,18)***; for this purpose, it can be used either on samples from the patient after drug treatment, or on purged bone marrow prior to transplant. The method is very sensitive, being capable of detecting ≤ 10 lymphoma cells ***(3)***. In a minority of patients with follicular lymphoma, the translocation breakpoint is not in the MBR or the MCR (e.g., it can be in the 5'-flank of *bcl-2*); the method described here does not detect such less prevalent types of rearrangements.

The general approach of the protocol detailed below is to use the primer sets specific for the MBR and MCR in PCR reactions with genomic DNA from the tumor to be tested for *bcl-2* rearrangement. A third primer set representing a nonrearranged gene is also used as a control to check that the PCR reaction is working and that the tumor DNA is capable of supporting PCR. In addition to being used on the tumor DNA to be tested, these primer sets are used in PCR with positive as well as negative control DNAs. The positive control consists of DNA from a cell line that contains a rearrangement in the MBR (SU-DHL-6). The negative control consists of normal human genomic DNA. After PCR, the reaction products are subjected to agarose gel electrophoresis, followed by blotting to a membrane ***(19)***. The membrane is then hybridized with a probe representing *bcl-2* (an oligonucleotide probe that recognizes a sequence interior to the primers). Hybridization of this probe to the PCR products is detected nonisotopically. The presence of a rearrangement of *bcl-2* in the sample DNA being tested can be detected as a band on the agarose gel, and is confirmed if this band hybridizes with the *bcl-2*-specific probe. As described in **Subheading 3.**, this protocol is designed to be carried out within one day. This requires certain specialized equipment such as a Gel Transfer Apparatus and a UV Crosslinker. If this equipment is not available, alternate, somewhat more time-consuming, procedures are described (*see* **Subheading 3.**).

2. Materials

2.1. Polymerase Chain Reaction

1. Oligonucleotide primers (*see* **Notes 1–3**):
   a. MBR (ER388, a 22-mer used as the upstream primer for MBR rearrangements):
   5'-CATTTCCACGTCAACAGAATTG-3' Mol wt ~6700
   Ext. coeff.: 33.7 μg/A_{260} nm
   b. MCR (ER08, a 21-mer used as the upstream primer for MCR rearrangements):
   5'-GATGGCTTTGCTGAGAGGTAT-3' Mol wt ~6600
   Ext. coeff.: 33.7 μg/A_{260} nm
   c. J_H (a 19-mer used as the downstream primer for both MBR and MCR rearrangements):
   5'-ACCTGAGGAGACGGTGACC-3' Mol wt ~5900
   Ext. coeff.: 32.7 μg/A_{260} nm
   d. PC04 (a 20-mer control oligonucleotide used for PCR of β-globin):
   5'-CAACTTCATCCACGTTCACC-3' Mol wt ~6000
   Ext. coeff.: 35.1 μg/A_{260} nm
   e. GH20 (a 20-mer control oligonucleotide used with PC04):
   5'-GAAGAGCCAAGGACAGGTAC-3' Mol wt ~6200
   Ext. coeff.: 30.9 μg/A_{260} nm

   When these primers are initially purchased, high concentration stocks are prepared (400–500 pmol/μL in distilled, deionized water). These are aliquoted (100-μL aliquots) and stored at –20°C. From these, working stocks (100 pmol/μL in distilled, deionized water) are prepared and stored aliquoted (25–50-μL aliquots) at –20°C. Thawing and refreezing of the aliquots is kept to a minimum.
2. AmpliTaq DNA polymerase: Stoffel fragment (Applied Biosystems, formerly Perkin-Elmer, cat. no. N808-0038, supplied at 10 U/μL) and the 10X Stoffel buffer (10X = 100 m*M* KCl, 100 m*M* Tris-HCl, pH 8.3) supplied with the enzyme (*see* **Note 2**). The enzyme is stored at –20°C in the storage buffer in which it is supplied by the supplier. It is useful to store the enzyme in a Stratacooler II Benchtop Cooler (Stratagene, La Jolla, CA) or similar container to insulate the enzyme from possible temperature fluctuations.
3. 10X dNTP solution: 2 m*M* dATP, 2 m*M* dCTP, 2 m*M* dGTP, 2 m*M* dTTP in distilled, deionized water. We use a set of dATP, dCTP, dGTP, and dTTP (Boehringer Mannheim, Indianapolis, IN, cat. no. 1 277 049). In this set, the dNTPs are supplied at 100 m*M* each. They are thus diluted 1:50 using distilled, deionized water (e.g., 100 μL of each dNTP to a final vol of 5 mL) to yield the solution containing 2 m*M* of each dNTP. This dNTP solution is then aliquoted into 200-μL aliquots and stored at –20°C. Aliquots should not be repeatedly frozen and thawed (i.e., they can be frozen and thawed 2–3 times at maximum).
4. 25 m*M* $MgCl_2$.

5. Genomic DNAs:
   a. Negative control DNA (Normal human genomic DNA [e.g., CLONTECH, Palo Alto, CA, cat. no. 6550-1]).
   b. Positive control DNA (genomic DNA from the SU-DHL-6 cell line, obtained from Dr. Alan Epstein, University of Southern California).
   c. Genomic DNA from the sample to be tested for *bcl-2* rearrangement.

   These can be isolated using the Puregene DNA Isolation Kit from Gentra Systems, Inc. (Minneapolis, MN). All genomic DNAs are diluted to a concentration of 50 ng/μL in TE buffer (10 m*M* Tris-HCl, pH 8.0, 1 m*M* ethylenediamine tetraacetic acid [EDTA]).
6. Tubes for carrying out the reaction (e.g., 0.2-mL tubes from Marsh Biomedical, Rochester, NY, cat. no. T-5002, or equivalent).

2.2. Agarose Gel Electrophoresis of the PCR Product

1. 10X Loading buffer: 50% glycerol, 0.4% bromophenol blue in distilled, deionized water.
2. 1X TBE buffer: 90 m*M* Tris-borate, 2 m*M* EDTA. A 5X stock is made up; this consists of 54 g Tris base, 27.5 g boric acid, and 20 mL 0.5 *M* EDTA (pH 8.0) per L (*see* **ref. *13***, p. B.23). The 0.5 *M* EDTA is made up by dissolving 186.1 g of EDTA, disodium dihydrate to 1 L and adjusting the pH to 8.0 with NaOH pellets (*see* **ref. *13***, p. B.11).
3. 10 mg/mL ethidium bromide in distilled, deionized water. This agent is stored protected from light at room temperature (**ref. *13***, p. B.11). **This agent is a mutagen. It should be handled with gloves only and should be kept contained. A mask should be worn when weighing out the chemical** (*see* **ref. *13***, p. 1.49 for methods of decontamination of solutions containing ethidium bromide).

2.3. Transfer of the PCR Product to Nylon Membrane

1. Denaturing solution: 0.5 *N* NaOH, 1.5 *M* NaCl.
2. Neutralizing solution: 0.1 *M* Tris-HCl, pH 7.5, 1.5 *M* NaCl.
3. Duralon UV nylon membrane (Stratagene, cat. no. 420101; *see* **Note 3**).
4. Whatman #1 filter paper.
5. Transfer buffer, either 10X SSPE or 10X SSC: The 10X SSPE is made up by 1:1 dilution of a 20X SSPE stock. The 20X SSPE stock consists of 3 *M* NaCl, 0.2 *M* NaH_2PO_4, and 0.02 *M* EDTA at pH 7.4; this is made up by bringing 175.3 g NaCl, 27.6 g $NaH_2PO_4 \cdot H_2O$, and 7.4 g EDTA, disodium dihydrate to ~900 mL, and then bringing the pH to 7.4 and the final volume to 1 L ***(13)***. 10X SSC can be substituted for 10X SSPE for the transfer, where 20X SSC consists of 3 *M* NaCl, and 0.3 *M* sodium citrate, pH 7.0 ***(13)***.
6. Perkin-Elmer 9600 thermal cycler (*see* **Note 4**).
7. Transfer apparatus (either a nucleic acid transfer apparatus such as the PosiBlot 30–30 Pressure Blotter with Pressure Control Station from Stratagene, or sponges/paper towels) (*see* **Note 5**).
8. Stratalinker 1800 UV Crosslinker (Stratagene) or a vacuum oven for baking the blot (*see* **Note 6**).

2.4. Hybridization of the Membranes with bcl-2-Specific Probes

1. Oligonucleotide/horseradish peroxidase probes:
   a. MBR Probe (ER132):
      5'-X-ATTGTGACAGTTATATCTG-3' Mol wt = 6445 L/mol/cm
      Mol. ext. coeff. $_{260}$ = 186,876 L/mol/cm
   b. MBCR Probe (ER125):
      5'-X-CTAAGCCAGCCAGTCA-3' Mol wt = 54505 L/mol/cm
      Mol. ext. coeff. $_{260}$ = 155,282 L/mol/cm

   The molecular weights and molar extinction coefficients listed are calculated for the biotinylated oligonucleotide before conjugation to the streptavidin-HPR complex. The absorbance readings at 260 nm will reflect the concentration of the oligonucleotide present with accuracy. These probes represent portions of *bcl-2* internal to the primer sets. For purposes of detection, the oligonucleotides are linked at their 5' ends to horseradish peroxidase through streptavidin/biotin binding (the biotinylated oligonucleotide is conjugated to streptavidin-HPR). These HRP-conjugated oligonucleotide probes are prepared and HPLC purified by OPERON Technologies, Inc. (Alameda, CA); purification involves a two-step procedure, purification of the biotinylated oligonucleotide before the conjugation and a second purification after the conjugation reaction to remove unreacted material. When the purified probes are received from the supplier, they are diluted in distilled, deionized water to a concentration of 5 pmol/μL, using the pmol amount of material as listed by the company. (If the pmol amount of material received from the company is not clear, *see* **Note 7**). The diluted probes are aliquoted into 20-μL aliquots and stored at –20°C protected from light.
2. Hybridization oven (e.g., Marsh Biomedical, Rochester, NY; cat. no. H-9300), or sealed bag system for hybridizing the membrane with the probe. Accessories for the hybridization oven include roller bottles (Marsh Biomedical; cat. no. H-9082, 3.5 cm in diameter, 30 cm long) and membrane mesh (Marsh Biomedical; cat. no. H-9088).
3. Prehybridization/hybridization solution: 5X SSPE, 0.5% SDS. This solution can be made up by mixing together 10 mL 20X SSPE, 1 mL 20% SDS, and 29 mL distilled, deionized water. The 20X SSPE is made up as described above (*see* **Subheading 2.3., step 5**).
4. Wash solution: 2X SSPE, 0.1% SDS. This solution can be made up by mixing together mL 20X SSPE, 0.2 mL 20% SDS, and 358 mL distilled, deionized water.
5. Final wash solution: 2X SSPE.

2.5. Development of the Membrane Using Enhanced Chemiluminescence

1. New England Nuclear Renaissance nucleic acid chemiluminescence reagents (Oxidizing Reagent and Enhanced Luminol Reagent, cat. no. NEL-202).
2. Autoradiographic film (e.g., XAR-5).
3. Processor for autoradiographic film.

Table 1
Experimental Design (*see* Methods, Subheading 3.1.)

	Primer set		
Genomic DNA	A. ER388/ER13 for MBR	B. ER08/ER13 for MCR	C. PCO4/GH20 for β-globin (control)
1. No DNA	Tube # A1	Tube # B1	Tube # C1
2. Negative control DNA[a]	Tube # A2	Tube # B2	Tube # C2
3. Positive control DNA[b]	Tube # A3	Tube # B3	Tube # C3
4. Tumor DNA to be tested[c]	Tube # A4	Tube # B4	Tube # C4

[a]The negative control is normal human DNA.

[b]The positive control is DNA from the SU-DHL-6 cell line.

[c]One tumor DNA is to be tested in the example illustrated. Each additional tumor sample to be tested will require an additional three reactions to be set up.

Table 2
Preparation of the Master Mixes[a] (*see* Methods, Subheading 3.2.)

Reagent	Volume (µL)[b]	Final concentration[c]
Distilled, deionized water	310	
10X Stoffel buffer	50	1X
$MgCl_2$ (25 m*M*)	30	1.5 m*M*
dNTP solution (2 m*M* each)	50	200 µ*M* each
AmpliTaq DNA polymerase (Stoffel fragm.)	5	5 U/50-µL reaction
Primer 1[d] (100 pmol/µL)	2.5	25 pmol/25-µL reaction
Primer 2[d] (100 pmol/µL)	2.5	25 pmol/25-µL reaction

[a]Three master mixes are made up; they are identical except each contains a different primer set (*see* footnote "d" below). The recipe shown is for 10 reactions.

[b]Final volume of each master mix = 450 µL.

[c]This will be the final concentration once the genomic DNA to be used in PCR has been added (step B4).

[d]In master mix A, primers 1 and 2 are ER388 and J_H. In master mix B, primers 1 and 2 are ER08 and J_H, and in master mix C, primers 1 and 2 are PC04 and GH20.

3. Methods

3.1. Polymerase Chain Reaction

1. PCR reactions with three different primer sets (primer sets A, B, and C) will be carried out according to the design outlined in **Table 1**.
2. To this end, three master mixes are prepared, one master mix for each primer set (e.g., master mixes A, B, and C). The master mixes are identical except that each contains a different primer set. These master mixes are prepared as shown in **Table 2**.

3. Aliquots of the master mixes are dispensed into the reaction tubes. Each reaction tube receives 45 µL of master mix. Thus, reaction tubes # A1-A4 (*see* **Table 1**) each receive 45 µL of master mix A. Reaction tubes # B1-B4 each receive 45 µL of master mix B. Reaction tubes C1–C4 each receive 45 µL of master mix C.
4. The DNA to be subjected to PCR is added to the reaction tubes. Each tube receives 5 µL of DNA (at a concentration of 50 ng/µL in TE buffer); this brings the final reaction vol to 50 µL and the final amount of DNA in the reaction to 250 ng. Thus, to tubes A1, B1, and C1 (*see* **Table 1**), TE buffer alone (5 µL) is added. To tubes A2, B2, and C2, the negative control DNA (5 µL) is added. To tubes A3, B3, and C3, the positive control DNA (5 µL) is added. To tubes A4, B4, and C4, the sample DNA to be tested for *bcl-2* rearrangement (5 µL) is added.
5. The reaction tubes are subjected to PCR in a Perkin-Elmer 9600 thermal cycler as follows: An initial denaturation is carried out at 94°C for 1 min. Forty cycles are carried out as follows: 94°C for 20 s, 60°C for 20 s, 72°C for 30 s. A final extension is carried out at 72°C for 1 min. If a Perkin-Elmer 9600 thermal cycler is not available, *see* **Note 4**.

3.2. Agarose Gel Electrophoresis of the PCR Products

1. Mix 20-µL aliquots of the products of the above PCR reactions with 2 µL of 10X loading buffer and loaded onto an agarose gel for electrophoresis. The gel consists of 2% agarose in 1X TBE buffer, with 0.5 µg/mL ethidium bromide being added after the agarose is melted and cooled and before the gel is poured (*see* **ref.** ***13***, p. 6.9 for detailed instructions on how to prepare and run an agarose gel). Reaction tubes #A1–A4 (*see* **Table 1**) should be loaded in adjacent lanes on the gel and reaction tubes #B1–B4 should similarly be adjacent to each other (as the "A" tubes will be hybridized with one probe and the "B" tubes with another in **Subheading 3.4.**). An additional lane is loaded with DNA mol wt markers (e.g., a 100 bp ladder, Life Technologies, Gaithersburg, MD).
2. Electrophoresis is carried out at 100 V until the bromophenol blue dye has migrated approximately 10 cm.
3. The gel is placed on a UV light box for visualization of the ethidium bromide stained bands representing PCR products. With all tubes containing DNA, but not those not containing DNA, a band of 269 bp is expected with primer set C (beta-globin control). With the positive control DNA, but not with no DNA or Negative Control DNA, a band of 0.3–0.4 kb is normally seen with primer set A (MBR). If these controls behave appropriately, one can then interpret the presence of a similar (not necessarily identical) sized band in the test sample as indicative of the likely presence of a rearrangement. Unfortunately, there is no cell line that serves as a control for primer set B. However, should the sample DNA have a rearrangement involving the MCR, a band of 0.2–0 3 kb is normally seen on the ethidium bromide stained agarose gel. Candidate bands seen on agarose gels will be confirmed by hybridization with a *bcl-2* specific probe, as described in **Subheadings 3.3.** and **3.4.** If no bands are visible, it is still advisable to hybridize with the *bcl-2*-specific probe, as it is possible that a band is present at levels not detected by visual inspection of the stained gel.

3.3. Transfer of the PCR Product to Nylon Membrane

1. The gel from **Subheading 3.1.** above is soaked in denaturing solution for 30–60 min.
2. The gel is then soaked in neutralizing solution for 30–60 min.
3. A piece of Duralon UV nylon membrane is cut to a size just larger than the gel. The membrane is wet by floating it on distilled, deionized water in a baking dish or similar container; it is then soaked in transfer buffer (10X SSPE or 10X SSC). Two pieces of Whatman filter paper are cut just larger than the membrane.
4. The PCR products in the gel are transferred to the membrane. For rapid transfer, a transfer apparatus such as a PosiBlot Pressure Blotter can be used. If such a transfer apparatus is not available, *see* **Note 5**. With the PosiBlot, the transfer is set up as follows. The two pieces of Whatman filter paper are placed on the support of the PosiBlot. The membrane is placed on the Whatman filters, with no air bubbles between the filters and the membrane. The "mask" of the PosiBlot is placed on top of the membrane, making sure that the mask covers the membrane completely. The gel is laced on top of the mask, right side up, and the sponge of the Posiblot (soaked in transfer buffer) is placed on top of the gel. The lid of the PosiBlot is put in place, and the pressure station is attached and powered up. Transfer is carried out for 1 h at 70–80 psi.
5. After transfer, the membrane is placed (face up) on a damp piece of Whatman paper and the PCR products are crosslinked to the membrane in a UV crosslinker (UV Stratalinker 1800 using the "Auto Cross Link" setting). If a UV crosslinker is not available, *see* **Note 6**. The membrane is cut to separate the portion that derives from reaction tubes A1–A4 and the portion that derives from reaction tubes B1–B4. At this point, the membranes can be wrapped in plastic wrap with damp filters around it and stored at 4°C.

3.4. Hybridization of the Membranes with bcl-2-Specific Probes

1. To prehybridize, the membranes are incubated in approx 10 mL of prehybridization/hybridization solution at 50°C for 10 min (with gentle agitation). The membrane derived from reaction tubes A1–A4 is prehybridized separately from the membrane derived from reaction tubes B1–B4. This step is most easily performed in roller bottles in a hybridization oven (e.g., MINIOVEN MKII, using a rotation speed of 9). A piece of mesh lies between the membrane to be hybridized and the wall of the roller bottle, to facilitate even wetting of the membrane. When rolling the membrane to insert it into the roller bottle, the side of the membrane containing the transferred PCR products is placed facing inward, away from the mesh and the walls of the roller bottle. Other types of hybridization apparatus will also work (e.g., heat sealed bags in a water bath; *see* **ref.** ***13***, p. 9.52).
2. The above solution is removed from the roller bottle and 10 mL of fresh prehybridization/hybridization solution is added, along with approx 10 pmol (2–5 µL, *see* **Note 6**) of the *bcl-2*-specific oligonucleotide/HRP probe. The MBR probe (ER132) is used for reactions A1–A4. The MCR probe (ER125) is used for the reactions B1–B4. This step and all subsequent steps should be carried out with a minimum of exposure of the probes to light (e.g., cover vessels with alu-

minum foil, cover hybridization oven door with aluminum foil, and plan each step ahead, being prepared to work rapidly through each step). Hybridization is carried out at 50°C for 1 h with rotation of the roller bottle containing the membrane.

3. Washing of the membrane to remove unhybridized probe is carried out as follows: The wash solution is prewarmed to 50°C. The hybridization solution (from **step 2** above) is removed and the membrane is then washed twice at 50°C with wash solution (10 min each wash). Changes of solutions should be carried out rapidly, such that the membrane remains wet and does not cool down. This is facilitated by the use of the roller bottles in the hybridization oven. With the roller bottles used in this laboratory, a volume of about 50 mL is used for each wash. Washing can also be carried out in sealed Tupperware-type containers (with the membrane right side up) in a water bath with shaking, in which case additional wash solution might be necessary. A final wash of the membrane is carried out at room temperature in ~200 mL final wash solution for 10 min (in a small dish covered with aluminum foil).

3.5. Development of the Membrane Using Enhanced Chemiluminescence (ECL)

1. The following steps are carried out rapidly in order to minimize exposure of the membrane or the ECL reagents to light.
2. The membrane is blotted on Whatman paper to remove excess wash solution and placed right side up into a small square dish.
3. The two ECL reagents (oxidizing reagent and enhanced luminol reagent) are mixed 1:1 (generally 5 mL each or approx 0.125 mL per cm^2 of membrane). The mixture is pipeted onto the membrane; it should cover the membrane completely but it will not roll off the membrane (owing to surface tension).
4. The dish containing the developing membrane is covered with aluminum foil and incubated in the dark for 1 min.
5. Excess developing reagent are blotted off the membrane and the membrane is wrapped in Saran Wrap and exposed to X-ray film. An initial 1–5-min exposure is carried out, with further exposures being carried out as necessary.

4. Notes

1. In this laboratory, oligonucleotides are purchased from OPERON TECHNOLOGIES, Inc. The oligonucleotides purchased for PCR are unmodified and have been desalted but have not been purified (e.g., no HPLC purification). It is not essential that the concentration of the working stocks of primers be exactly 100 pmol/μL, as the primers are present in excess.
2. Use of the Stoffel fragment of AmpliTaq DNA polymerase is highly recommended as preferable to the use of the entire intact polymerase.
3. The choice of membrane is critical to the success of this method. The Duralon UV is strongly recommended as a membrane that will yield successful results.
4. If a Perkin-Elmer 9600 thermal cycler is not available, an MJ thermal cycler (MJ Research, Inc., Watertown, MA) or equivalent can be used. An initial de-

naturation is carried out at 94°C for 1 min. Forty cycles are carried out as follows: 96°C for 1 s, 94°C for 20 s, 58°C for 1 s, 60°C for 20 s, 74°C for 1 s, 72°C for 30 s. A final extension is carried out at 72°C for 1 min. The additional 1-s pulses (underlined) used with the MJ thermal cycler relate to the fact that the Perkin-Elmer 9600 thermal cycler does not begin timing until the liquid in the tubes reaches the desired temperature. The MJ thermal cycler does not have this feature; it begins timing when the block is at temperature.

5. If a transfer apparatus is not available, the transfer can be carried out using sponges, a baking dish, and paper towels as described in **ref. *13*** (p. 9.34). Transfer buffer is placed in a baking dish and two sponges (usually at least 1.5-in. thick) are placed side by side in the transfer buffer, with the buffer level such that the surface of the sponges is not below the level of buffer. A piece of Whatman filter paper is placed on top of the sponges, with the filter paper curving over the sponges into the buffer at the top and bottom of the sponges (to act as a wick for the buffer). The gel is placed onto the Whatman filter paper and a double layer of Saran Wrap is formed around the outside of the gel (to prevent "shortcircuiting" of the gel by movement of the buffer through the sponge area around the edges of the gel). The membrane (after wetting) is placed on top of the gel, avoiding wrinkles or bubbles in the membrane. One of the two pieces of Whatman paper is wet with transfer buffer and placed on top of the membrane. The other piece of Whatman paper (not wet) is placed on top of the first. Paper towels are placed in a stack (approx 3 in. tall or more) on top of the assembled gel to be transferred and a glass plate is placed on top of the paper towels. A weight is placed (evenly distributed) on top of the stack. This assembly is left overnight. Transfer buffer is drawn up into the paper towel stacks and as it passes through the gel it transfers the PCR products to the membrane lying on top of the gel.
6. If a UV crosslinker apparatus is not available, the more traditional method of baking the membrane at 80°C in a vacuum oven for 1–2 h can be used as an alternative (*see* **ref. *13***, p. 9.46).
7. If the pmol amount of oligonucleotide/HRP probe received from the company is not known, it can be determined as follows: The OD_{260} of the material is measured. The volume to be used to resuspend the material (to a concentration of 5 pmol/µL) is then calculated as follows: For ER132, the volume to be used is equal to 0.934/(total OD_{260} units). For ER125, this volume is equal to 0.775/(total OD_{260} units). With increasing storage time, the oligonucleotide/HRP probes may appear to give a lesser signal. In this laboratory, 2 µL of the probe stock solution has been found to be sufficient for 10 mL of hybridization buffer when using probes that have been prepared recently. These probes, if stored properly, can be used for several years. However, as they are stored for increasing times, it may be necessary to use an increased amount of probe (up to 5 µL of the 5 pmol/µL solution for the 10 mL of hybridization buffer).

Acknowledgment

Ruth Craig's research is supported by a grant from the National Institutes of Health (CA57359).

References

1. Cleary, M. L., Smith, S. D., and Sklar, J. (1986) Cloning and structural analysis of cDNA for bcl-2 and a hybrid bcl-2/immunoglobulin transcript resulting from the t(14;18) translocation. *Cell* **47,** 19–28.
2. Kawasaki, E. S. (1992) The polymerase chain reaction: its use in the molecular characterization and diagnosis of cancers. *Cancer Invest.* **10,** 417–429.
3. Kawasaki, E. S. (1994) The polymerase chain reaction: Its role in the future of molecular and diagnostic oncology, in *Cancer Therapy in the Twenty-First Century, Vol. I: Molecular and Immunologic Approaches* (Huber, B. E., ed.) Futura, Mount Kisco, NY, 119–141.
4. Larsen, C. J., Mecucci, C., and Leroux, D. (1990) t(2: 18) and t(18;22) variant chromosomal translocations and bc1-2 gene rearrangements in human malignant lymphomas. *Nouv. Rev. Fr. Hematol.* **32,** 401–403.
5. Tsujimoto, Y. and Croce, C. M. (1986) Analysis of the structure, transcripts, and protein products of bcl, the gene involved in human follicular lymphoma. *Proc. Natl. Acad. Sci. USA* **83,** 5214–5218.
6. Craig, R. W. (1995) The bc1-2 gene family. *Semin. Oncol.* **6,** 35–44.
7. Chen-Levy, Z., Nourse, J., and Cleary, M. L. (1989) The bc1-2 candidate proto-oncogene is a 24-kilodalton integral-membrane protein highly expressed in lymphoid cell lines and lymphomas carrying the t(14;18) translocation. *Mol. Cell. Biol.* **9,** 701–710.
8. Cotter, F., Price, C., Zucca, E., and Young, B. D. (1990) Direct sequence analysis of the 14q+ and 18q-chromosome junctions in follicular lymphoma. *Blood* **76,** 131–135.
9. Segal, G. H., Jorgensen, T., Scott, M., and Braylan, R. C. (1994) Optimal primer selection for clonality assessment by polymerase chain reaction analysis: II Follicular lymphomas. *Hum. Pathol.* **25,** 1276–1282.
10. Kerrigan, D. P., Irons, J., and Chen, I. M. (1990) bc1-2 gene rearrangement in salivary gland lymphoma. *Am. J. Surg. Pathol.* **14,** 1133–1138.
11. Corbally, N., Grogan, L., Keane, M. M., Devaney, D. M., Dervan, P. A., and Carney, D. N. (1994) Bcl-2 rearrangement in Hodgkin's disease and reactive lymph nodes. *Am. J. Clin. Pathol.* **101,** 756–760.
12. Ji, W., Qu, G. Z., Ye, P., Zhang, X. Y., Halabi, S., and Ehrlich, M. (1995) Frequent detection of bcl-2/JH translocations in human blood and organ samples by a quantitative polymerase chain reaction assay. *Cancer Res.* **55,** 2876–2882.
13. Limpens, J., de Jong, D., van Krieken, J. H., Price, C. G., Young, B. D., van Ommen, G. J., and Kluin, P. M. (1991) Bcl-2/J_H rearrangements in benign lymphoid tissues with follicular hyperplasia. *Oncogene* **6,** 2271–2276.
14. Poppema, S., Kaleta, J., and Hepperle, B. (1992) Chromosomal abnormalities in patients with Hodgkin's disease: evidence for frequent involvement of the 14q chromosomal region but infrequent bcl-2 gene rearrangement in Reed-Sternberg cells. *J. Natl. Cancer Inst.* **84,** 1789–1794.
15. Alkan, S., Lehman, C., Sarago, C., Sidawy, M. K., Karcher, D. S, and Garrett, C. T. (1995) Polymerase chain reaction detection of immunoglobulin gene rearrange-

ment and bc1-2 translocation in archival glass slides of cytologic material. *Diagn. Mol. Pathol.* **4,** 25–31.
16. Liu, J., Johnson, R. M., and Traweek, S. T. (1993) Rearrangement of the Bcl-2 gene in follicular lymphoma. Detection by PCR in both fresh and fixed tissue samples. *Diagn. Mol. Pathol.* **2,** 241–247.
17. Gribben, J. G., Freedman, A., Woo, S. D., Blake, K., Shu, R. S., Freeman, G., Longtine, J. A., Pinkus, G. S., and Nadler, L. M. (1991) All advanced stage non-Hodgkin's lymphomas with a polymerase chain reaction amplifiable breakpoint of bc1-2 have residual cells containing the bc1-2 rearrangement at evaluation after treatment. *Blood* **78,** 3275–3280.
18. Lee, M.-S., Chang, K.-S., Cabanillas, F., Freireich, E. J., Trujillo, J. M., and Stass, S. A. (1987) Detection of minimal residual cells carrying the t(14;18) by DNA sequence amplification. *Science* **237,** 175–178.
19. Sambrook, J., Fritsch, E. F., and Maniatis, T. (1989) *Molecular Cloning: A Laboratory Manual,* 2nd ed., Cold Spring Harbor Press, Cold Spring Harbor, NY.

IV

Molecular Markers

17

Applications of Tissue Microdissection in Molecular Pathology

Principles and Guidelines

Michael R. Emmert-Buck, Irina A. Lubensky, Rodrigo F. Chuaqui, Larisa V. Debelenko, Cathy D. Vocke, Maria J. Merino, Paul H. Duray, W. Marston Linehan, Lance A. Liotta, and Zhengping Zhuang

1. Introduction

The study of human disease processes is an evolving field that is closely intertwined with the development of technology. The advent of the polymerase chain reaction (PCR) allows investigators new opportunities for genetic analysis of pathological processes. DNA and RNA analysis of small numbers of cells is now possible, allowing for study of specific defined cell populations or lesions. For example, application of tissue microdissection and PCR technology to human tumor samples represents a powerful method to study genetic alterations in cancer cells as they exist in vivo. The study of human tumor samples is complex, and can in fact be hampered by the exquisite sensitivity of PCR. A typical histologic field of cancer contains inflammatory, stromal, premalignant, and normal epithelial cells in addition to invasive tumor cells. PCR amplification of DNA or RNA from these "contaminating cells" interferes with accurate determination of tumor-specific genetic changes. Tissue microdissection provides a method to procure specific cell types from a human tumor sample, e.g., a pure population of tumor cells can be analyzed without interference from neighboring nontumor cells. Additionally, investigators can recover select subpopulations of cells such as premalignant lesions that cannot be studied in bulk tissue specimens. Our laboratory and others have developed and applied various microdissection approaches to human tissue samples. The

From: *Methods in Molecular Medicine, Vol. 14: Tumor Marker Protocols*
Edited by: M. Hanausek and Z. Walaszek © Humana Press Inc., Totowa, NJ

focus of the current chapter is to review a detailed protocol of the technique which our laboratory has employed. Specific applications of the technique applied to basic research studies as well as applied surgical pathology studies are described. The chapter finishes with a section on a new laser capture microdissection system developed at the National Cancer Institute (NCI).

2. Methods

2.1. Basic Technique

The concept of tissue microdissection is quite simple, however, the method is technically challenging particularly when a large number of samples are to be studied. Several approaches to microdissection have been described in the literature including: gross dissection of frozen tissue blocks to enrich for tumor cells *(1,2)*; irradiation of manually ink stained sections to destroy unwanted genetic material *(3)*; and microdissection with manual tools *(4–7)*. Our approach has been to directly procure cell populations of interest from tissue sections by manual hand microdissection using a 30-gage needle. With practice, this approach allows investigators to selectively recover only the cell populations of interest, including very small premalignant lesions. DNA, RNA, or enzyme activity can be recovered from the microdissected tissue using slight modifications of the technique. Use of a new, sterile 30-gage needle for each microdissection minimizes inadvertent PCR contamination.

2.2. Approach to Tissue Specimens

The tissue source utilized in a given study is of critical importance. Factors such as time and type of fixation of paraffin-embedded material, or the interval between surgery and freezing of tissue impacts heavily on subsequent quality of isolated proteins or nucleic acids. Investigators are often unaware of the fixation or freezing interval status of cases in a study, therefore, the quality of the tissue should be determined empirically. As a practical approach we often start with a gross razor blade scraping of one tissue section from each case, and assess the quality of the recovered DNA, RNA, or enzymes using the assay system to be utilized in the study. In this way specific cases appropriate for analysis are identified prior to investing the time and effort involved in microdissection. It is important that the tissue sections be properly cut and prepared. Use of clean, disposable microtome, or cryostat blades minimizes the potential for cross-contamination of cases. Additionally, cutting of sections with new, sharp blades decreases smearing of proteins or nucleic acids across the tissue section as it is being cut. There are unique considerations when performing studies with fresh frozen tissue or formalin fixed, paraffin-embedded tissue. The following are general guidelines.

1. Paraffin-embedded tissue: The processes of tissue fixation and processing damage both DNA and RNA, and denature proteins. However, PCR amplification of DNA recovered from paraffin-embedded tissue works quite well and allows investigators to perform studies on archival patient material. Studies on paraffin tissue have the general advantages of abundant samples for study, high quality of histologic detail for studying dysplastic and premalignant lesions, and frequent availability of follow-up clinical information on patients. There are several potential difficulties when using archival paraffin-embedded material that should be considered. Investigators should be prepared to discard cases and perform repeat analysis when necessary, often several times during the course of a study. The type of fixative used and the length of fixation impact heavily on the quality of the DNA recovered after microdissection, thus PCR amplification signals may be widely disparate among samples in a study, even when roughly equivalent amounts of tissue are microdissected and processed. If a sample does not initially give a PCR product, a 10-fold dilution of the template may yield a strong PCR product owing to dilution of tissue inhibitors of *Taq* polymerase. In our hands approx 10–20% of cases in which we are unaware of fixation conditions will not yield amplifiable DNA. However, if an entire series of cases have been properly processed and fixed, then recovery and excellent PCR amplification of close to 100% of the cases is possible.

   We have not systematically studied the effects of various fixatives on the quality of recovered DNA for PCR, however, in our experience standard 10% neutral buffered formalin works reasonably well. Fixatives containing heavy metals or low pH should be avoided. Nonformalin-based fixatives such as alcohols generally result in recovery of better quality DNA. Routine care in tissue processing is also helpful including immediate processing of samples after surgery and slicing of tissue samples into thin sections to allow rapid penetration of fixative. Over fixation (>8 h) should be avoided. We recommend selecting PCR primer sets that produce products in the 100–250 bp range since the DNA is often crosslinked and/or fragmented, and may not reliably amplify larger products.

   In our hands studies of RNA recovered from formalin fixed, paraffin-embedded tissue are often problematic. Investigators can recover and amplify RNA by reverse transcription (RT)-PCR, however, the quality of the RNA is variable and difficult to quantitate. We prefer to use frozen tissue samples for RNA studies, especially when interested in quantitative differences between two cell populations. If RT-PCR studies are to be attempted with paraffin-embedded material it is recommended to design PCR primers that amplify small PCR products in the 80–120 bp range.
2. Frozen tissue: Compared to formalin-fixed, paraffin-embedded tissue, freshly frozen tissue samples allow for recovery of active enzymes, as well as high-quality DNA and RNA. Drawbacks include the scarcity of material for study compared to archival tissue samples, and the less detailed histology that can be discerned in the tissue.

2.3. Microdissection

There is no absolute right or wrong method to perform microdissection. Each member in our group dissects a little differently. Speed, precision, and avoidance of contamination are the most important parameters, and any method that achieves these to the satisfaction of the dissector is adequate. However, we have found manual microdissection by hand to be superior to microdissection with mechanical micromanipulators especially in terms of speed. Microdissection by hand requires an initial investment in time and practice, however, usually 10–20 cases is enough to begin to feel comfortable with the approach. It is helpful to dissect on an inverted microscope that has more room to work in the vicinity of the tissue section. We rarely find it necessary to dissect at >200× final magnification.

2.3.1. Preparation of Slides

Microdissection can be performed on standard histology slides. Sections 10–15 µm thick placed on noncoated glass slides are optimal.

2.3.2. Paraffin-Embedded Tissue

The following are general guidelines.

1. Recut sections should be stored at or below room temperature, and should not be deparaffinized until immediately prior to microdissection.
2. Slide deparaffinization
   a. Soak twice for 5 min each in xylene.
   b. Soak twice for 2 min each in 95% ethanol.
   c. Soak twice for 2 min each in 80% ethanol.
   d. Soak twice for 2 min each in 50% ethanol.
   e. Soak twice for 2 min each in distilled H_2O.
   f. Soak twice for 5 min each in 3% glycerol prepared in distilled H_2O.
3. The final soak in 3% glycerol is particularly helpful in preparing the tissue for microdissection. This step renders the tissue less brittle, and dissected tissue fragments are easier to procure with the needle.
4. After removing the slide from the 3% glycerol step, shake the slide in the air to remove the layer of glycerol/water. The next 5–10 min are the optimal time for microdissection. The tissue is dry, but retains a soft consistency. If the dissection takes more than a few minutes the tissue will become increasingly brittle, and the dissected fragments may be repelled as the needle is brought in proximity to the tissue. If the tissue becomes overly dry, a 1–2 min resoak in the 3% glycerol/water solution is advised. It is important to ensure that the thin coat of fluid that covers the slide after removal from the glycerol/water is removed. Dissection with this fluid layer present results in diffusion of tissue fragments with potential for "contamination" of samples. Additionally, dissection of the tissue under fluid pro-

duces large strips of tissue which are not adequately homogenized in preparation for the one-step DNA extraction buffer.

5. Staining of tissues is optional. Some members of our group prefer to briefly stain slides in eosin prior to dissection. Hematoxylin and eosin (H&E) staining is also acceptable, however, this can result in diminished PCR amplification. The optimal approach is to dissect from unstained slides. An adjacent H&E-stained section can be used as a guide to direct the microdissection. Lowering the intensity and/or altering the refraction of the light source is helpful in visualizing the unstained tissue while dissecting. Microdissection can also be performed on tissue sections after immunohistochemical or *in situ* hybridization staining. This approach can be utilized for several purposes. For example, in loss of heterozygosity (LOH) studies of parathyroid tumors we found the scoring of allelic loss was confounded by "contamination" from normal endothelial cells within the tumors. Staining of the tumors with anti-CD34 antibody identified the vessels and allowed dissections that minimized or eliminated endothelial cell contamination, and resulted in clear, nonequivocal LOH scoring. Additional uses of prestaining tissue sections include: identification of cells for dissection which produce (or do not produce) a specific mRNA transcript or protein, identification of cells containing or associated with organisms, and assessment of the degree of inflammation within a tumor that is not apparent with standard H&E staining.
6. Dissection: Typically, we place a sterile 30-gage needle on a pencil sized syringe. The dissector should prop his or her elbow on a solid surface adjacent to and at the same height as the stage of the microscope to stabilize the dissecting hand. It is helpful to rest the ulnar aspect of the dissecting hand on the stage of the microscope and move the needle into the microscopic field, a few millimeters above the tissue. In this way the dissecting arm and hand can be rested on solid support surfaces, and microdissection can be performed by minute movements of the fingers. While viewing the tissue through the microscope, the cell population of interest should be gently scraped with the needle. The dissected cells will become detached from the slide and form small dark clumps of tissue that can be collected on the needle by electrostatic attraction. Several small tissue fragments can be procured simultaneously. Collection of an initial fragment on the tip of the needle will assist in procuring subsequent tissue.
7. The tip of the needle with the procured tissue fragments should be carefully placed into a small PCR tube with a minimum of 10 µL solution. Gentle shaking of the tube will ensure the tissue detaches from the tip of the needle. Pressing down on the shaft of the syringe to inject an air bubble into the extraction solution also helps to detach the tissue from the needle, and prevents any fragments from remaining lodged in the barrel of the needle.

2.3.3. Frozen Tissue

The following are general guidelines.

1. Dissection of frozen section slides is similar to paraffin sections with a few minor modifications. Sections should be cut and immediately frozen at –70°C. One sec-

tion at a time should be removed from the freezer and microdissected. Allow the section to warm at room temperature for approx 1 min. The frozen tissue sections are similar to paraffin-embedded slides after removal from glycerol/H_2O in that there is a window of approx 5–10 min in which the tissue will have dried sufficiently, but will remain soft enough for dissection. The remainder of the dissection is similar to that for paraffin-embedded tissue described in **Subheading 2.3.2.**

2. RNA for RT-PCR can similarly be extracted from frozen tissue sections. The dissections should be performed as rapidly as possible to minimize RNA degradation, and brief pretreatment of the slides prior to microdissection in ethanol or formalin can improve RNA recovery. In general, the recovery of RNA from frozen sections works well if the tissue has been properly processed, however, the quality of the RNA is variable. Investigators need to empirically determine the optimal conditions for a given RT-PCR study by considering quality of the tissue, level of endogenous RNases, size of RT-PCR product desired, and abundance of particular mRNA transcripts of interest.
3. Active enzymes can also be recovered from microdissected frozen tissue sections. We have employed two separate approaches to enzyme studies. Microdissection of standard frozen tissue sections as described in **steps 1** and **2** can be used to recover proteins in their native conditions. In our hands, this approach has worked very well for tissue inhibitors of metalloproteinases (TIMPs) which are stable, small molecular weight proteinase inhibitors. Less stable enzymes may require approaches that protect the integrity of the enzymes during the dissection procedure, particularly if the investigator requires accurate quantitation of enzyme activity. Placement of frozen tissue sections directly on agarose-coated slides can be helpful in maintaining enzyme stability ***(4)***. Additionally, the agarose gels can be prepared or soaked in custom buffers that will bathe the frozen section prior to and during the microdissection (e.g., pH, salt concentration, proteinase inhibitors, etc. can be varied specifically for a given enzyme). Some members of our group also prefer to use the agarose-coated slide microdissection approach for recovery of RNA.

2.3.4. Gel Slide Microdissection

Slides for microdissection are prepared by placing 200 µL of warm agarose on standard uncoated glass slides, covering with a glass slip, and allowing the gel to polymerize. Remove the glass slip from the slide and immediately place the frozen tissue section onto the agarose gel. The freshly cut section should be transferred directly from the cryostat to the agarose coated slide. Microdissection can be performed similar to the method described in **Subheading 2.3.2.** and **2.3.3.**, however, the dissector may find it easier to "tease" the tissue apart since the tissue remains bathed in the fluid from the gel, and can be gently pulled apart. The tissue will also separate along tissue planes (e.g., stroma and epithelium will easily separate from each other). The dissected tissue can be gently picked up from the slide, or alternatively the dissector can use the needle to physically cut the agarose and procure both the agarose and the tissue fragment together.

2.4. Processing of Microdissected Tissue

2.4.1. DNA

If the amount of microdissected material is substantial (>10,000 cells), then any of the standard procedures for isolating DNA are acceptable. However, if the number of cells procured is minimal, e.g., dissection of premalignant lesions, then a simple one-step DNA preparation for PCR is recommended. The resulting DNA preparation is not "clean," but is sufficient for PCR-based analyses. For example, a premalignant lesion in prostate may contain 200–500 cells, thus, dissection of the identical lesion from four consecutive frozen section recuts procures 800–2000 cells in total. Procured cells are immediately resuspended in 20 μL of solution containing 10 m*M* Tris-HCl, 1 m*M* ethylenediamine tetra-acetic acid (EDTA), 1% Tween-20, 0.1 mg/mL proteinase K, pH 8.0, and incubated overnight at 37°C. Longer incubation times and/or higher concentrations of proteinase K have been reported to improve the quality of DNA recovered from fixed tissue sections. The mixture is boiled for 8 min to inactivate the proteinase K and 0.2–1% of this solution is used for PCR analysis. If a lesion or particular cell population containing very small numbers of cells is desired, it is recommended to dissect the identical lesion from as many consecutive recuts as possible to maximize the total number of cells procured. If this is not possible, the few procured cells can be placed into 10 μL of extraction buffer, and 5–10 μL of this solution used in the PCR reaction. There is no optimal number of cells that should be collected from a microdissection since results vary significantly depending on the tissue source. For frozen tissue, approx 50–100 cells/μL of extraction buffer is recommended as a good starting point. Storage of frozen tissue after processing in the extraction buffer should be at 4°C for a short-term working solution, or –70°C for long-term storage. Freezing and thawing should be minimized. Excellent PCR amplification can be achieved with the samples stored at 4°C for a year or more after microdissection.

Amplification of DNA from paraffin-embedded sections after storage is variable, presumably related to the initial fixation conditions of the tissue. If possible, we perform the majority of the PCR reactions within the first 7–10 d after microdissection. Short-term storage should be at 4°C, and the solution should be covered with 50–100 μL of mineral oil. Long-term storage should be at –70°C. The majority of cases will continue to amplify well, however, a significant number of cases will no longer produce a strong PCR product and will need to be remicrodissected.

2.4.2. RNA

Recovery of RNA from microdissected tissue can be obtained by standard methods. In our laboratory we utilize the RNA microisolation kit from Stratagene (La Jolla, CA). Dissection of a minimum of several thousand cells is recommended, however, procurement of 10,000–20,000 cells is preferred for reliable RNA recovery. For example, in prostate tissue we routinely microdissect a minimum of 10,000 cells of normal epithelium or tumor that provides ample RNA for several RT-PCR studies. The microdissections for RNA are more difficult than that for DNA since a larger number of cells are required, and the dissections must be carefully performed to avoid contamination by cell types other than those desired by the dissector. High-copy mRNA transcripts from adjacent contaminating cells can produce erroneous results.

2.4.3. Enzymes

Optimal recovery solutions for enzymes must be determined empirically by individual investigators. In our studies of proteinases and inhibitors, standard buffers suitable to the individual enzymes have worked well. The number of cells required to analyze levels of a given enzyme must also be determined empirically. Assays based on activity of an enzyme can often be performed on very small numbers of cells because of the amplification of the enzyme substrate. For example, cathepsin B or telomerase activity can be determined from minute quantities of microdissected tissue.

2.5. PCR Interpretation

Since many studies utilizing microdissected tissue require PCR, investigators should be aware of the pitfalls that can accompany this technique. Proper controls and care to avoid contamination are essential. The advantage of PCR-based assays is that the starting material required is minimal so assays can be performed in duplicate or triplicate to ensure reproducibility. It is also recommended to perform dilutions of samples to assess optimal PCR conditions of individual samples, and to determine the sample concentration that is in the linear range for PCR amplification. The number of PCR amplification cycles should be kept as minimal as possible. For RNA studies, comparison of a gene of interest to a known housekeeping gene using the same tissue sample is an excellent way to obtain a relative level of gene expression.

Tissue microdissection is an especially powerful technique to examine loss of heterozygosity in tumors. The availability and abundance of microsatellite markers throughout the genome combined with the large number of PCR reactions that can be performed from a single microdissection allow investigators to carefully map allelic deletions in tumors. Also, the purity of the tumor sample results in clear, nonequivocal LOH scoring.

Care should be taken to avoid conditions that will result in artifactual allelic dropout when performing LOH studies. This is particularly a problem with microdissection of small numbers of cells from paraffin-embedded tissue, especially if the tissue is not optimally fixed. In our laboratory, we control for this phenomenon by carefully analyzing multiple normal cell samples from the same tissue block that contains the tumor. If reliable, consistent heterozygosity is not obtained from the normal tissue, the sample is discarded.

Another consideration when microdissecting tumors is the issue of tumor heterogeneity. Microdissection allows the investigator to selectively sample small regions of interest that potentially could lead to erroneous conclusions regarding the status of a given molecule if only one area of the tumor is assessed. Sampling from multiple, separate regions of the tumor can avoid this problem.

3. Applications

Tissue microdissection combined with PCR allows investigators to study select cell populations in normal and diseased tissue encompassing diverse pathological processes. In cancer research for example, study of normal, dysplastic, *in situ*, and invasive tumor cells as they exist in vivo can be performed, providing a unique opportunity to examine the nature and sequence of genetic alterations that occur during tumor progression. The technique described in this chapter has been particularly useful in the study of prostate and breast cancers. Both tumors typically contain an admixture of tumor and nontumor cells, and have been proposed to initially develop and progress as small premalignant lesions prior to developing into overt cancers.

The following section provides an overview of some of the work in our laboratory applying tissue microdissection to human tissue samples, starting with studies on prostate and breast cancer. The work is presented in brief summaries and highlights the concepts of using microdissection. Additional studies of renal cancer, colon cancer, neuroendocrine tumors, esophageal cancer, melanoma, patients with synchronous tumors, and infectious diseases are presented to highlight unique applications of the technique. Special emphasis is placed on premalignant lesions of prostate, breast, kidney, and esophagus.

We highly recommended that a pathologist be involved in studies utilizing tissue microdissection. Without exception all studies we have undertaken have been significantly enhanced or have progressively evolved based on a thorough understanding and appreciation of the histopathological appearance of various disease processes. The interplay among pathological assessment, clinical disease behavior, and genetic analysis has been especially informative.

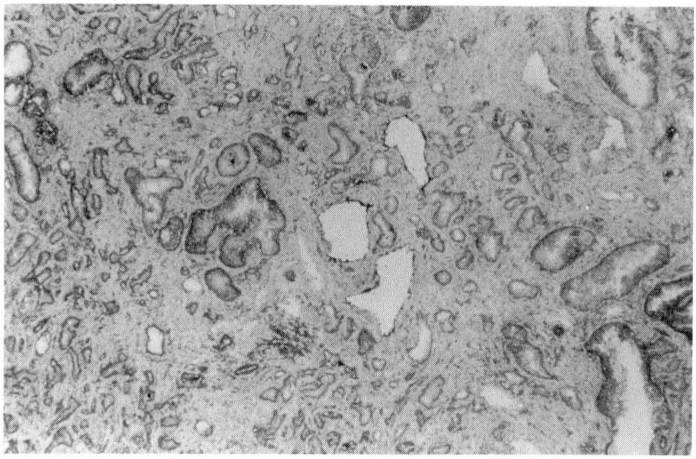

Fig. 1. Histologic field of prostate cancer before and after microdissection of three invasive glands. Normal epithelium, stroma and inflammatory cells were excluded from analysis (H&E stain, magnification ×40). Reprinted with permission from **ref.** ***16***.

3.1. Prostate Cancer

3.1.1. Analysis of Microdissected Prostate Carcinomas Reveals High Frequency of Allelic Loss on Chromosome 8p21 (see ***ref. 8****)*

In order to investigate tumor suppressor genes in prostatic neoplasms we performed a comprehensive LOH study on 99 tumors from prostate cancer patients. Previous reports have suggested that chromosomes 7q, 8p, 10q, and 16q may harbor tumor-suppressor genes important in prostate cancer, however, the relative frequency of LOH at these regions has not been clearly determined ***(9–15)***. In this study, we used tissue microdissection and 45 microsatellite markers, spanning the short arm of chromosome 8, the long arm of chromosome 7, the long arm of chromosome 10, and the long arm of chromosome 16, to examine LOH in prostate cancer (**Figs. 1** and **2**). Additional loci examined included known tumor suppressor genes DCC, p53, and BRCA1. The highest rate of LOH was found on chromosome 8p. The overall rate of loss was 85.9%. Two separate regions of loss were identified. Allelic loss was highest in the proximal region of the chromosome at 8p21 with >80% of the cases exhibiting a deletion in this region. Highest LOH was in the vicinity of microsatellite markers D8S136 and D8S137. No correlation was observed between LOH of this region and grade or stage of disease. All other areas of the genome studied showed significantly lower rates of LOH than 8p21. Chromosome 8p22 showed LOH of 25%, chromosome 16q24 showed LOH of 28%,

LOSS BY LOCUS:

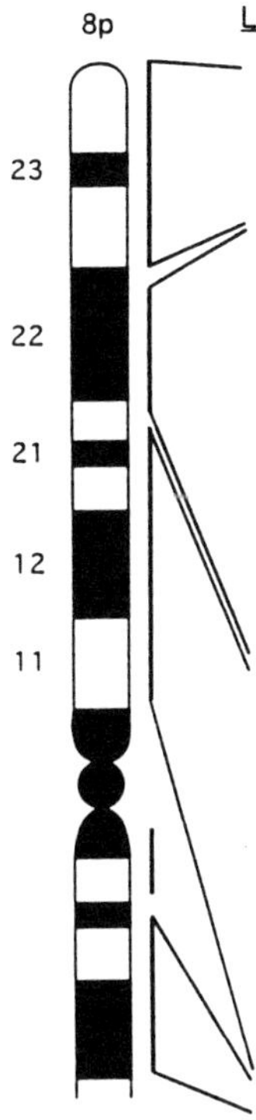

Locus	LOH/ Inf. cases	%LOH
264	10/80	13%
277	1/82	1.2%
349	6/74	8.1%
351	9/54	17%
552	4/60	6.7%
511	5/69	7.2%
549	8/48	17%
602	13/71	18%
254	9/51	18%
261	14/73	19%
258	8/73	11%
LPL5GT	3/34	8.8%
LPLTET	25/66	38%
298	17/52	33%
LPL3GT	24/65	37%
133	39/64	61%
136	53/75	71%
NEFL	35/56	63%
137	43/58	74%
339	27/67	40%
87	22/60	37%
259	18/58	31%
283	16/75	21%
278	6/45	13%
ANK	14/54	26%
285	3/70	4.3%

Fig. 2. Illustration of LOH rates on the short arm of chromosome 8. Allelic loss is highest in the vicinity of microsatellite markers D8S133, D8S136, NEFL, and D8S137 in band 8p21. A second region of LOH is seen in band 8p22.

and the highest rate of LOH on chromosome 10q was 8%. Allelic loss on chromosome 7q and tumor-suppressor genes DCC, p53, and BRCA1 was <10%. The present comprehensive study of LOH of 99 prostate cancer patients clearly identifies chromosome 8p21 as a region of high allelic loss in prostate cancer. Use of tissue microdissection and PCR analysis of LOH allows for definitive scoring of allelic loss.

*3.1.2. Allelic Loss on Chromosome 8p21 in Microdissected Prostatic Intraepithelial Neoplasia (PIN) (see **ref. 16**)*

Human prostate cancer has been proposed to progress through an *in situ* tumor phase known as PIN prior to evolving into overtly invasive cancer. PIN is frequently found in association with prostate carcinoma, and the cells of PIN have several histologic features similar to those of invasive prostate cancer cells *(17)*. However, there are currently few molecular studies examining the relationship of PIN and invasive prostate cancer. The classification of PIN as a neoplastic entity has relevance both to the understanding of the fundamental genetic alterations that occur early in the development of human prostate cancer as well as to the potential use of PIN as a clinical marker of malignant transformation prior to the development of prostate cancer. In this study, we utilized tissue microdissection to examine allelic loss on chromosome 8p21 in microdissected samples of normal prostatic epithelium, high-grade PIN, and invasive prostate carcinoma from the same patients. Among 30 patients with concomitant cancer and PIN, we found loss of heterozygosity on chromosome 8p21 in 63% (34/54) of loci of PIN examined, suggesting that abnormalities on chromosome 8p21 may be important in the early stages of prostatic carcinoma development. Several cases in which multiple loci of PIN from the same patient were sampled showed different patterns of allelic loss, including loss of opposite alleles. Fifty-five percent (16/29) of the prostate carcinomas contained a potential precursor PIN focus based on allelic loss pattern. Substantially lower rates of LOH in PIN were observed on chromosomes 10q and 16q. Based on chromosome 8p21 LOH, our results are consistent with the hypothesis that PIN is a neoplastic element that arises multilocally within the prostate gland, and that a subset of these lesions progress to become carcinoma. Combined with our previous study of LOH in a large series of prostate tumors, we conclude that chromosome 8p21 contains a tumor-suppressor gene fundamental to the early development of prostate cancer.

*3.1.3. Identification of a Novel Zinc Finger Containing Gene Upregulated in Prostate Tumors (see **ref. 18**)*

Evaluation of differential gene expression between normal and tumor cells is an important aspect of cancer research. Several methods have been established to compare gene expression between separate cell populations, and studies with cell lines have resulted in identification of several differentially regulated genes in human tumors *(19–23)*. However, less work has been done assessing and comparing levels of gene expression of normal epithelium and corresponding tumors as they exist in vivo. In this study we examined differential gene expression in microdissected populations of normal epithelial and corresponding tumor cells from the same patients. Differences in expression of zinc finger genes in normal and tumor RNA was detected using RT-PCR with an arbitrary and a degenerate zinc finger PCR primer set. All experiments were performed in duplicate to minimize misinterpretation of PCR artifact bands. A 130-bp product was identified selectively in a prostate tumor sample. Extraction, sequencing, and basic local alignment search tool (BLAST) analysis of this band (PB-39) revealed it to be a cDNA clone from the expressed sequence tag (EST) database. Preliminary analysis and cloning has shown PB-39 to be a novel zinc finger gene upregulated in prostate tumors. In conclusion, tissue microdissection and RT-PCR with degenerate primers has identified a novel zinc finger containing a gene upregulated in prostate cancer. Analysis of mRNA expression in microdissected tissue represents a method to quantitate expression of known genes in defined cell populations, as well as search for novel tumor-specific genetic alterations.

*3.1.4. Construction of a Representative cDNA Library from PIN (see **ref. 24**)*

cDNA libraries have proven essential to gene cloning experiments, positional gene cloning efforts, differential gene expression studies, and EST sequencing/mapping projects. Traditionally, cDNA libraries are constructed from large amounts of purified mRNA obtained from either large tissue samples or tissue-culture cell lines ***(25,26)***. We investigated the feasibility of combining microdissection of specific cell populations with cDNA library construction to generate specific cDNA libraries corresponding to cells that derive exclusively from a distinct tumorigenic stage. The cell type chosen for this study was the microscopic prostatic lesion, PIN.

Total RNA was extracted from microdissected cells and converted to blunt-ended, double-stranded cDNA by oligo(dT)-mediated reverse transcription followed by linker addition. A linker-specific primer was used to amplify the cDNA and the resulting PCR product subcloned. A total of 154 clones were sequenced and results indicate 81.5% of the clones derived from either known genes, anonymous ESTs, or unique transcripts, indicating a library of high quality. There was very little redundancy of known genes and ESTs demonstrating a high degree of library complexity. These results demonstrate the feasibility of constructing representative cDNA libraries from specific microdissected cell populations, even from small populations such as PIN. This method will allow for identification of transcripts specifically expressed in cells of a distinct origin and tumorigenic stage.

3.2. Breast Cancer

3.2.1. Identical Allelic Loss on Chromosome 11q13 in Microdissected In Situ *and Invasive Human Breast Cancer (see **ref. 27**)*

The current model for the development of breast cancer proposes that invasive breast tumors originate as atypical hyperplasias of the epithelium, progress to *in situ* lesions, and eventually develop into invasive cancers ***(28,29)***. However, little molecular evidence exists that supports this model. Previous methods of study have not allowed investigators to specifically examine genetic alterations in preinvasive breast lesions. In this study, we used tissue microdissection to investigate LOH on chromosome 11q13 in *in situ* and invasive breast tumor from the same patients. We examined chromosome 11q13 LOH in both *in situ* and invasive lesions of the breast, as compared to normal breast epithelium in 41 cases of sporadic breast cancer. LOH on chromosome 11q13 was found in 24 of 36 (67%) of the informative invasive breast cancer cases using two polymorphic DNA markers specific for this region (INT2 and PYGM). Twenty-one of the cases which demonstrated LOH in the invasive tumor also contained *in situ* carcinoma in the same tissue section. Seventy-one percent (15 of 21) of the microdissected *in situ* tumor showed LOH, and each case showed loss of the identical allele in the corresponding invasive tumor cells. The results of this study suggest that a tumor-suppressor gene located on chromosome 11q13 may play an important role in the early stages of development of sporadic human breast cancer. This finding provides molecular genetic support for the hypothesis that invasive breast cancer arises from *in situ* lesions.

3.2.2. Genetic Analysis of Atypical Ductal Hyperplasia (ADH) and In Situ *Breast Carcinoma (*see ***ref.* 30***)*

Demonstration of identical allelic loss on chromosome 11q13 in synchronous in situ and invasive ductal breast carcinoma has provided molecular evidence of the progression of ductal carcinoma *in situ* (DCIS) to invasive carcinoma. Very little is currently known of the molecular events that characterize other putative premalignant lesions such as ADH. In this study, we investigated the pattern of deletion on chromosome 11q13 in ADH, and various histologic types of *in situ* carcinomas. Twenty-nine cases of *in situ* carcinoma, and 12 cases of pure ADH were studied in patients without concomitant invasive breast cancer. Tissue microdissection of tumor/hyperplasia and normal cells was performed from paraffin-embedded sections. DNA was extracted and used for PCR analysis with polymorphic markers INT2 and PYGM on chromosome 11q13.

LOH was identified in 7/28 (25%) informative *in situ* carcinoma samples and in 0/10 informative ADH cases (**Fig. 3**). LOH was identified in 6/17 (35%) informative DCIS with at least moderate atypia (3 comedo, 2 solid with necrosis, 1 solid). In contrast, only 1/12 *in situ* tumors with mild atypia showed LOH (one lobular carcinoma). The present results show that LOH at 11q13 is identi-

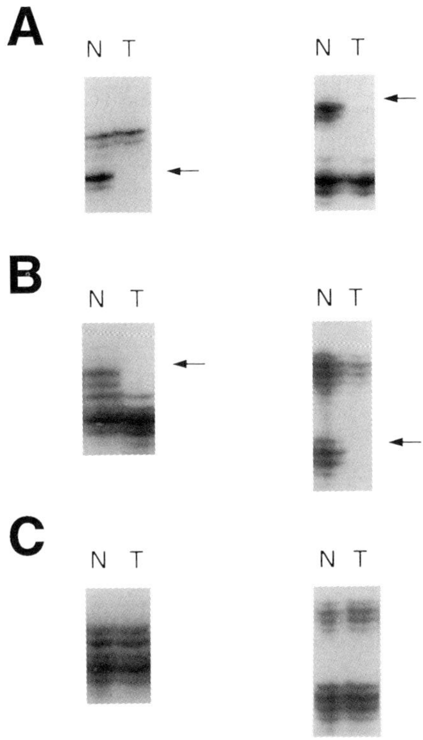

Fig. 3. Denaturing gel electrophoresis analysis of microsatellite markers on chromosome 11q13. **(A)** Two cases of *in situ* breast carcinoma with allelic loss at marker INT-2. Deleted allele in the tumor samples is indicated by the arrows. **(B)** LOH in two *in situ* breast cases analyzed at marker PYGM. **(C)** Retention of heterozygosity in two cases of ADH at markers INT-2 (left) and PYGM (right). Reprinted with permission from **ref. *30***.

fied in a significant proportion of pure *in situ* breast carcinomas, suggesting that a tumor-suppressor gene located at this locus may play a role in early stages of breast cancer progression. Higher rates of LOH at 11q13 were identified in high-grade DCIS. However, no significant LOH was identified at 11q13 in ADH, a putative premalignant lesion that has been proposed to occur prior to *in situ* carcinomas.

3.2.3. Detection of LOH in Microdissected Fine Needle Aspiration (FNA) Specimens of Breast Carcinomas (see **ref. 31**)

Improvement in screening programs have allowed mammographies to detect small breast lesions. FNA of these lesions has assumed an expanding role in the detection and diagnosis of early stage breast cancer. In this study, we evaluated the feasibility of detecting LOH in cytologic specimens obtained by FNA of breast carcinomas. Using tissue microdissection applied to cytologic specimens, representative cells were dissected from cytologic slides for LOH analysis. The results were compared to those identified in microdissected samples from the corresponding tissue blocks available in each case. Twenty archival cases of invasive ductal breast carcinomas with an available representative histologic block and a cytologic slide from a FNA were studied. A stepwise dissection of the cytologic specimen was performed. The area surrounding a cluster of tumor cells was first gently scraped to remove contaminating normal epithelial or inflammatory cells, then the cluster of malignant cells was dissected. From 3–6 tumor cell clusters, each consisting of 10–30 cells, were dissected in each case. Nontumor cells were also procured from the cytologic slides. Identical allelic patterns were identified in the dissected sample from the cytologic smear and from the histologic section in each case studied.

The present study confirms that FNA of breast lesions provide adequate samples for genetic analysis using tissue microdissection. A high reproducibility of LOH patterns in microdissected FNA samples compared to histological sections was observed. With the expanding knowledge of molecular changes and identification of genetic markers in breast cancer, the application of genetic analysis to FNA specimens will help in providing future diagnostic, as well as prognostic information for patients.

3.3. Renal Cancer

3.3.1. LOH on 3p25–26 in Premalignant Renal Lesions in Von Hippel-Lindau (VHL) Patients (see **ref. 32**)

VHL disease is an autosomal dominant disorder characterized by the development of multiple tumors in different organs. VHL patients develop a spectrum of multifocal bilateral renal lesions which include benign cysts, atypical cysts, *in situ* renal cell carcinoma (RCC), and RCC. Histologically, cysts with a single cell layer lining are considered benign, those with lining two to three cell layers are classified as atypical, and those lesions with more than three cell layers are classified as RCCs. These lesions are often microscopic and surrounded by normal somatic tissue which makes it technically difficult to obtain specific cells of interest for genetic evaluation. Previous studies suggested that renal cysts in VHL patients might represent precursor lesions of RCC, however, there is no direct molecular evidence of the relationship between RCC and benign or atypical cysts. We analyzed 21 renal lesions from two informative patients for VHL gene deletions (3p25–26 LOH) using tissue microdissection of formalin-fixed, paraffin-embedded tissue and PCR-based single-stranded conformational polymorphism (SSCP) analysis. Tissue microdissection allowed procurement of specific cell populations from lesions less than one high-power field in size, and from a single layer lined cysts in formalin-fixed, paraffin-embedded tissue. DNA was PCR-amplified and analyzed by SSCP. Chromosome 3p LOH was detected in nine RCC, five microscopic RCC *in situ*, five atypical cysts, and one benign cyst. The study shows for the first time a chromosome 3p deletion in *in situ* RCC as well as in atypical and benign renal cysts in VHL patients. Our findings suggest that atypical and even some benign cysts may represent early stages in the development of RCC. Thus, the histologic classification of renal lesions in VHL is arbitrary, and microscopic renal cysts have malignant potential.

3.3.2. Identical Genetic Changes Are Detected in Different Components of Wilms' Tumors (WTs)

WT is an embryonal malignancy of the kidney that affects approx 1 in 10,000 infants and young children. The typical histologic picture is a triphasic pattern consisting of blastema, epithelium, and stroma. It is generally assumed that the tumor arises from metanephric blastema, but the occasional presence of various heterologous elements such as cartilage, bone, or striated muscle has produced controversy over its histogenesis. A genetic locus on chromosome 11p13 has been linked to the initiation of Wilms' tumorigenesis. We studied the WT-1 locus in the various histologic elements seen in WT using tissue microdissection. Eighteen cases of sporadic WT showing the broad range of morphologies characteristic of these tumors were studied. We examined LOH using polymorphic markers D11S904 and D11S1392 on chromosome 11p13. Nine of 18 cases (50%) showed LOH at WT-1. In the nine cases showing LOH, three cases were triphasic tumors, one of which contained a fully differentiated striated muscle component. The different histologic elements (blastema, stroma, epithelium, and, in one case, striated muscle) of these three triphasic tumors were separately microdissected and analyzed for LOH at WT-1. LOH was seen in all the histological components, including striated muscle elements (**Fig. 4**). In addition, the identical allele was deleted in all components. These results show that when LOH at WT-1 is detected in WT it is present in all elements of the tumor, and suggest that heterologous elements, if present, are of tumor origin and do not represent normal cells trapped within tumor.

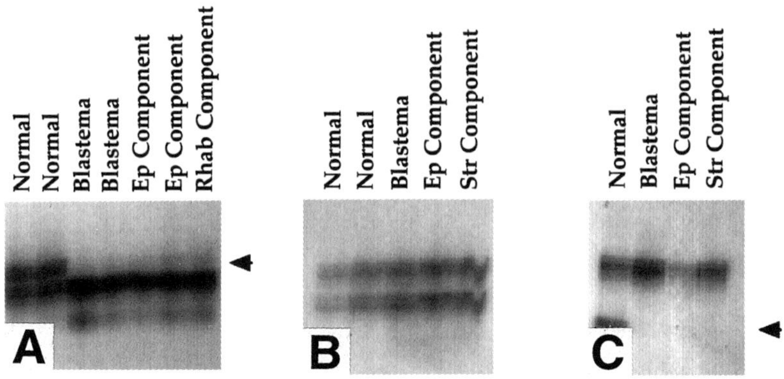

Fig. 4. Denaturing gel electrophoresis analysis of three cases of WT. **(A)** Identical LOH at WT1 on chromosome 11p13 in all the components of the tumor. Two blastemal regions, two epithelial region and an area of rhabdoid differentiation were analyzed. **(B)** A case of WT analyzed at the WT1 gene in blastemal, epithelial, and stromal components. No LOH was observed in this case. **(C)** A case of WT demonstrating identical allelic loss in blastemal, epithelial, and stromal components.

3.4. Colon Cancer

3.4.1. Increased Gelatinase A (MMP-2) and Cathepsin B Activity in Invasive Tumor Regions of Human Colon Cancer Samples (see **ref. 4**)

Gelatinase A (MMP-2) and cathepsin B are proteinases that have been proposed to participate in human tumor invasion and metastasis. Precise quantitation of the activity of these enzymes in invading tumors has not been previously described. We utilized tissue microdissection to determine levels of enzyme activity in specific microscopic areas of invasive human colon cancer. Tissue specimens smaller than one high-power field were extracted from the samples and analyzed. Increased levels of proenzyme and active enzyme forms of gelatinase A (MMP-2), as well as increased cathepsin B activity were localized in regions of tumor invasion as compared to a matched number of normal epithelial cells from the same patient. Levels of progelatinase B (MMP-9) were also increased in the tumors, however, we did not observe activation of this enzyme. To investigate the mechanism of gelatinase A activation, we amplified DNA of specific, microdissected tumor cell populations using PCR. We did not detect a mutation in the activation locus of the enzyme in any of the tumors studied, suggesting activation may be the result of upregulation of a tumor-associated gelatinase A activating species. Our results indicate that gelatinase A and cathepsin B activity are significantly upregulated in fields of invasive colon tumors. This is the first study that compares the enzyme activity of these two proteinases in specific regions of tumor invasion with a matched number of corresponding normal epithelial cells from the same patient. Tissue microdissection of frozen tissue sections may prove valuable in the study of enzymes in human tumor samples.

3.4.2. Detection of VHL Gene Deletion in Microdissected Sporadic Human Colon Carcinoma Specimens (see **ref. 33**)

Previous studies have suggested that inactivation of tumor-suppressor genes on chromosomes 5q, 17p, 18q, and 8p play a role in the development of colorectal carcinoma. However, chromosome 3p at the VHL disease gene locus (3p25–26) has not been previously implicated in the development or progression of sporadic colorectal carcinoma. We analyzed VHL gene alterations on chromosome 3p in sporadic human colon carcinomas and adenomas using tissue microdissection of both paraffin-embedded and frozen human tumor specimens. VHL disease gene deletion was detected by PCR and SSCP in microdissected colon carcinoma specimens. Allelic loss of VHL gene was detected in 7 of 11 (64%) informative patients who underwent colectomy for primary sporadic colon carcinoma. However, no allelic loss of VHL gene was demonstrated in colonic adenomas of eight informative patients. Our results indicate that VHL disease gene deletion frequently occurs in sporadic colon carcinoma. Since this deletion was not present in adenomas, VHL gene may play a role in colonic carcinogenesis and represent a relatively late event in colonic neoplasia progression.

3.5. Neuroendocrine Tumors

3.5.1. Detection of Differential Patterns of 11q13 LOH in Multiple Parathyroid, Pancreatic, and Duodenal Tumors from Individual Multiple Endocrine Neoplasia Type 1 (MEN1) Patients (see **ref. 34**)

Familial MEN1 (FMEN1) is an autosomal dominant hereditary disorder characterized by tumors of multiple parathyroid glands (90–97%), endocrine pancreas (30–82%) and duodenum (25–60%), and adenomas of anterior pituitary gland (60%). The putative MEN1 tumor-suppressor gene has been linked to chromosome 11q13. MEN1 studies to date, with one exception, have been restricted to analysis of a single parathyroid gland or a single pancreatic endocrine neoplasm from individual members of MEN1 kindreds. It has been previously unknown whether the LOH pattern on the wild-type allele in the same patient varies between individual MEN1 tumors or among tumors of different histologic origins. We studied 44 histologically different microdissected tumors from nine unrelated FMEN1 patients with eight polymorphic markers spanning the area of the putative MEN1 gene on 11q13 to investigate a novel approach to gene mapping using multiple, small, archival endocrine lesions from the same patient. Because contaminating small vessels in microscopic sections of parathyroid lesions confound definitive LOH scoring, we utilized CD34 immunostain-guided microdissection of tumors to avoid endothelial contamination. Significantly higher LOH was detected in parathyroid hyperplasia (100%) and endocrine tumors of pancreas (83%), compared to gastrinomas of duodenum (21%). X Chromosome inactivation analysis of multiple regions within seven individual hyperplastic parathyroid glands demonstrated one to several independent tumor clones within each gland suggesting a multifocal origin for MEN1-associated parathyroid disease. Differential LOH patterns in multiple tumors in the same patient constitutes a new approach to tumor-suppressor gene mapping in familial disorders. Tissue microdissection facilitates this approach by allowing procurement of multiple small tumors from individual patients.

*3.5.2. Detection of Frequent Allelic Deletions of MEN1 Gene Locus (11q13) in Gastric Enterochromaffin-Like (ECL)-Cell Carcinoids in Patients with Zollinger-Ellison Syndrome (ZES) and MEN1 (see **ref. 35**)*

It is currently not known if nonclassical tumors in MEN1 patients such as adrenal adenomas and gastric carcinoids occur secondary to inactivation of the putative MEN1 tumor-suppressor gene. Only three gastric ECL-cell carcinoids in ZES/MEN1 patients were previously reported, and results are not conclusive. We studied 20 ECL-cell gastric carcinoids from six ZES/MEN1 patients for LOH on 11q13 to assess whether the MEN1 tumor-suppressor gene is involved in the genesis of these lesions. Five duodenal microgastrinomas from three of the patients were also evaluated. All tumors were immunostained with gastrin antibody to differentiate gastrinomas from ECL-cell carcinoids. Tumor and corresponding normal tissue were microdissected from routine histological sections from endoscopic biopsies. Extracted genomic DNA was PCR-amplified with seven polymorphic markers on 11q13. LOH on 11q13 was detected in 15 of 20 ZES/MEN1 ECL-cell carcinoids (75%). Four of five MEN1-related microgastrinomas demonstrated 11q13 LOH (80%). The three patients with ECL-cell carcinoids and gastrinomas showed loss of the identical allele in both tumor types. The study shows that gastric ECL-cell carcinoids in ZES/MEN1 patients have a high rate of allelic loss in MEN1 gene locus (11q13). The frequency and patterns of allelic deletions are similar in MEN1-associated ECL-cell gastric carcinoids and duodenal gastrinomas implicating the MEN1 gene in their tumorigenesis. Thus, gastric ECL-cell carcinoid is an independent tumor type of MEN1 that shares a common developmental mechanism with pancreatic and parathyroid MEN1 tumors.

3.6. Esophageal Cancer

3.6.1. Barrett's Esophagus: Metaplastic Cells with LOH at the Adenosis Polyposis Coli (APC) Gene Locus Are Clonal Precursors to Invasive Adenocarcinoma (see **ref. 36***)*

Barrett's esophagus is associated with an increased risk of developing invasive adenocarcinoma. It is difficult to detect those patients at high risk, and most methods rely on the histologic recognition of high-grade dysplasia from surveillance endoscopic biopsy specimens. The goal of this study was to investigate APC gene alterations in different histologic regions representative of the putative stages of progression from Barrett's epithelium (BE) to carcinoma. In addition, clonality studies were performed to assess whether BE, dysplasia, and carcinoma are derived from the same cellular origin. Twelve resection specimens from the Massachusetts General Hospital were selected for the study. From all 12 cases, at least 3 areas of invasive adenocarcinoma were microdissected directly from H&E-stained sections and analyzed for LOH at the APC gene locus. Five cases showed APC gene deletions in the adenocarcinomatous component, thus additional microdissection of the following areas of interest were performed on these five cases; normal control tissue (esophageal squamous epithelium, gastroesophageal junctional epithelium, lymphoid tissue), BE distant from dysplasia (more than one 10× power field, e.g., 2 mm, separated from dysplastic area), BE adjacent to dysplasia (less than one 10× power field, e.g., 2 mm, separated from dysplastic area), dysplasia, and invasive carcinoma. Deletion of the identical allele at the APC gene locus was detected both in invasive adenocarcinoma and dysplasia in all five cases. Clonality analysis using X chromosome inactivation in the two female cases showed the dysplasia and invasive adenocarcinoma to be clonal. BE adjacent to dysplasia showed LOH of the APC genc in two of five cases, and the deleted allele was the same as that in the adjacent dysplastic epithelium in both cases. One of the female patients with LOH of the APC gene in BE adjacent to dysplasia showed monoclonality in the BE, and the clonal pattern was the same as that of the adjacent dysplasia and invasive adenocarcinoma. All five cases of BE distant from dysplasia, and all normal tissues (esophageal squamous epithelium, gastroesophageal junctional epithelium, lymphoid cells) showed no LOH of the APC gene. Clonality analysis showed these lesions, as well as normal tissue, to be polyclonal. These data show that genotypic alterations in the APC gene precede the histopathologic changes of carcinoma in Barrett's esophagus syndrome (**Fig. 5**). Therefore, genotyping of Barrett's metaplastic epithelium may supplement the histopathologic evaluation of Barrett's esophagus.

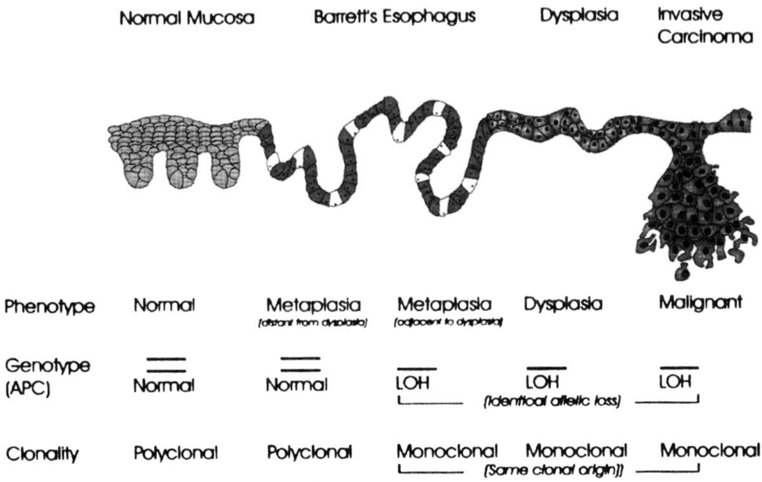

Fig. 5. Schematic illustrating phenotypic and genotypic changes seen in Barrett's esophagus and adjacent cancer. Invasive tumor, dysplasia, and metaplasia adjacent to dysplasia all showed identical allelic loss and clonality patterns. Reprinted with permission from **ref.** ***36***.

3.7. Melanoma

3.7.1. Genetic Progression of Human Melanoma Based on Tissue Microdissection and Comparative Genomic Hybridization (CGH) (see *ref.* 37)

Human cutaneous malignant melanoma progresses through a series of well-defined clinical and histopathological stages. It has been assumed that neoplastic progression of this disease advances from a common acquired nevus or dysplastic nevus through the primary radial growth phase (RGP), primary vertical growth phase (VGP), and finally to distant metastasis. However, it has never been directly shown that VGP is clonally derived from RGP. Furthermore, it has not been possible previously to conduct a detailed genetic analysis on pure tumor cells from archival material because the lesions are a heterogeneous mixture of normal and neoplastic cells, and the entire specimen must be excised and fixed for clinical diagnosis.

In this study we microdissected normal cells, RGP, and VGP from three archival paraffin-embedded specimens, and analyzed DNA copy number changes in tumor cells using CGH. Similar loss of genetic material on chromosomes 1p, 16, 17q, and 22 was observed in the RGP and VGP cells in the three cases. The results of the present study imply that VGP cells are derived from RGP during progression of melanoma to a metastatic phenotype. Tissue microdissection and CGH are helpful in identifying the sequence of genetic changes which occur in progressive melanoma stages.

3.8. Patients with Synchronous Tumors

*3.8.1. Molecular Analysis of Concomitant Uterine and Ovarian Endometrioid Tumors (see **ref. 38**)*

The presence of simultaneous carcinomas involving both the ovary and uterine corpus presents a diagnostic challenge, particularly if the tumors have a similar histology. The classification of these lesions as either two separate primary tumors, or as a single primary tumor with a metastasis has significant implications with respect to patient prognosis and recommendations for therapy. Several morphologic criteria have been proposed as guidelines for the classification of these lesions, however, certain cases remain difficult to evaluate. The application of current molecular biology techniques to pathologic specimens can provide genetic information that can be helpful in establishing the relationship between concurrent neoplasms. In this study, we used tissue microdissection and polymorphic DNA markers on chromosomes 17q21.3–22 and 11q13 to study LOH in 13 patients who presented with endometrioid tumors in both the uterus and ovary. Ten of the 13 cases showed LOH in one or both tumors. In 8 of the 13 cases the detected LOH on either chromosome 17q21.3-22 or 11q13 occurred selectively in only one of the two tumor sites. The results of this study suggest that the 8 cases with LOH selective for one tumor site represent patients with two separate primary tumors. Allelic loss studies may serve as helpful clinical adjunct studies in the assessment and management of patients with synchronous cancers. Continued analysis of the usual patterns of LOH in individual tumor types, as well as the identification and characterization of additional tumor suppressor genes may provide useful markers in the future assessment of cancer patients.

3.9. Infectious Diseases

3.9.1. Kaposi's Sarcoma-Associated Herpes-Like Virus (KSHV) Detected in Visceral Kaposi's Sarcoma (KS) from HIV+ Patients

Recently, Moore and Chang amplified DNA from KS and showed it to be a new member of the human herpes virus family ***(38a)***. Subsequently, they and others have shown that KSHV is associated with cutaneous KS in both HIV+ and HIV– patients, as well as with body cavity-based lymphomas in HIV+ patients. No studies have been published to date detecting KSHV in visceral KS lesions or other vascular lesions in HIV+ patients. Using tissue microdissection of formalin-fixed, paraffin-embedded tissue, we tested for the presence of KSHV in selected vascular lesions in HIV+ patients. Five cases of KS were positive for KSHV DNA and consisted of the following: one case of skin KS, one of duodenal KS, and two of lung KS. In addition, surprisingly, one case of cavernous hemangioma of the skin in an HIV+ patient was positive. One of the KSHV bands was cut out of the polyacrylamide gel, eluted into TE, reamplified using the KSHV primers, and electrophoresed on an agarose gel. A single, 233-bp product was cut out of this gel and sequenced using the cycle sequencing method. A total of 163 bp were sequenced of this 233-bp product and was found to differ by only 3 bp from the Genbank sequence submission for KSHV minor capsid gene (e.g., 98% homology with KSHV). In the current study, we have shown that KSHV DNA was associated with cutaneous and visceral KS in HIV+ patients. Through the use of tissue microdissection, we have further shown this association to be specifically linked to lesional tissue, and not adjacent normal tissue. This study is the first that has detected KSHV DNA in visceral KS lesions as well as cavernous hemangioma of the skin in HIV+ patients; thus, KSHV might be tropic to endothelial cells, in both HIV+ and HIV– patients.

3.9.2. Pneumocystis carinii (PC) DNA Identified in a Spleen Showing Dystrophic Calcification and Sclerosis of White Pulp in a HIV+ Patient

A 4-yr-old HIV+ patient was noted to have multiple, minute splenic calcifications of unknown etiology 6 mo premortem. The patient died of PC pneumonia and an autopsy was performed. Silver stained sections of lung and mediastinal lymph node showed PC. Histologic examination of the spleen revealed depletion of white pulp with sclerosis and dystrophic calcification. No PC was identified in spleen, abdominal lymph nodes, or in other organs. Since PC is known to be associated with calcification and this patient died of PC pneumonia, molecular studies aimed at identifying the presence of PC DNA in the splenic lesions were undertaken. Sections of formalin-fixed, paraffin-embedded lung (positive control), mediastinal lymph node, spleen, and kidney (negative control) were selected from the autopsy material. Areas of spleen showing sclerosis and dystrophic calcification were microdissected from an H&E-stained slide and the DNA extracted. In addition, DNA was extracted from splenic red pulp, as well as the additional tissue sections above. PCR was performed using primers specific for PC mitochondrial ribosomal RNA large subunit, and DNA sequencing was performed. PCR reactions of DNA from lung, hilar lymph node, abdominal lymph node, sclerotic spleen, and splenic red pulp showed a 191-bp product. PCR reaction of DNA from uninvolved kidney repeatedly showed no product. A 104-bp fragment of the 191-bp product from the sclerotic spleen was sequenced, and a GenBank search found 101 bp of this sequence to be homologous to the mitochondrial PC ribosomal

RNA large subunit (e.g., 97% homology). In conclusion, PC DNA has been identified in the spleen of an HIV+ pediatric patient showing sclerosis and dystrophic calcification of white pulp. This case study supports previous findings that PC may be involved in the histogenesis of this lesion. This case study also demonstrates the utility of the microdissection and PCR techniques in identifying the presence (or past presence) of microorganisms in specific microscopic lesions in formalin-fixed, paraffin-embedded archival tissue.

3.10. Future Technology: NCI Laser Capture Microdissection

As the human genome project identifies all the human genes, the medical diagnostic lab will be transformed. As more and more genes become linked to the cause of, predisposition for, or clinical behavior of specific diseases, there will be a growing need to routinely screen the assigned genes for mutations or altered expression patterns. Using one or more of rapid screening methods, it is expected that genetic testing in the future will consist of panels of tests rather than limited tests for only a few genes ***(39)***. Microchips will allow for simultaneous testing of multiple DNA mutations ***(40)***. The future examination of mRNA expression needs to be adapted to examine hundreds to thousands of genes simultaneously rather than determining expression of an individual gene. Serial analysis of gene expression and microarray analysis are approaches for assessment of multiple mRNA transcripts ***(41,42)***.

However, even the most sophisticated genetic testing methods applied to tissue specimens will be of limited value if the input DNA or mRNA is derived from or contaminated by the wrong cells. Microscopic regions of normal, inflammatory, or reactive host cells present in even the smallest biopsy can produce erroneous results, especially if PCR based assays are utilized. The genetic signature of the disease is lost in the background amplification noise.

The microdissection method presented in this chapter solves the cell sampling problem and provides a reliable means to procure pure populations of selected tissue cells for analysis. The power of the technique ranges from tissue-based genetic diagnosis to research studies on premalignant lesions. However, the currently described sampling method is inadequate for widespread clinical and research applications. The manual microdissection technique is simple to perform, but is labor intensive and requires a high degree of manual dexterity. Consequently, the future expansion of tissue microdissection to routine use in the research or diagnostic laboratory will be greatly facilitated by a simple, inexpensive, automated microdissection method.

We have developed a laser capture microdissection system at the NCI that greatly increases the speed and efficiency of tissue microdissection ***(43,44)***. The system works by placing an ultrathin transparent composite film on top of a tissue section, and activating the film with a pulse from a focused laser beam

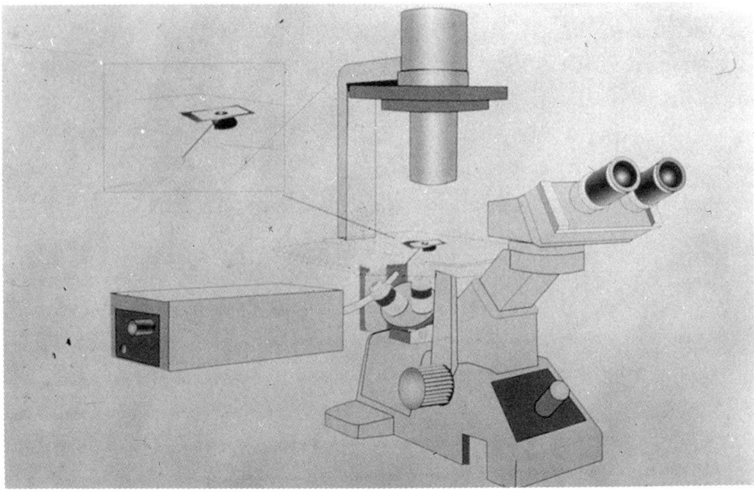

Fig. 6. Schematic of laser capture microdissection system. A laser attached to a standard inverted microscope allows for rapid and precise microdissection of select cell populations from tissue sections.

(**Fig. 6**). The activated film adheres tightly to the underlying cells that are then selectively procured from the tissue section when the film is removed. The laser capture is performed while the investigator observes the tissue through the microscope, thus precise and selective transfer is ensured. The investigator can recover a single cell population of interest from the tissue section, or alternatively, the investigator can move the tissue section and procure several different regions of the tissue with laser pulses, e.g., many separate loci of invasive tumor can be transferred to the same film and procured together. The film is then placed into DNA, RNA, or protein extraction buffer and the genetic material is recovered for analysis.

The laser capture approach has several features that make it preferable to manual microdissection. Speed and precision are greatly improved, and the tissue transfers are easy to perform. The system is very simple and no moving parts are necessary. The laser energy required to transfer tissue to the film is minimal and can be accomplished with small, inexpensive low-power lasers that can be adapted to routine microscopes. Sterile, disposable transfer films minimize potential contamination. Additionally, one tissue section can be utilized for procurement of different cell populations. For example, we typically utilize three films for microdissection of a prostate tumor section. The first film is used to procure all the normal epithelium present in the section, the second film is used to procure all the loci of *in situ* tumor, and the third film is used to procure all the loci of invasive tumor. The transfer of all three cell types takes no more than a few minutes to perform.

A potential disadvantage of manual microdissection is that contamination can occur during dissection if small clumps of tissue from one microdissected region are inadvertently procured while microdissecting a second region from the same tissue section, e.g., dissection of normal and tumor from the same slide. Although this is fairly infrequent and not much more than a nuisance for research studies, this could potentially have grave consequences in a clinical setting where an accurate genetic assessment is needed. This potential difficulty is not encountered with the laser capture system.

In summary, the ability to genetically analyze human tumors is advancing rapidly because of advancements in technology, particularly the widespread application of PCR. Unique opportunities to study the development and progression of human cancers as they exist in the patient are now available. Tissue microdissection is an important tool that is necessary to properly assess specific genetic alterations occurring in tumors. The challenge of basic researchers, pathologists, and clinicians is to utilize the new research opportunities and expanding base of genetic information for improved diagnostic, prognostic, and therapeutic benefit of patients.

Acknowledgments

The development of tissue microdissection techniques, and research studies summarized in this chapter were performed by M. R. Emmert-Buck and Z. Zhuang while in anatomic pathology residency training in the Laboratory of Pathology (LP), National Cancer Institute. Drs. Emmert-Buck and Zhuang gratefully acknowledge the support and encouragement of the entire LP staff. The authors thank the following collaborators who participated in the research studies: Rudy O. Pozzatti, Charles D. Florence, Stephen E. Strup, David G. Bostwick, Scott B. Jennings, David B. Krizman, Jeffrey M. Trent, Rhonda A. Weiss, Alex Lash, Robert F. Bonner, Settara Chandrasekharappa, Francis S. Collins, Allen M. Spiegel, Stephen J. Marx, Elias Campo, and Bonnie F. Sloane.

References

1. Fearon, E., Hamilton, S. R., and Vogelstein, B. (1987) Clonal analysis of human colorectal tumors. *Science* **238,** 193–197.
2. Radford, D., Fair, K., Thompson, A. M., Ritter, J. H., Holt, M., Steinbrueck, T., Wallace, M., Wells, S. A., and Donnis-Keller, H. R. (1993) Allelic loss on chromosome 17 in ductal carcinoma in situ of the breast. *Cancer Res.* **53,** 2947–2949.
3. Shibata D., Hawes, D., Li, Z.-H, Hernandez, A., Spruck, C. H., and Nichols, P. W. (1992) Specific genetic analysis of microscopic tissue after selective ultraviolet radiation, fractionation and polymerase chain reaction. *Am. J. Pathol.* **141,** 539–543.

4. Emmert-Buck, M., Roth, M. J., Zhuang, Z., Campo, E., Rozhin, J., Sloane, B. F., Liotta, L. A., and Stetler-Stevenson, W. G. (1994) Increased gelatinase A and cathepsin B activity in invasive tumor regions of human colon cancer samples. *Am. J. Pathol.* **145,** 1285–1290.
5. Zhuang, Z., Berttheau, P., Emmert-Buck, M. R., Liotta, L. A., Gnarra, J., Linehan, W. M., and Lubensky, I. A. (1995) A microdissection technique for archival DNA analysis of specific cell populations in lesions less than one millimeter in size. *Am. J. Pathol.* **146,** 620–625.
6. Noguchi, S., Motomura, K., Inaji, H., Imaoka, S., and Koyama, H. (1994) Clonal analysis of predominantly intraductal carcinoma and precancerous lesions of the breast by means of polymerase chain reaction. *Cancer Res.* **54,** 1849–1853.
7. Park, T.-W., Felix, J. C., and Wright, T. C. (1995) X Chromosome inactivation and microsatellite instability in early and advanced bilateral ovarian carcinomas. *Cancer Res.* **55,** 4793–4796.
8. Vocke, C., Pozzatti, R. O., Bostwick, D. G., Florence, C. D., Jennings, S. B., Strup, S. E., Duray, P. H., Liotta, L. A., Emmert-Buck, M. R., and Linehan, W. M. (1996) Analysis of 99 microdissected prostate carcinomas reveals high frequency of allelic loss on chromosome 8p12–21. *Cancer Res.* **56,** 2411–2416.
9. Bova, G., Carter, B. S., Bussemakers, J. G., Emi, M., Fujiwara, Y., Kyprianou, N., Jacobs, S. C., Robinson, J. C., Epstein, J. I., Walsh, P. C., and Isaacs, W. B. (1993) Homozygous deletion and frequent loss of chromosome 8p22 loci in human prostate cancer. *Cancer Res.* **53,** 3869–3873.
10. Cunningham, C., Dunlop, M. G., Wyllie, A. H., and Bird, C. C. (1992) Deletion mapping in colorectal cancer of a putative tumor suppressor gene in 8922–21.3. *Oncogene* **8,** 1391–1396.
11. Devilee, P., van Vliet, M., van Sloun, P., Dijkshoorn, N. K., Hermans, J., Pearson, P. L., and Cornelisse, C. J. (1991) Allelotype of human breast carcinoma: a second major site for loss of heterozygosity is on chromosome 6q. *Oncogene* **6,** 1705–1711.
12. Emi, M., Fujiwara, Y., Nakajima, T., Tsuchiya, F., Tsuda, H., Hirohashi, S., Maeda, Y., Tsurute, K., Miyaki, M., and Nakamura, Y. (1992) Frequent loss of heterozygosity for loci on chromosome 8p in hepatocellular carcinoma, colorectal cancer, and lung cancer. *Cancer Res.* **52,** 5368–5372.
13. Fujiwara, Y., Emi, M., Ohata, H., Kato, Y., Nakajima, T., Mori, T., and Nakamura, Y. (1993) Evidence for the presence of two tumor suppressor genes on chromosome 8p for colorectal carcinoma. *Cancer Res.* **53,** 1172–1174.
14. MacGrogan, D., Levy, A., Bostwick, D. G., Wagner, M., Wells, D., and Bookstein, R. (1994) Loss of chromosome arm 8p loci in prostate cancer: mapping by quantitative allelic imbalance. *Genes Chromosomes Cancer* **10,** 151–159.
15. Trapman, J., Sleddens, H. F. B. M., van der Weiden, M. M., Dinjens, W. N. M., Konig, J. J., Schroder, F. H., Faber, P. W., and Bosman, F. T. (1994) Loss of heterozygosity of chromosome 8 microsatellite loci implicates a candidate tumor suppressor gene between the loci *D8S87* and *D8S133* in human prostate cancer. *Cancer Res.* **54,** 6061–6064.

16. Emmert-Buck, M., Vocke, C. D., Pozzatti, R. O., Duray, P. H., Jennings, S. B., Florence, C. D., Zhuang, Z., Bostwick, D. G., Liotta, L. A., and Linehan, W. M. (1995) Allelic loss on chromosome 8p12–21 in microdissected prostatic intra-epithelial neoplasia (PIN). *Cancer Res.* **55,** 2959–2962.
17. Bostwick, D. and Brawer, M. K. (1987) Prostatic intra-epithelial neoplasia and early invasion in prostate cancer. *Cancer* **59,** 788–794.
18. Chuaqui, R., Englert, C. R., Strup, S., Vocke, C. D., Zhuang, Z., Duray, P. H., Bostwick, D. G., Linehan, W. M., Liotta, L. A., and Emmert-Buck, M. R. (1997) PB39: identification of a novel gene upregulated in clinically aggressive human prostate cancer. *Urology* **50,** 302–307.
19. Liang, P., Averboukh, L., Keyomarsi, K., Sager, R., and Pardee, A. B. (1992) Differential display and cloning of messenger RNAs from human breast cancer versus mammary epithelial cells. *Cancer Res.* **52,** 6966–6968.
20. Mok, S., Wong, K. K., Chan, R. K. W., Lau, C. C., Tsao, S. W., Knapp, R. C., and Berkovitz, R. S. (1994) Molecular cloning of differentially expressed genes in human epithelial ovarian cancer. *Gynecol. Oncol.* **52,** 247–252.
21. Kocher, O., Cheresh, P., Brown, L. F., and Lee, S. W. (1995) Identification of a novel gene, selectively up-regulated in human carcinomas, using the differential display technique. *Clin. Cancer Res.* **1,** 1209–1215.
22. Stone, B. and Wharton, W. (1994) Targeted RNA fingerprinting: the cloning of differentially-expressed cDNA fragments enriched for members of the zinc finger gene family. *Nucleic Acids Res.* **22,** 2612–2618.
23. Watson, M. and Fleming, T. P. (1994) Isolation of differentially expressed tags from human breast cancer. *Cancer Res.* **54,** 4598–4602.
24. Krizman, D., Chuaqui, R. F., Meltzer, P. S., Trent, J. M., Duray, P. H., Linehan, W. M., Liotta, L. A., and Emmert-Buck, M. R. (1996) Construction of a representative cDNA library from prostatic intraepithelial neoplasia. *Cancer Res.* **56(23),** 5380–5383.
25. Okayama, H. and Berg, P. (1982) High-efficiency cloning of full-length cDNA. *Mol. Cell. Biol.* **2,** 161–170.
26. Gubler, U. and Hoffman, B. J. (1983) A simple and very efficient method for generating cDNA libraries. *Gene* **25,** 263–269.
27. Zhuang, Z., Merino, M. J., Chuaqui, R., Liotta, L. A., and Emmert-Buck, M. R. (1995) Identical allelic loss on chromosome 11q13 in microdissected *in situ* and invasive human breast cancer. *Cancer Res.* **55,** 467–471.
28. Page, D., Dupont, W. D., Rogers, L. W., and Rados, M. S. (1985) Atypical hyperplastic lesions of the female breast. A long term follow up study. *Cancer* **55,** 2698–2708.
29. Page, D. and Dupont, W. D. (1990) Anatomical markers of human premalignancy and risk of breast cancer. *Cancer* **66,** 1326–1335.
30. Chuaqui, R., Zhuang, Z., Emmert-Buck, M. R., Liotta, L. A., and Merino, M. J. (1997) Analysis of loss of heterozygosity (LOH) on chromosome 11q13 in atypical ductal hyperplasia and *in situ* carcinoma of the breast. *Am. J. Pathol.* **150(1),** 297–303.
31. Chuaqui, R., Vargas, M. P., Castiglioni, T., Elsner, B., Zhuang, Z., Emmert-Buck, M. R., and Merino, M. J. (1996) Detection of heterozygosity loss in microdissected fine needle aspiration specimens of breast carcinoma. *Acta Cytologica* **40,** 642–648.

32. Lubensky, I., Gnarra, J., Bertheau, P., Waither, M., Linehan, W. M., and Zhuang, Z. (1997) Allelic deletions of the VHL gene detected in multiple microscopic clear cell renal lesions in von Hippel-Lindau disease patients. *Am. J. Pathol.* **149(6),** 2089–2094.
33. Zhuang, Z., Emmert-Buck, M. R., Roth, M. J., Gnarra, J., Linehan, W. M., Liotta, L. A., and Lubensky, I. A. (1996) Von Hippel-Lindau disease gene deletion detected in microdissected sporadic human colon carcinoma specimens. *Hum. Pathol.* **27,** 152–156.
34. Lubensky, I., Debelenko, L. V., Zhuang, Z., Emmert-Buck, M. R., Dong, Q., Chandrasekharappa, S., Guru, S., Manickam, P., Olufemi, E.-S., Marx, S. J., Spiegel, A. M., Collins, F. S., and Liotta, L. A. (1996) Tissue specific patterns of 11q13 LOH in multiple parathyroid, pancreatic, and duodenal tumors from individual MEN1 patients. *Cancer Res.* **56(22),** 5272–5278.
35. Debelenko, L. V., Emmert-Buck, M. R., Zhuang, Z., Epshteyn, E., Moskaluk, C., Jensen, R. T., Liotta, L. A., and Lubensky, I. A. (1997) The MEN1 gene locus is involved in the pathogenesis of gastric ECL-cell carcinoids in MEN1-ZES patients. *Gastroenterology* **113,** 773–781.
36. Zhuang, Z., Vortmeyer, A. O., Mark, E. J., Odze, R., Emmert-Buck, M. R., Merino, M. J., Moon, H., Liotta, L. A., and Duray, P. H. (1996) Barrett's esophagus: metaplastic cells with loss of heterozygosity at the APC gene locus are clonal precursors to invasive adenocarcinoma. *Cancer Res.* **56,** 1961–1964.
37. Wiltshire, R., Duray, P. H., Bittner, M. L., Visakorpi, T., Meltzer, P. S., Tuthill, R. J., Liotta, L. A., and Trent, J. M. (1995) Direct visualization of the clonal progression of primary cutaneous melanoma: application of tissue microdissection and comparative genomic hybridization. *Cancer Res.* **55,** 3954–3957.
38. Emmert-Buck, M. R., Chuaqui, R., Zhuang, Z., Nogales, F., Liotta, L. A., and Merino, M. J. (1997) Molecular analysis of concomitant uterine and ovarian endometrioid tumors. *Int. J. Gynecol. Pathol.* **16(2),** 143–148.

38a. Chang, Y., Cesarman, E., Pessin, M. S., Lee, F., Culpepper, J., Knowles, D. M., and Moore, P. S. (1994) Identification of herpesvirus-like DNA sequences in AIDS-associated Kaposi's sarcoma. *Science* **266(5192),** 1865–1869.

39. Nowak, R. (1995) Entering the postgenome era. *Science* **270,** 368–371.
40. Abbott, A. (1996) DNA chips intensify the sequence search. *Nature* **379,** 392.
41. Schena, M., Shalon, D., Davis, R. W., and Brown, P. (1995) Quantitative monitoring of gene expression patterns with a complementary DNA microarray. *Science* **270,** 467–469.
42. Velculescu, V., Zhang, L., Vogelstein, B., and Kinzler, K. (1995) Serial analysis of gene expression. *Science* **270,** 484–487.
43. Emmert-Buck, M. R., Bonner, R. F., Smith, P. D., Chuaqui, R., Goldstein, S. R., Zhuang, Z., Weiss, R. A., and Liotta, L. A. (1996) Laser capture microdissection (LCM), *Science* **274,** 998–1001.
44. Bonner, R. F., Emmert-Buck, M. R., Cole, K. A., Pohida, T., Chuaqui, R. F., Goldstein, S. R., and Liotta, L. A. Laser capture microdissection: molecular analysis of tissue. *Science,* in press.

18

EWS Gene Fusions as Diagnostic Markers in Sarcomas

Principles and Guidelines

Marc Ladanyi

1. Introduction

Highly specific chromosomal translocations are found in several primitive sarcomas ***(1–3)***. Biologically, the breakpoints of these translocations involve various putative or confirmed transcription factor genes, some of which appear to participate in normal mesenchymal development and differentiation. These genes are rearranged by the translocations, resulting in the formation of chimeric genes, encoding novel tumor-specific transcription factors that are presumed to disrupt normal differentiation and lead to sarcomagenesis. Several lines of evidence suggest that the fusion genes encoded by these translocations are likely to be either necessary or sufficient for sarcomagenesis. The area has been the subject of several recent reviews ***(1–3)***. Besides their biological significance, these translocations are of great diagnostic interest as tumor markers. The morphological diagnosis of sarcomas is often problematic. The possibility of detecting tumor-type-specific translocations represents an extremely useful diagnostic modality. The *EWS* gene, located at 22q12, is the single most commonly involved gene in these translocations and thus plays a pivotal role in several different types of sarcomas, including Ewing's sarcoma, clear cell sarcoma, desmoplastic small round cell tumor, and extraskeletal myxoid chondrosarcoma (*see* **Table 1**). We expect that the usefulness of *EWS* rearrangements as tumor markers in sarcomas will only continue to grow as additional *EWS* gene fusions are identified.

Although chromosomal translocations in sarcomas are extremely specific, they may be apparently absent in rare cases of the tumor type with which they

From: *Methods in Molecular Medicine, Vol. 14: Tumor Marker Protocols*
Edited by: M. Hanausek and Z. Walaszek © Humana Press Inc., Totowa, NJ

Table 1
Major *EWS* Gene Fusions in Sarcomas: RT-PCR Detection

Chromosomal translocation	Fusion product	*EWS* forward primer[a]	Reverse primer	Product size (bp)	Internal and fusion junction probes
Ewing's sarcoma/PNET					
t(11;22) (q24;q12)	*EWS/FLI1*	ex. 7	*FLI1* ex. 9: ACTCCCCGTTGGTCCCCTCC	type 1: 120	*EWS* ex. 7: TATAGCCAACAGAGGAGCAG
				type 2: 183	*FLI1* ex. 6: CAAGCTCCTCTTCTGACTGAG
		ex. 7	*FLI1* ex. 6: GTTGAGGCCAGAATTCATGTTA	type 1: 327	*EWS/FLI1* type 1: ACGGGCAGCAGAACCCTTCTTATG
				type 2: 390	*EWS/FLI1* type 2: ACGGGCAGCAGAGTTCACTGCTGG
t(21;22) (q22;q12)	*EWS/ERG*	ex. 7	*ERG* ex. 9: AAAGCTGGATCTGGCCACTG	Various	*ERG*: AGTCGAAAGCTGCTCACCATCT
Clear-cell sarcoma (MMSP)					
t(12;22) (q13;q12)	*EWS/ATF1*	ex. 8	*ATF1*: TCTCCGTCTCCTTTTCTGC	126	*EWS/ATF1:* CGGTGGAATGGGAAAAATTTTGAA
Desmoplastic SRCT					
t(12;22) (p13;q12)	*EWS/WT1*	ex. 7	*WT1* ex. 9: GACCAGGAGAACTTTCGCTGAC	197	*EWS/WT1:* ACGGGCAGCAGAGTGAGAAACCAT
Extraskeletal myxoid CS					
t(9;22) (q22;q12)	*EWS/CHN*				
	type 1	ex. 12	*CHN:* CCTGGAGGGGAAGGGCTAT	109	*EWS/CHN* type 1: AATGGTTTGATGATATGCCCTGCG
	type 2	ex. 7	*CHN:* CCTGGAGGGGAAGGGCTAT	275	*EWS/CHN* type 2: ACGGGCAGCAGAAGCCCACTGCGG

[a]EWS forward primers: exon 7: TCCTACAGCCAAGCTCCAAGTC; exon 8: GGGAAGAGGGGGATTTGA; exon 12: AAGGCGATGCCACAGTGTC.

are associated. Whether these negative cases represent instances of morphologic misdiagnosis or the existence of as-yet undefined translocation variants is presently unclear. For instance, in a recent large study of Ewing's sarcoma/peripheral neuroectodermal tumor (ES/PNET), cases that lacked the diagnostic fusion product were shown on review to have pathological features atypical for ES/PNET ***(4)***.

This chapter will review two molecular diagnostic methods for the detection of *EWS* rearrangement in sarcomas, namely Southern blotting and reverse transcription-polymerase chain reaction (RT-PCR). Other diagnostic approaches, such as fluorescent *in situ* hybridization on interphase nuclei ***(5,6)*** and long-range DNA polymerase chain reaction (PCR) ***(7)*** will not be discussed.

1.1. Ewing's Sarcoma and Peripheral Neuroectodermal Tumor

ES is a small round cell tumor showing occasional neural differentiation, typically arising within a bone of a child or adolescent. It is closely related pathologically to PNET, which occurs mainly in soft tissues and shows more definite neural features. That the ES/PNET group of tumors represented the same tumorigenic process occurring at different sites was supported by the finding of the specific translocation t(11;22) (q24;q12) in over 90% of cases ***(8)***. The cloning of its breakpoints in 1992 inaugurated the era of molecular diagnosis of sarcomas. Molecular data now confirm that this clinicopathologic group also includes the "Askin tumor" (a clinical variant localized to the chest wall). Some biphenotypic sarcomas with myogenic and neural differentiation, sometimes also called malignant ectomesenchymomas, may also be pathogenetically related to this group of tumors ***(9)***. Accurate morphologic diagnosis of this complex group of primitive small round cell tumors can be notoriously difficult, creating a special need for objective diagnostic approaches.

The cloning of the t(11;22) breakpoints in 1992 led to the identification of the *EWS* gene ***(10,11)***. Since then, our understanding of the genes involved in the ES/PNET group of tumors has grown tremendously. It is now clear that almost all cases of ES/PNET have either of two highly homologous gene fusions: *EWS/FLI1*, corresponding to the t(11;22) translocation, or *EWS/ERG*, derived from the t(21;22) translocation (*see* **Table 1**). The normal function of EWS is presently unclear, although the carboxyl terminal portion encodes a functional RNA-binding domain ***(10,12,13)***. The *EWS* gene is ubiquitously expressed. Its promoter region, which drives this constitutive expression, is thus responsible for the inappropriate high-level expression of *EWS* fusion genes in ES/PNET (and in the other sarcomas below). The amino terminal region of EWS, included in the chimera, can function as a regulatory domain for the portions of FLI1 or ERG to which it is fused ***(13)***. The genomic breakpoints in *EWS* are conveniently clustered within a small 7-kilobase (kb)

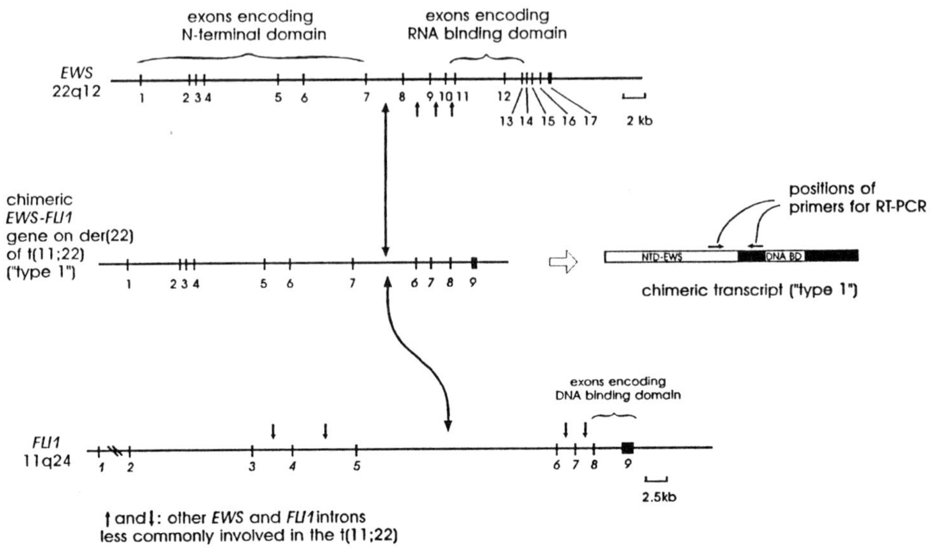

Fig. 1. Schematic diagram of t(11;22) (q24;q12) of ES/PNET showing the structure of the normal *EWS* and *FLI1* genes, the most common type of fusion gene resulting from the translocation ("type 1"), and the resulting chimeric RNA transcript. The introns are drawn to scale, except where crosshatched. Exons encoding some known functional domains are indicated, and the *FLI1* exons are in italics to distinguish them from the *EWS* exons in the chimeric gene. In the t(11;22) (q24;q12), the breakpoints may occur in one of four *EWS* introns and one of six *FLI1* introns, resulting in a great diversity of fusion genes, although not all combinations have been reported to date. The intron between exons 8 and 9 of FLI1 can also be involved (not indicated). (Reproduced from **ref. *2*** with permission).

region, with most occurring in introns 7 and 8 ***(11,14)*** (**Fig. 1**). This allows *EWS* rearrangements to be readily detected by Southern blot analysis, as discussed below ***(11,15,16)*** (*see* **Fig. 2**).

1.1.2. EWS/FLI1

In the t(11;22) (q24;q12), the breakpoint within the *FLI1* gene at 11q24 can occur in one of six introns, encompassing a 40-kb region (**Fig. 1**). The *FLI1* rearrangement is thus poorly suited to Southern blot detection. FLI1 is in the ETS proto-oncogene family and encodes a transcription factor with an ETS-type sequence-specific DNA-binding domain ***(17,18)***. The genes normally regulated by the FLI1 transcription factor remain unknown. In the *EWS/FLI1* fusion gene, the 3' portion of the *FLI1* gene, including the region coding for DNA binding, replaces the RNA-binding domain of *EWS*, thus encoding a protein consisting of the amino terminal portion of the EWS protein and the car-

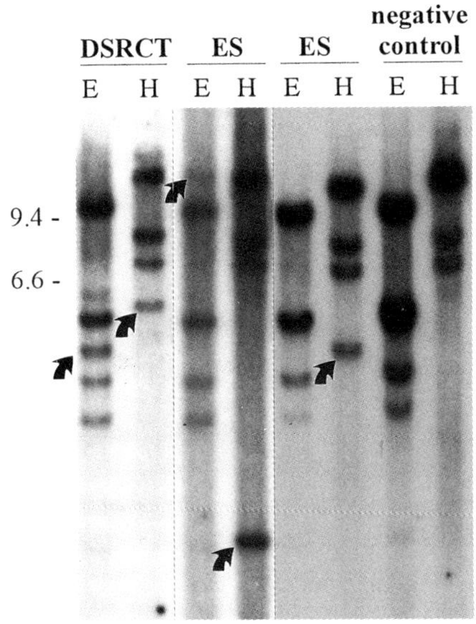

Fig. 2. Southern-blot detection of *EWS* rearrangement in genomic DNA samples extracted from two ES cases and one DSRCT case, digested with *Eco*RI (E) and *Hin*dIII (H), hybridized with the *EWS* cDNA probe described in the text. The normal germline band pattern is seen in the negative control lane and consists of 8.2-, 4.7-, 3.5-, 2.9-, and 2.0- (faint) kb bands in the *Eco*RI digest and 9.5-, 7.0-, and 6.0-kb bands in the *Hin*dIII digest. Rearranged bands are indicated by arrows. The positions of two size marker bands (in kb) are shown on the left.

boxyl terminal portion of the FLI1 protein (**Fig. 1**). Other *EWS* gene fusions follow the same structural pattern, whereby the EWS RNA-binding domain is replaced by a heterologous DNA-binding domain.

The occurrence of molecular variants is highly relevant for molecular diagnosis. In this and other *EWS* gene fusions, it is important to distinguish recurrent variants, that result from normal splicing of the chimeric gene from unique variants that result from abnormal splicing events because of unusual genomic breakpoints. So far, 10 recurrent variants of the *EWS-FLI1* chimeric transcript have been described (**Table 2**), representing different combinations of exons from *EWS* and *FLI1*. Eight of these were reported by Zucman et al. ***(19)***. Additionally, two other recurrent variant fusions have been identified, *EWS* exon 7 to *FLI1* exon 9 and *EWS* exon 7 to *FLI1* exon 7 ***(20,21)***. This heterogeneity reflects the remarkable coincidence that the splice junctions of exons 7, 9, and, 10 of *EWS* and exons 4–9 of *FLI1* occur at the same codon nucleotide position ***(14,19)***. Thus, the potential combinatorial diversity of fusion types could rise to 18. Fortunately for molecular diagnosis, the two most common fusions, *EWS*

Table 2
Molecular Heterogeneity of the *EWS/FLI1* Fusion Product

Junction of *EWS* exon	To *FLI1* exon	Estimated frequency
7	4	<2%
7	5	20–25%[a]
7	6	60–70%[b]
7	7	<2%
7	8	2%
7	9	<2%
9	4	2%
9	7	2%
10	5	6%
10	6	4–5%

[a]"Type 2" fusion.
[b]"Type 1" fusion.

exon 7 to *FLI1* exon 6 and *EWS* exon 7 to *FLI1* exon 5, respectively, types 1 and 2 ***(10)***, account for about 80% of all cases with chimeric *EWS-FLI1* RNA transcripts (**Table 2**). All *EWS/FLI1* fusion transcripts encode the essential structural elements of the chimeric gene product; namely, the entire DNA-binding domain of *FLI1* (represented by *FLI1* exon 9) and the entire *N*-terminal domain of *EWS* (encoded by *EWS* exons 1–7). Correspondingly, *EWS* exon 7 and *FLI1* exon 9 PCR primers should amplify all *EWS-FLI1* molecular variants. In several reported instances, a given tumor has expressed two or more different *EWS/FLI1* transcripts, presumably because of alternative splicing of a single rearranged allele ***(19,20,22)***. This phenomenon is illustrated in **Fig. 3**.

Mapping of the rearrangements finds about 90% of *EWS* genomic breakpoints evenly distributed between introns 7 and 8 ***(14,19)***. Correlation of these data with the structure of the resulting chimeric RNAs shows that breakpoints in either of these introns result in chimeric RNAs which include exon 7 but not exon 8 of *EWS*, because the usage of exon 8 would result in out-of-frame fusion with *FLI1* ***(19)***.

1.1.3. EWS/ERG

Molecular studies have established that approx 5% of ES/PNET contain the translocation t(21;22) (q22;q12) instead of the t(11;22). The t(21;22) rearranges *EWS* with *ERG*, an ETS family gene highly homologous to *FLI1*, located at 21q22 ***(19,23,24)***. Functionally, this *EWS* gene fusion is analogous to *EWS/FLI1*. The exon structure of *ERG* appears to follow that of *FLI1*, and,

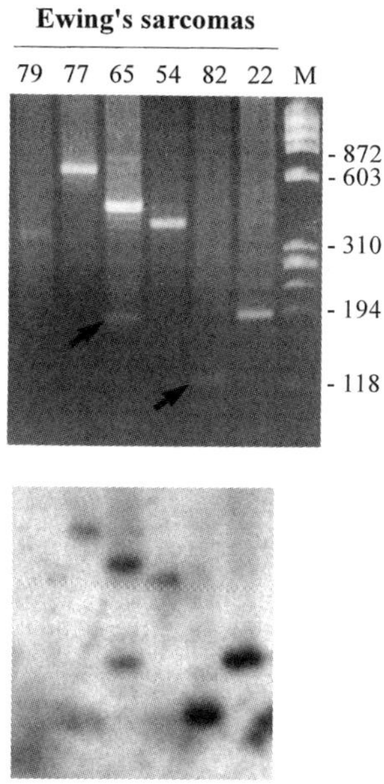

Fig. 3. RT-PCR analysis of chimeric *EWS/FLI1* mRNA in a panel of six ES/PNET using *EWS* exon 7 and *FLI1* exon 6 primers (*see* **Table 1**). **(Top)** The ethidium bromide-stained agarose gel of the RT-PCR products. Arrows indicate two of the fainter products. The sizes of selected bands (in bp) of the size marker (M) are shown on the right. **(Bottom)** The blot of the same gel hybridized with the *FLI1* exon 6 internal probe (**Table 1**). Case 79 is negative. The *EWS/FLI1* fusion types (*see* **Table 2**) in the remaining cases are as follows: case 77, *EWS* exon 9 to *FLI1* exon 4 junction; case 54, *EWS* exon 10 to *FLI1* exon 6 junction; case 82, *EWS* exon 7 to *FLI1* exon 6 junction ("type 1"), and case 22, *EWS* exon 7 to *FLI1* exon 5 junction ("type 2"). Case 65 contains two fusion transcripts, one joining *EWS* exon 10 to *FLI1* exon 5, the other joining *EWS* exon 7 to *FLI1* exon 5. These two transcripts presumably were derived by alternative splicing of *EWS* exons 8 to 10, following transcription of a single fusion gene.

similarly, four molecular variants of *EWS/ERG* have been described so far ***(19,24,25)***. The high degree of homology between *FLI1* and *ERG* complicates the molecular diagnosis of these two gene fusions. For instance, the commonly used *FLI1* exon 9 primer (primer 11.3, ***10***) has only two mismatches with *ERG* and may thus hybridize to *ERG*, depending on PCR stringency.

1.1.4. Rare EWS *Fusion Transcripts in ES/PNET*

Two other rare variant ES/PNET translocations, t(7;22) (p22;q12) and t(17;22) (q12;q12), have recently been cloned, each in a single case. Both involve novel ETS domain transcription factor genes. In the former, *EWS* fuses with a novel gene designated *ETV1* ***(26)***. The latter represents a rearrangement of *EWS* with the *E1AF* gene at 17q12 ***(27)***. Interestingly, *E1AF* and *ETV1* are more homologous to each other than to *FLI1* or *ERG*. It is still unclear whether these two gene fusions are associated with particular morphologic variants of ES/PNET.

1.2. Clear Cell Sarcoma: EWS/ATF1

Clear-cell sarcoma (CCS) is a deep soft-tissue tumor that produces melanin, also known as malignant melanoma of soft parts (MMSP). It typically occurs between the ages of 20 and 40, and the limbs are the most common location. Cytogenetically, at least 65% of CCS analyzed have been found to harbor the translocation t(12;22) (q13;q12) ***(28)***. In this translocation, the *EWS* gene is rearranged with the *ATF-1* gene on chromosome 12 ***(29)***. The latter encodes a transcription factor with a leucine zipper dimerization domain and a basic DNA-binding domain ***(30)***. In the fusion mRNA transcript, the RNA-binding domain of EWS is replaced by the DNA-binding domain of ATF-1. RT-PCR results have been published in only a small number of CCS samples, two cell lines and three primary tumors ***(29,30)***. In all five instances, the identical fusion was detected between exon 8 of *EWS* and codon 65 of the *ATF-1* transcript. Thus, molecular diagnostic experience with this gene fusion is still at an early stage.

1.3. Desmoplastic Small Round-Cell Tumor: EWS/WT1

The desmoplastic small round-cell tumor (DSRCT) is a primitive sarcoma showing widespread abdominal serosal involvement, a peculiar histologic appearance with prominent desmoplasia, and striking divergent, multilineage differentiation. It typically occurs in young males and is highly lethal. We showed that the t(11;22) (p13;q12) translocation seen in this sarcoma represents a rearrangement between the *EWS* gene at 22q12 and the Wilms' tumor gene, *WT1*, at 11p13, generating a fusion gene that encodes a chimeric RNA resulting from an in-frame junction of *EWS* exon 7 to *WT1* exon 8 ***(31,32)*** (**Fig. 4**). Thus, this chimeric RNA encodes a putative protein in which the RNA-binding domain of EWS is replaced by a functional portion of the WT1 DNA-binding domain.

From the molecular diagnostic viewpoint, it is notable that, like the native *WT1* transcript, which undergoes alternative splicing ***(33)***, the chimeric *EWS/WT1* transcripts include both alternatively spliced forms of the zinc-finger domain of WT1, "+KTS" and "–KTS" ***(32)***. This phenomenon is apparent only if the 3' primer site is located in *WT1* exon 10, 3' to the KTS codons. No recurrent molecular variants of the *EWS/WT1* gene fusion have so far been described, although a unique variant resulting from an unusual genomic breakpoint in *WT1* has recently been reported ***(34)***.

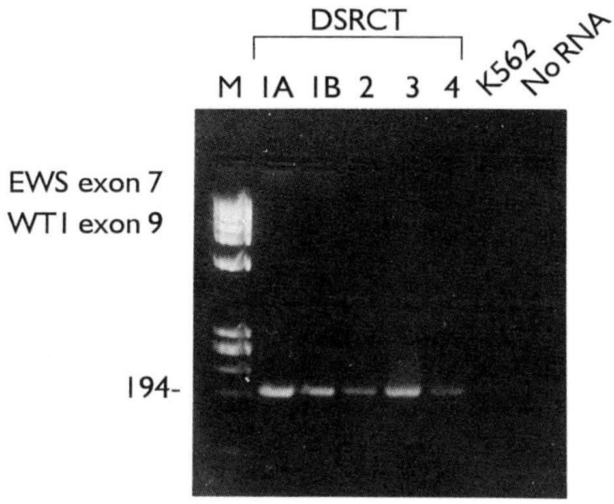

Fig. 4. RT-PCR analysis of chimeric *EWS/WT1* mRNA in four cases of DSRCT using the primers shown in **Table 1**. The 197-bp product seen in the positive lanes corresponds to the fusion of *EWS* exon 7 to *WT1* exon 8. The size of the closest marker (M) band, 194 bp, is indicated on the left. No RT-PCR products are seen with RNA from an unrelated tumor cell line, the K562 acute myeloid leukemia, or in the absence of RNA (no RNA control).

1.4. Extraskeletal Myxoid Chondrosarcoma: EWS/CHN

The recurrent translocation t(9;22)(q22-31;q12) of extraskeletal myxoid chondrosarcoma (CS) represents a rearrangement of the *EWS* gene with a novel gene at 9q22 designated *CHN* or *TEC* ***(35,36)***. *CHN* encodes a novel orphan nuclear receptor with a zinc finger DNA-binding domain most homologous to the DNA-binding domains of members of the retinoid receptor family. It is the human homolog of the rat *nor1* gene ***(37)***. Unlike other *EWS* translocations, the coding region of the translocation partner in this case is not truncated, but is fused in its entirety to the N-terminal domain of EWS. Three fusion variants have so far been described. In the most common, type 1, *EWS* exon 12 is fused to position -2 of the *CHN* cDNA (the exon structure of *CHN* is presently unknown) ***(35)***. The *EWS* fusion point in this variant represents the most 3' exon included in any *EWS* chimera to date and results in the inclusion of a portion of the EWS RNA-binding domain within the predicted protein. The type-2 variant fuses *EWS* exon 7 to position –176 of the *CHN* cDNA, resulting in a novel open-reading frame of 59 amino acids ***(35,36)***. The type-3 variant appears to be a unique, probably nonrecurrent variant resulting from an unusual genomic breakpoint within *EWS* exon 12 ***(35)***. So far, at least a quarter of extraskeletal myxoid CS cases appear negative for the *EWS/CHN* gene fusion and may harbor a different genetic lesion.

1.5. *Myxoid Liposarcoma:* EWS/CHOP

Myxoid liposarcoma is characterized by the t(12;16)(q13;p11) translocation, which represents a fusion of the *TLS* or *FUS* gene at 16p11 with the *CHOP* gene at 12q13. The *CHOP* gene belongs to the CCAAT/enhancer binding protein family and may be involved in the differentiation of adipose tissue ***(38,39)***. TLS is a ubiquitously expressed RNA-binding protein that shows highest homology to EWS ***(40,41)***. TLS and EWS are closely related structurally and functionally. In in vitro transformation and transcription assays, the N-terminal domain of EWS can functionally substitute for the TLS portion of *TLS-CHOP* ***(42)***. Remarkably, these in vitro studies appear to have presaged the discovery of a naturally occurring *EWS/CHOP* gene fusion in rare cases of myxoid liposarcoma. This rearrangement, described in two cases by Panagopoulos et al. ***(43)***, results at the mRNA level in a fusion of *EWS* exon 7 to *CHOP* exon 2, corresponding to complex variants of a t(12;22) (q13;q12).

2. Southern Blot Detection of *EWS* Rearrangement

2.1. *Principles*

Southern blotting is a well-established technique, and excellent detailed protocols are available in several manuals ***(44,45)***. This section reviews the application of Southern blotting to the detection of *EWS* rearrangements.

Southern blot detection of *EWS* rearrangements is possible because of the clustering of genomic breakpoints within a 7-kb region of the *EWS* gene, as described above ***(11,14)***. The genomic breakpoints in most translocation partners of *EWS* have not been extensively mapped, with two exceptions: *FLI1*, where breakpoints are spread out over 40 kb, and *WT1*, where the genomic breaks involve a relatively short intron.

Southern blotting, although complicated by the requirement for a significant amount of frozen tissue for DNA extraction, nonetheless remains uniquely useful in some settings. For instance, Southern blotting can reliably detect *EWS* rearrangement, regardless of the translocation partner or molecular variation in the fusion gene. It can also be used as a starting point in the isolation of novel translocation partners of *EWS* ***(31)***.

2.2. *Probe Selection*

The breakpoint cluster region in *EWS*, sometimes called EWSR1, can be covered by several genomic probes ***(11)*** or by a single cDNA probe. Using a panel of at least three genomic probes, different groups have shown excellent detection of *EWS* rearrangement in cases positive for the t(11;22) by cytogenetics or RT-PCR ***(4,11,16)***. We have used a single partial *EWS* cDNA probe to obtain the same level of detection of *EWS* rearrangements ***(15)***. This partial *EWS* cDNA probe is generated by PCR using normal human placental cDNA (Clontech, Palo Alto, CA) as a template, although cDNA from almost any tissue should contain the ubiquitously expressed *EWS* transcript. This 741 base-pair probe spans nucleotides 527–1267 of the *EWS* cDNA and is synthesized by PCR with the primers: AGCCTAGGA TATGGACAGA (*EWS* exon 6 forward) and CTTTCCTGTTTCCTTGTCC (*EWS* exon 12 reverse). The probe thus hybridizes to exons 6–12 and covers in *Eco*RI- and *Hin*dIII-digested DNA the entire genomic breakpoint cluster region in *EWS*, including the breaks between exons 12 and 13 in some extraskeletal myxoid CS, which are 3' to the originally defined "EWSR1" breakpoint cluster region in ES/PNET ***(10,14)***. The normal germline band pattern with this partial *EWS* cDNA on Southern blots is relatively simple and is illustrated in **Fig. 2**.

The use of a cDNA probe instead of a genomic probe precludes restriction mapping of the breakpoints but also obviates the need for multiple genomic probes. This *EWS* cDNA probe described above contains no repetitive sequences (which complicate the use of some genomic probes) and does not crosshybridize at high stringency with closely related sequences in pseudogenes or in the *TLS* gene.

2.3. Precautions

Although Southern blotting is a very robust technique, there are, nevertheless, some pitfalls. cDNA probes usually hybridize to several exons and often generate a complex pattern of germline bands. With multiple germline bands, the risk of comigration of rearranged and germline bands is obviously greater. Therefore, at least three different enzyme digests are recommended to establish the absence of rearrangement. Another problem with cDNA probes is that some exons may be quite short (<50 bp) and may not hybridize reliably with the probe, resulting in an evanescent germline band susceptible to variations in probe quality and hybridization conditions. A thorough characterization of the normal pattern of bands in a good-quality control DNA is thus required.

Southern blotting is not entirely immune to contamination. This is a less frequently discussed pitfall. Bacterial contamination of DNA either because of in vivo infection of the source tissue or because of improper storage of the DNA sample, can give strong nongermline bands resulting from the presence of bacterial plasmid DNA hybridizing with traces of pBR322-related vector sequences that are often copurified with the probe insert ***(46)***. Obviously, this problem arises only with probes cloned in plasmids, not PCR-generated probes.

3. RT-PCR: Detection of *EWS* Gene Fusions

3.1. Principle

RT-PCR is the mainstay in the detection of the chimeric transcripts resulting from *EWS* rearrangements. The general principles of RT-PCR are widely available in manuals ***(44,45,47)*** and kit inserts (GeneAmp RNA PCR kit, Perkin–Elmer, Norwalk, CT). RT-PCR requires good-quality snap-frozen tissue for RNA extraction and may be complicated by the molecular variability of some rearrangements described above (for example, 10 molecular variants of *EWS/FLI1*).

Both single-step and nested RT-PCR approaches have been used. The latter is usually needed to achieve the sensitivity required for minimal residual disease detection. In general, however, a conventional single-step RT-PCR protocol, followed by transfer and hybridization of the products, is preferred because of the significantly greater contamination risks inherent in nested PCR methods and because hybridization with an internal probe provides confirmation of the results.

Unlike Southern blotting, RT-PCR assays can also be multiplexed, allowing the detection of multiple possible targets in a single reaction. The application of this type of approach to the molecular diagnosis of sarcomas is illustrated by Downing et al. ***(48)***.

3.2. Primer Selection and Methods

We use the GeneAmp RNA PCR protocol and reagents (Perkin–Elmer) the only modification being the use of Superscript II RT (Gibco-BRL, Gaithersburg, MD), a modified RNase H^- MMLV RT enzyme. The RT reaction is performed on 500 ng–2 μg of total RNA, extracted by the acid guanidinium thiocyanate–phenol–chloroform method (Tel-Test, Friendswood, TX) ***(49)***.

For the RT step, we do not recommend oligo-dT priming, because the 3' end of many transcripts may be at a considerable distance from the chimeric RNA fusion point. Since RNA extracted from surgical specimens of sarcomas is of variable quality, such large fragments are not reliably reverse-transcribed. On the other hand, random primers are more prone to generate nonspecific products because, unlike oligo-dT, they bind to ribosomal RNAs as well. The use of the downstream primer may produce the best results.

In the PCR step, as a general rule, shorter target sequences (<500 bp) are easier to amplify, because of the variable quality of primary tumor RNA samples and because PCR efficiency is greater for short products. For instance, for first-line detection of *EWS/FLI1*, we use an *EWS* exon 7 primer (primer 22.3 ***[10]***) with a *FLI1* exon 6 primer, which detect approx 90–95% of *EWS/FLI1* fusion RNAs, including the two most common types (*see* **Table 1**) ***(19)***. The use of a *FLI1* exon 6 primer results in shorter products and therefore more

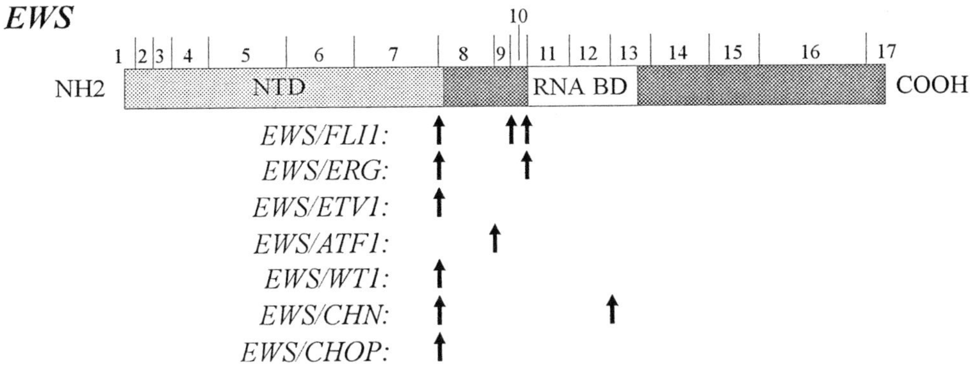

Fig. 5. Schematic diagram of the different fusion points reported so far in *EWS* chimeric transcripts, in relation to the functional domains encoded by the *EWS* exons (numbered above). For the *EWS/E1AF* rearrangement, the hybrid transcript has not yet been characterized. NH2, amino-terminal; COOH, carboxyl terminal; NTD, NH_2-terminal domain; RNA BD, RNA-binding domain.

efficient amplification than the originally described *FLI1* exon 9 primer (primer 11.3 ***[10]***). ES cases negative for *EWS/FLI1* using these primers are studied for the rarer *EWS/FLI1* types, using the *FLI1* exon 9 primer instead of the exon 6 primer. To assist in *EWS* primer selection, the positions of the different *EWS* fusion points reported so far in *EWS* chimeric transcripts are schematized in **Fig. 5**.

3.3. Precautions

Poor PCR efficiency can exponentially reduce the final amount of PCR product. It is therefore essential to start with optimal PCR parameters. There are many approaches to PCR optimization ***(50)***. We use empiric $MgCl_2$ optimization and "touchdown" annealing. In the latter approach, the PCR annealing temperature is gradually decreased from about 8°C above to 2°C below the predicted optimal annealing temperature over the first 10 cycles ("touchdown PCR") ***(51)***. This favors specific over misprimed products and largely eliminates the need for empiric optimization of the annealing temperature.

If the RT-PCR is negative for an *EWS* chimeric transcript, the quality of the RNA sample should be confirmed, regardless of the absorption spectrophotometer reading. The RNA sample may be significantly degraded or may contain enzyme inhibitors, for example, heparin or residual proteinase K. The use of *EWS* or β-actin transcripts as controls for RNA quality is not advised because there are processed pseudogenes of both that could result in false-positive results because of contaminating genomic DNA in RNA samples ***(52,53)***.

Validation of the PCR products relies on their size and their hybridization with internal oligonucleotide probes. For *EWS* gene fusions showing little or no molecular variability, e.g., *EWS/WT1* or *EWS/ATF1*, a band of the correct size is usually sufficient evidence (**Fig. 4**). For *EWS* gene fusions showing extensive molecular variability, such as *EWS/FLI1* or *EWS/ERG*, it is wise to blot and hybridize the products with an oligonucleotide spanning the fusion junction (or serially with two oligonucleotides, each internal to one primer) (**Table 1**, **Fig. 3**). Transfer and hybridization usually also provide a 10-fold increase in sensitivity. Alternately, the product can be sequenced. If the RT-PCR assay is being used for minimal residual disease detection, the sensitivity of each run should be monitored by the inclusion of sensitivity controls (for example, containing $1{:}10^4$ and $1{:}10^5$ tumor RNA).

Contamination can occur either in the RT-PCR reagents or in the RNA samples. The former problem is easier to spot because most, if not all, lanes will show a contaminating band. Thus, all runs should include a no-RNA control to assess possible reagent contamination. To exclude contamination of RNA samples, it is prudent to repeat the RT-PCR in all positive cases omitting the RT enzyme. Since all of the RT-PCR target sequences in *EWS* chimeric transcripts span the fusion junction, which invariably represents an exon boundary, there is little or no risk that contaminating genomic DNA will yield products.

As general precautions, the following well-known approaches to avoid false positives are useful ***(54)***: the use of pipet tips with aerosol barriers and ultraviolet irradiation of pipeters, tips, racks, and work area prior to PCR reagent assembly, in a dedicated PCR reaction assembly hood (e.g., Template Tamer; Oncor, Gaithersburg, MD). RNA extraction, RT-PCR reagent assembly, and product analysis should be kept strictly physically separate and the materials stored apart as well. In terms of laboratory workflow, on any given day, the handling and electrophoresis of PCR products prior to setting up the next batch of RT-PCR or RNA extractions should be avoided.

Note Added in Proof

Peter et al. have recently described yet another member of the ETS family fused to *EWS* in two cases of Ewing's sarcoma. The gene was designated *FEV* for fifth Ewing variant ***(55)***. Speleman et al. have recently reported a novel molecular variant of the *EWS-ATF1* fusion in clear cell sarcoma. In their case, the fusion occurred between *EWS* exon 10 and codon 110 of *ATF1* instead of the previously observed fusion between *EWS* exon 8 and *ATF1* codon 65 ***(56)***. We have recently observed two molecular variants of the *EWS-WT1* fusion of desmoplastic small round cell tumor. Instead of the usual fusion involving EWS exon 7, the *EWS* exon involved in the *EWS-WT1* fusion was exon 9 in one case and exon 10 in another (M. Ladanyi and W. Gerald, unpublished observations).

These *EWS-ATF1* and *EWS-WT1* cases illustrate that the molecular variability observed in *EWS-FLI1* will extend to other *EWS* fusions as well. Finally, an important polymorphism has been described in the *EWS* gene ***(57)*** that is of great significance as a possible pitfall in the interpretation of *EWS* gene rearrangements by Southern blotting. The polymorphism is a deletion of 2.5 kilobases within a stretch of ALU repeats in intron of the *EWS* gene. Thus far it has only been observed in individuals of African origin. Because this is a substantial length polymorphism, it is detectable in multiple restriction enzyme digests. As corresponding normal tissue is not always available in tumors studied for *EWS* rearrangement, it is now important to become familiar with the position of these polymorphic bands before rendering diagnoses based on Southern blot analysis of the *EWS* gene.

References

1. Rabbitts, T. H. (1994) Chromosomal translocations in human cancer. *Nature* **372,** 143–149.
2. Ladanyi, M. (1995) The emerging molecular genetics of sarcoma translocations. *Diag. Mol. Pathol.* **4,** 162–173.
3. Cooper, C. S. (1996) Translocations in solid tumours. *Curr. Opin. Genet. Dev.* **6,** 71–75.
4. Delattre, O., Zucman, J., Melot, T., Garau, X. S., Zucker, J. M., Lenoir, G. M., Ambros, P. F., Sheer, D., Turc-Carel, C., Triche, T. J., Aurias, A., and Thomas, G. (1994) The Ewing family of tumors—a subgroup of small-round-cell tumors defined by specific chimeric transcripts. *N. Engl. J. Med.* **331,** 294–299.
5. Taylor, C., Patel, K., Jones, T., Kiely, F., De Stavola, B. L., and Sheer, D. (1993) Diagnosis of Ewing's sarcoma and peripheral neuroectodermal tumour based on the detection of t(11;22) using fluorescence in situ hybridisation. *Br. J. Cancer* **67,** 128–133.
6. Desmaze, C., Zucman, J., Delattre, O., Melot, T., Thomas, G., and Aurias, A. (1994) Interphase molecular cytogenetics of Ewing's sarcoma and peripheral neuroepithelioma t(11;22) with flanking and overlapping cosmid probes. *Cancer Genet. Cytogenet.* **74,** 13–18.
7. Ladanyi, M. and Gerald, W. L. (1995) PCR analysis of genomic junction fragments in the EWS-WT1 gene rearrangement in DSRCT. *Mod. Pathol.* **8,** 144A.
8. Turc-Carel, C., Aurias, A., Mugneret, F., Lizard, S., Sidaner, I., Volk, C., Thiery, J. P., Olschwang, S., Philip, I., Berger, M. P., Philip, T., Lenoir, G. M., and Mazabraud, A. (1988) Chromosomes in Ewing's sarcoma. I. An evaluation of 85 cases and remarkable consistency of t(11;22)(q24;q12). *Cancer Genet. Cytogenet.* **32,** 229–238.
9. Sorensen, P. H. B., Shimada, H., Liu, X. F., Lim, J. F., Thomas, G., and Triche, T. J. (1995) Biphenotypic sarcomas with myogenic and neural differentiation express the Ewing's sarcoma EWS/FLI1 fusion gene. *Cancer Res.* **55,** 1385–1392.

10. Delattre, O., Zucman, J., Plougastel, B., Desmaze, C., Melot, T., Peter, M., Kovar, H., Joubert, I., de Jong, P., Rouleau, G., Aurias, A., and Thomas, G. (1992) Gene fusion with an ETS DNA-binding domain caused by chromosome translocation in human tumours. *Nature* **359,** 162–165.
11. Zucman, J., Delattre, O., Desmaze, C., Plougastel, B., Joubert, I., Melot, T., Peter, M., De Jong, P., Rouleau, G., Aurias, A., and Thomas, G. (1992) Cloning and characterization of the Ewing's sarcoma and peripheral neuroepithelioma t(11;22) translocation breakpoints. *Genes Chromosom. Cancer* **5,** 271–277.
12. Burd, C. G. and Dreyfuss, G. (1994) Conserved structures and diversity of functions of RNA-binding proteins. *Science* **265,** 615–621.
13. Ohno, T., Ouchida, M., Lee, L., Gatalica, Z., Rao, V. N., and Reddy, E. S. P. (1994) The EWS gene, involved in Ewing family of tumors, malignant melanoma of soft parts and desmoplastic small round cell tumors, codes for an RNA binding protein with novel regulatory domains. *Oncogene* **9,** 3087–3097.
14. Plougastel, B., Zucman, J., Peter, M., Thomas, G., and Delattre, O. (1993) Genomic structure of the EWS gene and its relationship to EWSR1, a site of tumor-associated chromosome translocation. *Genomics* **18,** 609–615.
15. Ladanyi, M., Lewis, R., Garin-Chesa, P., Rettig, W. J., Huvos, A. G., Healey, J. H., and Jhanwar, S. C. (1993) EWS rearrangement in Ewing's sarcoma and peripheral neuroectodermal tumor. Molecular detection and correlation with cytogenetic analysis and MIC2 expression. *Diag. Mol. Pathol.* **2,** 141–146.
16. Sorensen, P. H. B., Liu, X. F., Delattre, O., Rowland, J. M., Biggs, C. A., Thomas, G., and Triche, T. (1993) Reverse transcriptase PCR amplification of EWS/FLI-1 fusion transcripts as a diagnostic test for peripheral primitive neuroectodermal tumors of childhood. *Diag. Mol. Pathol.* **2,** 147–157.
17. Zhang, L., Lemarchandel, V., Romeo, P. H., Ben-David, Y., Greer, P., and Bernstein, A. (1993) The Fli-1 proto-oncogene, involved in erythroleukemia and Ewing's sarcoma, encodes a transcriptional activator with DNA-binding specificities distinct from other Ets family members. *Oncogene* **8,** 1621–1630.
18. Prasad, D. D., Rao, V. N., and Reddy, E. S. (1992) Structure and expression of human fli-1 gene. *Cancer Res.* **52,** 5833–5837.
19. Zucman, J., Melot, T., Desmaze, C., Ghysdael, J., Plougastel, B., Peter, M., Zucker, J. M., Triche, T. J., Sheer, D., Turc-Carel, C., Ambros, P., Combaret, V., Lenoir, G., Aurias, A., Thomas, G., and Delattre, O. (1993) Combinatorial generation of variable fusion proteins in the Ewing family of tumours. *EMBO J.* **12,** 4481–4487.
20. Zoubek, A., Pfleiderer, C., Salzer-Kuntschik, M., Amann, G., Windhager, R., Fink, F. M., Koscielniak, E., Delattre, O., Strehl, S., Ambros, P. F., Gadner, H., and Kovar, H. (1994) Variability of EWS chimaeric transcripts in Ewing tumours: a comparison of clinical and molecular data. *Br. J. Cancer* **70,** 908–913.
21. Bhagirath, T., Abe, S., Nojima, T., and Yoshida, M. C. (1995) Molecular analysis of a t(11; 22) translocation junction in a case of Ewing's sarcoma. *Genes Chromosom. Cancer* **13,** 126–132.
22. May, W. A., Gishizky, M. L., Lessnick, S. L., Lunsford, L. B., Lewis, B. C., Delattre, O., Zucman, J., Thomas, G., and Denny, C. T. (1993) Ewing sarcoma

11;22 translocation produces a chimeric transcription factor that requires the DNA-binding domain encoded by FLI1 for transformation. *Proc. Natl. Acad. Sci. USA* **90,** 5752–5756.
23. Sorensen, P. H. B., Lessnick, S. L., Lopez-Terrada, D., Liu, X. F., Triche, T. J., and Denny, C. T. (1994) A second Ewing's sarcoma translocation, t(21;22), fuses the EWS gene to another ETS-family transcription factor, ERG. *Nature Genet.* **6,** 146–151.
24. Giovannini, M., Biegel, J. A., Serra, M., Wang, J. Y., Wei, Y. H., Nycum, L., Emanuel, B. S., and Evans, G. A. (1994) EWS-erg and EWS-Fli1 fusion transcripts in Ewing's sarcoma and primitive neuroectodermal tumors with variant translocations. *J. Clin. Invest.* **94,** 489–496.
25. Ida, K., Kobayashi, S., Taki, T., Hanada, R., Bessho, F., Yamamori, S., Sugimoto, T., Ohki, M., and Hayashi, Y. (1995) EWS-FLI-1 and EWS-ERG chimeric mRNAs in Ewing's sarcoma and primitive neuroectodermal tumor. *Int. J. Cancer* **63,** 500–504.
26. Jeon, I.-S., Davis, J. N., Braun, B. S., Sublett, J. E., Roussel, M. F., Denny, C. T., and Shapiro, D. N. (1995) A variant Ewing's sarcoma translocation (7;22) fuses the EWS gene to the ETS gene ETV1. *Oncogene* **10,** 1229–1234.
27. Kaneko, Y., Yoshida, K., Handa, M., Toyoda, Y., Nishihira, H., Tanaka, Y., Sasaki, Y., Ishida, S., Higashino, F., and Fujinaga, K. (1996) Fusion of an ETS-family gene, E1AF, to EWS by t(17;22)(q12;q12) chromosome translocation in an undifferentiated sarcoma of infancy. *Genes Chromosom. Cancer* **15,** 115–121.
28. Sreekantaiah, C., Ladanyi, M., Rodriguez, E., and Chaganti, R. S. K. (1994) Chromosomal aberrations in soft tissue tumors. Relevance to diagnosis, classification, and molecular mechanisms. *Am. J. Pathol.* **144,** 1121–1134.
29. Zucman, J., Delattre, O., Desmaze, C., Epstein, A., Stenman, G., Speleman, F., Fletchers, C. D. M., Aurias, A., and Thomas, G. (1993) EWS and ATF-1 gene fusion induced by t(12;22) translocation in malignant melanoma of soft parts. *Nature Genet.* **4,** 341–345.
30. Brown, A. D., Lopez-Terrada, D., Denny, C., and Lee, K. A. W. (1995) Promoters containing ATF-binding sites are de-regulated in cells that express the EWS/ATF1 oncogene. *Oncogene* **10,** 1749–1756.
31. Ladanyi, M. and Gerald, W. (1994) Fusion of the EWS and WT1 genes in the desmoplastic small round cell tumor. *Cancer Res.* **54,** 2837–2840.
32. Gerald, W. L., Rosai, J., and Ladanyi, M. (1995) Characterization of the genomic breakpoint and chimeric transcripts in the EWS-WT1 gene fusion of desmoplastic small round cell tumor. *Proc. Natl. Acad. Sci. USA* **92,** 1028–1032.
33. Haber, D. A., Sohn, R. L., Buckler, A. J., Pelletier, J., Call, K. M., and Housman, D. E. (1991) Alternative splicing and genomic structure of the Wilms tumor gene WT1. *Proc. Natl. Acad. Sci. USA* **88,** 9618–9622.
34. de Alava, E., Ladanyi, M., Rosai, J., and Gerald, W. L. (1995) Detection of chimeric transcripts in desmoplastic small round cell tumor and related developmental tumors by reverse transcriptase polymerase chain reaction. A specific diagnostic assay. *Am. J. Pathol.* **147,** 1584–1591.

35. Labelle, Y., Zucman, J., Stenman, G., Kindblom, L. G., Knight, J., Turc-Carel, C., Dockhorn-Dworniczak, B., Mandahl, N., Desmaze, C., Peter, M., Aurias, A., Delattre, O., and Thomas, G. (1995) Oncogenic conversion of a novel orphan nuclear receptor by chromosome translocation. *Hum. Mol. Genet.* **4,** 2219–2226.
36. Clark, J., Benjamin, H., Gill, S., Sidhar, S., Goodwin, G., Crew, J., Gusterson, B. A., Shipley, J., and Cooper, C. S. (1996) Fusion of EWS gene to CHN, a member of the steroid/thyroid receptor gene superfamily, in a human myxoid chondrosarcoma. *Oncogene* **12,** 229–235.
37. Huang, A., Campbell, C. E., Bonetta, L., McAndrews-Hill, M. S., Chilton-MacNeill, S., Coppes, M. J., Law, D. J., Feinberg, A. P., Yeger, H., and Williams, B. R. (1990) Tissue, developmental, and tumor-specific expression of divergent transcripts in Wilms tumor. *Science* **250,** 991–994.
38. Ron, D. and Habener, J. F. (1992) CHOP, a novel developmentally regulated nuclear protein that dimerizes with transcription factors C/EBP and LAP and functions as a dominant negative inhibitor of gene transcription. *Genes Dev.* **6,** 439–453.
39. Ron, D., Brasier, A. R., McGehee, R. E., Jr., and Habener, J. F. (1992) Tumor necrosis factor-induced reversal of adipocytic phenotype of 3T3-L1 cells is preceded by a loss of nuclear CCAAT/enhancer binding protein (C/EBP). *J. Clin. Invest.* **89,** 223–233.
40. Crozat, A., Aman, P., Mandahl, N., and Ron, D. (1993) Fusion of CHOP to a novel RNA-binding protein in human myxoid liposarcoma. *Nature* **363,** 640–644.
41. Rabbitts, T. H., Forster, A., Larson, R., and Nathan, P. (1993) Fusion of the dominant negative transcription regulator CHOP with a novel gene FUS by translocation t(t(t(12;16) in malignant liposarcoma. *Nature Genet.* **4,** 175–180.
42. Zinszner, H., Albalat, R., and Ron, D. (1994) A novel effector domain from the RNA-binding protein TLS or EWS is required for oncogenic transformation by CHOP. *Genes Dev.* 8, 2513–2526.
43. Panagopoulos, I., Höglund, M., Mertens, F., Mandahl, N., Mitelman, F., and Åman, P. (1996) Fusion of the EWS and CHOP genes in myxoid liposarcoma. *Oncogene* **12,** 489–494.
44. Sambrook, J., Fritsch, E. F., and Maniatis, T. (1989) *Molecular Cloning. A Laboratory Manual* (2nd ed.) Cold Spring Harbor Laboratories, Cold Spring Harbor, NY.
45. Ausubel, F. M., Brent, R., Kingston, R. E., Moore, D. D., Seidman, J. G., Smith, J. A., and Struhl, K., eds. (1992) *Short Protocols in Molecular Biology* (2nd ed.) John Wiley & Sons, NY.
46. Howell, M. D. and Kaplan, N. O. (1987) Spurious DNA blot hybridization resulting from bacterial contamination of primary tissue preparations. *Anal. Biochem.* **161,** 311–315.
47. Kawasaki, E. S. (1990) Amplification of RNA, in *PCR Protocols. A Guide to Methods and Applications* (Innis, M. A., Gelfand, D. H., Sninsky, J. J., and White, T. J., eds. San Diego, CA, Academic, 21–27.
48. Downing, J. R., Khandekar, A., Shurtleff, S. A., Head, D. R., Parham, D. M., Webber, B. L., Pappo, A. S., Hulshof, M. G., Conn, W. P., and Shapiro, D. N. (1995) Multiplex RT-PCR assay for the differential diagnosis of alveolar rhabdomyosarcoma and Ewing's sarcoma. *Am. J. Pathol.* **146,** 626–634.

49. Chomczynski, P. and Sacchi, N. (1987) Single-step method of RNA isolation by acid guanidinium thiocyanate–phenol–chloroform extraction. *Anal. Biochem.* **162,** 156–159.
50. Roux, K. H. (1995) Optimization and troubleshooting in PCR. *PCR Methods. Appl.* **4,** S185–S194.
51. Don, R. H., Cox, P. T., Wainwright, B. J., Baker, K., and Mattick, J. S. (1991) "Touchdown" PCR to circumvent spurious priming during gene amplification. *Nucleic Acids Res.* **19,** 4008.
52. Bovee, J. V., Devilee, P., Cornelisse, C. J., Schuuring, E., and Hogendoorn, P. C. (1995) Identification of an EWS-pseudogene using translocation detection by RT-PCR in Ewing's sarcoma. *Biochem. Biophys. Res. Commun.* **213,** 1051–1060.
53. Menon, R. S., Chang, Y.-F., St. Clair, J., and Ham, R. G. (1991) RT-PCR artifacts from processed pseudogenes. *PCR Methods Appl.* **1,** 70,71.
54. Kwok, S. and Higuchi, R. (1989) Avoiding false positives with PCR. *Nature* **339,** 237,238.
55. Peter, M., Couturier, J., Pacquement, H., Michon, J., Thomas, G., Magdelenat, H., and Delattre, O. (1997) A new member of the ETS family fused to EWS in Ewing tumors. Oncogene 14, 1159–1164.
56. Speleman, F., Delattre, O., Peter, M., Hauben, E., Van Roy, N., and Van Marck, E. (1997) Malignant melanoma of the soft parts (clear-cell sarcoma): confirmation of EWS and ATF-1 gene fusion caused by a t(12;22) translocation. *Mod. Pathol.* **10,** 496–499.
57. Zucman-Rossi, J., Batzer, M. A., Stoneking, M., Delattre, O., and Thomas, G. (1997) Interethnic polymorphism of EWS intron 6: genome plasticity mediated by Alu retroposition and recombination. *Hum. Genet.* **99,** 357–363.

19

p53 Detection in Breast Cancer

Penelope L. Davis and J. Dirk Iglehart

1. Introduction

p53 is a nuclear phosphoprotein whose function is classified as a tumor suppressor *(1)*. Mutations in the p53 gene are currently regarded as the most common genetic alteration in human cancer *(2)*, including breast cancer (reviewed in **ref.** *3*). Most of these are point mutations within highly conserved regions of the gene that produce altered protein with increased stability *(4)*, allowing easy detection in affected cells by immunohistochemical and immunoblotting techniques. The presence of elevated levels of mutant p53 may itself be a prognostic factor in human breast cancer *(5,6)*. Furthermore, a significant association between high levels of p53 and established pathological criteria (e.g., tumor stage, estrogen, and progesterone receptor levels) have been described in a number of studies *(5,6)*. This chapter will describe two procedures routinely used in our laboratory for the detection of p53 protein in mammary epithelial cell lines and in breast-tumor tissue. The methods have been used to analyze p53 levels in breast tumors *(5–7)* and used in many studies to monitor p53 expression following DNA damage in cell lines *(8–10)*.

1.1. Immunohistochemical Analysis

Immunohistochemical staining of p53 involves the well-established application of avidin–biotin technology for localization of proteins *(7)*. A p53-specific primary antibody is used, followed by a biotinylated secondary antibody; reactive protein is subsequently localized using a chromogenic form of avidin (alkaline phosphatase [AP], horseradish peroxidase [HRP]) such that reaction of the avidin–enzyme conjugate with a suitable substrate produces an insoluble colored product in the cell. Use of an enzyme label as opposed to a fluorescent label allows subsequent counterstaining of the cell nuclei and/or cytoplasm; in addition, slides can be stored without any appreciable loss of signal, whereas

From: *Methods in Molecular Medicine, Vol. 14: Tumor Marker Protocols*
Edited by: M. Hanausek and Z. Walaszek © Humana Press Inc., Totowa, NJ

fluorescent staining is more liable to fade in time. For a more detailed discussion of general immunohistochemical methods and applications, *see* vol. 10 of the *Methods in Molecular Medicine* (Chapters 10, 11, and 14) published by Humana Press. For specific consideration of p53 detection, *see* **ref. *11***.

1.2. Immunoblotting Analysis

Immunohistochemical analysis of breast tumors has proven to be an accurate method of screening for the presence of mutant and/or wild-type p53. An alternative method that allows a more quantitative assessment of p53 protein levels, as well as providing information to back up immunohistochemical data, is Western blotting. This technique is more widely used to analyze p53 expression in cell lines; however, its application in breast tumors is also valuable.

Western blotting is a sensitive assay for the detection and characterization of proteins ***(12)***. The technique involves solubilization and electrophoretic separation of proteins, followed by electrophoretic transfer to nitrocellulose membrane. Immunological identification of bound proteins involves sequential probing of the membrane with a primary p53-specific antibody and an HRP-linked secondary antibody. Visualization of specific polypeptides is achieved using enhanced chemiluminescence (ECL) ***(13)***, a nonradioactive, light-emitting method of detection whereby HRP-catalyzed oxidation of luminol in the presence of chemical enhancers causes the emission of light, which is then detected by short exposure to X-ray film.

There are various different immunoblotting procedures available: The methodology described here has been optimized for the detection of p53 in breast tissue and cell lines. For a more extensive account of western blotting procedures, *see* vol. 10 of the *Methods in Molecular Medicine* series (Chapter 24) (*see* **Subheading 1.1.**). A detailed description of sodium dodecyl sulfate polyacrylamide electrophoresis (SDS-PAGE) is not relevant for this chapter; for details *see* vol. 1 of the same series (Chapter 6). The most significant advance in western blotting has been in the visualization step using ECL detection reagents. The main advantages of ECL over other detection methods (radioactive, chromogenic) are:

1. High sensitivity—approximately 10 times more sensitive than other systems;
2. A stable, permanent signal is recorded on film, which can be quantitated by densitometry; and
3. Reprobing—the membrane can be sequentially reprobed with other antibodies. Also, antibodies can be easily stripped from membranes without antigen damage.

2. Materials

2.1. Immunohistochemistry in Paraffin-Embedded Specimens

1. Archived paraffin-embedded tumor blocks.
2. Microtome suitable for cutting thin sections from paraffin blocks.

3. Charged or "plus" glass microscope slides, which are precleaned and designed to adhere to paraffin sections.
4. Appropriate size cover slips.
5. Permanent mounting media to weld the cover slip to the histology slide; we use ACCU-Mount from Baxter Labs, Inc. (West Chester, PA).
6. Deparaffinization reagents: ethanol, xylene, and acetone.
7. Staining racks to hold the slides during application of reagents, and a humidity chamber to cover the slides during addition of aqueous reagents. There are commercially available slide holders with a reservoir for water and a light-proof lid to hold in the moisture.
8. Conventional laboratory oven capable of heating over a range of temperatures from 37 to 90°C.
9. Microwave oven—note the maximum wattage (*see* **Note 1**).
10. Phosphate buffered saline (PBS) and PBS with 2% bovine serum albumin (PBS + 2% BSA): Dissolve 360 g NaCl, 8 g KCl, 8 g $KHPO_4$, 45.6 g Na_2HPO_4 in 4 L of distilled water. For PBS + 2% BSA, add 2 g of powdered BSA/100 mL of buffer for daily use. BSA must be ultrapure or crystalline and IgG-free.
11. Primary antibody against p53. PAB1801 (Ab-2; Oncogene Science, Manhasset, NY) is a murine IgG_1 monoclonal antibody (MAb) that recognizes a denaturation-resistant epitope in the human p53 protein located between amino acids 32 and 79. Dilute PAB1801 to a working concentration of 1.0 μg/mL in PBS + 2% BSA (*see* **Note 2**).
12. Nonimmune mouse IgG_1, diluted to 1.0 μg/mL in PBS + 2% BSA.
13. Biotinylated, affinity-purified horse antimouse secondary antibody against mouse IgG (Vector Laboratories, Burlingame, CA).
14. 10% Normal horse serum, diluted into PBS + 2% BSA.
15. Elite Universal ABC Kit (Vector Laboratories) or other streptavidin/biotin-based detection systems that are available commercially. These kits come with enzyme-conjugated streptavidin (in this case, streptavidin is conjugated to peroxidase), biotinylated secondary antibody, and color-development systems. These kits contain prediluted reagents, instructions, and information about troubleshooting difficulties.
16. DAB (3,3'-diaminobenzidine tetrahydrochloride). DAB can be purchased as a powder or in ready-to-use kits. In the authors' laboratory, a stock solution is prepared by dissolving 10 g of DAB into 500 mL of 0.05*M* Tris-HCl, pH 7.6. The stock is filtered, divided into 2.5-mL and 5.0-mL aliquots, and frozen. For use, the stock solution is thawed and diluted into 0.05*M* Tris-HCl, pH 7.6 buffer (2.5 mL stock into 100 mL buffer or 5 mL stock into 200 mL buffer) to give a final working concentration of 0.5 mg DAB/1 mL of solvent. Just prior to use, 1 mL of 0.6% hydrogen peroxide is added to the working solution (*see* **Note 3** for precautions).
17. Appropriate counterstain if desired. The authors use 1% methyl green, particularly if black and white photographs are taken.
18. Positive and negative control sections (*see* **Notes 4** and **5**).

2.2. Western Blotting (see Note 6)

1. Snap frozen surgical specimens of breast cancers from biopsies or mastectomies. Store at –120°C.
2. Breast-cancer cell lines can be obtained from the American Type Culture Collection (Rockville, MD).
3. Solubilization buffer: 50 m*M* Tris-HCl, pH 8.0, 5 m*M* ethylenediamine tetraacetic acid (EDTA), 150 m*M* NaCl, 0.5% Nonidet P-40. Filter and store at 4°C. Add the following protease and phosphatase inhibitors from stock solutions immediately before use (*see* **Note 6**)
   a. 0.5 m*M* Phenylmethylsulfonyl fluoride (PMSF): 100 m*M* stock solution in 100% isopropanol. Store at –20°C.
   b. 25 µg/mL Leupeptin: 5 mg/mL stock solution in distilled water. Store at –20°C.
   c. 25 µg/mL Aprotinin: 5 mg/mL stock solution in 0.01*M* HEPES pH 8.0. Store at –20°C.
   d. 10 µg/mL Soybean trypsin inhibitor: 5 mg/mL stock solution in distilled water. Store at –20°C.
   e. 1 m*M* Benzamidine: 1 *M* stock solution. Store at 4°C.
   f. 10 µg/mL Pepstatin A: Prepare from a 1 mg/mL stock solution in 100% ethanol. Store at –20°C.
   g. 80 m*M* β-Glycerophosphate: Prepare from a 1 *M* stock solution, pH 7.3. Store at 4°C.
   h. 20 m*M* EGTA: Prepare from a 0.5 *M* stock solution, pH 8.0. Store at room temperature.
   i. 15 m*M* MgCl: Prepare from a 1*M* stock solution. Store at room temperature.

   For cell culture extraction, add **items a–e**; for tissues add **items a–i.**
4. Plastic, sterile, disposable cell scrapers.
5. Brinkmann homogenizer.
6. Sonicator with a microtip adaptor.
7. Mini PROTEAN II electrophoresis cell (BioRad, Hercules, CA).
8. Mini Trans-blot electrophoretic transfer cell (BioRad).
9. Rainbow-colored protein molecular weight markers (Amersham, Arlington Heights, IL).
10. Stock acrylamide solution; ultrapure PROTOGEL—a stabilized, premixed solution of 30% (w/v) acrylamide and 0.8% (w/v) *bis*-acrylamide (National Diagnostics, Boston, MA) (*see* **Note 8**).
11. Resolving gel buffer (Buffer R): 1.5*M* Tris-HCl, 0.4% (w/v) SDS, pH 8.8.
12. Stacking gel buffer (Buffer M): 0.5*M* Tris-HCl, 0.4% (w/v) SDS, pH 6.8.
13. 10X Running buffer: 30.3 g Tris base, 144 g glycine, 10 g SDS dissolved in 1 L distilled water. Make a 1X solution just before use.
14. Ammonium persulfate (APS), 5% (w/v); will remain effective for up to 3 wk at room temperature.
15. *N,N,N',N'*-Tetramethylene-ethylenediamine (TEMED). Store at 4°C.
16. 2X Sample loading buffer: 50 m*M* Tris-HCl, pH 6.8, 25% (v/v) glycerol, 6% (w/v) SDS, 0.01% (w/v) bromophenol blue. Make up 20 mL and store

in 1-mL aliquots at –20°C; add 40 µL 2-mercaptoethanol to each aliquot after thawing.

17. Electroblotting buffer: 0.05*M* Tris base, 0.19*M* glycine, 20% (v/v) methanol. Store at 4°C.
18. Nitrocellulose membrane: 0.45 µm pore size.
19. 10X PBS with Tween (PBST): Dissolve 360 g NaCl, 8 g KCl, 8 g $KHPO_4$, 45.6 g Na_2HPO_4 in 4 L distilled water. Add Tween-20 to a final concentration of 0.1%.
20. Heat-sealable plastic bags and heat sealer.
21. Blocking buffer (blotto): 5% instant nonfat dried milk in PBST. Preferably make fresh, but can be stored at 4°C for 3–5 d.
21. Primary antibody: Anti-p53 DO1 mouse monoclonal antisera (Santa Cruz Biotechnology, Inc., Santa Cruz, CA) diluted 1/1000 in PBST/blotto (*see* **Note 9**). Can be reused 2 to 3 times within 5 d; store diluted antisera at 4°C.
22. Secondary antibody: goat antimouse HRP-conjugated antisera (Amersham) diluted 1/4000 in PBST/blotto.
23. ECL detection reagents 1 and 2 (Amersham).
24. Autoradiography cassettes.
25. X-ray film: Kodak X-Omat AR5.
26. Stripping buffer: 100 m*M* 2-mercaptoethanol, 2% SDS, 62.5 m*M* Tris-HCl, pH 6.7.
27. Glass hybridization jars and hybridization oven.

3. Methods

3.1. Immunohistochemistry for p53 in Formalin-Fixed Tissue Sections

1. Cool paraffin blocks in ice water and cut sections 4–6-µm thick.
2. Mount the sections on coated slides by heating at 42°C for 15 min and air dry for several hours or overnight.
3. Place the slides in staining racks and deparaffinize by soaking them in Coplin jars filled with xylene. Soak for 5 min in three changes of fresh xylene.
4. Rehydrate by immersion in graded alcohol solutions as follows: 5 min in two changes of 100% ethanol, 5 min in two changes of 95% ethanol, and 5 min in distilled water.
5. To quench endogenous peroxidase activity, soak the hydrated slides in 0.3–3% hydrogen peroxide in ethanol prepared just prior to use. For instance, we prepare a 100-mL H_2O_2/ethanol solution by adding 1–10 mL of 30% H_2O_2 to 95% ethanol prior to using. Slides should be soaked for 5–10 min in a 3% solution; longer times are used for lower peroxide concentrations.
6. p53 antigen detection in formalin-fixed tissue benefits from an antigen-retrieval method. In this protocol, we recommend heating in a 700-W oven for 5 min in a 0.1 *M* citrate buffer, cooling for 3 min at room temperature, and repeating the microwave heating cycle for 3 min. Allow the slides to come to room temperature (*see* **Note 1** for alternatives and details).
7. Rinse the slides in PBS with 2% BSA by placing them in carriers and dipping them in Coplin jars containing buffer.

8. Incubate the slides with 10% normal horse serum for 15 min by applying enough reagent to cover the tissue section completely (about 100 μL). It is advisable to incubate the slides within the humidity chamber to prevent them from drying.
9. Shake or blot off the excess normal serum; it is not necessary to wash the slides at this point. Add the primary antibody diluted in PBS with 2% BSA as above (**Subheading 2.1., item 11**). Incubate in the humidity chamber for 30 min at room temperature. As an alternative, slides can be left overnight at 40°C in a humidity chamber with the primary antibody.
10. Rinse the slides well in PBS.
11. Incubate the slides with secondary antibody (biotinylated, affinity-purified horse antimouse IgG provided in the Vector Elite kit). Incubate 30 min at room temperature in the humidified chamber.
12. Rinse well in PBS.
13. Add the Elite ABC reagent (Vectastain), which is made up 30 min prior to use according to instructions provided with the Elite ABC kit. This reagent is essentially biotin-conjugated HRP, which is complexed with avidin and retains exposed biotin-binding sites on the avidin. Allow the ABC reagent to incubate for 30 min at room temperature in the humidity chamber.
14. Wash extensively in PBS.
15. Add DAB diluted as recommended above and incubate for 2–5 min until the solution bathing the tissue section turns dark and opaque (*see* **Note 3** for precautions).
16. Rinse copiously in gently running tap water.
17. Counterstain as desired. In the authors' laboratory, methyl green counterstaining provides adequate tissue coloration to discriminate histology and makes the p53-positive cells easy to identify or quantify.
18. Since DAB is alcohol-insoluble, a permanent-mounting medium (ACCU-Mount) can be used. The tissue is first dehydrated in three changes of acetone and then cleared with xylene rinses. One drop of mounting medium is centered on the tissue section and placed at the top of the slide. A cover slip is picked up by the inverted slide, which is turned over, the edge blotted, and trapped bubbles allowed to settle out to the sides.
19. Slides are examined under light microscopy; *see* **Note 4** for help with interpretation.

3.2. Western Blotting

3.2.1. Extraction of Protein from Tissue Samples and Cell Layers

1. Cell cultures: Wash adherent cell monolayers twice with cold PBS. Add 0.3 mL solubilization buffer/5X 10^5 cells and rock gently for 5 min before scraping cells with cell scraper. Collect the resulting whole-cell extract and sonicate for 15 s on ice at setting 4 of a Branson microtip sonicator. Spin lysates in Eppendorf tubes at high speed for 10 min at 4°C to remove cell debris. Transfer supernatant to a fresh tube and store at –20°C (*see* **Notes 7** and **10**).
2. Breast tumors: Thaw tumor and allow to sit on ice in solubilization buffer (0.2–1.0 mL depending on the size of tumor) for 10 min to soften the tissue.

Homogenize for 1 min to fully disrupt the tissue and sonicate on ice for 45 s. Spin briefly to pellet the bulk of the insoluble material, transfer to Eppendorf tubes, and spin extracts for a minimum of 30 min at 4°C. Store supernatant at –20°C (*see* **Note 11**).

3.2.2. Electrophoresis of Protein Samples

1. Prepare a 10% polyacrylamide-resolving gel in the mini-electrophoresis cell apparatus (*see* **Note 12**) by mixing the following: 14.2 mL distilled water, 8.8 mL buffer R, 11.6 mL PROTOGEL acrylamide, 250 µL 10% APS, 30 µL TEMED. Once this has set, pour the stacking gel: 9.0 mL distilled water, 3.8 mL buffer M, 2.3 mL PROTOGEL acrylamide, 125 µL APS, 15 µL TEMED.
2. Dilute protein lysates (50 µg, *see* **Note 13**) and 8 µL rainbow markers with 2X sample loading buffer, i.e., mix 1 vol of sample with 1 vol of buffer. Heat samples at 95–100°C for 2–5 min to fully denature the proteins and load onto the gel.
3. Electrophorese samples in 1X running buffer at 25–30 mA constant current until bromophenol blue dye reaches the bottom of the gel (*see* **Note 14**). A detailed description of the theory and methodology involved in SDS-PAGE is given in **ref.** ***11***.

3.2.3. Assembly of the Western Blot "Sandwich"

1. Cut two pieces of Whatman 3MM filter paper to the same size as the fiber pads and one piece of nitrocellulose cut to the same size as the gel (*see* **Note 14**). Prewet pads, paper, and membrane in electroblotting buffer. Soak gel in buffer for 1–2 min.
2. Place the gel-holder cassette on the bench with the gray side down (negative, cathode) and lay components onto this surface in the following order: pad, filter paper, gel, membrane, filter paper, pad. Exclude any air bubbles that may interfere with the transfer by gently rolling a 10-mL glass pipet over the surface of the second filter paper.
3. Close the cassette by lowering the clear panel (positive, anode) and clamp together. Insert the assembly vertically into the buffer tank such that the gray panel faces the cathode (–) electrode (*see* **Note 16**). Insert the Bio-ice cooling unit (prepared in advance by filling it with distilled water and storing it in the freezer; *see* **Note 17**) and fill the tank with electroblotting buffer. Place a stir bar in the tank.
4. Carry out the transfer (stirring gently) at 60-V constant voltage for 2 h at 4°C (*see* **Note 18**).

3.2.4. Development of the Western Blot

1. Remove the membrane from the "sandwich" and check to see if the colored rainbow markers are visible, indicating efficient transfer (*see* **Note 19**). Rinse the blot briefly in PBST and seal in a heat-sealable plastic bag with blocking buffer (5–10 mL). Incubate for 1 h on an orbital shaker or rocking platform (to block any nonspecific antibody sites on the nitrocellulose). All subsequent incubations

are carried out in this way; all washes are carried out in a plastic container in PBST with vigorous shaking.

2. Remove the blot from the bag and wash for 5 min.
3. Seal the blot in a fresh bag with primary p53 antibody (*see* **Note 9**) and incubate for 1–2 h.
4. Wash blot with 5 to 6 changes of PBST (approx 100 mL).
5. Seal the blot in a fresh bag with secondary antibody and incubate for 1 h (*see* **Note 20**).
6. Wash blot as in **step 4**.
7. Mix ECL detection reagents in a 1:1 ratio to give sufficient volume to cover the membrane. Drain excess buffer from the membrane and place in a small container. Add the ECL reagent to the protein side of the membrane (i.e., colored markers facing up) and incubate for 1 min. Ensure that the membrane surface remains completely covered.
8. Drain off excess fluid and wrap membrane in plastic film (Saran Wrap). Smooth out air pockets.
9. Place the membrane in a film cassette and expose to film for 15 s (*see* **Note 21**).
10. Rinse off detection reagents in PBST and store membrane wet-wrapped in plastic film (Saran Wrap) at 4°C.
11. The membrane is routinely stripped of bound antibodies and reprobed with a mouse antiactin antisera (1 µg/mL dilution in blotto) to control for protein loading errors (*see* **Note 22**). Place the membrane in a small glass hybridization jar with 15 mL of stripping buffer and incubate rotating at 50°C in a hybridization oven for 30–40 min.
12. Wash the membrane twice for 10 min in PBST at room temperature using large vol of buffer (200 mL). Carry out immunodetection with the actin antibody as in **steps 1–9**.

4. Notes

1. Antigen retrieval, by heating sections in buffered citrate solutions to temperatures of over 100°C, has been shown to be effective for formalin-fixed and paraffin-archived tissue blocks. Microwave ovens provide quick heating of the buffers and are widely available. These ovens come with maximal power ratings between 500 and 1000 W. The authors use a 700-W oven and set to maximum, heat for 5 min, cool for 5 min, and repeat the microwave heating for 3 min. For more powerful ovens, adjust the power settings to provide intermittent cycles, during which boiling occurs for about 15 s. The total times the slides are boiling is about 2 min. Slides can be placed in a suitable holder and placed in a container with buffer, or Coplin jars can be used for smaller procedures.
2. A variety of antibodies against p53 are now commercially available. The authors have used other antibodies in paraffin tissues with equal success.
3. **DAB is a potent carcinogen and should be used under a barrier hood or a fume hood. Disposal of DAB must be done according to approved handling of carcinogenic chemicals in each institution.**

4. In most circumstances, p53 is detected as a nuclear antigen. We have used a murine IgG_1 as a negative control. Control sections should be devoid of staining ("0" in a 0–3 scale where "3" is the maximum intensity). Sections are scored positive when there is widespread nuclear staining (>25% of cells or the majority of nuclei staining) and when the intensity is considered at least a "2" in the 0–3 scale. In this case, the authors have consistently found underlying p53 mutations in the "positive" cases. Using less stringent criteria will lead to false-positive estimations of p53 mutations. For instance, normal p53 is regulated during the cell cycle and is strongly induced by agents that damage genomic DNA. In addition, a given tumor has a signature p53 mutation and is clonal with respect to this mutation. Therefore, the entire tumor should demonstrate high-level nuclear staining. In human breast cancer, about 30% of clinical isolates will have a p53 mutation. In contrast, 60% of human colon and lung cancers will display p53 mutations, and nearly 90% or more of human small-cell carcinomas of the lung will carry a mutant p53 allele.
5. Positive control sections are easily found. Cell blocks can be made from a variety of tissue cultured cell lines known to carry p53 mutations. In human breast cancer, these lines include BT-20, BT-474, T-47D, and MDA-MB-468. HBL-100 is a normal human mammary cell line immortalized with SV40 T-antigen and that sequesters large amounts of wild-type p53 protein in the nucleus. In contrast, the MCF-7 human breast cancer cell line harbors a wild-type p53. Cytospins can be made and fixed in formalin or pellets imbedded in paraffin and cut onto slides.
6. A detailed account of common variables of Western blotting, as well as explanations of the theory behind certain steps, are covered comprehensively in the Notes section in vol. 10 of the Methods in Molecular Medicine series (Chapter 24): For example, the use of different acrylamide concentrations; determination of correct antibody dilutions; reasons for including methanol in the transfer buffer, use of different membrane supports, water cooling of apparatus, and alternative blocking agents and detection systems. Therefore, for the sake of simplicity, notes relating to the Western blotting techniques described here will not address these points, and the reader is referred to the aforementioned chapter.
7. Most methods of solubilization that use mechanical disruption procedures release intracellular proteases that can digest the target protein. Denatured proteins are more likely to be degraded than native proteins. To minimize proteolytic activity, it is important to keep solubilized extracts cold and to include the various protease inhibitors listed here. Many are considered to be very toxic and should be handled with care; for example, PMSF is extremely destructive to the mucous membranes of the respiratory tract, eyes, and skin. For further information on protease inhibitors, *see* **ref. *14***.
8. PROTOGEL is a premixed acrylamide solution; therefore, its use reduces risk of exposure to neurotoxic and carcinogenic acrylamide dust while weighing and filtering reagents.
9. We have found the p53-specific DO1 mouse MAb (Santa Cruz) to be the most sensitive antisera to detect p53 protein in Western blots. Other antisera are avail-

able that recognize linear epitopes in the amino or carboxy terminal of p53; for example, PAB 1801 (Oncogene Science and Santa Cruz). Exceptions are PAB1620 and PAB246, both of which recognize conformational epitopes of wild-type p53, and PAB240, which recognizes a "mutant"-specific conformation; since these will only recognize p53 under nondenaturing conditions, they are more widely employed in immunoprecipitation analysis of p53.

10. Extracts should always be kept on ice and stored at –20 or –100°C for long-term storage. Freeze/thawing of extracts does not appear to affect the stability of p53 protein, although other proteins may be sensitive to such treatment. Extracts (both cell and tissue lysates) are routinely aliquoted into small volumes before freezing.
11. Tumor samples tend to vary in size and fat content. Larger tumors (~1 cm in diameter) may need to be cut into smaller pieces with a sterile scalpel blade to allow more efficient homogenization of tissue; often they also require longer centrifugation to pellet insoluble material (up to 1 h). After the final spin, a fatty layer occasionally forms on the surface of the lysate, which is easily dispersed, causing the extract to look cloudy. This cannot be avoided, but samples of the extract can be withdrawn as required by carefully lowering the Eppendorf tip slowly beneath the fatty layer.
12. Larger gels can also be used; however, the minigels are easily assembled, can take as little as 45 min to run, give good resolution, and use fewer reagents.
13. As little as 20 µg of protein extract is sufficient to detect p53 on a Western blot, but 50 µg is more routinely loaded to ensure detection of lower, wild-type p53 levels.
14. This normally takes ~1 h, but the current can be lowered to run gels more slowly if convenient.
15. Wear gloves for all manipulations, because oil from hands will block efficient transfer.
16. The nitrocellulose must always face the anode (+), i.e., current forces proteins out of the gel and onto the membrane from (–) to (+).
17. The frozen water acts as a heat sink for heat generated during the transfer process. It will maintain an appropriate buffer temperature for approx 1 h if carried out at 4°C.
18. Sometimes an overnight transfer is more convenient; use 25 V constant voltage at 4°C.
19. The membrane can be stained with reversible stains (e.g., ponceau red or fast green) to show transferred proteins and to allow marking the position of different protein markers if rainbow markers are not available. It also gives some indication of the accuracy of equal protein loading.
20. The use of bags instead of an open container allows the membrane to be incubated in a small volume, thereby eliminating excess waste of antisera. The blocking and antibody incubation steps can be interrupted at any stage by storing the sealed membrane at 4°C.
21. For p53 detection, ECL exposure times vary from 5 s to 5 min. In some mutant p53 cell lines, a 2-s exposure is sufficient. If overexposure occurs, blots can be left in the cassette for 5–10 min before re-exposing to film.

22. The blot can be stripped and reprobed several times. The use of a hybridization oven for this step is more convenient than floating a sealed bag in a heated water bath. There is an alternative method of probing the same blot with two different antibodies, provided that their molecular weights are sufficiently different to allow their resolution. For example, following the blocking step, the membrane can be cut in half such that the upper half retains polypeptides of >50 K and the lower half <50 K; each half can be probed separately for p53 and actin (42 K), respectively, at the same time. This is easier to do with blots derived from larger gels that result in greater resolution of proteins, but it is feasible with the miniblots. Protein levels can be quantitated by densitometry; actin levels should not vary between samples and are used to correct for loading errors.

References

1. Levine, A. J., Momand, J., and Finlay, C. A. (1991) The p53 tumour supressor gene. *Nature* **351,** 453–456.
2. Hollstein, M., Sidransky, D., Vogelstein, B., and Harris, C. C. (1991) P53 mutations in human cancer. *Science* **235,** 49–53.
3. Harris, A. L. (1992) P53 expression in human breast cancer. *Adv. Cancer Res.* **59,** 69–88.
4. Caron de Fromentel, C. and Soussi, T. (1992) P53 tumour supressor gene: a model for investigating human mutagenesis. *Genes Chrom. Cancer* **4,** 1–15.
5. Marks, J. M., Humphrey, P. A., Wu, K., Berry, D., Bandarenko, N., Kerns, B. M., and Iglehart, J. D. (1994) Overexpression of p53 and HER-2*neu* proteins as prognostic markers in early stage breast cancer. *Ann. Surg.* **219,** 332–341.
6. Davidoff, A. M., Herndon, J. E., Glover, N. S., Kerns, B. M., Pence, J. C., Iglehart, J. D., and Marks, J. R. (1991) Relation between p53 overexpression and established prognostic factors in breast cancer. *Surgery* **110,** 259–264.
7. Bayer, E. A., Skutelsky, E., and Wilchek, M. (1979) The avidin–biotin complex in affinity cytochemistry. *Meth. Enzymol.* **62,** 308–315.
8. Elbendary, A., Berchuck, A., Davis, P. L., Havrilesky, L., Bast, J. C., Iglehart, J. D., and Marks, J. R. (1994) Transforming growth factor β1 can induce CIP1/WAF1 expression independent of the p53 pathway in ovarian cancer cells. *Cell Growth Differ.* **5,** 1301–1307.
9. Gudas, J., Nguyen, H., Li, T., Hill, D., and Cowan, K. H. (1995) Effects of cell cycle, wildtype p53 and DNA damage on p21 expression in human breast epithelial cells. *Oncogene* **11,** 253–261.
10. Kastan, M. B., Onyekwere, O., Sidransky, D., Voglestein, B., and Craig, R. (1991) Participation of p53 protein in the cellular response to DNA damage. *Cancer Res.* **51,** 6304–6311.
11. Kerns, B. M., Jordan, P. A., Moore, M. H., Humphrey, P. A., Berchuck, A., Kohler, M. F., Bast, R. C., Iglehart, J. D., and Marks, J. R. (1992) P53 overexpression in formalin-fixed, paraffin-embedded tissue detected by immunohistochemistry. *J. Histochem. Cytochem.* **40,** 1047–1051.

12. Towbin, H., Staehlin, T., and Gordon, J. (1979) Electrophoretic transfer of proteins from polyacrylamide gels to nitrocellulose sheets: procedure and some applications. *Proc. Natl. Acad. Sci. USA* **76,** 4350–4354.
13. Whitehead, T. P. (1979) Analytical luminescence: its potential in the clinical laboratory. *Clin. Chem.* **25,** 1531–1546.
14. Barret, A. J. and Salvesen, G., eds. (1986) Proteinase inhibitors, in *Research Monographs in Cell and Tissue Physiology,* vol. 12. Elsevier, Amsterdam, The Netherlands, pp. 3–18.

20

Use of the Polymerase Chain Reaction Technique to Detect the t(14;18) Translocation in Lymphoid Tissue

Maryalice Stetler-Stevenson and Megan S. Lim

1. Introduction

The t(14;18) (q32;21) chromosomal translocation is characteristic of follicle center cell lymphomas, involving 95% of cases ***(1,2)***. In addition, it has been found in a variety of other malignant lymphomas, including 20% of all diffuse B-cell lymphomas ***(3)***, diffuse small cleaved-cell lymphomas ***(4)***, and Hodgkin's disease ***(5–7)***. These observations have led to the theory that the t(14;18) (q32;21) translocation is a critical oncogenic event in the multiple-hit model of lymphomagenesis. The t(14;18) (q32;21) translocation juxtaposes the *bcl-2* gene on chromosome 18 with the immunoglobulin heavy-chain joining region (J_H) gene on chromosome 14 ***(8,9)***, resulting in deregulation and overexpression of the *bcl-2* gene ***(10)***. Programmed cell death (apoptosis) is inhibited by bcl-2 protein, thus conferring a survival advantage to the lymphoma cells ***(11)***. The breakpoints on chromosome 14 and 18 are not random, but occur in specific, identified regions. The breakpoint on chromosome 18 is, in the majority of cases, clustered at two main regions in the *bcl-2* gene: 50–60% in the 150-bp major breakpoint region (MBR) ***(12)*** and 25% in the minor cluster region (MCR) ***(13)***. The breakpoint on chromosome 14 always occurs within J_H, a small portion of the heavy-chain gene.

The t(14;18) (q32;21) translocation was initially detected by traditional cytogenetic techniques and then by restriction enzyme Southern blot analysis. However, because of clustering of the breakpoints in well-defined regions of chromosomes 14 and 18, screening for the translocation is optimally performed by the polymerase chain reaction (PCR) technique. The DNA sequence at the *bcl-2*/J_H join is amplified using PCR and oligonucleotide primers specific for the regions of chromosomes 18 and 14 flanking the translocation. The

From: *Methods in Molecular Medicine, Vol. 14: Tumor Marker Protocols*
Edited by: M. Hanausek and Z. Walaszek © Humana Press Inc., Totowa, NJ

extremely sensitive PCR assay, which allows the detection of one cell containing the translocation present among 1 million normal cells, is ideally suited for detection of minimal disease ***(14,15)***, and is more rapid than restriction enzyme studies. In addition, unlike traditional methods, this technique can be used to detect the t(14;18) (q32;21) in a variety of tissue sources, including fresh, frozen, and formalin-fixed paraffin-embedded tissues, as well as cytological specimens ***(7,16,17)***. Therefore, the evaluation of tissues for the t(14;18) translocation using PCR can play a key role in diagnosis and, because of its greater sensitivity, the determination of minimal residual disease in patients undergoing workup for bone-marrow transplantation ***(18)***. Important caveats to keep in mind in the interpretation of PCR assays for t(14;18) in lymphoid tissues are its presence in benign, reactive hyperplastic lymphoid tissue as well as in peripheral blood lymphocytes ***(19,20)*** and the various causes for false positivity. Modifications of the PCR method to increase the sensitivity of detection, such as nested primers and booster PCR, increase the risk of contamination with amplicons or DNA from another patient. This may create a higher rate of false positivity. In addition, Segal et al. ***(21)*** have reported that standard primers frequently used for detecting the t(14;18) mbr breakpoint also amplify a 167-bp sequence within the Epstein-Barr virus (EBV) DNA, resulting in a false-positive result in EBV infection. These observations underscore the importance of utilizing a specific detection system in analysis of PCR products and correlating the results with the clinical history, as well as immunohistochemical and morphological features in the interpretation of a PCR assay for t(14;18) translocations in lymphoproliferative disorders. In this chapter, we will briefly outline methodology for PCR detection of the mbr t(14;18) using a 5' primer homologous to the *bcl-2* gene on chromosome 18 and a 3' primer complementary to the consensus sequence in the J_H region on chromosome 14.

2. Materials

2.1. DNA

1. DNA extraction kit.
2. Proteinase K (10 mg/mL) (for an extraction method based on proteinase K digestion).
3. TBE buffer: 0.045 *M* Tris-borate, 0.001 *M* EDTA, pH 8.0.
4. Tris-EDTA (TE) buffer: 1 *M* Tris-HCl, pH 7.4, 0.5 *M* EDTA, pH 8.0.
5. Ethanol.
6. Equilibrated phenol.
7. Chloroform.

2.2. PCR Setup and Reaction

1. PCR tubes (0.5-mL sterile PCR tubes, Perkin reaction tubes).
2. Pipet tips (aerosol-resistant tips).

3. Dedicated pipeters.
4. Automated thermocycler (e.g., Perkin–Elmer Cetus, Emeryville, CA).
5. PCR buffer (10X PCR buffer, Perkin–Elmer).
6. *Taq* polymerase (AmpliTaq or other commercially available source).
7. dNTPs, 25 m*M* (Promega or other commercially available source).
8. Mineral oil (chemical grade).
9. Primers:
   a. MBR primer: 5'-TTAGAGAGTTGCTTTACGTG-3';
   b. J_H consensus primer: 5'-ACCTGAGGAGACGGTGACCAGGGT-3';
   c. β-actin primer: sense 5'-AGGCCAACCGCGAGAAGATGACC-3';
   d. β-actin primer: antisense 5'-GAAGTCCAGGGCGACGTAGCAC-3'.

   Primers can be obtained from a local oligonucleotide synthesis laboratory or a commercial source.

2.3. Detection of PCR Products

1. Gel-loading buffer: 1X TBE buffer. Prepare from 10X TBE buffer: 121 g Tris-base, 68 g boric acid, 7.6 g EDTA; make up to 1 L.
2. Agarose: specialty agarose for preparative electrophoresis of nucleic acids, e.g., SeaKem GTG (FMC BioProducts, Rockland, ME).
3. TAE buffer: 0.04*M* Tris-acetate, 0.001*M* EDTA, or TBE buffer: 0.045*M* Tris-borate, 0.001*M* EDTA.
4. Gel-electrophoresis apparatus.
5. Ethidium bromide: 10 mg/mL (Life Technologies, Gaithersburg, MD).
6. DNA marker (1-kb ladder, Life Technologies).
7. Oligonucleotide-specific probe internal to *bcl-2* primer: 5'-CAACACAGACCC ACCCAGAGCCCTCCTGCCCTCCTTCCGCGGGGC-3'.

   Probe can be obtained from a local oligonucleotide synthesis laboratory or a commercial source.

3. Methods

3.1. DNA Extraction

The DNA is extracted in an area free of PCR products or high copies of target DNA sequences (e.g., plasmid preparations). In addition, the space is irradiated with UV light after each DNA extraction to destroy any possible DNA contamination.

3.1.1. Isolation of DNA from Fresh and Frozen Tissue

Genomic high-mol-wt DNA can be isolated from a variety of fresh and frozen tissue sources using rapid DNA isolation kits from various manufacturers. Cell suspensions can be directly lysed and DNA isolated. With solid blocks of tissue, pulverization of tissue frozen in liquid nitrogen to form a powder or fine mincing with disposable scalpels, is desirable.

1. Follow manufacturers' instructions when using DNA isolation kits. For example, using the DNA Stratagene extraction kit, DNA is obtained by a modification of a procedure based on separating contaminating protein from DNA by salt precipitation. It requires 2–3 h to complete and does not require phenol or chloroform. It is capable of processing numerous samples and is limited only by the available amount of centrifuge space.
2. Alternatively, use a method based on proteinase K digestion followed by chloroform/phenol extraction and then isopropanol precipitation and rinsing in 75% ethanol (*see* vol. 2 of the *Methods in Molecular Biology* series published by Humana Press for details).

   Dissolve extracted DNA in a buffer or water and keep at 4°C for up to 3 mo. Long-term storage should be at –20°C.
3. Determine concentration of DNA using a UV spectrophotometer. DNA should be stored at an optimal concentration of 1–2 µg/µL.

3.1.2. Isolation of DNA from Paraffin-Embedded Tissue

DNA can be isolated from paraffin-embedded tissue using commercially available kits or simple proteinase K digestion as described below.

1. Cut two 6-mm sections using disposable blades to prevent cross-contamination.
2. Deparaffinize the sections in xylene and wash in ethanol. For tissues fixed in B5, the procedure is modified to include a 4-min incubation in 1% (w/v) iodine in xylene to remove the mercury introduced by the fixative.
3. Digest the deparaffinized tissue in 200 µg/mL of proteinase K in 1X PCR buffer. Digestion is followed by debris removal by centrifugation.
4. Continue DNA isolation as described in **Subheading 3.1.1., step 2**.

The integrity of the isolated DNA is evaluated in a PCR-based assay using the following primers to the human β-actin gene: sense 5'-AGGCCAACCGCG AGAAGATGACC-3' and antisense 5'-GAAGTCCAGGGCGACGTAGCAC-3'. The reactions are set up in an identical manner to the t(14;18) PCR assay described below with substitution of the human β-actin gene primers for the *bcl-2* and J_H primers. The specimens are considered suitable for amplification only if a specific control fragment is discernible after electrophoresis through a 2% agarose gel.

3.2. PCR Assay

Reactions are carried out in 50- or 100-µL volumes (*see* **Note 1**).

1. Quality assurance: Because of the extreme sensitivity of PCR and the risk of false positives, set up the actual PCR reactions in a dedicated area, such as a hood or room. This area is kept free of PCR products and high-copy DNA, such as plasmids containing target sequence. In addition, the space, pipeters, tips, and racks are irradiated with UV light after each PCR setup to destroy any possible DNA contamination. Equipment (pipeters, tips, buffers) should be kept in PCR

Table 1
PCR Master Mix Volumes

Component	Volume, μL/reaction[a]	Final concentration
Water	68.5	
10X PCR buffer	10.0	1.64 m*M* (Mg^{2+})
MBR primer, 1.0 μ*M*	5.0	1.0 m*M*
J_H primer, 1.0 μ*M*	5.0	1.0 m*M*
dNTPS, 25 m*M* each: dATP, dCTP, dGTP, and dTTP	1.0	250 m*M*
Taq DNA polymerase, 5 U/mL	0.5	2.5 U/tube

[a]Total volume: 90 μL/reaction.

area and used solely for PCR setup. Pipeters with ejectors should be used in conjunction with aerosol-resistant tips to avoid carryover from tube to tube. Tubes containing DNA should be centrifuged prior to opening to reduce accidental spraying of area with DNA. Lab coats and gloves should be worn at all times. Gloves should be changed frequently during the procedures, especially after handling any tube containing DNA. Lab coats should be dedicated to the sample-preparation area. In each assay, DNA from the lymphoma cell line SUDHL-6 (ATCC, Rockville, MD) is used as a positive control. Placental DNA is used as a negative DNA control and distilled water as a negative amplification control. A great deal of scrutiny should be exercised in setting up various parts of the experiment to minimize sources of contamination that can be detrimental.

2. Labeling of PCR tubes: In the PCR area, assemble and label the 0.5-mL sterile PCR tubes and place them in a rack. Tubes to be labeled are the assay bank (water), positive control, and DNA of test specimen.
3. Master mix preparation (*see* **Note 2**): DNA is amplified in a reaction mixture containing 1X PCR buffer (10 m*M* Tris-HCl, 1.5 m*M* $MgCl_2$, 50 m*M* KCl, 0.1% porcine skin gelatin) and 0.2 n*M* each dNTP, 2.5 U *Taq* DNA polymerase, and 0.5 m*M* of each MBR and the J_H consensus primer. Prepare a master mix containing the buffer, primers, nucleotides, and often *Taq* polymerase in the PCR setup area for all of the PCR reaction tubes. This is done prior to handling of DNA. The volume of each component in the master mix is obtained by multiplying the volumes in **Table 1** by the number of PCR reaction tubes.

   In the case of hot-start PCR (*see* **Note 3**), an anti-*Taq* antibody (e.g., Taqstart antibody) is added to the master mix to prevent *Taq* activity until high heat denatures the antibody, releasing the enzyme. Preparation of a master mix before aliquoting out DNA samples not only adds convenience but also decreases chances of contamination during the setup process. Master mix solutions can be prepared at 10X final concentration, aliquoted, and stored at either –20 or 4°C for easy and convenient use. This allows for experiment-to-experiment consistency, reduces chances of unsuspected contamination, and removes random error resulting from calculation and pipet variation.

4. PCR setup: In the PCR setup area, add 8 µL of H_2O to all except the test blank tube. Then add 2 µL of test DNA (0.5–1.0 µg) to appropriately labeled tubes. Then add 10 µL of sterile H_2O to the tube labeled as the assay blank. The blank is set up last so that any contamination that occurs will be detected. Starting with the test samples, aliquot 90 µL of PCR master mix to each tube to bring the final volume of each reaction to 100 µL. Add 90 µL master mix last to the test blank. Add 100 µL of mineral oil to all tubes (test blank last) using disposable pipet tips. Change tips between each tube. Vortex each tube and briefly centrifuge.
5. Thermal cycling: Place tubes into the preheated thermal cycler and begin the cycling according to the following schedule:
   a. Denaturation: 2 min at 95°C;
   b. Annealing: 1 min at 55°C;
   c. Extension: 2 min at 72°C;
   for 35–40 cycles. Then continue as follows:
   d. Time delay: 10 min at 72°C
   e. Soak: 10 min at 4°C (this can be left overnight).

3.3. Detection of PCR Products

1. Preparation of 2% agarose gel (ideal for identification of PCR products with an estimated size range of 100–350 bp): Using TAE or TBE buffer, prepare 2% agarose solution (for example, 4 g agarose/200 mL buffer) in a glass flask or bottle. The solution should not fill >50% of the container. Loosely plug the container and heat in a boiling water bath or microwave oven until the agarose is dissolved. Cool the solution to 60°C and add ethidium bromide to final concentration of 5 µg/mL. Poor into the gel mold containing a comb for sample wells. The gel should be approx 3–5-mm thick. Allow the gel to set at room temperaturc for 30 min and 10 min at 4°C. Place the gel in an electrophoresis apparatus, cover with buffer, and carefully remove the comb.
2. Loading of gel: Mix 20–40 µL of PCR-amplified samples with 4 µL of gel-loading buffer (containing bromphenol blue and xylene cyanol dyes) and load into individual sample wells, taking care not to tear the bottom of the wells. DNA marker (1 kb ladder, Life Technologies) is used to provide a size estimate in base pairs.
3. Electrophoresis: Shut the electrophoretic apparatus lid and attach electrodes so that DNA will migrate toward the anode (red electrode). To check that the system is working, look for bubbles being generated at electrodes and advancement of the dyes in the loading buffer. The gel is electrophoresed at 8–9 V/cm (120 V for a 14-cm gel) until the bromphenol blue (fastest running dye) is 75% advanced along the gel. This usually takes 1–2 h.
4. Product visualization: Visualize products by UV illumination. A picture is taken with a ruler alongside the DNA size marker for future reference in analyzing radiographs of Southern blots (*see* **Note 4**).

5. Southern blot: Wash the gel for 5 min in water and for 10 min in denaturation/transfer solution. The gel is then blotted onto a nylon membrane overnight. Depurination by weak acid is not necessary because the DNA fragments are small and transfer efficiently.
6. Hybridization: Hybridize the Southern blot with the labeled *bcl-2* probe and confirm the identity of the visualized products. Radioactive or other labeling methods can be used.

4. Notes

1. Strict quality control and constant surveillance for contamination is necessary in a diagnostic setting. Our policy is to set up PCR reactions first thing in the morning, before working with DNA, and especially before working with amplicons (the PCR-amplified target sequence contained in the PCR products). DNA extraction should be performed before working with amplicons, but after PCR setup. We use color-coded coats for the amplicon-handling area. These colored coats identify a "contaminated" individual and are never worn into the PCR setup area.
2. Hot-start PCR is helpful with poor-quality DNA, such as is usually obtained from paraffin-embedded tissue. Traditional hot-start PCR involves adding the *Taq* polymerase after the tubes have been brought to 95°C. There is great risk of contamination in this procedure in that tubes containing hot DNA are opened in the PCR area. On insertion of the pipet tip to add the *Taq* polymerase, spattering and spraying of liquid can occur. To avoid the risk of contamination, an antibody to *Taq* is added with the enzyme so that the enzyme is kept in an inactive form until high temperatures are obtained. This removes the need to open the tubes and further manipulate the specimen.
3. When there is doubt regarding their identity, PCR products can be eluted from gels and the DNA sequence determined.
4. Simple observation of bands on ethidium gels is often not adequate, because EBV contains a sequence that is amplified by this system, producing a 167-bp product. Therefore, unless the PCR product differs significantly in size from 167 bases, confirmation of the t(14;18) by another method is necessary.

References

1. Ynuis, J. J., Oken, M. M., Kaplan, M. E., Ensrud, K. M., Howe, R. R., and Theologiodes, A. (1982) Distinctive chromosomal abnormalities in histologic subtypes of non-Hodgkin's lymphoma. *N. Engl. J. Med.* **307,** 1231–1234.
2. Fukuhara, S., Rowley, J. D., Variakojis, D., and Golomb, H. M. (1979) Chromosomal abnormalities in poorly differentiated lymphoma. *Cancer Res.* **39,** 3119–3123.
3. Aisenberg, A. C., Wilkes, B. M., and Jacobson, J. O. (1988) The bcl-2 gene is rearranged in many diffuse B-cell lymphomas. *Blood* **71,** 969–972.
4. Leith, C. P., Willman, C. L., Spier, C. M., Miller, T. P., and Grogen, T. M. (1994) The presence of bcl-1 and bcl-2 gene rearrangements in diffuse small cleaved-cell lymphoma. A disease with diverse molecular and immunophenotypic findings. *Diagn. Mol. Pathol.* **3,** 178–183.

5. Stetler-Stevenson, M. (1992) The t(14;18) translocation in Hodgkin's disease. *J. Natl. Cancer Inst.* **84,** 1770,1771.
6. Bhagat, S. K. M., Medeiros, L. J., Weiss, L. M., Wang, J., Raffeld, M., and Stetler-Stevenson, M. (1993) Bcl-2 expression in Hodgkin's disease. Correlation with the t(14;18) translocation and Epstein-Barr virus. *Am. J. Clin. Pathol.* **99,** 604–608.
7. Reid, A. H., Cunningham, R. E., Frizzera, G., and O'Leary, T. J. (1993) Bcl-2 rearrangement in Hodgkin's disease. Results of polymerase chain reaction, flow cytometry and sequencing on formalin-fixed, paraffin-embedded tissue. *Am. J. Pathol.* **142**, 395–402.
8. Tsujimoto, Y., Finger, L. R., Yunis, J., Nowell, P., and Croce, C. (1984) Cloning of the chromosome breakpoint of neoplastic B cells with the t(14;18) chromosomal translocation. *Science* **226**, 1097–1099.
9. Tsujimoto, Y., Cossman, J., Jaffe, E., and Croce, C. (1985) Involvement of the bcl-2 gene in human follicular lymphoma. *Science* **228**, 1440–1443.
10. Graninger, W. B., Seto, M., Boutain, B., Goldman, P., and Korsmeyer, S. J. (1987) Expression of bcl-2 and bcl-2-Ig fusion transcripts in normal and neoplastic cells. *J. Clin. Invest.* **80**, 1512–1515.
11. Hockenberry, D., Nunez, G., Milliman, C., Schreiber, R. D., and Korsmeyer, S. J. (1990) Bcl-2 is an inner mitochondrial membrane protein that blocks programmed cell death. *Nature* **348**, 334–336.
12. Tsujimoto, Y., Louie, E., Bashir, M. M., and Croce, C. M. (1988) The reciprocal partners of both the t(14;18) and the t(11;14) translocations involved in B-cell neoplasms are rearranged by the same mechanism. *Oncogene* **2**, 347–351.
13. Cleary, M. L., Smith, S. D., and Sklar, J. (1986) Cloning and structural analysis of cDNAs for bcl-2 and a hybrid bcl-2/immunoglobulin transcript resulting from the t(14;18) translocation. *Cell* **47**, 19–28.
14. Stetler-Stevenson, M., Raffeld, M., Cohen, P., and Cossman, J. (1988) Detection of occult follicular lymphoma by specific DNA amplification. *Blood* **72**, 1822–1825.
15. Lee, M.-S., Chang, K.-S., Cabanillas, F., Freireich, E. J., Trujillo, J. M., and Stass, S. A. (1987) Detection of minimal residual cells carrying the t(14;18) by DNA sequence amplification. *Science* **237**, 175–178.
16. Limpens, J., Beelen, M., Stad, R., Haverkort, M., van Krieken, J. H. J. M., van Ommen, G. B., and Kluin, P. M. (1993) Detection of the t(14;18) translocation in frozen and formalin-fixed tissue. *Diagn. Mol. Pathol.* **2**, 99–107.
17. Alkan, S., Lehman, C., Sarago, C., Sidawy, M. K., Karcher, D., and Garrett, C. (1995) Polymerase chain reaction detection of immunoglobulin gene rearrangement and bcl-2 translocation in archival glass slides of cytologic material. *Diagn. Mol. Pathol.* **4**, 25–31.
18. Berinstein, N. L., Jamal, H. H., Kuzniar, B., Klock, R. J., and Reis, M. D. (1993) Sensitive and reproducible detection of occult disease in patients with follicular lymphoma by PCR amplification of t(14;18) both pre- and posttreatment. *Leukemia* **7,** 113–119.

19. Limpens, J., De Jong, D., Van Krieken, J. H., Price, C. G., Young, B. D., van Ommen, G. J., and Kluin, P. M. (1991) bcl-2/JH rearrangements in benign lymphoid tissues with follicular hyperplasia. *Oncogene* **6**, 2271–2276.
20. Liu, Y., Hernandez, A. M., Shibata, D., and Cortopassi, G. A. (1994) Bcl-2 translocation frequency rises with age in humans. *Proc. Natl. Acad. Sci. USA* **91**, 8910–8914.
21. Segal, G. H., Scott, M., Jorgensen, T., and Braylan, R. C. (1994) Primers frequently used for detecting the t(14:18) major breakpoint also amplify Epstein-Barr viral DNA. *Diagn. Mol. Pathol.* **3**, 15–21.

21

Detection of *ras* Gene Mutations Using Oligonucleotide Ligation Technology

Faye A. Eggerding

1. Introduction

The human *ras* genes (H-, K-, and N-*ras*) are members of a superfamily of low-mol-wt GTP-binding proteins that function as G proteins in signal transduction pathways controlling cell proliferation and differentiation ***(1,2)***. *Ras* genes acquire oncogenic potential primarily as a result of point, missense mutations in codons 12, 13, or 61, producing single amino-acid substitutions that alter the ability of the protein to bind or hydrolyze GTP. The net consequence of somatic *ras* missense mutations is to lock the protein in a GTP-bound, active conformation, thus perturbing cellular physiology and contributing to tumorigenesis. Different tumor types show specificity of *ras* oncogene activation. For example, activated H-*ras* occurs most often in bladder cancers ***(3,4)***; K-*ras* mutations are found predominantly in pancreatic (90% cases), colon (40–50% cases) and lung carcinomas ***(5–7)***, and N-*ras* mutations are most often associated with hematopoietic malignancies, particularly myeloid leukemias ***(8,9)***.

Somatic mutations in one of the three *ras* genes are among the most common genetic abnormalities in human cancers, exceeded only by mutations in the *p53* gene ***(2)***. Because *ras* activation is so common and may occur both early in the development of some cancers and during tumor progression, detection of mutant *ras* alleles is an important procedure in the molecular oncology laboratory. Detection of single nucleotide substitution mutations in K-*ras* genes may be of diagnostic significance as an early tumor marker for colorectal and pancreatic carcinomas ***(10–12)***. *Ras* mutations may also serve as prognostic markers. The presence of point mutations in codon 12 of K-*ras* is a biomarker of poor prognosis in adenocarcinoma of the lung ***(13)***. *Ras* gene mutations, such as N-*ras* mutations in myeloid leukemia, are potential tumor markers for

From: *Methods in Molecular Medicine, Vol. 14: Tumor Marker Protocols*
Edited by: M. Hanausek and Z. Walaszek © Humana Press Inc., Totowa, NJ

determining minimal residual disease. In addition, rapid and sensitive identification of *ras* mutations in tumors should have expanded clinical application as anticancer therapies, directed at inhibition of oncogenic *ras* activity, become available.

1.1. Strategy of Oligonucleotide Ligation

In this chapter, a strategy to detect point mutations in *ras* oncogenes will be presented that combines the sensitivity of target amplification by polymerase chain reaction (PCR) with the specificity of sequence distinction by oligonucleotide ligation ***(14–17)***. Allelic discrimination in the oligonucleotide ligation assay (OLA) is based on the ability of thermostable DNA ligase to discriminate a single base-pair mismatch at the ligation junction (**Fig. 1**). The ligation reaction requires only that the terminal and penultimate nucleotides on both sides of the junction of the two probes be base-paired correctly; mismatches located elsewhere in the probe do not prevent ligation ***(15,16)***.

Mutation-specific probes are synthesized for each possible single-base, nonsilent mutation in codons 12, 13, and 61 of *ras* oncogenes. Normal and mutant allelic probes hybridize upstream and in immediate juxtaposition to a common, downstream probe; if sequences at the probe junctions are not perfectly base-paired, the probes will not be joined by ligase. Common probes are labeled with fluorophores, and allelic probes each have different lengths. Genotype-specific ligation products are therefore distinguished from other components in the reaction mixture by both fluorescence and size after electrophoresis on an automated DNA sequencer ***(18)***.

A modification of the PCR and OLA protocol, coupled amplification and ligation (CAL), combines DNA amplification by PCR with DNA genotyping by OLA ***(19)*** in one, single-tube assay. DNA amplification and DNA genotyping can be sustained in one reaction by virtue of differences in melting behavior of PCR primer-template and OLA probe-template DNA hybrids. *Ras* gene targets are amplified by multiplex PCR using primers with high melting temperatures during the initial stage of the reaction to provide templates for ligation; allelic discrimination occurs simply by lowering the reaction temperature to facilitate hybridization and competitive oligonucleotide ligation of low-melting, allele-specific OLA probes to the amplified target sequences.

2. Materials

2.1. Materials for Sample Preparation

1. Digestion buffer: 100 m*M* NaCl, 10 m*M* Tris-HCl, pH 8.0, 25 m*M* EDTA, 0.5% sodium dodecyl sulfate (SDS), 0.1 mg/mL proteinase K.
2. Proteolytic digestion buffer: 50 m*M* Tris-HCl, pH 8.5, 1 m*M* EDTA, 0.5% Tween-20, 2 mg/mL proteinase K.

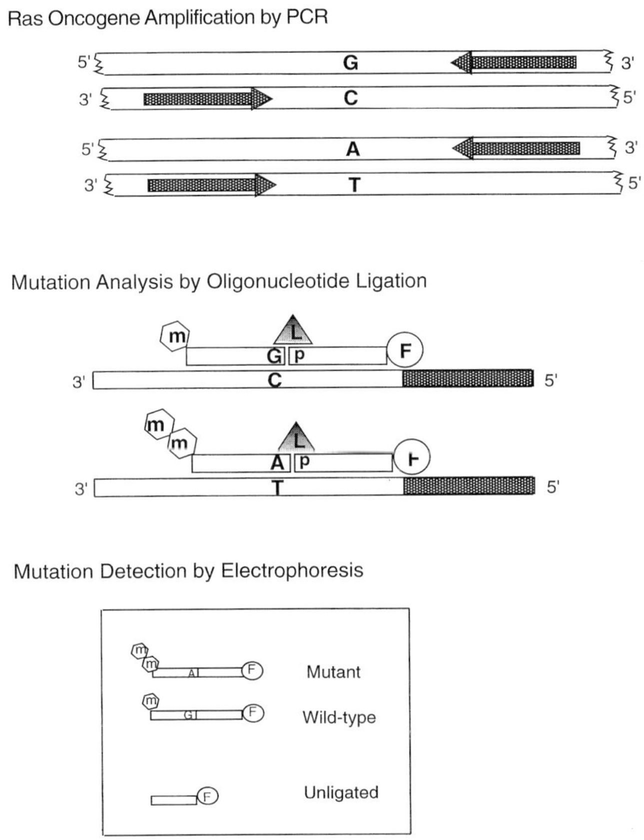

Fig. 1. Detection of *ras* gene mutations using PCR and oligonucleotide probe ligation. The hatched arrows represent PCR primers for synthesis of target *ras* gene PCR products. Allelic variants (G → A transition) in the amplified target are distinguished by oligonucleotide ligation. Probes hybridizing in juxtaposition to amplified target are covalently joined by DNA ligase (L); probes mismatched to the target by a single nucleotide at their junctions are not ligated. Upstream allelic probes contain 5'-mobility modifiers (M) for size separation; the downstream reporter probe is 5'-phosphorylated (p) and labeled with a fluorescent dye at the 3' end (F). Ligation products are detected by denaturing polyacrylamide gel electrophoresis on a fluorescent DNA sequencer.

3. 25:24:1 (v/v/v) Phenol/chloroform/isoamyl alcohol equilibrated with 50 m*M* Tris-HCl, pH 8.0.
4. TE buffer: 10 m*M* Tris-HCl, pH 8.0, 1 m*M* EDTA.
5. 3 *M* sodium acetate, pH 5.2.
6. 100% Ethanol, ice-cold; 70% ethanol, room temperature.
7. Glycogen (Boehringer-Mannheim [Mannheim, Germany], cat. no. 901393).

8. Sephadex G50 column equilibrated with TE buffer containing 100 m*M* NaCl.
9. SpeedVac evaporator (Savant, Instruments, Inc., Holbrodi, NY).

2.2. Materials for Amplification of ras Oncogenes

1. AmpliTaq DNA polymerase, 5 U/µL (Perkin–Elmer Co., Norwalk, CT).
2. 10X PCR amplification buffer: 500 m*M* KCl, 100 m*M* Tris-HCl, pH 8.3, 15 m*M* $MgCl_2$, 0.1% (w/v) gelatin.
3. 25 m*M* $MgCl_2$ solution.
4. 2.5 m*M* dNTP mix (consists of dATP, dCTP, dTTP, and dGTP).
5. Oligonucleotide PCR primers, 20 µ*M* (20 pmol/µL in sterile water; store at –20°C).
6. Template DNA, 100 ng/µL (human genomic DNA in sterile water).
7. 3% MetaPhor agarose (FMC BioProducts, Rockland, ME).
8. Automated thermal cycler (GeneAmp PCR System 9600, Perkin–Elmer; Applied Biosystem Div., Foster City, CA).

2.3. Materials for Analysis of ras Mutations by Oligonucleotide Ligation

1. *Thermus aquaticus (Taq)* DNA ligase, 40 U/µL (New England BioLabs Inc., Beverly, MA).
2. 10X oligonucleotide ligation assay (OLA) buffer: 200 m*M* Tris-HCl, pH 7.6, 100 m*M* DTT, 10 m*M* NAD^+, 1% Triton X-100 (New England BioLabs).
3. 10X CAL buffer: 200 m*M* Tris-HCl, pH 7.6, 500 m*M* KCl, 100 m*M* $MgCl_2$, 100 m*M* DTT, 10 m*M* NAD^+, 0.1% Triton X-100.
4. Allele-specific oligonucleotide probes modified in the 5' position by addition of a mobility modifier group, 20 µ*M* (20 pmol/µL in sterile water; store at –20°C).
5. Reporter oligonucleotides modified by addition of a 5' phosphate group and by a fluorophore in the 3' position, 20 µ*M* (20 pmol/µL in sterile water; store at –20°C). Fluorophores are stable indefinitely at –20°C.
6. 10X OLA probe mix (250 n*M*): prepare by diluting 20 µ*M* stock solutions of appropriate allelic and reporter oligonucleotide probes 1/80 in water.
7. Automated thermal cycler (GeneAmp PCR System 9600, Perkin–Elmer).

2.4. Materials for Automated Fluorescent Detection of Ligation Products

1. Fluorescent DNA sequencer (Model 373A with Genescan 672 software, Applied Biosystems Division, Perkin–Elmer).
2. 10X Tris-Borate-EDTA (TBE): 0.89 *M* Tris base, 0.89 *M* boric acid, 30 m*M* EDTA.
3. Polyacrylamide gel: 8% acrylamide, 8.3 *M* urea sequencing gel, 1X TBE.
4. Gel loading solution: 10 m*M* EDTA in deionized formamide.
5. Internal electrophoresis lane standard, 50 n*M* (30–70 nucleotide oligomers labeled with the fluorophore ROX, 6-carboxy-X-rhodamine).

3. Methods

3.1. Sample Preparation

Sample preparation is crucial for insuring the quality and reproducibility of PCR, particularly when working with clinical samples, such as small tumor-biopsy specimens. Although the use of purified DNA is highly preferable, PCR and oligonucleotide ligation will work directly on cells lysed only by boiling ***(18,19)***. Following DNA extraction, and before initiating PCR, it is useful to assess the quantity of DNA recovered by removing a small aliquot for fluorometry. The methods outlined below have been used successfully in this laboratory for preparation of genomic DNA from tissue culture cells, fresh biopsy material, and microdissected tissue sections (*see* **Note 1**).

3.1.1. Tissue Culture Cells, Biopsy Tissue, and Whole Blood

1. Starting with tissue-culture cells: Collect cells from the flask by trypsinization and centrifuge 5 min at 500*g*. Wash cells in a small volume of ice-cold PBS, centrifuge 5 min at 500*g*, and resuspend the cells in digestion buffer (1 mL buffer/10^8 cells) ***(20)***.
2. Starting with whole-biopsy tissue: After excision, mince the tissue with scalpel blades and snap-freeze in liquid nitrogen. Grind tissue to a fine powder with a chilled mortar and pestle. Suspend the powdered tissue in digestion buffer (1 mL buffer/100 mg tissue) ***(20)***.
3. Starting with whole blood: Isolate mononuclear cells and lymphocytes from whole blood using a Ficoll-Hypaque gradient and wash the cells twice with ice-cold PBS ***(21)***. Resuspend the cells in digestion buffer (5×10^3 cells/µL).
4. Lyse the tissue culture cells, whole-tissue cells, or blood mononuclear cells by overnight (12–18 h) incubation at 50°C with gentle agitation.
5. Extract the samples in a microfuge tube with an equal volume of phenol/chloroform/isoamyl alcohol equilibrated with 50 m*M* Tris-HCl, pH 8.0. Microfuge briefly at room temperature and collect the aqueous phase. Re-extract the aqueous phase two to three times as necessary.
6. Precipitate the DNA by addition of 1/10 vol of 3 *M* sodium acetate, pH 5.2, and 2–2.5 vol of ice-cold 100% ethanol. Glycogen (2 µg) may be added as a carrier molecule to enhance the efficiency of the DNA precipitation. Mix and place in a –20°C freezer for 30 min or longer.
7. Pellet the DNA by spinning at high speed in a fixed-angle microfuge for 5 min and remove the supernatant. Wash the pellet with room-temperature 70% ethanol, microfuge, and remove the supernatant. Dry the DNA pellet in a Speed-Vac evaporator.
8. Dissolve the dry DNA pellet in sterile distilled water or TE, pH 8.0, and store at 4°C or at –20°C.

3.1.2. Paraffin Sections

1. Sections (5–50 μm) are cut and fixed onto glass slides.
2. Tissue sections are deparaffinized by incubating the slides in Coplin jars with xylene for 5–20 min, depending on the thickness of the tissue section. The slides are then placed in 100% ethanol for 5 min and air-dried.
3. Scrape microdissected tissue from the glass slides and transfer to 1.5-mL Eppendorf tubes containing enough proteolytic digestion buffer so that the microdissected tissue occupies about 30% of the volume of the buffer (typically about 100 to 200 μL of buffer).
4. Incubate microdissected tissue fragments in proteolytic digestion buffer overnight or until tissue is dissolved. Large samples may need longer incubation times and additional proteinase K ***(22–24)***.
5. Following proteinase K digestion, extract the samples at least two times with an equal volume of phenol/chloroform/isoamyl alcohol equilibrated with 50 m*M* Tris-HCl, pH 8.0.
6. Precipitate the DNA and resuspend the pellet as described in **Subheading 3.1.1.**
7. Crude DNA extraction from microdissected tissues may be performed by incubating the tissue in TE buffer containing 500 μg–2 mg/mL proteinase K at 50°C. When digestion is complete, microfuge the sample and incubate at 100°C for 10 min to inactivate proteinase K before initiating PCR.

3.1.3. Tissue Fluids and Scrapings

1. Cells from tissue fluids or scrapings are collected by centrifugation, washed with gentle agitation in ice-cold PBS, and pelleted by centrifugation at 500*g* for 5 min.
2. The cell pellet is resuspended in sterile distilled water (10^6–10^7 cells/mL), boiled for 30 min–1 h, and briefly microcentrifuged to remove cell debris ***(18,19)***.
3. Before PCR, the supernatant containing the crudely extracted DNA is purified by passage over a Sephadex G50 column equilibrated with TE buffer containing 100 m*M* NaCl.
4. Alternatively, more reliable results can be obtained by scaling down the procedure described in **Subheading 3.1.1.** for small sample volumes and by adding glycogen to improve the efficiency of precipitation of small amounts of DNA present in tissue fluids or smears that may contain few viable cells.

3.2. Amplification of ras Oncogenes

The purity of DNA templates and the design of the oligonucleotide primers for PCR are the two factors most crucial to the overall efficacy of amplification reactions. Long (36–51 bases), high-melting-temperature (T_m) primers flanking codons 12 and 13 and codon 61 in exons 1 and 2, respectively, of *ras* oncogenes were used at an annealing and elongation temperature of 72°C. Under these conditions, primer-directed amplification at each *ras* gene locus was optimally efficient with high yields of correct-sized PCR products uncontaminated with nonspecific spurious-sized fragments. To maximize product yields in multiplex amplifications, the magnesium concentration can be

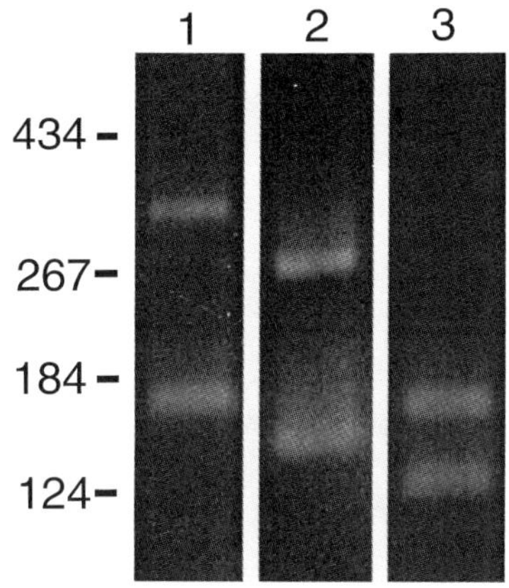

Fig. 2. Multiplex amplification of exons 1 and 2 of H-, K-, and N-*ras* oncogenes. PCR reaction products were analyzed on a 3.0% MetaPhor agarose gel. H- (lane 1), K- (lane 2), and N-*ras* (lane 3) amplification product sizes for exon 1 are 169, 260, and 126 base pair, and for exon 2 are 339, 133 and 172 base pair, respectively. Size markers represent the sizes of *Hae*III-digested pBR322 DNA fragments.

increased from 1.5 to 3.0 m*M* or greater without evident mispriming at nontarget sites. Long PCR primers are more gene-specific, and they can be annealed and extended at a higher temperature, which maximizes *Taq* polymerase activity and enables simultaneous amplification and genotyping utilizing the CAL method (*see* **Subheading 3.3.2.**). The PCR protocol used varies depending on whether the sequential PCR and OLA (**Subheading 3.2.1.** and **3.3.1.**) or the CAL protocol (**Subheading 3.3.2.**) is employed.

3.2.1. PCR Protocol

1. Prepare the PCR reaction in Perkin–Elmer GeneAmp thin-walled reaction tubes (N801-0537) as follows: 2.5 µL 10X PCR buffer, 2.0 µL 25 m*M* $MgCl_2$, 2 µL (250 µ*M* each dNTP) 2.5 m*M* dNTP mix, 0.5 µL (400 n*M* each primer) PCR primers (20 µ*M*), 0.5 µL (2.5 U) AmpliTaq polymerase, and 1.0–5.0 µL (100 ng/µL) genomic DNA sample. Sterile distilled water is added to a final volume of 25 µL.
2. Close the GeneAmp tube and amplify by PCR in a thermal cycler (GeneAmp PCR system 9600, Perkin–Elmer) using the following cycling parameters: initial denaturation, 94°C, 5 min; 30 two-step cycles: 94°C, 30 s; and 72°C, 2 min; and final extension, 72°C, 10 min.
3. To assess the quality of a PCR, the amplification products can be visualized by UV transillumination following electrophoresis on 3% MetaPhor agarose gels stained with 0.5 µg/mL ethidium bromide. A representative result is shown in **Fig. 2** (*see* **Note 2**).

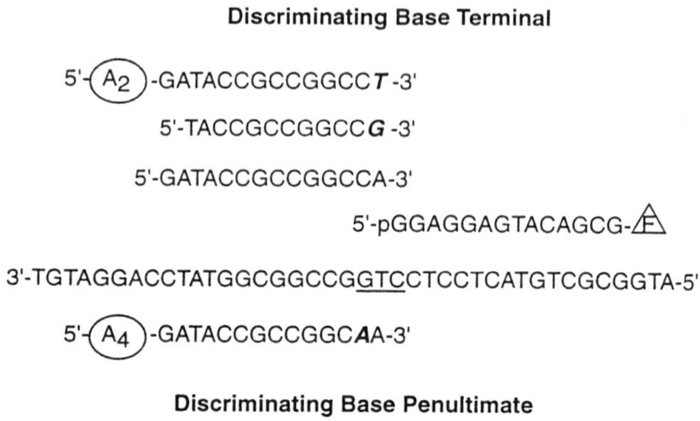

Fig. 3. Oligonucleotide ligation probe design. The sequences of oligonucleotide probes used to distinguish normal and mutant alleles at H-*ras* codon 61 are shown in the 5' to 3' orientation. The nucleotide sequence of exon 2 surrounding codon 61 (underlined) of the normal H-*ras* allele is also shown. A set of three probes is required to analyze each alternative sequence, one for each allele and one reporter 5'-phosphorylated probe labeled with a fluorophore (F) at its 3' end. Allele-specific oligonucleotides are designed such that the distinguishing base is the 3'-terminal base, or in one case, the 3'-penultimate base. Allelic sequences are indicated in bold italic. Noncomplementary 5'-poly(A) extensions are added to the 5' ends of some allelic probes as mobility modifiers for sizing. On ligation to the fluorescently labeled reporter probe, each ligation product has a characteristic electrophoretic mobility, differing in size by two bases.

3.3. Analysis of ras Mutations by Oligonucleotide Ligation

Oligonucleotide probes used for detection of *ras* mutations are 12- to 22-mers with similar melting properties for performing anneal-ligation reactions at 55°C. Allelic oligonucleotides are designed to have a 3' sequence (the terminal or penultimate base) specific for either the wild-type allele or one of the *ras*-mutant alleles (**Fig. 3**). Noncomplementary poly(A) extensions or nonnucleotide oligomers (for example, oligomers of pentaethylene-oxide, PEO) may be added to their 5' ends as mobility modifiers for electrophoretic separation. Oligomeric PEO mobility modifiers can be added onto the allelic probes during automated oligonucleotide synthesis using PEO phosphoramidite monomers ***(25)***. Reporter (downstream) oligonucleotides, hybridizing immediately 3' to the allelic oligonucleotides, have a 5' phosphate group, required for ligation, and a fluorophore introduced in the 3' position. Allele-specific ligation products have a unique electrophoretic mobility determined by the length of the ligated oligonucleotides and the mobility modifier, and they each contain a fluorophore label for detection on a fluorescent DNA sequencer.

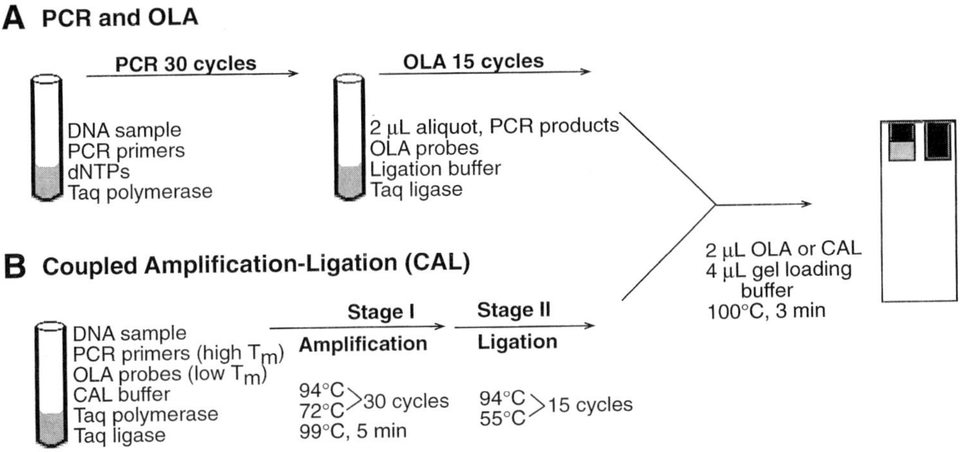

Fig. 4. Schematic representation of alternative oligonucleotide ligation assay formats. **(A)** A DNA sample is amplified by PCR and an aliquot of the PCR reaction is transferred to a second tube for analysis by oligonucleotide ligation. **(B)** A DNA sample is analyzed in a single tube, one-step coupled amplification-ligation format (CAL). DNA amplification with high T_m PCR primers occurs in stage I; genotyping by oligonucleotide ligation occurs in stage II by lowering the temperature to allow hybridization of low T_m ligation probes (*see* **Note 3**). Aliquots of the ligation products from either (A) or (B) are analyzed on an automated fluorescent DNA sequencer.

The sensitivity of oligonucleotide ligation is greatly increased by combining it with a primary amplification of template DNA by PCR. Here, two strategies for ligase-mediated mutation detection in conjunction with PCR will be described (**Fig. 4**) (*see* **Notes 3–8**).

3.3.1. Oligonucleotide Ligation Protocol

Oligonucleotide ligation may be carried out as a two-step procedure in combination with an initial PCR amplification as follows:

1. Prepare the OLA reaction in GeneAmp thin-walled reaction tubes as follows: 1.0–2.0 µL PCR-amplified DNA sample (*see* **Subheading 3.2.1.**), 2.0 µL 10X OLA buffer, 1.0–2.0 µL 10X OLA probe mix, and 1.0 µL (40 U) Taq DNA ligase.
   Sterile distilled water is added to a final volume of 20 µL.
2. Close the GeneAmp tube and initiate anneal-ligation cycles in a thermal cycler using the following cycling parameters: 94°C, 5 min; initial denaturation, 20 anneal-ligation cycles: 94°C, 45 s; and 55°C, 2 min 30 s.

3.3.2. Coupled Amplification and Ligation Protocol

1. Prepare the CAL reaction in GeneAmp thin-walled reaction tubes as follows: 2.5 µL 10X CAL buffer, 2.0 µL 2.5 m*M* each of four dNTPs (250 µ*M* each dNTP),

0.5 μL PCR primers (20 μ*M*) (400 n*M* each primer), 0.5 μL (2.5 U) AmpliTaq polymerase, 1.0–2.0 μL 10X OLA probe mix, 5.0 μL (200 U) *Taq* DNA ligase, and 1.0–5.0 μL (100 ng/μL) genomic DNA sample.

Sterile distilled water is added to a final volume of 25 μL.

2. Close the GeneAmp tube and begin CAL in a thermal cycler using the following cycling parameters: initial denaturation, 94°C, 5 min; 30 two-step PCR cycles: 94°C, 30 s; 72°C, 2 min 30 s; 99°C, 5–10 min (inactivate AmpliTaq); and 20 anneal-ligation cycles: 94°C, 30 s; 55°C, 2 min.

Initially, conditions are weighted in favor of PCR to allow amplification of *ras* oncogenes to provide templates for the ligation reaction. A reaction buffer system was developed to support both PCR and oligonucleotide ligation. Mutation detection by competitive oligonucleotide ligation occurs when the temperature is lowered to 55°C to permit hybridization of the OLA probes. Reporter oligonucleotide probes contain a 3' fluorophore label for the detection of ligation products and for blockage of 3' end extension by *Taq* polymerase. Extension from the 3' ends of allelic oligonucleotides may occur but would be undetectable because it prevents ligation.

3.4. Automated Fluorescent Detection of Ligation Products

1. Add 1.0–2.0 μL aliquots of the PCR-OLA reaction (**Subheading 3.3.1.**) or the CAL reaction (**Subheading 3.3.2.**) to 4.5 μL of gel-loading solution containing 0.5 μL of a mixture of ROX-labeled synthetic oligonucleotide size markers.
2. Heat the samples to 98°C for 2–5 min to denature before loading onto a pre-electrophoresed 8% polyacrylamide, 8.3 *M* urea sequencing gel. Start the 373A DNA sequencer and carry out the automated run according to standard procedure for 3–4 h.
3. Perform the fluorescent fragment analyses with the Applied Biosystems Genescan 672 software. Results of experiments using the PCR-OLA and CAL procedures for analysis of H-, K-, and N-*ras* mutations are illustrated in **Fig. 5**. The assay is capable of detecting *ras* oncogene mutations in a DNA sample in the presence of a 100-fold excess of normal DNA (**Fig. 6**).

4. Notes

1. The methods used for sample preparation may be simplified or additional steps added as determined empirically. In general, purification of DNA samples by phenol-chloroform extraction and ethanol precipitation give the most consistent results. Instead of SDS, a nonionic detergent, such as Tween-20 (0.5%), which is more compatible with *Taq* polymerase, can be used in the digestion buffer ***(21)***. Be careful not to aspirate the genomic DNA precipitate after the 70% ethanol wash since it may not stick to the walls of the tube. Crude DNA preparations must be stored at –20°C since they are more susceptible to degradation by endo-

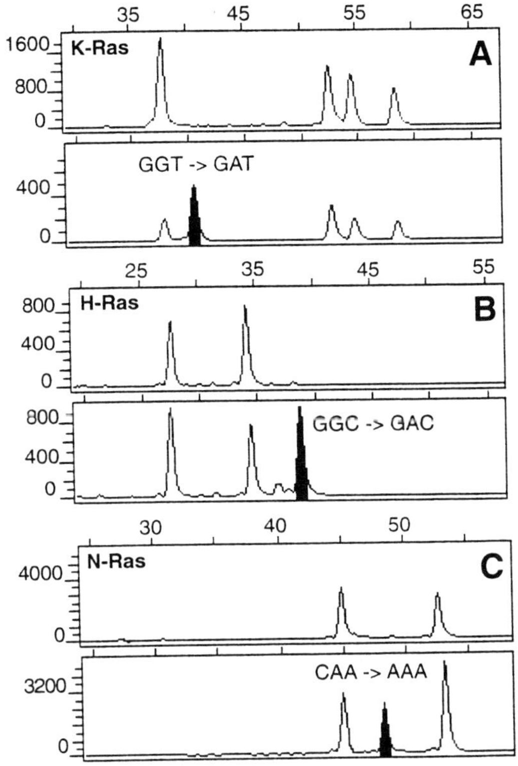

Fig. 5. Data from oligonucleotide ligation analyses of *ras* alleles in human tumor cell lines and normal DNA. Ligation products were separated on 8% polyacrylamide gels in a fluorescent DNA sequencer. Electrophoretogram displays representing real-time fluorescence detection of ligation products are shown. Y-axes display peak heights measured by fluorescence intensity in arbitrary units, and X-axes display the computed sizes of the ligation products in bases. Mutant alleles are indicated as filled peaks. **(A)** K-*ras* exon 1 amplification products were analyzed by multiplex oligonucleotide ligation to detect normal and mutant alleles at codons 12 and 13. The top profile is from normal DNA and the lower one shows a G → A transition in codon 12 in a human lung-cancer cell line (A427). **(B)** Five common mutant alleles in codons 12 and 61 of the H-*ras* gene were analyzed in a multiplex CAL format. Sequences of ligation probes for codon 61 alleles are shown in **Fig. 2**. Wild-type alleles are shown in the top panel and a G → A transition in codon 12 in a human breast-cancer cell line (Hs578T) is shown in the lower panel. **(C)** N-*ras* exons 1 and 2 amplification products were analyzed by multiplex oligonucleotide ligation to detect normal and mutant alleles at codons 12 and 61. A C → A transversion in codon 61 of human neuroblastoma DNA (Sk-N-Sh cells) is shown in the lower profile.

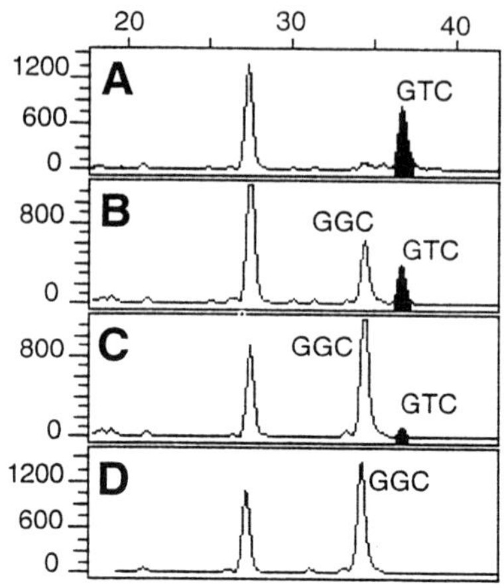

Fig. 6. Detection of H-*ras* mutations in a background of normal DNA. H-*ras* alleles in wild-type and T24 human bladder carcinoma cells were analyzed in multiplex CAL reactions as described in the legend to **Fig. 5**. **(A)** T24 cell DNA homozygous for a G → T transversion (GGC → GTC). **(B)** T24 and wild-type DNA in a 1:10 ratio, **(C)** T24 and wild-type DNA in a 1:100 ratio. **(D)** T24 and wild-type DNA in a 1:200 ratio.

nucleases than samples purified by phenol-chloroform extraction, which may be stored at 4°C or frozen.

2. If the assay gives ambiguous signals, the quality of the PCR amplification of *ras* oncogenes should be determined by agarose-gel electrophoresis. The two-temperature stringent PCR format (94°C, 30 s; 72°C, 2 min) with long, high-melting primers markedly improves PCR specificity and results in robust product yields (**Fig. 2**).
3. Inclusion of a 5–10 min 99°C incubation step after stage I of the CAL reaction (**Fig. 4B**) to inactivate *Taq* polymerase improves ligation efficiency and results in considerably increased yields of specific ligation products. This step blocks any 3' end extension of allelic OLA probes by the *Taq* polymerase that may otherwise occur during the 55°C anneal-ligation stage of the reaction ***(19)***.
4. PCR contamination is always a concern, especially when working with small amounts of material (e.g., DNA from microdissected tissue sections). Work in a PCR-product-free area, preferably a separate room with positive pressure ventilation. Meticulously clean the equipment used for sectioning with 10% bleach followed by ethanol rinses. Always dissect normal tissues first to prevent contaminating the germline sample with tumor tissue. In the CAL protocol, all reagents are added simultaneously in one tube, thus minimizing the risk of PCR carryover contamination. To check for contamination, a negative control (sterile distilled water in place of template DNA) should be done with each PCR reaction.
5. The ligation of oligonucleotides, mismatched at the junction (for example, resulting from crossligation of a wild-type *ras* probe to a mutant *ras* target or vice versa), may be minimized or prevented by adjusting the ligation conditions

(concentration of monovalent cations, magnesium ions, ligase, or OLA probes in the ligation buffer). OLA probe concentrations of 1.25–2.50 × $10^{-8}M$ resulted in efficient and allele-specific ligation of *ras* alleles. High concentrations of fluorochrome-labeled reporter probes are to be avoided because they can obscure the detection of ligation-specific products during gel electrophoresis. The specificity of the ligation can also be enhanced by performing ligations at or close to the melting temperature of the OLA probes so that hybridization of imperfectly matched probes is destabilized ***(18)***. Of the eight possible base-pair mismatches (A-A, A-C, G-T, A-G, C-C, C-T, G-G, T-T), G-T mismatches are known to be the least well-discriminated by thermostable ligase ***(26)***.

6. Allelic OLA probes should be designed such that the ligation product from the normal allele is smaller and therefore migrates faster electrophoretically than that of the mutant allele. In this way, minor peaks representing failure sequences cannot be erroneously interpreted as mutant alleles. The 5' poly(A) residues present on allelic probes for sizing are particularly vulnerable to depurination after alkali deprotection during oligonucleotide synthesis, resulting in the formation of shorter failure sequences able to ligate with fluorescent probes. Use of PEO phosphoramidite monomers instead of poly(A) residues as mobility modifiers should alleviate this problem.
7. Salient features of the PCR-OLA and CAL assays are multiplexing potential and low signal-to-noise ratios. OLA probes differing in their denaturation temperatures can be readily multiplexed because specificity is primarily determined by probe ligation. By using different fluorescent dyes to label oligonucleotide reporter probes, the multiplexing capacity can be further enhanced. Because OLA probes are complementary to only one DNA strand, and the amplification of ligation products is linear and not exponential, there is absolutely no background, target-independent ligation.
8. The combination of PCR and OLA or CAL adapts naturally to the analysis of activating point mutations in *ras* genes in human tumors. Over 20 specimens microdissected from paraffin-embedded tissue sections from patients with primary lung carcinomas and 15 biopsy specimens from patients with pancreatic cancer were analyzed for K-*ras* mutations; 14 codon 12 single nucleotide substitution mutations were detected by allele-specific oligonucleotide ligation, and the results were confirmed by sequencing of cloned PCR products or PCR-based direct sequencing ***(27)***. The assay can detect *ras* mutations in tumor cells making up <5% of the clinical sample; to identify *ras* mutations by PCR-based direct sequencing the mutation must be present in over 12% of the cells in the tumor sample ***(4)***. The ability to simultaneously test for common *ras* oncogene mutations in one reaction and in a single electrophoresis lane without the use of radioactive reagents makes this method well-suited for clinical diagnostic testing.

Acknowledgments

I am grateful to Eric Shulse and my colleagues at the Applied Biosystems Division of Perkin–Elmer Corporation for supporting this work.

References

1. Pronk, G. J. and Bos, J. L. (1994) The role of p21^{ras} in receptor tyrosine kinase signalling. *Biochim. Biophys. Acta* **1198,** 131–147.
2. Bos, J. L. (1989) Ras oncogenes in human cancer: a review. *Cancer Res.* **49,** 4682–4689.
3. Capon, D. J., Chen, E. Y., Levinson, A. D., Seeburg, P. H., and Goeddel, D. V. (1983) Complete nucleotide sequences of the T24 human bladder carcinoma oncogene and its normal homologue. *Nature* **302,** 33–37.
4. Burchill, S. A., Neal, D. E., and Lunec, J. (1994) Frequency of H-ras mutations in human bladder cancer detected by direct sequencing. *Br. J. Urol.* **73,** 516–521.
5. Almoguera, C., Shibata, D., Forrester, K., Martin, J., Arnheim, N., and Perucho, M. (1988) Most human carcinomas of the exocrine pancreas contain mutant c-K-ras genes. *Cell* **53,** 549–554.
6. Bos, J. L., Fearon, E. R., Hamilton, S. R., Verlaan-de Vries, M., van Boom, J. H., van der Eb, A. J., and Vogelstein, B. (1987) Prevalence of ras gene mutations in human colorectal cancers. *Nature* **327,** 293–303.
7. Shimizu, K., Birnbaum, D., Ruley, M. A., Fasano, O., Suard, Y., Edlund, L., Taparowsky, E., Goldfarb, M., and Wigler, M. (1983) Structure of the K-ras gene of the human lung carcinoma cell line Calu-1. *Nature* **304,** 497–500.
8. Taparokwsky, E., Shimizu, K., Goldfarb, M., and Wigler, M. (1983) Structure and activation of the human N-ras gene. *Cell* **34,** 581–586.
9. Farr, C. J., Saiki, R. K., Erlich, H. A., McCormick, F., and Marshall, C. J. (1988) Analysis of RAS gene mutations in acute myeloid leukemia by polymerase chain reaction and oligonucleotide probes. *Proc. Natl. Acad. Sci. USA* **85,** 1629–1633.
10. Sidransky, D., Tokino, T., Hamilton, S. R., Kinzler, K. W., Levin, B., Frost, P., and Vogelstein, B. (1992) Identification of ras oncogene mutations in the stool of patients with curable colorectal tumors. *Science* **256,** 102–104.
11. Jen, J., Powell, S. M., Papadopoulos, N., Smith, K. J., Hamilton, S. R., Vogelstein, B., and Kinzler, K. W. (1994) Molecular determinants of dysplasia in colorectal lesions. *Cancer Res.* **54,** 5523–5526.
12. Caldas, C., Hahn, S. A., Hruban, R. H., Redstron, M. S., Yeo, C. J., and Kern, S. E. (1994) Detection of K-ras mutations in the stool of patients with pancreatic adenocarcinoma and pancreatic ductal hyperplasia. *Cancer Res.* **54,** 3568–3573.
13. Slebos, R. J. C., Kibbelaar, R. E., Dalesio, O., Kooistra, A., Stam, J., Meijer, C. J. L. M., Wagenaar, S. S., Vanderschueren, R. G., van Zandwijk, N., Mooi, W. J., Bos, J. L., and Rodenhuis, S. (1990) Kirsten ras oncogene activation as a prognostic marker in adenocarcinoma of the lung. *N. Engl. J. Med.* **323,** 561–565.
14. Saiki, R., Gelfand, R., Stoffel, S., Scharf, S., Higuchi, R., Horn, G., Mullis, K., and Erlich, H. (1988) Primer-directed enzymatic amplification of DNA with a thermostable DNA polymerase. *Science* **239,** 487–491.
15. Landegren, U., Kaiser, R., Sanders, J., and Hood, L. (1988) A ligase-mediated gene detection technique. *Science* **241,** 1077–1080.
16. Whiteley N. M., Hunkapiller, M. W., and Glaxer, A. (1989) Detection of specific sequences in nucleic acids. U. S. Patent No. 4,883,750.

17. Nickerson, D. A., Kaiser, R., Lappin, S., Stewart, J., Hood, L., and Landegren, U. (1990) Automated DNA diagnostics using an ELISA-based oligonucleotide ligation assay. *Proc. Natl. Acad. Sci. USA* **87,** 8923–8927.
18. Eggerding, F. A., Iovannisci, D. M., Brinson, E., Grossman, P., and Winn-Deen, E. S. (1995) Fluorescence-based oligonucleotide ligation assay for analysis of cystic fibrosis transmembrane conductance regulator gene mutations. *Hum. Mutation* **5,** 153–165.
19. Eggerding, F. A. (1995) A one-step coupled amplification and oligonucleotide ligation procedure for multiplex genetic typing. *PCR Meth. Appl.* **4,** 337–345.
20. Strauss, W. M. (1994) Preparation of genomic DNA from mammalian tissue, in *Current Protocols in Molecular Biology,* vol. 1, Wiley, Secaucus, NJ, Section 2.2.1.–2.2.3.
21. Higuchi, R. (1989) Preparation of samples for PCR, in *PCR Technol.* (Ehrlich, H. A., ed.), Stockton, NY, pp. 31–38.
22. Schubert, E. L., Bischoff, F. Z., Whitaker, L. L., Pleasants, L. M., and Hansen, M. F. (1993) A method to isolate DNA from small archival tissue samples for p53 gene analysis. *Hum. Mutation* **2,** 123–126.
23. Greer, C. E., Wheeler, C. M., and Manos, M. M. (1994) Sample preparation and PCR amplification from paraffin embedded tissues. *PCR Meth. Appl.* **3,** S113–S122.
24. Stratton, M. R., Collins, N., Lakhani, S. R., and Sloane, J. P. (1995) Loss of heterozygosity in ductal carcinoma *in situ* of the breast. *J. Pathol.* **175,** 195–201.
25. Grossman, P. D., Bloch, W., Brinson, E., Chang, C. C., Eggerding, F. A., Fung, S., Iovannisci, D. A., Woo, S., and Winn-Deen, E. S. (1994) High-density multiplex detection of nucleic acid sequences: oligonucleotide ligation assay and sequence-coded separation. *Nucleic Acids Res.* **22,** 4527–4534.
26. Barany, F. (1991) Genetic disease detection and DNA amplification using cloned thermostable ligase. *Proc. Natl. Acad. Sci. USA* **88,** 189–193.
27. Eggerding, F. A., Sun, Q., and Chapman, P. B. (1995) Rapid detection of H-, K-, and N-Ras oncogene point mutations by automated genetic analysis: application to human cancers. *Am. J. Hum. Genet.* Suppl., vol. 57 no. 4, A63.

22

Detection of Prostate-Cancer Cells in Blood and Bone Marrow by RT-PCR

Robert L. Vessella and Eva Corey

1. Introduction

As cancer treatment options broaden, there is a corresponding increase in the need for improved methods of detecting various stages and forms of the disease. Only through the use of highly sensitive and discriminating diagnostic tests can a rational and informed decision be made from among the clinical interventions available to the physician. The case of prostate cancer is an ideal paradigm of this challenging problem. As with many other types of solid tumors, complete cure can frequently be achieved through established methods of treatment if the malignancy is organ-confined. However, accurate staging of the disease when gross metastases are not evident is problematic because the techniques in use have poor sensitivities for micrometastatic disease. Consequently, since the optimal treatment strategy is different for primary and metastatic disease, the patient may not receive the best initial treatment.

In prostate cancer, aggressive early detection programs instituted over the past 5 yr have reduced the proportion of patients presenting with discernible metastatic disease. The detection rate of primary disease has increased simultaneously, and these factors have combined to raise the frequency of radical prostatectomy (removal of the prostate gland) as the treatment of choice for organ-confined disease. A few of these patients will be found to have disease outside their prostate at the time of surgery; however, for the majority, the procedure concludes with physician and patient believing that the disease was limited to the prostate. Unfortunately, 15–30% of these patients will subsequently develop clinically evident metastatic disease, indicating that occult micrometastases were in fact present at the time of surgery, and had this been known, other treatment options would have been more strongly considered.

From: *Methods in Molecular Medicine, Vol. 14: Tumor Marker Protocols*
Edited by: M. Hanausek and Z. Walaszek © Humana Press Inc., Totowa, NJ

Current models of metastasis hold that cells "shed" by primary tumors are borne by circulation (blood and lymphatics) to other organs, where they are deposited and form secondary tumors. Therefore, there is likely to be a very strong correlation between the appearance of cancer cells in blood and the advent of metastasis. In practice, it is found that both circulation and the bone marrow, which is a major site for metastasis of prostate cancer, are fruitful sources of specimens for early detection of shed cancer cells. The need to diagnose micrometastasis in its earliest stages has spurred the development of more sensitive techniques for detecting these migrating cells.

Over the past few years, we and others have attempted to develop such a detection method for micrometastatic disease. Ours is a two-part approach: first, to use the exquisite sensitivity of the reverse transcription-polymerase chain reaction (RT-PCR) to detect prostate epithelial cells in circulation and/or bone marrow, and, second, to find the correlation between the presence of these cells and other specific disease attributes, such as virulence or metastatic potential.

RT-PCR offers significant promise of improvements in the detection of prostate epithelial cells in the peripheral circulation, bone marrow, and lymph nodes. These epithelial cells are presumed to be malignant cells, since normal epithelial cells are not thought to "shed" or disseminate from the source organ. However, with the phenomenal sensitivity of the RT-PCR reaction, the possibility exists that a positive reaction may be obtained from nonmalignant cells (*see* **Subheading 1.2.**). Moreover, one cannot assume that the detected cells, even if malignant, are identical to those responsible for metastatic disease. Therefore, based on our current knowledge, it is potentially incorrect to infer that a positive RT-PCR result coincides with micrometastatic disease; however, for the purposes of this chapter, we hereafter assume that the cells detected are cancer cells with the potential to develop into micrometastases.

The premise of the RT-PCR approach is that the cells to be detected express a messenger RNA that is not found in other cells within the environment being tested. For detection of prostate-cancer cells in the peripheral blood, bone marrow, or lymph nodes, the mRNA for prostate-specific antigen (PSA) is presently the target transcript of choice. PSA is a 32-kDa serine protease produced almost exclusively by prostate epithelial cells, the cells of origin for prostate cancer. The use of PSA as a serum tumor marker for early diagnosis and monitoring of treatment has been the subject of well over 3000 scientific articles and is beyond the scope of this discussion. PSA is unquestionably the best serum marker available today in human oncology ***(1–4)***. Although the serum PSA level is not helpful in determining whether the disease is organ-confined or micrometastatic, we have been able to show that detection of PSA mRNA signals the presence of prostate-cancer cells in blood or bone marrow.

1.1. General Technical Remarks on RT-PCR

The first use of RT-PCR for detecting circulating cells shed by solid tumors utilized transcripts of tyrosinase, a tissue-specific gene in melanocytes ***(5)***. Among other transcripts, tyrosine hydroxylase gene transcript has been used for detection of neuroblastoma cells ***(6)***, the α-fetoprotein mRNA transcripts for hepatocellular cancer ***(7)***, and the cytokeratin mRNA for breast cancer ***(8)***. This broad spectrum indicates why the interest in the RT-PCR approach is so high: It offers nearly limitless opportunities, not only for detection of cells from all types of human solid tumors, but also the potential of further molecular phenotyping in regard to virulence, drug resistance, and so on.

The process of developing a RT-PCR protocol thus begins with the identification of an appropriate target gene. As indicated previously, the transcript must be highly specific for the cell of interest within an environment of other cells that lack the transcript. Even if the transcript is expressed by other cell types within the body, as long as it is not expressed within the environment being tested, it may be a suitable target (e.g., cytokeratin) ***(8)***. The RT-PCR will give rise to a band of a distinct size in gel electrophoresis specific to the cell type of interest.

Conversion of the specimen RNA to cDNA is most frequently accomplished by one of two general approaches. In the first, primer specific for the RNA of interest is used to convert only the desired target RNA transcript to cDNA. Alternatively, all of the mRNA in the specimen extract is converted to cDNA using oligo-$T_{12–18}$ or random hexamers, and target specificity arises from primers used in the PCR amplification of the cDNA. Although the first approach may result in higher specificity, we prefer to use the second approach since it yields a "bank" of cDNA representing all of the mRNA transcripts in the specimen, which is stable in storage and may be invaluable in follow-up studies with other primers (yet to be defined) for markers of virulence and the like.

The size of the band in gel electrophoresis is governed by the number of nucleotides between the primers that bind to the cDNA template following reverse transcription. Although the extraction method is designed to yield high-purity total RNA from the specimen, which is then converted to cDNA for PCR amplification, some trace of genomic DNA contamination is unavoidable. Thus, in designing the primers to be used in the PCR amplification process, it is critical to have them span exon splice junctions so that the genomic DNA will not be amplified. If this is not possible, the primers should be designed to bind to different exons, so that the product from amplified genomic DNA can be easily distinguished by its larger size from the product of the spliced mRNA (both products will be generated). In either approach, the primer pair must be designed to react only with the cDNA target sequence of interest.

Thus, not only must the target mRNA transcript be chosen for cell specificity, but the primers must also anneal specifically to that target cDNA to yield the expected product. Presumed specificity in design is followed by multiple rounds of testing negative control specimens derived from the environment of interest (e.g., peripheral blood) as well as appropriate positive control cells mixed with the negative control cells.

1.2. Use of RT-PCR to Detect Prostate Cancer

The sensitivity of detection by RT-PCR is several-fold better than any other method in current use, such as flow cytometry or immunohistochemistry, which have lower limits of detection in the range of one tumor cell/250,000–500,000 white blood cells. Recent progress and fine-tuning of RT-PCR utilizing PSA mRNA as the target has yielded several RT-PCR protocols with high sensitivity. It should be noted that each of these techniques is optimized for each reagent and step of the reaction, including the specific DNA thermocycler dedicated to the analysis. As many of us have learned by experience, altering any aspect of the standard procedure may yield inferior sensitivity or spurious results until the procedure is reoptimized. Using the RT-PCR procedure presented here, we achieved a sensitivity of approximately one LNCaP (prostate-cancer cell line)/10–100 million lymphoid cells (or 2–10 copies of PSA cDNA).

In addition to the spurious results that can occur when care is not given to optimization of the reaction, there are other sources of false negatives and false positives. The most common source of false negatives is a subtle technical error. Specimen sampling errors, although not technically a false negative, do result in reports that no cells were detected. To help eliminate technical errors, it is good practice to include in the analysis the parallel detection of mRNA from a housekeeping gene. In addition, a sensitivity control should be considered, such as a plasmid cDNA corresponding to the target transcript. This could be run in parallel, or, if engineered to be longer or shorter than the amplified portion of the native cDNA following RT, may be spiked directly into the specimen reaction. False positives can also present problems. For example, some normal cells within the test environment may produce a few "illegitimate" copies of the mRNA, which on several million-fold amplification, could result in a positive reaction ***(9,10)***. Should this occur, the sensitivity of the procedure may need to be lowered, for example, by decreasing the number of cycles. A false positive can also arise from contamination of the specimen with a source of the target transcript. This is a critical issue that requires constant attention. A third situation that may lead to false positives can occur when the PCR process is not optimal for specific annealing of the primers to the target cDNA. Although it is unlikely that this would result in a product indistinguishable in size from the desired product, a standard safeguard is to verify the integrity of every positive product either by restriction digestion (details provided) or by DNA sequencing. Very slight changes in the optimized procedure, specimen acquisition, specimen processing, or analysis can also lead to such spurious results.

Despite these potential difficulties, many of which may be overcome with sufficient emphasis on detail, design, and controls, RT-PCR remains a promising and exciting approach to molecular staging. For example, we and others have shown a good correlation between the detection of cells in the peripheral circulation in patients with prostate cancer and an increasing known stage *(11–14)*. In other words, the percentage of patients with detectable prostate epithelial cancer cells in circulation increases as the clinical staging progresses from organ-confined disease to known distant metastasis. However, there is a relatively broad range of frequency of detection at each of the stages among laboratories. This is attributed to the different procedures and patient populations. For this reason, a RT-PCR consortium was established to assess several different RT-PCR methods in use for targeting PSA mRNA and to initiate a blind multisite clinical trial *(14)*. What follows is a detailed methodology for a third-generation RT-PCR protocol developed in our laboratory for the detection of prostate-cancer cells in blood and bone marrow using PSA mRNA as the target transcript. We have analyzed over 1000 specimens using this technique and have confidence in its accuracy and reproducibility.

2. Materials

2.1. Drawing of Blood

1. Vacutainer cell preparation tube (CPT) with sodium citrate (Baxter Scientific, West Chester, PA).
2. Tourniquet 21G Vacutainer needle and Vacutainer needle holder.
3. Alcohol swab and Band Aid™.

2.2. Drawing of Bone Marrow

1. Histo-paque-1077 (Sigma, St. Louis, MO).
2. Sodium citrate solution (6%) (Drug Services).
3. 20-mL Syringe.

2.3. Isolation of Mononuclear Cells

1. Sterile Dulbecco's phosphate saline buffer.
2. 50-mL Sterile conical centrifuge tubes.
3. 10-mL Sterile pipet.
4. 5-mL Sterile pipet.

2.4. Cell Count

1. Hemacytometer and coverslip.
2. Microscope.

3. Pipetmen.
4. Hand counter.
5. 3% Acetic acid in water (v/v) and 0.4% Tryptan blue in phosphate-buffered saline (PBS).

2.5. RNA Extraction

1. 1.5- and 0.65-mL Presiliconized microcentrifuge tubes, RNase- and DNase-free.
2. Pipetmen (e.g., Gilson P-10, P-20, P-200, P-1000).
3. Barrier tips (200–1000, 30–300, 0.5–20, and 0.2–10 µL).
4. Vortex.
5. RNA STAT 60 Total RNA Isolation Reagent (Tel-Test "B," Inc., Friendswood, TX). Store at 4°C.
6. Chloroform, molecular-biology grade.
7. 75% (v/v) Ethanol, molecular-biology grade.
8. Isopropanol, molecular-biology grade.
9. AquaNase-free water, ultrapure RNase- and DNase-free, diethyl pyrocarbonate (DEPC)-treated, sterile, pyrogen-free, free of enzyme inhibitors.
10. UV spectrophotometer.
11. Quartz cuvets.

2.6. Reverse Transcription

1. 0.65-mL Presiliconized microcentrifuge tubes, RNase- and DNase-free.
2. Pipetmen (e.g., Gilson P-10, P-20, P-200, P-1000).
3. Barrier tips (200–1000, 30–300, 0.5–20, and 0.2–10 µL).
4. 0.05 µg/µL Random hexamers. Store at –20°C.
5. SuperScript II RNase H^- reverse transcription kit (Gibco-BRL, Gaithersburg, MD).
   a. SuperScript II reverse transcriptase.
   b. 5X First-strand buffer: 250 m*M* Tris-HCl, pH 8.3 at room temperature, 375 m*M* KCl, 15 m*M* $MgCl_2$, 0.1 *M* dithiothreitol (DTT). Store at –20°C.
6. 10 U/µL RNase inhibitor (–20°C). Store at –20°C.
7. AquaNase-free water, ultrapure RNase- and DNase-free, DEPC-treated, sterile, pyrogen-free, free of enzyme inhibitors.
8. 10 m*M* Deoxynucleotide triphosphate mixture (dNTP). Store at –20°C.
9. Microcentrifuge.
10. Thermal cycler.

2.7. PCR

1. Ultrathin-walled PCR tubes.
2. Pipetmen (e.g., Gilson P-10, P-20, P-200, P-1000).
3. Barrier tips (200–1000, 30–300, 0.5–20, and 0.2–10 µL).
4. 10X PCR buffer: 200 m*M* Tris-HCl, 500 m*M* KCl, pH 8.4, at room temperature. Store at –20°C.
5. 50 m*M* $MgCl_2$. Store at –20°C.
6. 2 m*M* dNTP. Store at –20°C.

7. β-2-Microglobulin (MIC) and PSA 5' and 3' oligonucleotides.
8. *Taq* DNA polymerase, store at –20°C.
9. TAQSTART antibody (TaqAb) (Clontech, Palo Alto, CA). Store at –20°C.
10. AquaNase-free water, ultrapure RNase- and DNase-free, DEPC-treated, sterile, pyrogen-free, no enzyme inhibitors.
11. Thermal cycler (HYBAID Omnigene with heated lid, Labnet).
12. Mineral oil, molecular-biology grade.

2.8. Gel Electrophoresis

1. Gel-electrophoresis box.
2. Power supply.
3. Microwave oven.
4. Agarose.
5. 10X TBE buffer: 121 g Tris base, 68 g boric acid, 7.6 g ethylenediamine tetraacetic acid (EDTA). Make up to 1 L.
6. 10 mg/mL Ethidium bromide solution. Handle with gloves as a potential carcinogen.
7. Orange loading dye: 925 μL 25% Ficoll + 75 μL 2% Orange G.
8. DNA molecular-weight marker.
9. UV box.
10. Camera and film.

2.9. ClaI Digestion

1. 0.5 mL Microcentrifuge tubes.
2. *Cla*I enzyme solution and 10X *Cla*I enzyme reaction buffer. Store at –20°C.
3. Water bath, heating block, or incubator at 37°C.

3. Methods

3.1. Collection of Blood Samples and Isolation of Mononuclear Cells

Safety consideration: handle the blood samples with precautions appropriate for biohazardous materials.

1. Draw aseptically 8 mL of venous blood into a Vacutainer CPT.
2. Process blood within 2 h after drawing. If processing is delayed for more than 2 h, the tube can be centrifuged as described later and stored at 4°C. Do not store longer than overnight.
3. Centrifuge the Vacutainer CPT at 1500–1800*g* for 30 min at room temperature.
4. Gently resuspend the resulting mononuclear cell layer into the plasma by inverting the tube.
5. Pipet the resuspended mononuclear cells and plasma into a 50-mL sterile, conical centrifuge tube. Rinse the tube with 5 mL Dulbecco's PBS and transfer to the tube with resuspended mononuclear cells.
6. Bring volume up to 40 mL with PBS.

7. Remove a 10-μL aliquot and dilute with 40 μL 3% acetic acid for a cell count (*see* **Subheading 3.3.**).
8. Spin the 50-mL centrifuge tube with the white blood cells in PBS at 300*g* for 20 min at 4°C.
9. Remove the supernatant and let tube drain for 1–2 min.
10. Resuspend the cell pellet in STAT60 for RNA isolation by pipeting the solution with a 5-mL pipet (1 mL of STAT60 for every 5×10^6 mononuclear cells).
11. If not used immediately for RNA isolation, this solution can be stored up to 2 wk at –80°C.

3.2. Bone Marrow Sample Collection and Mononuclear Cell Isolation

1. Draw aseptically 10 mL of bone marrow into a 20-mL syringe containing 10 mL 6% sodium citrate solution.
2. Transfer the bone marrow mixture from **step 1** into a 50-mL sterile centrifuge tube.
3. Using a pipet, carefully underlay the tube contents from **step 2** with 15 mL Histopaque-1077.
4. Centrifuge this mixture at 400*g* for 30 min at room temperature.
5. Remove the upper (plasma) layer with a 5-mL pipet to within 1 cm of the opaque interface containing the mononuclear cells. Discard the upper layer.
6. Using a new 5-mL pipet, transfer the opaque interface into a sterile 50-mL centrifuge tube.
7. Bring the volume to 40 mL with PBS.
8. Continue as described in **Subheading 3.1., steps 7–11**.

3.3. Determination of Mononuclear Cell Count

1. Use 50 μL of mononuclear cell suspension in 3% acetic acid from **step 7** and **step 8**, respectively (**Subheadings 3.1.** and **3.2.**).
2. Add 50 μL Tryptan blue solution; mix by pipeting up and down.
3. Use pipetman to transfer about 10 μL of solution to coverslip. Let the cell suspension flow under the grid area. Do not overfill.
4. Under the microscope, count all cells contained in each of the four large squares.
5. The number of cells in each square should be between 50 and 100. If there are >100 or <50 cells, repeat with appropriate dilution.
6. Determine the average number of cells in one large square. Count the number of cells per milliliter using following equation:

$$\text{cells/mL} = (\text{average number of cells per one large square} \times 10^4/\text{mL}) \times (1/\text{dilution})$$

3.4. RNA Isolation

1. Transfer 1 mL of the STAT60 suspension from **step 10** (**Subheading 3.1.**) to a 1.7-mL microcentrifuge tube (equivalent to 5×10^6 cells, since the volume of the cell pellet is negligible) (*see* **Note 1**). Incubate at room temperature for 7 min to dissociate nucleoprotein complexes.

2. Add 0.2 mL of chloroform and shake vigorously for 15 s. Incubate for 5 min at room temperature.
3. Spin in microcentrifuge at 12,000*g* (max) for 15 min at 4°C.
4. Transfer the colorless upper phase (aqueous) containing RNA to a clean presiliconized microcentrifuge tube. Do not transfer the interface, which contains proteins. The volume of the aqueous phase should be about 0.6 mL.
5. Add 0.6 mL of isopropanol to the aqueous phase and briefly vortex. Incubate the sample for 10 min at room temperature to allow the RNA to precipitate.
6. Centrifuge the sample at 12,000*g* (max), for 10 min at 4°C. The RNA appears as a white pellet at the bottom of the tube.
7. Remove the supernatant with a pipetman. Add 1 mL of 75% ethanol to the RNA pellet. Invert the tube several times to wash the pellet.
8. Spin for 5 min at 7500*g* at 4°C.
9. Remove the supernatant. Dry the RNA pellet for 5 min at room temperature. Avoid drying the pellet completely, which can lead to difficulty in redissolving the RNA.
10. Dissolve the RNA in 15 µL of AquaNase-free water. Vortex briefly or pass the pellet a few times through a pipet tip.
11. Store RNA solutions at –80°C (*see* **Note 2**).

3.5. Determination of RNA Concentration

1. Mix 498 µL of water and 2 µL of the RNA solution in a 1.7-mL microcentrifuge tube. Transfer the solution to a quartz cuvette.
2. Turn on the UV spectrophotometer and wait 10–15 min to allow the lamp to warm up.
3. Pipet 500 µL of water into another quartz cuvet and calibrate the instrument on water at 260 and 280 nm.
4. Transfer a cuvet with the RNA solution to the spectrophotometer and read the absorbance at 260 and 280 nm.
5. Calculate the 260/280 ratio to determine the purity of the RNA. Total RNA should be free of DNA and protein contamination.
6. Calculate the RNA concentration assuming: 1 OD = 40 µg RNA.

$$\text{Original RNA conc.}(\mu g/mL) = A_{260} \times 40 \text{ (Total volume of RNA solution, } \mu L)/\text{(volume of RNA solution added, } \mu L)$$

3.6. Reverse Transcription

1. Keep RNA and all reagents on ice. Wear gloves throughout the procedure.
2. Prepare one 0.5-mL presiliconized, RNase-free microcentrifuge tube for each sample. Add 5 µg of total RNA (*see* **Note 3**) and 1 µL of random hexamers (0.05 µg/µL) (*see* **Note 4**). Bring volume to 10 µL with AquaNase-free water.
3. Using a thermal cycler, denature this RNA-primer solution at 70°C for 10 min.
4. Immediately transfer the tube to ice for 5 min.
5. Spin the tube for 5 s to collect the condensate in the bottom of the tube.
6. Prepare a reverse transcription master mixture in a 1.7-mL microcentrifuge tube.

For multiple samples, prepare 10% extra master mixture to allow for pipeting losses. For each sample use 4 µL 5X first-strand buffer, 2 µL 0.1 *M* DTT, 1 µL 10 m*M* dNTP mix, 1 µL RNase inhibitor (10 U/µL), 1 µL Superscript II RT, and 1 µL H_2O.

7. Mix the master mix by inverting the tube.
8. Add 10 µL of master mix to each of the RNA-random hexamer mixtures (20 µL total reaction volume). Mix gently but thoroughly.
9. Transfer each tube to the thermal cycler and incubate the samples for 5 min at 25°C, followed by 1 h at 42°C.
10. Incubate at 99°C for 5 min to inactivate the RT.
11. Spin the reactions briefly in a microcentrifuge to collect the condensate.
12. Store cDNA at –80°C.

3.7. PCR

3.7.1. MIC PCR (see **Note 5**): Housekeeping Gene Control for RT

1. Mix 0.5 µL *Taq* DNA polymerase (5 U/µL) with 0.5 µL TaqAb for each PCR sample. Mix gently by pipeting and incubate for 5 min at room temperature to allow the TaqAb to bind and inactivate *Taq* (*see* **Note 6**).
2. While the above reaction is incubating, prepare a PCR master mixture in a 1.7-mL microcentrifuge tube. For multiple samples, prepare 10% extra master mixture to allow for pipeting losses.
   For each sample, add 5 µL 10X PCR buffer, 2 µL 50 m*M* $MgCl_2$, 1 µL 5' MIC primer (5 pmol/µL) (*see* **Note 7**), 1 µL 3' MIC primer (5 pmol/µL) (*see* **Note 7**), 4 µL 2 m*M* dNTP mix, and 36 µL H_2O.
3. Pipet the *Taq*/TaqAb mixture into an aliquot of master mixture and mix by inverting the tube.
4. Pipet 49 µL of master mixture into each labeled ultrathin-walled PCR tube.
5. Add 1 µL cDNA of the sample of interest to each reaction 50 µL total volume of PCR reaction.
6. Close the tubes and transfer them into a thermal cycler with a heated lid.
7. Run the following PCR program:
   a. 80°C for 3 min, 1 cycle (*see* **Note 8**);
   b. 94°C for 3 s (*see* **Note 9**);
      69°C for 1 min (*see* **Note 10**) for 25 cycles;
   c. 72°C for 7 min (*see* **Note 11**) for 1 cycle.
8. These samples will be used for agarose-gel-electrophoretic determination of the presence of the 550-bp MIC band to determine RT performance (*see* **Notes 12–14**).

3.7.2. Prostate-Specific Antigen PCR (see **Notes 15–19**)

1. Mix 0.5 µL *Taq* DNA polymerase (5 U/µL) with 0.5 µL TaqAb for each PCR sample. Mix gently by pipeting and incubate for 5 min at room temperature to allow the TaqAb to inactivate *Taq*.

2. While the reaction is incubating, prepare a PCR master mixture in a 1.7-mL microcentrifuge tube. For multiple samples, prepare 10% extra master mixture to allow for pipeting losses.
   For each sample, add 5 µL 10X PCR buffer, 2 µL 50 m*M* $MgCl_2$, 1 µL 5' PSA primer (5 pmol/µL) (*see* **Note 16**), 1 µL 3' PSA primer (5 pmol/µL) (*see* **Note 16**), 4 µL 2 m*M* dNTPs mix, 32 µL H_2O.
3. Pipet the *Taq*/TaqAb mixture into the master mixture and mix by inverting the tube.
4. Pipet 45 µL of master mixture into each labeled ultrathin-walled PCR tube.
5. Add 5 µL of the cDNA of interest 50 µL total PCR reaction volume.
6. Close the tubes and transfer them to a thermal cycler with a heated lid under "tube-based" temperature control.
7. Program the following PCR:
   a. 80°C for 3 min (*see* **Note 15**), 1 cycle;
   b. 94°C for 3 s;
      69°C for 1 min, 40 cycles; and
   c. 72°C for 7 min, 1 cycle.
8. These samples will be used for agarose-gel-electrophoretic determination of the presence of the 460-bp PSA band and for restriction analysis.

3.8. Gel Electrophoresis

3.8.1. Preparation of 1.5% Agarose Gel

1. Weigh 0.75 g of agarose, transfer into a 300-mL Erlenmeyer flask, and add 50 mL of 0.5X TBE.
2. Microwave the flask twice for 1 min or until the agarose is dissolved. Swirl the suspension between heating steps.
3. Incubate the flask in a 50°C water bath for 5 min.
4. Add 1 µL of ethidium bromide solution (10 mg/mL), mix, and pour the agarose solution into the gel-casting apparatus.
5. Allow the gel to solidify at room temperature for 30–50 min.

3.8.2. Loading Samples and Running Electrophoresis

1. In a microcentrifuge tube or on a piece of Parafilm, mix 5 µL of loading dye and 10 µL of a PCR reaction mixture (*see* **Note 20**). Mix by pipeting.
2. Prepare a DNA molecular weight marker sample (*see* **Note 21**), using 5 µL of loading dye and an appropriate amount of DNA ladder.
3. Remove the comb from the gel and overlay the gel with 0.5X TBE buffer.
4. Using a pipetman, load samples slowly into the wells. Loading dye causes the sample to sink into the well.
5. Run the gel at 40–60 mA until the orange dye reaches the bottom of the gel.
6. Remove the gel from the gel box (use gloves when handling the gel; ethidium bromide is a potential carcinogen) and place it on a UV box.
7. Take a picture of the gel. The DNA fragments will appear as white bands on a black background.

3.9. Restriction Analysis (see Note 22)

1. Pipet 1.5 µL of 10X *Cla*I enzyme reaction buffer into a 0.5-mL microcentrifuge tube.
2. Add 0.5 µL of *Cla*I enzyme solution and 13 µL of the PSA PCR sample.
3. Close the tube and place it into a 37°C water bath or incubator for 1 h.
4. Run a 1.5% agarose gel for analysis.

4. Notes

1. Higher cell densities in the STAT60 suspension cause contamination of the RNA preparation with high-molecular weight DNA. Other RNA isolation solutions (e.g., RNAZOL B) can be used according to the manufacturers' instructions.
2. RNA preparations can degrade over time. If you need to work with an RNA preparation older than 2 mo, first run a gel to check the RNA integrity.
3. We use relatively large amounts of input total RNA in the RT reaction to obtain high sensitivity in the PSA RT-PCR step. A smaller amount of total RNA can be used, but the sensitivity of the RT-PCR will be affected.
4. A higher concentration of random hexamers usually results in a higher yield of shorter cDNA products.
5. To check the performance of the RT reaction, a PCR of the housekeeping gene MIC is run as an internal standard for each sample.
6. Hot-start conditions with the TAQSTART antibody are used to ensure high specificity of the PCR.
7. MIC primers: positions 148–2560 (148–401, 1018–1045, 2290–2560) of MIC DNA; GenBank accession number M17987; product: 550 bp.
   a. 5' MIC primer (28-mer, GC content 50%, T_m = 79°C): 5'-CACGTCATCCAG CAGAGAATGGAAAGTC-3'.
   b. 3' MIC primer (28-mer, GC content 53.8%, T_m = 39.2°C): 5'-TGACCAAGA TGTTGATGTTGGATAAGAG-3'.

The 28-mer PCR primers are designed for high specificity and compatibility with the two-step PCR.

8. The purpose of the first cycle of the PCR program is to deactivate TaqAb and liberate Taq DNA polymerase to start DNA synthesis.
9. Using ultrathin-walled PCR tubes and "tube-based" temperature control in the thermal cycler, 94°C for 3 s is entirely sufficient for DNA denaturation.
10. The 69°C step incorporates both annealing and extension.
11. The 72°C step is included to allow completion of all DNA chains.
12. If a thermal cycler with a heated lid is not available, add 50 µL of molecular-biology grade mineral oil to each tube. Oil prevents evaporation of the reaction mixture and helps to maintain stable concentrations in the PCR reaction.
13. Once it has been optimized, any changes in the procedure, even changing to another thermocycler of the same model, may require reoptimization.
14. The choice of housekeeping gene as a control for RT performance and RNA integrity is an important issue. Two common housekeeping genes in use as RT

controls, β-actin and glyceraldehyde-3-phosphate-dehydrogenase, have known "processed pseudogenes" in the genome. These pseudogenes have nearly the same sequences as the transcribed mRNA. If primers can anneal to a pseudogene, PCR can potentially amplify genomic DNA contamination in addition to the target cDNA. The two products would be indistinguishable in size. This is a serious problem, because amplification of the pseudogene present in DNA contamination of the RNA preparation leads to a false assurance that the RT was successful, when in fact the RT may have failed. It is therefore necessary to check for the presence of a pseudogene before settling on a particular housekeeping gene for use as a positive RT control. First, check Genbank for primer sequence matches. Second, run a PCR on genomic DNA with the same primers; no product of expected size should be generated. Finally, run RT reactions with and without the RT enzyme, generating the expected product and no product, respectively. (RNA is present in both samples.) These three experiments will ensure that the primers will anneal only to the cDNA of interest, and that the presence of a band of the right size after RT-PCR amplification is a proper control of RT performance.

15. A major obstacle to sensitive and specific PCR is usually competing side-reactions, such as an amplification of nontarget sequences in background DNA (mispriming) and primer oligomerization. These reactions take place mainly during PCR setup when all reactants have been mixed, but before thermal cycling. Hot-start conditions using the TAQSTART antibody are used to ensure high specificity of PCR. *Taq* polymerase is liberated from *Taq*/TaqAb complex at 80°C, which is high enough to suppress annealing of primers to nontarget sequences.
16. PSA primers: position 110–569 of PSA cDNA, Genbank access number M21895; product: 460 bp.
    a. 5' PSA primer (25-mer, GC content 64%, $T_m = 77°C$): 5'-TTGTGGCCTCTC GTGGCAGGGCAGT-3'.
    b. 3' PSA primer (26-mer, GC content 53.8%, $T_m = 75°C$): 5'-TGGTCACCTTCT GAGGGTGAACTTGC-3'.
17. To obtain the highest specificity, PSA primers were chosen from regions of the PSA gene exhibiting the high diversity between PSA and human glandular kallikrein (hK2).
18. It is generally necessary to optimize the cycling parameters for each thermal cycler:
    a. If possible, begin with a PCR of a positive control sample, for example, a plasmid containing PSA cDNA or cDNA obtained by reverse transcription of RNA from cells that are PSA-positive (LNCaP, prostate-cancer cell line).
    b. If the positive control does not yield the proper electrophoretic PSA-band (460 bp):
        i. Discard water, magnesium solution, buffers, and primer solutions; repeat the PCR.
        ii. If the band still does not appear, try adjusting the cycling parameters: Extend denaturation and/or annealing/extension times and/or decrease the annealing/extension temperature.

   c. If unwanted bands are observed, the PCR conditions are not sufficiently selective. Raise the annealing/extension temperature in 2–5°C increments in subsequent optimization runs, or try a different concentration of Mg^{2+} (especially if primers other than those recommended are used).
19. If a thermal cycler with a heated lid is not available, add 50 µL of molecular-biology grade mineral oil to each tube. Oil prevents evaporation of the reaction mixture and helps to maintain stable concentrations in the PCR.
20. If there is oil in the PCR reaction, avoid contaminating the electrophoresis sample with oil, which may prevent the sample from sinking into the well.
21. The DNA molecular weight marker allows determination of the size of the amplified products and, consequently, identification of the PSA fragment. Occasionally, weak bands of other molecular weights are present in the samples.
22. The *Cla*I digest is performed to ensure the specificity of the PSA PCR. A *Cla*I restriction site is present in the middle of the PSA fragment that is amplified with the above mentioned PSA primers; the *Cla*I site is not present in the hK2 gene. A positive PSA digest will exhibit bands of 220 and 240 bp. If the 460-bp band does not change on digestion, it is not caused by the presence of PSA mRNA, and the sample cannot be considered PSA-positive. Using the primers and conditions described here, we have infrequently observed an amplified band that was not digested by *Cla*I (<2%).

Acknowledgments

We acknowledge the excellent technical assistance of Ed Arfman and Matt Oswin, and we are grateful to Michael Corey, for invaluable editorial assistance. This work was funded by the Lucas Foundation, the CaP CURE Foundation, Urocor Inc., the Department of Veterans' Affairs and a George M. O'Brien Center Award from NIDDK.

References

1. Oesterling, J. (1991) Prostate specific antigen: a critical assessment of the most useful tumor marker for adenocarcinoma of the prostate. *J. Urol.* **145,** 907–923.
2. Partin, A. W., Pound, C. R., Clemens, J. Q., Epstein, J. I., and Walsh, P. C. (1993) Serum PSA after anatomic radical prostatectomy: the Johns Hopkins Experience after 10 years. *Urol. Clin. N. Am.* **20,** 713–716.
3. Malm, J. and Lilja, H. (1995) Biochemistry of prostate specific antigen, PSA. *Scand. J. Clin. Lab. Invest.* **221(Suppl.),** 15–22.
4. Kabalin, J. N., McNeal, J. E., Johnstone, I. M., and Stamey, T. A. (1995) Serum prostate-specific antigen and the biologic progression of prostate cancer. *Urology* **46,** 65–70.
5. Wang, X., Heller, R., VanVoorhis, N., Cruse, C. W., Glass, F., Fenske, N., Berman, C., Leo-Messina, J., Rappaport, D., and Wells, K. (1994) Detection of submicroscopic lymph node metastases with polymerase chain reaction in patients with malignant melanoma. *Ann. Surg.* **220,** 768–774.

6. Miyajima, Y., Kato, K., Numata, S., Kudo, K., and Heribe, K. (1995) Detection of neuroblastoma cells in bone marrow and peripheral blood at diagnosis by the reverse transcriptase-polymerase chain reaction for tyrosine hydroxylase mRNA. *Cancer* **75,** 2757–2761.
7. Komeda, T., Fukuda, Y., Sando, T., Kita, R., Furukawa, M., Nishida, N., Amenemori, M., and Nakao, K. (1995) Sensitive detection of circulating hepatocellular carcinoma cells in peripheral venous blood. *Cancer* **75,** 2214–2219.
8. Datta, Y. H., Adams, P. T., Drobyski, W. R., Ethier, S. P., Terry, V. H., and Roth, M. S. (1994) Sensitive detection of occult breast cancer by the reverse-transcriptase polymerase chain reaction. *J. Clin. Oncol.* **12,** 475–482.
9. Chelly, J., Concordet, J. P., Kaplan, J. C., and Kahn, A. (1989) Illegitimate transcription: transcription of any gene in any cell type. *Proc. Natl. Acad. Sci. USA* **86,** 2617–2621.
10. Smith, M. R., Biggar, S., and Hussain, M. (1995) Prostate-specific antigen messenger RNA is expressed in non-prostate cells: implications for detection of micrometastases. *Cancer Res.* **55,** 2640–2666.
11. Moreno, J. G., Croce, C. M., Fischer, R., Monne, M., Vihko, P., Mulholland, S. G., and Gomella, J. G. (1992) Detection of hematogenous micrometastasis in patients with prostate cancer. *Cancer Res.* **52,** 6110–6112.
12. Katz, A. E., Olsson, C. A., Raffo, A. J., Cama, C., Perlman, H., Seaman, E., O'Toole, K. M., McMahon, D., Benson, M. C., and Buttyan, R. (1994) Molecular staging of prostate cancer with the use of an enhanced reverse transcriptase-PSA assay. *Urology* **43,** 765–775.
13. Wood, D. P., Jr., Banks, E. R., Humphreys, S., McRoberts, J. W., and Rangnekar, V. M. (1994) Identification of bone marrow micrometastases in patients with prostate cancer. *Cancer* **74,** 2533–2540.
14. Ghossein, R. A., Scher, H. I., Gerald, W. L., Kelly, W. K., Curley, T., Amsterdam, A., Zhang, Z. F., and Rosai, J. (1995) Detection of circulating tumor cells in patients with localized and metastatic prostatic carcinoma: clinical implications. *J. Clin. Oncol.* **13,** 1190–1200.

23

Quantitative, Competitive RT-PCR Analysis of Biomarkers in the Study of Neoplasia

Richard D. Hockett, Jr., Miriam D. Rogers, and Sudhir Srivastava

1. Introduction

The current status of the evaluation of many neoplasias (e.g., carcinoma of the prostate), relies on analysis of tumor stage and/or grade to predict the outcome of disease. Recent developments in screening practices with circulating markers, such as prostate-specific antigen (PSA), has led to the diagnosis of preneoplastic and early neoplastic lesions *(1–4)*. Although it is generally thought that these very early lesions will eventually progress to malignant disease, it is not always the case. In the early lesions, staging and grading of tumors has not predicted eventual outcome with any certainty. For this reason, other surrogate biomarkers have been sought to help distinguish potential bad outcomes from those that do not progress and therefore need no intervention. Surrogate biomarker analysis of neoplasia can involve measurement of protein species, and several markers have been associated with specific neoplastic conditions *(5–8)*. Alternatives to protein analysis include genetic-instability determinations *(9–12)* or mRNA analysis for molecules, such as growth factors/proto-oncogenes *(13–15)*. Assaying mRNA makes sense if the gene of interest does not encode for proteins against which antibodies are currently available, when those genes are controlled primarily at the level of transcription of mRNA, or if protein assays are insensitive. Analysis of mRNA has classically used Northern blot assays, which suffer from sensitivity problems, or RNA protection assays (S1 nuclease assay or RNase protection assay), which are more sensitive than Northern blotting, but are still relatively insensitive. With the advent of the polymerase chain reaction (PCR), attempts have been made to first measure DNA *(16–18)* and then RNA markers for disease *(13–15)*.

From: *Methods in Molecular Medicine, Vol. 14: Tumor Marker Protocols*
Edited by: M. Hanausek and Z. Walaszek © Humana Press Inc., Totowa, NJ

Utilization of PCR following reverse transcription (RT) of mRNA into cDNA has exceptional sensitivity for RNA measurement, but several factors have prevented its widespread use in a quantitative format ***(19–26)***. The critical modification developed to bypass quantitative problems was to perform competitive PCR, in which a known amount of a synthetic, competitive DNA construct was added to the PCR reaction. Typically, the DNA competitor fragment produced during PCR was of a slightly different size than the endogenous cDNA product, but utilized the same primer sequences as the analyte DNA and usually represented a deletion or addition subclone of the analyte cDNA ***(19–23)***. In the PCR, the same primers competed to amplify the competitor or the authentic cDNA species in the same reaction mixture, thus compensating for vagaries in the PCR synthesis rate during exponential amplification. After PCR, the two amplified products were distinguished and quantitated by performing ethidium bromide (EtBr) staining of agarose gel electrophoresis, followed by densitometry. A series of PCR reactions were run with different concentrations of competitor cDNA and a standard amount of authentic cDNA.

This type of analysis measured DNA concentrations quite accurately. However, in the case of RT-PCR, it did not correct for the inefficiencies of RNA reverse transcription. To address this problem, investigators generated RNA species from their DNA deletion or addition subclones and used this RNA to spike guanidine extracts with standard amounts of RNA competitor prior to the RT reaction ***(20–22,24)***. This type of analysis corrected for the inefficiencies of RT and PCR and gave accurate, reproducible results. Unfortunately, the gel-based analysis system usually employed for detection and quantitation of PCR products limited the number of samples that could be easily and practically analyzed. Furthermore, the analysis of several different analytes was hampered by the need to clone the cDNA of each analyte in question and subsequently make a deletion or addition subclone within the region to be analyzed by PCR. Therefore, in the design of an ideal, generally applicable quantitative, competitive RT-PCR (QC-RT-PCR) procedure, three things would be accomplished:

1. Control of all aspects of the procedure from RNA extraction through first-strand cDNA synthesis and during PCR;
2. Quantitation of multiple templates by a process that does not require cloning the cDNA for each analyte of interest; and
3. Analysis of PCR products by a readout system that does not utilize agarose-gel electrophoresis.

This chapter describes our attempt to incorporate the technology and principles of quantitative PCR into one such format.

2. Materials

2.1. Enzymes

The listed enzymes are available commercially from several different sources with a range of concentrations (Boehringer Mannheim, Indianapolis, IN; Stratagene, La Jolla, CA; Promega, Madison, WI; United States Biochemicals (USB/Amersham), Cleveland, OH; Gibco-BRL, Gaithersburg, MD). The protocols described use amounts based on standard concentrations supplied by most manufacturers.

1. Restriction endonucleases.
2. T3 RNA polymerase.
3. Human placental ribonuclease inhibitor (RNasin).
4. *Thermus aquaticus* DNA polymerase (*Taq* polymerase).
5. T4 DNA ligase.
6. Modified T7 DNA polymerase (Sequenase©).
7. Reverse transcriptase.
8. Terminal transferase.

2.2. Radionucleotides

^{3}H-UTP is available commercially from several different sources at several different specific activities. The purpose of using tritiated UTP is to trace label synthetic RNA molecules for determining concentration. Therefore, it is desirable to use as low a specific activity as possible and still be able to easily detect incorporation within the final product. We suggest using a specific activity of 30–40 Ci/mmol.

2.3. Enzyme Buffers

1. 10X Restriction endonuclease buffers (supplied by most manufacturers).
2. 10X Ligase buffer (supplied with enzyme).
3. 5X Tdt buffer (supplied with enzyme).
4. 10X Annealing buffer: 10 m*M* Tris-HCl, pH 7.4, plus 1 m*M* ethylenediamine tetra-acetic acid (EDTA).
5. 10X RNA transcription buffer (supplied with enzyme).
6. 100 m*M* Stock solutions of ultrapure grade deoxynucleotides: deoxyadenosine triphosphate (dATP), deoxyguanosine triphosphate (dGTP), deoxythymidine triphosphate (dTTP), and deoxycytidine triphosphate (dCTP). Dilute in H_2O to 1.25 m*M* each for use in PCR reactions or 2.5 m*M* each for use in reverse transcription (RT).
7. 100 m*M* Stock solutions of ultrapure grade ribonucleotides: ATP, GTP, CTP, and UTP. Generally supplied as 10 m*M* solutions. Use directly in RNA transcription reactions.
8. Reverse transcription mixture: 4 µL 5X RT buffer (manufacturer supplied), 2 µL 0.1 *M* DTT, 2 µL 2.5 m*M* dNTPs, and 50 U RT, in a volume of 9 µL.

9. PCR master mix: 1.6 μL 1.25 m*M* dNTPs, 1 μL 10X PCR reaction buffer (manufacturer supplied; the $MgCl_2$ final concentration [m*M*]) will vary (*see* **Notes 1–5**), 0.1–0.5 m*M* (final concentration) each primer, and 0.25 U *Taq* polymerase, in a volume of 6 μL. The PCR master mix can be prepared in quantities for 100 reactions, aliquoted in sialized tubes, and stored at –80°C for up to 3 mo. The optimal concentration of primers in the master mix varies between primer pairs and must be determined empirically for each primer set.

2.4. Other Buffers and Solutions

1. Phosphate-buffered saline (PBS): 0.15*M* NaCl, 3 m*M* KCl, 8 m*M* Na_2HPO_4, 2 m*M* KH_2PO_4, 0.1% NaN_3, pH 7.8.
2. Plate-washing buffer: PBS, pH 7.8, with 0.2% Tween-20.
3. PCR dilution buffer: plate-washing buffer with 0.1% tartrazine.
4. *Para*-Nitro phenylphosphate (pNPP) color solution: 1*M* diethanolamine plus 0.5 m*M* $MgCl_2$, pH 9.8 with 1 mg/mL pNPP added.
5. MS-2 purified phage RNA (MS-2 RNA available commercially from Boehringer Mannheim, Indianapolis, IN).
6. Solution D: 4 *M* Guanidine isothiocyanate (GITC), 1% sodium lauryl sarcosine (SLS), 15 m*M* Na citrate, pH 5.3, 0.1 *M* 2-mercaptoethanol (2-ME), and 10 μg/mL MS-2 phage RNA. Make 4 *M* GITC solution by adding 58 mL RNase free H_2O, 3 mL 3 *M* Na citrate, pH 5.3, and 1.5 mL 10% SLS to 50 g GITC; 4 *M* GITC can be stored at 4°C for up to 3 mo. Add 2-ME (7.2 μL of 14.4 *M* stock solution/mL GITC) and MS-2 RNA (10 μL/mL), just prior to use. Some guanidine solutions are available commercially.
7. Avidin: 5 mg/mL in plating buffer (0.1 *M* $NaHCO_3$, pH 9.8). The best source for avidin is a preparation called Avidin-DX (Vector Labs, Burlingame, CA).
8. EIA blocking solution: PBS with 1% bovine serum albumin (BSA).
9. EIA plate PCR DNA denaturing solution: 25 m*M* NaOH with 2 m*M* EDTA.
10. EIA plate probe solution: ddUTP-digoxigenin conjugated 25-bp DNA probes (50 ng/mL) in hybridization buffer (20% formamide, 6X SSPE [0.9 *M* NaCl, 1.2 *M* NaH_2PO_4, 6 m*M* EDTA], and 1 mg/mL sheared herring sperm DNA).
11. 0.1 *M* $CoCl_2$: Supplied with Tdt.
12. 25 m*M* Digoxigenin-conjugated ddUTP (available commercially from Boehringer Mannheim).
13. Oligo dT (available commercially as a lyophilized powder): Resuspend in RNase-free H_2O at 20 ng/μL.
14. Random hexamers (available commercially as a lyophilized powder): Resuspend in RNase-free H_2O at 2 μg/μL. Dilute just prior to use at 200 ng/μL.

3. Methods

3.1. Construction of a General Cloning Vector

A general cloning vector allows for a much easier building of future competitors. To generate such a general vector, two principles must be kept in mind: (a) the distance from the poly A+ tail, used for RT priming, to the 5' primer in the competitor must be close to the distance of the poly A+ tail to the 5' priming site in each analyte within the competitor; and (b) the distance between the primer pairs should be the same for analytes and competitors (*see* **Notes 1–5**).

1. Choose suitable plasmid vector with RNA transcription initiator sites flanking the polylinker and multiple cloning sites on each side of a blunt end restriction endonuclease (RE) site.
2. Clone "stuffer" segment into this centralized RE site.
3. Sequence-confirm the orientation of the stuffer so that the appropriate "sense" oligonucleotide can be ordered for use as detection probe.
4. Clone the "spacer/poly A+" segment into the 3' most RE site in the "stuffer" vector. A unique RE site must be positioned no more than 5 bp from end of poly A+ tail (include this unique RE site in "stuffer/poly A+" segment). This unique RE site is used to linearize the competitor plasmid before RNA transcription is performed.
5. Sequence-confirm the orientation of the "stuffer/poly A+" segment.

3.2. Construction of Specific Competitors

The choice of 5' and 3' primers for PCR should be given much attention. The best primers for this type of analysis are short (15–20 bases in length) and produce a single band after PCR of a complex cDNA. If the primers produce multiple bands on PCR, they may still be used, but the sensitivity of the assay may be affected.

1. Synthesize the sense and antisense PCR primer templates (both 5' and 3' templates, a total of four oligonucleotides) on a oligonucleotide synthesizer (several companies will do this commercially). Do not forget to add the appropriate restriction endonuclease overhangs to the end of each oligonucleotide.
2. Anneal 10 µg of each 5' template oligonucleotide and 10 µg of each 3' template oligonucleotide in a separate Eppendorf tube in 100 µL total volume (20 µg total DNA, 200 ng/µL annealed product).
3. Clone each annealed template (usually start with the 5' template) into the general cloning vector.
4. Isolate plasmids with inserts and sequence-confirm orientation and correct template sequences at the miniprep stage.
5. Large-scale purify the completed competitor by any of several plasmid isolation techniques (CsCl or column isolation).

3.3. Preparation of DNA Competitor (see Notes 3 and 4)

1. Digest 25 µg of completed, sequence-confirmed competitor with appropriate restriction endonuclease. This RE site must be within 5 bp of the poly A+ tail on the 3' end of competitor, because this linearized DNA will also be used for RNA synthesis.

2. Isolate linearized competitor on an agarose gel and purify by absorption to silica beads (available commercially as Geneclean©, BIO 101).
3. Take an OD_{260} reading using one-half of purified sample.
4. Calculate copies per microliter based on OD_{260} reading by the following formula:

$$\frac{[(OD_{260})\ (50\ \mu g/mL/OD_{260})\ (10^{-3}\ \mu L/mL)\ (6.023 \times 10^{17}\ \text{molecules}/\mu mol)]/}{[(660\ \mu g/\mu mol/bp)\ (\text{Size of entire competitor plasmid in bp})]} \quad (1)$$

5. Dilute to appropriate concentrations in H_2O with 10 µg/mL MS-2 RNA in sialized tubes. Use directly in PCR reactions.

3.4. Preparation of RNA Competitor (see Notes 3 and 4)

1. In a 1.5-mL Eppendorf tube, mix 10 µL of linearized DNA competitor (5 µg), 10 µL of 10X RNA transcription buffer, 10 µL 0.1*M* DTT, 3 µL RNasin (5 U/mL), 5 µL 10 m*M* ATP, 5 µL 10 m*M* GTP, 5 µL 10 m*M* CTP, 5 µL 10 m*M* UTP, 4 µL ^{3}H-UTP, 5 µL RNA polymerase (T3, T7, or SP6 depending on the site of the plasmid), and 38 µL H_2O.
2. Incubate at 37°C for 1 h.
3. Add 3 µL RNA polymerase.
4. Incubate at 37°C for 1 h.
5. Purify full-length synthetic RNA over oligo-dT column.
6. Quantitate copies per microliter using the following formula:

$$\frac{[(T_1)\ (5.0 \times 10^{-5}\ \text{mmol cold U})\ (6.023 \times 10^{20}\ \text{copies/mmol})]/}{[(T_o \times V_o)\ (\#\ \text{of Us in construct})]} \quad (2)$$

where T_o is the counts of entire reaction mixture, T_1 is the counts of isolated full-length product, and V_o is the volume of reaction mixture.
7. Dilute to the desired concentration in solution D in sialized tubes.

3.5. Preparation of Cell/Tissue Extract

Total RNA is isolated from cells or tissues by a modification of the acid/phenol extraction procedure of Chomczynski and Sacchi ***(27)***. Cells/tissues are lysed in solution D at a concentration of 1000–5000 cells/µL. Aliquots of solubilized RNA in GITC are stored in liquid N_2 until use. In order to minimize errors no reagent in the following steps should be pipeted in <2 µL volumes.

1. Add 2–10 µL RNA samples to solution D (up to 100 µL) and place in 0.5-mL Eppendorf tubes.
2. Add 11 µL 2 *M* Na acetate, pH 4.0, along with the appropriate amount of RNA competitor.
3. Add 110 µL of water-saturated phenol, pH 4.3, vortex, and place the sample on ice for 5–10 min.
4. Add 50 µL $CHCl_3$ to the sample, vortex again, and then spin in a microcentrifuge at high speed.

5. Transfer the upper (aqueous) phase to a new tube. For tissue preparations, repeat phenol/chloroform extraction. For single-cell preparations and after the second phenol extraction for tissues, add 110 µL $CHCl_3$, vortex, and spin.
6. Transfer the aqueous phase to another 0.5-mL Eppendorf tube and add an equal volume of isopropanol.
7. Place the tubes at –80°C for 15 min. After microcentrifugation, the pelleted sample is washed in 125 µL 80% EtOH, respun, decanted, and air-dried.

3.6. Reverse Transcription

3.6.1. Oligo dT Priming

1. Resuspend the nearly invisible pellet in 10 µL H_2O with 1 µL (20 ng/µL) oligo dT.
2. Heat the solution at 70°C for 10 min and then snap cool on ice for 5 min.
3. Add 9 µL of reverse transcription mixture and incubate the samples at 37°C for 1 h. This cDNA reaction mixture can be used directly in PCR or stored at –20°C for several weeks.

3.6.2. Random Priming

1. Resuspend the nearly invisible pellet in 10 µL H_2O with 1 µL (200 ng/µL) random hexamers.
2. Heat the solution at 70°C for 10 min and then slow cool to room temperature.
3. Add 9 µL of reverse transcription mixture and incubate the samples at 37°C for 1 h. This cDNA reaction mixture can be used directly in PCR or stored at –20°C for several weeks.

3.7. PCR

1. In a 0.5-mL reaction tube, place 4 µL of the above RT reaction product and 6 µL of a PCR master mix.
2. Add one drop of molecular biology grade oil.
3. Cycle the mixture in a thermocycler at 94°C for 30 s, 55°C for 1 min (this temperature to be determined empirically with each set of PCR primers), and 72°C for 30 s for the appropriate number of cycles (25–40 depending on the experiment; 35 is the routine number of cycles).

3.8. Plate-Based EIA

3.8.1. Preparation of EIA Plates

EIA plates are coated 1 d to 1 wk prior to use.

1. Add 150 µL of avidin (5 µg/mL) in plating buffer to each well, and incubate the plates at 37°C for 1 h, or alternatively at 4°C overnight.
2. Wash the plates twice in plate-washing buffer and then block the uncoated plate surface with 180 µL of blocking solution (PBS/1% bovine serum albumin [BSA]). The plates can be stored at 4°C for up to 3 wk in blocking solution. If plates are stored in 1% BSA for more than a few days, it is advisable to add NaN_3 (0.1%, w/v).
3. Just prior to using, wash the plates twice in plate-washing buffer.

3.8.2. Plating of PCR Product

After the PCR, the product is diluted and aliquots are placed into four separate, adjacent wells of an avidin-coated 96-well plate. Two wells (i.e., the PCR product placed in these wells) will be hybridized with digoxigenin-labeled analyte oligonucleotide, and two wells will be hybridized with digoxigenin-labeled stuffer oligonucleotide.

1. After PCR, dilute the 10-µL PCR reaction mixture 1:30 directly in the PCR tube by adding 290 µL PCR dilution buffer.
2. Mix the solution thoroughly, and place 30 µL of the diluted product into four adjacent avidin-coated microtiter wells that contain 130 µL of plate-wash buffer.
3. Incubate the plates at 42°C for at least 1 h.
4. Wash the plates twice with plate-washing buffer.
5. Add 160 µL of denaturing solution. Incubate plates at room temperature for 2 min.
6. Wash the plates twice with plate-washing buffer.
7. Add 160 µL/well of probe solution and incubate the plates at 42°C for a minimum of 30 min.
8. After hybridization, wash the plates three times with plate-washing buffer.
9. Add 160 µL of antidigoxigenin Fab fragments conjugated with (alkaline phosphatase) diluted 1:5000 in PBS/1% BSA.
10. Incubate the plates at 37°C for 1–2 h.
11. Wash plates four times with plate-washing buffer.
12. Add 180 µL pNPP color solution and incubate the plates at 37°C.
13. Read OD_{405} at various time points.

3.8.3. ddUTP Labeling of Detection Oligo T

The enzyme Tdt is used to modify detection oligonucleotides by placing digoxigenin-conjugated ddUTP on the 3' end.

1. In a 0.5-mL Eppendorf tube, place 5 µg of desired oligonucleotide (≤10 µL).
2. Add 4 µL 5X Tdt reaction buffer (supplied by the manufacturer).
3. Add 4 µL 0.1 *M* $CoCl_2$ (usually supplied by the manufacturer).
4. Add 1 µL 25 m*M* digoxigenin-conjugated ddUTP.
5. Add 1 µL Tdt (50 U/µL).
6. Add H_2O to 20 µL and incubate at 37°C for 20 min.
7. Dilute to 50 ng/µL with hybridization buffer (add 80 µL). Probes can be stored at 4°C up to 4 wk. Alternatively, large batches of detection oligonucleotides can be made (100 µg), diluted, aliquoted, and stored at –20°C for at least 1 yr.

4. Notes

1. Overview of procedure: **Fig. 1** shows a schematic overview of the procedure. To achieve the desired goals for a QC-RT-PCR procedure, a competitor RNA "minigene" was developed that would be spiked at different concentrations to guanidine extracts of cell populations prior to RNA extraction and cDNA synthe-

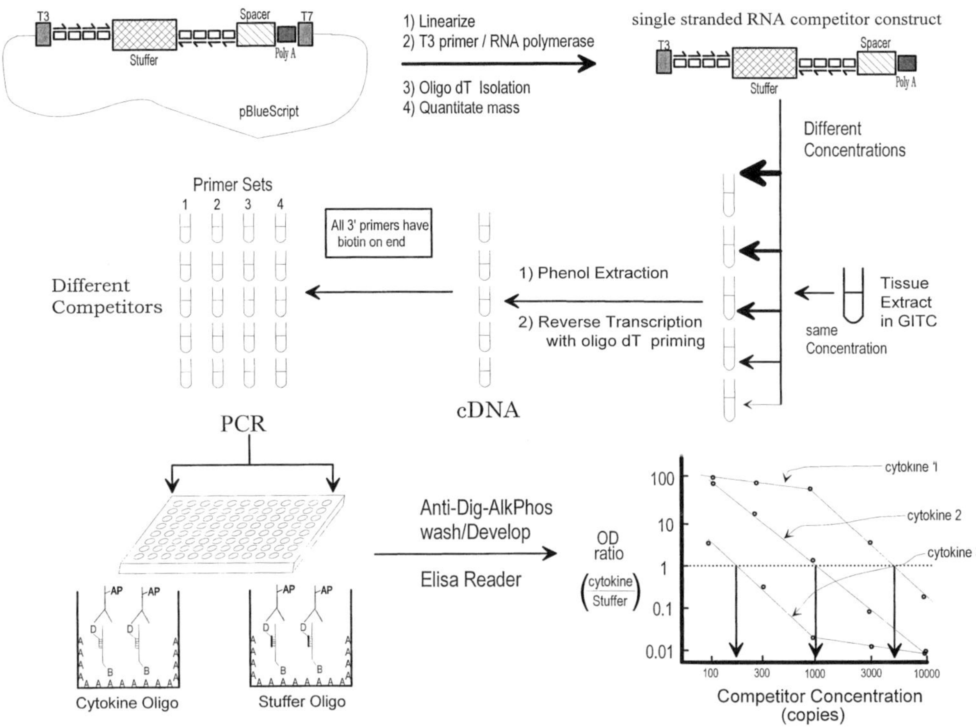

Fig. 1. Schematic representation of the steps in the described QC-RT-PCR assay. Reprinted with permission from Hockett et al. ***(28)***.

sis. Each competitor is constructed by preparing oligonucleotides corresponding to the primers of the relevant mRNAs and subcloning them into a general cloning vector, thus flanking the "stuffer" segment, which is used to detect the competitor PCR product. The vector also contains a "spacer" segment of approx 300–400 bp, flanked by a poly A+ tail of 15 or more base pairs. The purpose of the spacer segment is to maintain a similar distance between the poly A+ and the 3' primer cassette in any given vector to that of each analyte in question, in order to keep the distance that RT must synthesize cDNA nearly the same for the analyte and competitor. The poly A+ tail has a dual purpose. It is used to prime cDNA synthesis using oligo dT, but it also functions as a capture target for the purification of full-length competitor molecules with oligo dT latex beads. The DNA template must be constructed in a plasmid vector that has a primer site for one of the common DNA-dependent RNA polymerases (T3, T7, or SP6 polymerase). One such polymerase is used to transcribe the RNA competitor from this DNA template. After RNA transcription and purification, competitor concentration is assessed by trace-labeling techniques.

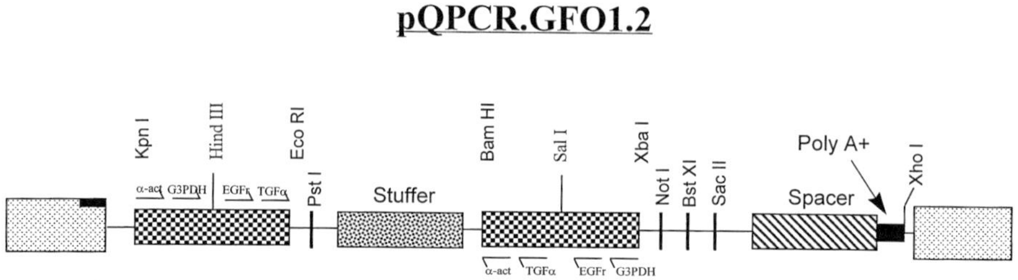

Analyte	Message Size	Competitor Size	Poly A⁺ to 3' Primer Distance Message	Poly A⁺ to 3' Primer Distance Competitor
α-actinin	291	345	600	397
EGFr	463	345	>5 kb	357
G3PDH	344	385	425	340
TGFα	369	305	500	380

Fig. 2. Schematic representation of the competitor GFO1.2. For the EGFr gene, the distance from the poly A+ tail to the 3' primer is not precisely known, but is several kilobases. For this reason, EGFr must use random priming instead of oligo dT priming for generation of first-strand cDNA.

The detection scheme for determining PCR products for competitor and endogenous message is EIA, i.e., specifically an enzyme-linked immunosorbent assay (ELISA) procedure and not agarose gel electrophoresis followed by densitometry. This method of detection was chosen because of the inherent limitations of agarose gel electrophoresis and densitometry when attempting to analyze multiple analytes from many samples. With ELISA, multiple analyses can be performed in a single 96-well plate. After PCR, products are captured in a 96-well plate precoated with avidin. This is possible because the antisense return primer (3' primer in most cases) is labeled with biotin on its 5' end (during primer synthesis). Both competitor and endogenous fragments are thus tagged with biotin on the antisense strand. Each PCR reaction product is diluted and aliquots placed into four wells: two wells for detection of competitor fragment concentration and two wells for detection of endogenous analyte message concentration. After capture on the avidin-coated plate, the DNA strands are denatured by alkali and the nonbiotinylated strand is washed away.

Oligonucleotide probes specific for each analyte and the "stuffer," 3' end labeled with digoxigenin, are hybridized in the appropriate wells, excess probe washed away, and alkaline phosphatase-conjugated antidigoxigenin antibody Fab fragments added. Subsequently, standard alkaline phosphatase ELISA with pNPP is performed. Absorbance measurements are obtained from an ELISA reader, and analyte levels are determined from standard plots of competitor RNA.

The outline of one such potential competitor is seen in **Fig. 2**. In this competitor, termed pGFO1.2, a 230-bp piece of murine genomic DNA (nontranscribed in any cell type) has been cloned into the plasmid vector pBluescript (Stratagene, La Jolla, CA), and designated as "stuffer" in **Fig. 2**. The utilization of an existing plasmid vector allows the maintenance of several cloning sites on each side of the stuffer in which the primer templates will be inserted. The orientation of the stuffer fragment must be determined by sequence analysis so that the appropriate detection oligonucleotide can be utilized (sense vs antisense). The other essential piece that must be placed downstream of the stuffer, has been designated "spacer/poly A+" segment in **Fig. 2**. In this instance, a 387-bp piece of a chicken cDNA, which contains an 18-bp poly A+ stretch, was subcloned downstream of the stuffer. Again, the proper orientation of each segment must be confirmed by sequence analysis. In addition, a unique restriction site must be placed adjacent to the poly A+ tail. In this instance, an *Xho*I site, inherent in the chicken cDNA clone, is maintained immediately next to the poly A+ tail. The distance for this restriction site must be kept to a minimum. In our experience, more than 5 bp away causes interference with oligo dT priming and cDNA synthesis by RT (R. D. Hockett et al., unpublished observations). Notice that the PCR product size differs slightly for each analyte and competitor. This was purposefully done so that the EIA procedure could be quality-controlled to insure that the read-out system behaves as designed (*see* **Note 5**).

2. Hints for primer selection. The choice of PCR primers is critical for efficient and reliable performance of this assay. Care taken in selecting and testing primer pairs before competitor construction will save many headaches during the quality-control aspects of competitor evaluation. In the initial choice of primers, any of the available computer programs can calculate "optimal" primers for PCR. The key parameters for efficient, reliable priming are Mg^{2+} concentration and annealing temperature. For logistical reasons, attempts should be made to keep the annealing temperature constant among all of the primer pairs. The optimal Mg^{2+} concentration needs to be empirically determined but can be easily adjusted within the 10X PCR buffer. Be advised that the authors have utilized several different primer-selection programs and have found that all of them have failed in choosing primer pairs for PCR, given the strict temperature constraints sought. For this reason, all potential primers should first be tested on a complex cDNA mixture and analyzed for the presence of a single, predicted band prior to cloning into a competitor construct. A typical example of not testing primers before competitor synthesis is shown in **Fig. 3**. In this figure, transforming growth factor α (TGFα) PCR products from a reaction using the prostate cancer cell line DU145 (known to make TGFα protein), with competitor GFO1 ("old competitor" in **Fig. 3**) were electrophoresed in an agarose gel, stained with EtBr, and photographed. It is easy to see the other extraneous bands primed during PCR. In contrast, TGFα PCR products from primers tested prior to competitor synthesis and cloned into the redesigned competitor GFO1.2 ("new competitor" in **Fig. 3**) demonstrate only the two expected bands (TGFα and competitor). Priming the extraneous bands decreases the sensitivity of the assay by consuming primers generating DNA fragments that will not be detected during hybridization in the EIA procedure.

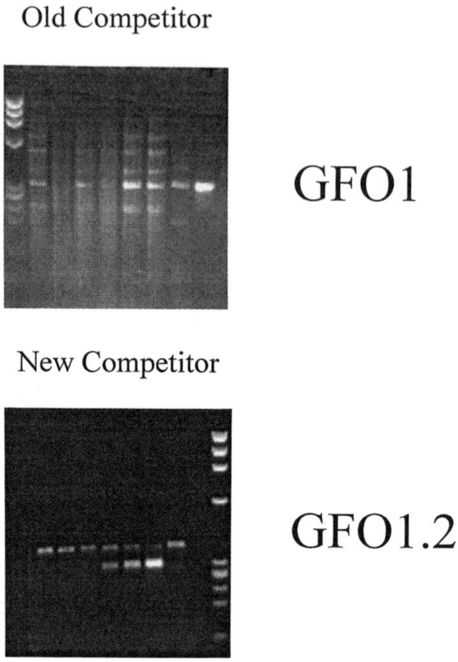

Fig. 3. Titration of TGFα in DU145 cells using two different competitors. Both pictures show EtBr-stained agarose gel electrophoresis of TGFα PCR products from titrations of the same DU145 cell extract. Note the extra bands seen in all lanes of the old competitor, GFO1, when compared to the new competitor, GFO1.2.

An additional important consideration to primer selection is the location within the cDNA to place both primers. Not only is the distance between primer pairs important, but so is the relation to the poly A+ tail. Although theoretically it is possible to place the primers anywhere within the mRNA, using random hexamers to prime cDNA synthesis, in mRNAs larger than 2 kb we have consistently found differences between upstream primer sets and those nearer the poly A+ tail. For example, in comparing competitors pGFO1 and pGFO1.2, not only do pGFO1.2 primers behave better during PCR (**Fig. 3**), they are also located in a different area of each gene (EGFr, TGFα, α-actinin, G3PDH). The pGFO1.2 primers are all located on the 3' end of each cDNA, instead of the middle of the protein-coding region (GFO1, data not shown). When comparing both sets of primers, the primers from the 3' end of genes consistently showed an increase in the apparent copy number readout from the QC-RT-PCR assay (**Table 1** and **Fig. 4**). For the erbB-2 gene, in SKBR3 cells, the apparent copy number increased two- to threefold; for TGFα, in DU145 cells, the apparent increase was eight- to ninefold. For EGFr, the increase was even more dramatic.

Table 1
Positional Effects on Copy Number Endpoint/5000 Cells

Analyte	5' End	3' End
TGFα (DU145)	158,000	1,300,000
ErbB-2 (SKBR3)	168,000	385,000
EGFr (DU145)	<100	2,500,000

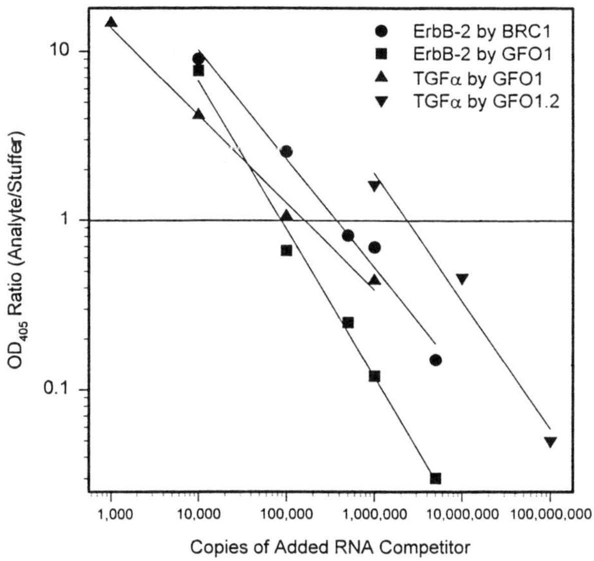

Fig. 4. Comparison of erbB-2 and TGFα end-point determination with two different RNA competitors in SKBR-3 cells (erbB-2) and DU145 cells (TGFα). ErbB-2 and TGFα determination from the same SKBR-3 and DU145 cell extracts using competitors with primers located in the protein-coding region (GFO1) compared to competitors with primer sequences located in the 3' untranslated regions (BRC1 and GFO1.2).

3. Cloning hints: Once three to five primer pairs for different analytes have been selected, large oligonucleotides corresponding to the head-to-tail arrangement of 5' and 3' primer sequences are synthesized. Appropriate restriction endonuclease overhangs must be placed on each end of the sense and antisense oligonucleotides for each template. Although the same restriction endonuclease site could be placed on each end of the primer template, the efficiency of correctly oriented primer sites is enhanced if different sites are placed on each end for "directional subcloning." Furthermore, utilizing two RE sites precludes the need to treat the vector with phosphatase to remove the 5'-PO_4 group, thus preventing religation of vector sequence without insert. It is also advisable to engineer a novel RE site in between the middle

Table 2
Effect of Methyl Mercury on EGFr and Cytokeratin 8 Measurements

Condition	EGFr equivalence point	Cytokeratin 8 equivalence point
No MeHgOH	1600	6400
MeHgOH at 45°C	63,500	984,000

two primer sites for each template. The purpose of this additional RE site is to prevent the need to resynthesize the entire template in case one primer template does not perform properly (*see* **Note 5**). Sequence analysis of miniprep DNA must be done to confirm the desired template (usually start with the 5' template first) was inserted, and the remaining template must be ligated into the appropriate sites, downstream of the stuffer, in similar fashion. Again, confirmation of the appropriate 3' template must be done by sequence analysis of miniprep DNA.

4. RNA secondary structure considerations: Large RNA molecules are known to have an extensive secondary structure that can inhibit the recognition and/or function of RNA enzymes ***(29–32)***. In order to fully denature some RNA molecules, we have employed the denaturant methyl mercury hydroxide (MeHgOH). Most RNA species we have measured by QC-RT-PCR show no change in copy-number endpoint after MeHgOH treatment (data not shown). However, some molecules, such as EGFr and cytokeratin 8, display major changes in apparent copy-number endpoint after MeHgOH treatment (**Table 2**). The best hypothesis for this effect is that these molecules exhibit tight secondary structure that is only denatured after reaction with MeHgOH. Relaxing of this structure allows RT to process the RNA more efficiently.
5. Quality control: The only consistent problem we have encountered with this procedure concerns the hybridization of detection oligonucleotides. We utilize strict parameters for choosing oligonucleotides: 30-bp length, 50–60% GC content, no hairpin loop structures present, and calculated T_m of 55–90°C. However, some oligonucleotides do not optimally hybridize at 42°C and 30% formamide concentration. In these instances, the percentage of formamide concentration can be changed or a new oligonucleotide selected. In order to determine if a problem exists, we routinely quantitate PCR products by alternate methodologies and compare results to the EIA procedure. The "gold standard" for this analysis is agarose gel electrophoresis followed by EtBr staining and densitometry. Keep in mind that for agarose electrophoresis to work the competitor and analyte PCR products must differ in size by at least 40–50 bp. We accomplish this by alternating the order of primer sites on the 5' and 3' primer cassettes before synthesis, insuring that all products differ by at least 50 bp. Once the detection oligonucleotides utilized in EIA have been compared favorably by these two techniques, the need to perform agarose electrophoresis is abrogated. To accomplish the comparison, choose a titration of analyte with four to five different competitor concentrations. Increase the PCR reaction mixture from 10–20 µL, in order to have enough prod-

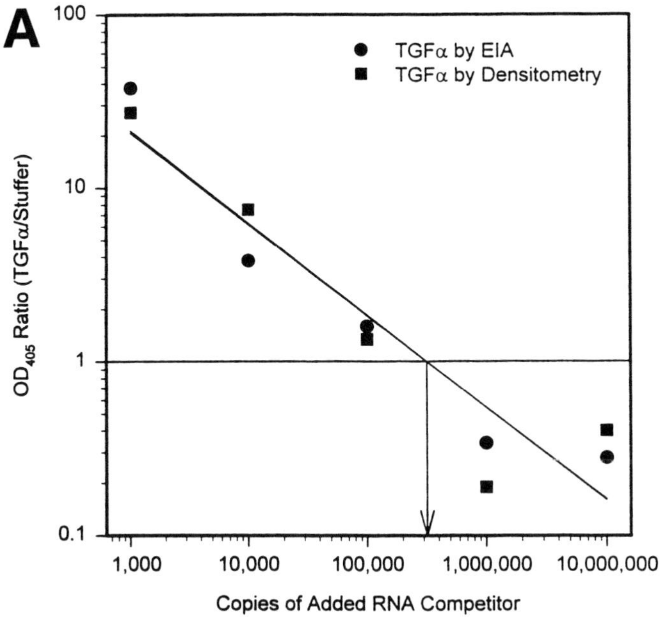

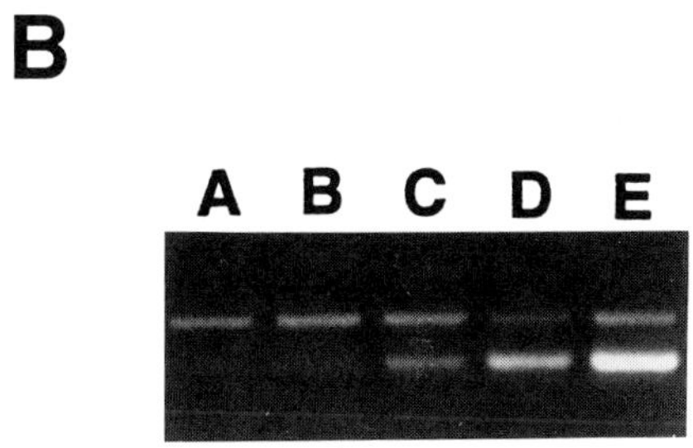

Fig. 5. Comparison of EIA and agarose gel densitometry endpoints. The same PCR titration of TGFα message from DU145 cells was analyzed by EIA and agarose gel electrophoresis followed by densitometry. **(A)** Illustration of the titration by EIA and gel densitometry; **(B)** The EtBr-stained agarose gel used for gel densitometry. The calculated endpoints from each analysis methodology are essentially identical. A, 100,000 copies of added RNA competitor. B, 300,000 copies of added RNA. C, 1,000,000 copies of added RNA. D, 3,000,000 copies of added RNA. E, 10,000,000 copies of added RNA.

uct to perform both techniques. After PCR, pipet 10 μL into an agarose gel for electrophoresis and staining, and then proceed with the EIA procedure as written. An example of a comparison of an analyte from competitor GFO1.2 (TGFα) is seen in **Fig. 5**. If the two procedures do not agree, then adjustments to the detection oligonucleotide or hybridization conditions can be made, and the new EIA procedure tested in the same fashion.

References

1. Lee, C. T. and Oesterling, J. E. (1995) Diagnostic markers of prostate cancer: utility of prostate specific antigen in diagnosis and staging. *Sem. Surg. Oncol.* **11,** 23–35.
2. Kardamakis, D. (1996) Tumor serum markers: clinical and economical aspects. *Anticancer Res.* **16,** 2285–2288.
3. Bangma, C. H., Blijenberg, B. G., and Schroder, F. H. (1995) Prostate specific antigen: its clinical use and application in screening for prostate cancer. *Scand. J. Clin. Lab. Invest.* **221,** 35–44.
4. Randrup, E. and Baum, N. (1996) Prostate specific antigen testing for prostate cancer. Practical interpretation of values. *Postgrad. Med.* **99,** 227–234.
5. Suresh, M. R. (1996) Classification of tumor markers. *Anticancer Res.* **16,** 2273–2277.
6. Pamies, R. J. and Crawford, D. R. (1996) Tumor markers. An update. *Med. Clin. North Am.* **80,** 185–199.
7. Grignon, D. J. and Hammond, E. H. (1995) College of American Pathologists Conference XXVI on clinical relevance of prognostic markers in solid tumors. Report of the prostate cancer working group. *Arch. Path. Lab. Med.* **119,** 1122–1126.
8. Berek, J. S. and Bast, R. C. (1995) Ovarian cancer screening. The use of serial complementary tumor markers to improve sensitivity and specificity for early detection. *Cancer* **76,** 2092–2096.
9. Tahara, E. (1995) Genetic alterations in human gastrointestinal cancers. The application to molecular diagnosis. *Cancer* **75,** 1410–1417.
10. Svendsen, L. B. (1993) Congenital genetic instability in colorectal carcinomas. *Danish Med. Bull.* **40,** 546–556.
11. Weinert, T. and Lydall, D. (1993) Cell cycle checkpoints, genetic instability and cancer. *Sem. Cancer Biol.* **4,** 129–140.
12. Thibodeau, S., Bren, G., and Schaid, D. (1993) Microsatellite instability in cancer of the proximal colon. *Science* **260,** 816–819.
13. Ciardiello, F., Kim, N., McGeady, M. L., Liscia, D. S., Saeki, T., Bianco, C., and Salomon, D. S. (1991) Expression of transforming growth factor alpha in breast cancer. *Annal. Oncol.* **2,** 169–182.
14. Klimpfinger, M., Zisser, G., Ruhri, C., Putz, B., Steindorfer, P., and Hofier, H. (1990) Expression of c-myc and c-fos mRNA in colorectal carcinoma in man. *Virchows Archiv. B Cell Pathol. Mol. Pathol.* **59,** 165–171.
15. Melo, J. V. (1996) The molecular biology of chronic myeloid leukemia. *Leukemia* **10,** 751–756.
16. Thein, S. L., Jeffreys, A. J., Gooi, H. C., Cotter, F., Flint, J., O'Connor, N. T. J., Weatherall, D. J., and Wainscoat, J. S. (1993) Detection of somatic changes in human cancer DNA by DNA fingerprint analysis. *Br. J. Cancer* **55,** 816–819.
17. Peinado, M. A., Malkhosyan, S., Velazquez, A., and Perucho, M. (1992) Isolation and characterization of allelic losses and gains in colorectal tumors by arbitrarily primed polymerase chain reaction. *Proc. Natl. Acad. Sci. USA* **89,** 10,065–10,069.
18. Meling, G. I., Lothe, R. A., Borresen, A. L., Hauge, S., Graue, C., Clausen, O. P., and Rognum, T. O. (1991) Genetic alterations within the retinoblastoma locus in

colorectal carcinomas. Relation to DNA ploidy pattern studied by flow cytometric analysis. *Br. J. Cancer* **64,** 475–480.

19. Becker-Andre, M. and Hahlbrook, K. (1989) Absolute mRNA quantification using the polymerase chain reaction (PCR). A novel approach by a PCR aided transcript titration assay (PATTY). *Nucleic Acids Res.* **17,** 9437–9446.
20. Piatek, M., Saag, M. S., Yang, L. C., Clark, S. J., Kappes, J. C., Luk, K. C. Hahn, B. H., Shaw, G. M., and Lifson, J. D. (1993) High levels of HIV-I in plasma during all stages of infection determined by competitive PCR. *Science* **259,** 1749–1754.
21. Piatek, M., Luk, K. C., Williams, B., and Lifson, J. D. (1993) Quantitative competitive polymerase chain reaction for accurate quantitation of HIV DNA and RNA species. *Biotechniques* **14,** 70–81.
22. Menzo, S., Bagnarelli, P., Giacca, M., and Varal, A. (1992) Absolute quantitation of viremia in human immunodeficiency virus infection by competitive reverse transcription and polymerase chain reaction. *J. Clin. Microbiol.* **30,** 1752–1757.
23. Scadden, D. T., Wang, Z., and Groopman, J. E. (1992) Quantitation of plasma human immunodeficiency virus type 1 RNA by competitive polymerase chain reaction. *J. Infect. Dis.* **165,** 1119–1123.
24. Wang, A. M., Doyle, M. V., and Mark, D. F. (1989) Quantitation of mRNA by the polymerase chain reaction. *Proc. Natl. Acad. Sci. USA* **86,** 9717–9721.
25. Gilliland, G., Perrin, S., Blanchard, K., and Bunn, H. F. (1990) Analysis of cytokine mRNA and DNA: detection and quantitation by competitive polymerase chain reaction. *Proc. Natl. Acad. Sci. USA* **87,** 2725–2729.
26. Bagnarelli, P., Menso, S., Valenia, A., Manzin, A., Giacca, M., Ancarani, F., Scalise, G., Varaldo, P. E., and Clementi, M. (1992) Molecular profile of human immunodeficiency virus type 1 infection in symptomless patients and in patients with AIDS. *J. Virol.* **66,** 7328–7335.
27. Chomczynski, P. and Sacchi, N. (1987) Single step method of RNA isolation by acid guanidinium thiocyanate phenol-chloroform extraction. *Anal. Biochem.* **162,** 156–159.
28. Hockett, R. D., Janowski, K. M., and Bucy, R. P. (1995) Simultaneous quantitation of multiple cytokine mRNAs by RT-PCR utilizing plate based EIA methodology. *J. Immunol. Meth.* **187,** 273–285.
29. Richardson, J. P. and Macy, M. R. (1981) Ribonucleic acid synthesis termination protein rho function: effects of conditions that destabilize ribonucleic acid secondary structure. *Biochemistry* **20,** 1133–1139.
30. Alford, R. L. and Belmont, J. W. (1990) Stable RNA secondary structure in a retroviral vector insert terminates reverse transcriptase elongation in vitro but not in cultured cells. *Hum. Gene Ther.* **1,** 269–276.
31. Forough, R., Engleka, K., Thompson, J. A., Jackson, A., Imamura, T., and Maciag, T. (1991) Differential expression in Escherichia coli of the alpha and beta forms of heparin binding acidic fibroblast growth factor-1: potential role of RNA secondary structure. *Biochim. Biophys. Acta* **1090,** 293–298.
32. Berkhout, B., Gatignol, A., Silver, J., and Jeang, K. T. (1990) Efficient trans activation by the HIV-2 Tat protein requires a duplicated TAR RNA structure. *Nucleic Acids Res.* **18,** 1839–1846.

24

Differential Display to Define Molecular Markers and Genes That Mediate Malignancy

Edward J. Pavlik, Katherine Nelson, Suseela Srinivasan, Thomas L. Johnson, and Paul D. DePriest

1. Introduction

Differential display is a powerful way to identify alterations in gene expression that can be contrasted by different states or treatments in side-by-side comparisons *(1)*. An inherent strength of the polymerase chain reaction (PCR)-based differential-display approach is that only a very modest amount of starting material is needed. For example, DNAs from untreated and treated matched preparations are run side-by-side so that bands that have been induced are absent in the untreated preparation and displayed prominently in the treated lane, while repressed gene bands are noted by their absence in the treated lane. As few as 2 μg of total RNA is theoretically sufficient for displaying all genes induced by any single treatment. Signals selected from differential-display gels are then used to screen an expression library for the full-length cDNA. Thus, the differential-display approach has the potential for identifying individual genes that correspond to different biological states or respond to different treatments. These identifications are then followed by efforts to determine if the full-length cDNAs that are displayed will introduce the correct responses when transfected to appropriate test cells. This approach has been used to successfully identify:

1. Genes expressed in human breast carcinoma vs mammary epithelial cells *(2)*;
2. A candidate tumor suppressor gene in human mammary epithelial cells *(3)*;
3. The gene for the M2 subunit of ribonucleotide reductase in premalignant breast epithelial lesions *(4)*;
4. A downstream target gene in the ras signaling pathway *(5)*;
5. Genes differentially expressed in human ovarian carcinoma *(6)*;

From: *Methods in Molecular Medicine, Vol. 14: Tumor Marker Protocols*
Edited by: M. Hanausek and Z. Walaszek © Humana Press Inc., Totowa, NJ

6. Genes expressed in the later stages of liver regeneration, including the ribosomal protein S24 ***(7)***;
7. Genes for the histocompatibility antigen (HLA-DR), laminin B2, melanoma inhibitory activity, and tissue inhibitor metalloproteinase-3 ***(8)***;
8. Homeobox genes in tobacco ***(9)***; and
9. The cytokeratin endo A gene and the α subunit of mitochondrial F_1 adenosine triphosphate (ATP) synthase in mouse preimplantation development ***(10)***.

1.1. Technical Overview

Differential display takes advantage of the poly A tail on mRNA so that poly T and two additional bases serve as an anchored primer in the reverse transcription (RT) reaction that generates cDNA. There are 12 possible two-base combinations adjacent to the poly A tail for the RT primer, which can be used to assemble sets of degenerate anchored oligo(dT) primers, each ending in one of the four DNA bases (N = A,T,G,C) with degeneracy in the penultimate position (M = A,G,C). Thus, a degenerate primer set is defined by the 3' base (N) with degeneracy in the penultimate position (M) as T_{12}MN, so that for T_{12}MG the degenerate primer set consists of 5'-TTTTTTTTTTTTGG-3', 5'-TTTTTTTTTTTTAG-3', and 5'-TTTTTTTTTTTTCG-3'. Using the four possible degenerate primer sets (T_{12}MN; N = A,G,C,T), the total mRNA can be divided into four different subpopulations through reverse transcription ***(11)***. Recently, it has been reported that one-base-anchored oligo(dT) primers provide excellent selectivity in subdividing mRNA into three populations ***(12)***. Next, each of the cDNA subpopulations is amplified using a 5' primer 10-mer of arbitrary sequence and the appropriate anchored primer from the RT reaction in the presence of [α-^{35}S]-dATP or [α-^{32}P]-dATP. With only six 10-mer arbitrary sequences, it is theoretically possible to amplify all possible eukaryotic genes ***(1)***. In fact, the use of 26 arbitrary 5' upstream primers with 12 degenerate anchored oligo(dT) primers in 312 PCR reactions resulted in ~38,000 display bands, which is more than twice the predicted number of genes expected to be expressed ***(13)***. Since the annealing positions to cDNAs for the 5' primer of arbitrary sequence should be randomly distributed in distance from the poly A tail, the amplified products from various mRNAs will differ in size, allowing the amplified products to be displayed as a ladder on 6% sequencing gels by autoradiography when [α-^{35}S]-dATP or [α-^{32}P]-dATP ***(14)*** is included in the PCR reaction. The use of [α-^{32}P]-dATP may be a better choice because [α-^{35}S]-dATP tends to volatilize and contaminate thermocyclers. Arbitrary primer length and PCR conditions have been worked out to maximize specificity for some individual genes ***(1,15)***. Any cDNA species would be amplified by PCR provided that the distance at which a second primer anneals is <2–3 kb from the beginning of the poly A tail. Since the average size of mRNA (~1.2 kb) is less than this distance, cDNA for virtually all expressed genes should be amplified by PCR. Because DNA sequencing gels can resolve cDNAs up to 500 nt, arbitrary primer annealing positions should be within 500 nt. Thus, arbitrary primers selected for use should have the ability to anneal within a 500-nt distance ***(1)***. After display on sequencing gel films, individual bands are then cut out, eluted, and reamplified with the originating 10-mer. The reamplification product is cloned, expression confirmed, and then sequenced. The cloned cDNA can then be used to screen a cDNA library to obtain the full-length cDNA. Ideally, two to four 48-lane sequencing gels can be used to display all possible genes induced by a single treatment.

1.2. Overall Approach for Differential Display, Reamplifications, Cloning, Screening, and Sequencing of Full-Length cDNAs

Differential-display gels demonstrate bands that may show expression altered by various treatments. A clear advantage of the differential-display approach is that it immediately visualizes bands that are exclusively related to different states or treatments and thereby allows only these bands to be pursued further. However, it must be emphasized that the displays require *considerably* more additional efforts, which can be quite challenging technically, before gene identification is obtained. Bands that are identified by differential display must be confirmed as expressed in the originating cells. Once confirmed, the differential display cDNA can be used to probe libraries for the full-length cDNA, and the full-length cDNA used for sequencing. It is wise to anticipate that any infidelities incorporated during PCR ***(16,17)*** need to be minimized by cloning the cDNA selected by differential display.

After reamplification and cloning, positive clones are screened for insertion and the insert reverified for expression in originating cells by Northern analysis, slot blots, or ribonuclease protection. Cloned inserts are then sequenced, restriction sites identified, and the insert fragment used for screening full-length cDNA from a cDNA library. Sequencing is then obtained from full-length cDNA.

1.3. Considerations Relevant to the Application of Differential Display

One of the strengths of differential display is that this approach is intuitively straightforward, providing clear demonstration early in the procedure of differences in expression related to treatment. However, there are also several critical aspects of differential display that should not be overlooked.

1.3.1. False Positives

The frequency of false positives on differential-display gels prevents the use of the displays to accurately estimate the number of genes that are expressed or repressed by a comparison. It is likely that at least some of the false positives originate from mispriming during PCR with the degenerate 10-mers. Second, it is possible that under certain conditions, the lower stringency of differential-display PCR may be more capable of revealing genes that are expressed in relatively high abundance. For example, bands representing abundant smaller gene fragments could be amplified better than those for larger rare genes. Such a situation would not arise owing to sensitivity of detection, since [^{32}P]-label or [^{35}S]-label is incorporated during amplification. However, it could occur when more abundant fragments out-compete rare fragments for primer and undergo more cycles of amplification. In any given preparation, the extent to which genes expressed with low abundance might be missed by differential display is never clear. To avoid missing these genes, complimentary approaches must be employed that achieve cDNA normalization in order to increase the frequency of rare cDNAs while simultaneously decreasing the percentage of abundant cDNAs ***(18)***.

1.3.2. Choice of cDNAs to Pursue from Display Gels

The problem of having too many genes to pursue can be reduced by expression verification and the elimination of false positives. However, it is possible that several display-gel cDNA bands represent different regions of the same mRNA. It is wise to use each selected cDNA as a crosshybridization probe against clones transfected with other cDNAs in order to assess overlap. Secondly, functional definitions for conditions regulating expression can inadvertently be made too broad. For example, if differential displays were to be used to identify genes that are expressed or repressed in response to treatment that stimulates growth, it is possible to select a mixed expression of both primary and secondary genes, depending on the duration of treatment. The extent to which secondary gene expression induced by treatment-regulated primary gene expression will obscure primary expression in the displays can be examined by shortening treatment times and looking for early changes in expression (i.e., treatment-regulated primary gene expression). For example, in the context of treatment-stimulated proliferation, care must be taken to avoid mistaking secondary gene expression related to proliferation *per se* for treatment-regulated primary gene expression because the number of secondary-expression events may overwhelm investigation. By abbreviating the conditions on which comparison is based, the experimental examination would be limited to a concise number of candidate cDNAs that are more likely to be related to gene expression primarily involved in mediating the condition under experimental examination. Lastly, by choosing the largest cDNAs identified on the display gels, the probability will be higher that the sequence extends beyond the 3' untranslated region selected for by the T_{12}MN primers in the PCR reaction. In summary, because differential display can present an unrealistically large number of cDNA bands for consideration, the investigator must be prepared to substantially reduce this number and narrow experimental focus to a small manageable number of cDNAs.

2. Materials

2.1. Reagents

1. Several commercial products are marketed specifically for differential display and are available from GenHunter (Nashville, TN) (RNAimage™ product, RNase-free MessageClean™ kit) and Clontech (Palo Alto, CA) (Delta™ product).
2. Nylon membrane: Colony Plaque Screen™, positively charged (DuPont, NEN, Boston, MA).
3. Denaturing solution: 4 *M* guanidine thiocyanate, 0.5% *N*-lauroyl sarcosine, 0.1 *M* β-mercaptoethanol in 25 m*M* citrate buffer; adjust pH to 6.0.
4. Phenol/chloroform/isoamyl alcohol mixture (25:24:1).
5. Promega (Madison, WI) or Perkin-Elmer (Applied Biosystems, Foster City, CA) PCR buffers and restriction enzymes.
6. Perkin-Elmer Ampli*Taq*.
7. DNA polymerases and TaqStart antibody (Clontech, Palo Alto, CA).
8. QIAEX kit (QIAGEN, Valencia, CA).
9. pCR-Script SK(+) vector and XL1-Blue competent *Escherichia coli* cells (Stratagene, La Jolla, CA).
10. All the reagents necessary for labeling nucleic acids probes are available from Amersham Co. (Arlington Heights, IL).
11. Ambion S1 ribonuclease protection (SNP) kit (Austin, TX).

2.2. Equipment

1. Brinkman (Westbury, NY) polytron sonicator.
2. Thermocycler.
3. Electrophoresis equipment.
4. Autoradiography equipment.

3. Methods

3.1. Differential-Display Procedures

Several commercial products are marketed specifically for differential display (*see* **Subheading 2.**). These products tend to emphasize the display aspect of this approach. Each of these products is capable of yielding different displays. The RNAmap™ product from GenHunter (Nashville, TN) uses anchored T_{12}MN primers for the RT reaction and 10-mer primers in a low-stringency PCR amplification (40°C for annealing). The RNAimage product from GenHunter uses three different 1-base anchored (A, G, C) oligo(dT) primers containing *Hin*dIII and 13-mer primers based on the 10-mers in the RNAmap™ product to which the *Hin*dIII site has been added. The Delta™ product from

Table 1
Experimental Outline

1. Choose primers and conditions, validate primer-display capacity, including reproducibility, reaction success, and microcentrifuge-tube suitability. (A reference source of RNA is useful.) Practice and master sequencing-gel transfers for autoradiography.
2. To keep efforts manageable, identify experimental conditions that are likely to yield a concise number of genes to pursue.
3. Run display gels and prioritize selection of differential bands by considering size and intensity.
4. Reamplify, clone, and screen for the expression of the reamplification fragment.
5. Verify the expression of cloned screen-positive reamplification fragments in originating cells by Northern analysis, SNP, or RT-PCR SNP.
6. Assess uniqueness and overlap with cross-hybridization analysis.
7. Obtain full-length cDNA by library screening or combined 5'- and 3'-RACE reactions.
8. Sequence the full-length cDNAs by primer walking or by making nested deletions.
9. To assess gene function, transfect suitable cell targets with the full-length cDNA.

Clontech makes use of oligo(dT)-primed, single-stranded cDNA synthesis and a combination of nine 29-mer "T" and ten 25-mer arbitrary "P" primers for PCR amplification that utilizes higher stringency annealing (22–25 cycles at 60°C) after three initial cycles annealing at 40°C. The "T" primers contain an 18-nt "anchor" followed by $(dT)_9N_{-1}N_{-1}$, where N_{-1}= A, G, or C. Thus, one product divides cDNA at the RT step and uses subsequent low-stringency PCR amplification, while the other uses nine "T" primers to effectively divide cDNA during PCR amplification under higher stringency conditions. In addition, the Delta™ product incorporates long-distance PCR using a mixture of *Taq* and $Vent_R$ DNA polymerases and TaqStart antibody ***(19,20)***. The higher fidelity long-distance PCR approach can increase yields and lower backgrounds by reducing the number of cycles and thereby prevent overcycling of abundant expression products. The degree to which either of these approaches offers an advantage in rare transcript sensitivity, expression-proven specificity, or comprehensive utility for detecting all putative transcripts, has not been established.

We present the methodologies that we have found to be effective. An experimental outline for identifying gene expression by differential display is presented in **Table 1**.

1. Harvest tumors and tumor cells.
2. Mince tumor fragments of up to 1 g and rinse with 1X PBS, add 8 mL of prechilled denaturing solution (*see* **Subheading 2.1.**). The denaturing solution is rotated over cells or tumor fragments so that lysis is achieved. Consolidate lysates as 12-mL volumes in 50-mL sterile conical tubes and subject to a final

disruption with a 15–30 s exposure in a Brinkman polytron sonicator set on "high." Each 12-mL preparation receives 2 *M* sodium acetate (1.2 mL, pH 4.0) and is mixed by inversion before adding 12 mL of phenol/chloroform/isoamyl alcohol mixture (25:24:1). After mixing by inversion and shaking vigorously for 10 s, the preparation is chilled on ice for 15 min. The mixture is then transferred to a diethyl pyrocarbonate (DEPC)-treated, thick-walled polypropylene tube and centrifuged for 20 min at 10,000*g* at 4°C. The interface and bottom organic phase containing protein and DNA are avoided, and the top aqueous phase is carefully removed.

3. Prepare total RNA (a minimum of 2 µg total RNA is sufficient for reactions involving all oligo(dT) primers in combination with 20 arbitrary 10-mers) and use it to make cDNA with reverse transcriptase using dinucleotide poly T primers. All materials are treated with 0.05% DEPC at room temperature for 1 h to destroy ribonucleases and then autoclaved to inactivate DEPC. Precipitate RNA by adding an equal volume of isopropanol for an overnight exposure at –20°C and pellet by centrifugation (10,000*g*, 15 min, 4°C). The RNA is resuspended in 5 mL of cold denaturing solution until it has dissolved, and an equal volume of isopropanol is added to again precipitate the RNA, as already described (brief heating to 65°C may be needed to completely dissolve the pellet). The RNA is again pelleted by microcentrifugation, washed with iced 75% ethanol (20 mL) and recentrifuged. The ethanol is removed, the pellet air-dried for 5–10 min, and the RNA then resuspended in 1–3 mL of RNase-free deionized water and stored at –20°C for up to 3 wk before use. Quality of these preparations has been satisfactory when A_{260}/A_{280} ratios have been >1.7. Whenever A_{260}/A_{280} ratios are <1.7, additional phenol/chloroform extraction is performed.
4. Because it is essential that the total RNA preparation be absolutely free of any DNA contamination, so that it accurately represents only mRNA available to the differential display, subject the total RNA preparation to DNase treatment (1 U/50 µL vol containing 50 µg RNA) using RNase-free MessageClean™.
5. After DNase treatment, extract the preparation with 40 µL of phenol/chloroform (3:1) to remove residual proteins, centrifuge, and collect the supernatant. The RNA is then precipitated by adding 5 µL 3 *M* Na acetate and 200 µL 95% ethanol, 5% DEPC-treated distilled water and kept overnight at –80°C. After centrifugation, the RNA is redissolved in 50 µL of DEPC water and 2 µL (0.2 µg) of DNA-free total RNA is used for reverse transcription.
6. Determine concentration at OD_{260} and adjust to 0.1 µg/µL with DEPC-treated water just before use (*see* **Note 1**).
7. Run the reverse transcription of mRNA with RNAmap reagents in 0.25-mL PCR tubes using anchored oligo(dT) primers degenerate at the second base from the 3' end, in a reaction mixture (19 µL) containing each deoxyribonucleoside triphosphate (dNTP) (250 µ*M*, 1.6 µL), $T_{12}MN$ (10 µ*M*, 2 µL), RNA (2 µL, 0.1 µg/µL DNA-free total RNA), reverse transcriptase buffer (5X, 4 µL), and water (9.4 µL). The thermocycler is programmed as follows: 65°C, 5 min; 37°C, 60 min; 95°C, 5 min, 4°C. After 10 min at 37°C, add 1 µL Moloney murine leukemia virus (MMLV) RT RNAmap reagent (100 U/µL) to each tube. The tubes

are spun briefly after reverse transcription to collect condensation, set on ice for immediate use or stored at –20°C for later use.

8. Run the PCR in a 20-µL reaction volume containing: dNTP (25 µ*M*, 1.6 µL), $T_{12}MN$ (10 µ*M*, 2 µL), one of the six amplification primers (AP) (2 µ*M*, 2 µL), reverse transcriptase mix containing the cDNA (from **step 7** above, with appropriate matching $T_{12}MN$ added earlier [2 µL]), PCR buffer (10X, 2 µL supplemented with $MgCl_2$ to a final concentration of 1.5 m*M* [α-^{32}P])-dATP (1200 Ci/mmol, 1 µL), Perkin-Elmer Ampli*Taq* (0.2 µL), and water (9.2 µL). Reagents are RNAmap unless otherwise specified (*see* **Note 2** for AP primers).

   Mix reactants well by pipeting up and down and cap with 25 µL of mineral oil, unless thin-walled reaction tubes are used. The thermocycler is programmed at 95°C for 5 min followed by 40 cycles as follows: 94°C, 45 s; 40°C, 2 min; 72°C, 5 min. Total time on a rapid cycler using thin-walled reaction tubes is ~3 h (*see* **Note 3**).

9. Combine the PCR samples (3 µL) with 98% deionized formamide, containing 10 m*M* EDTA, pH 8.0, 0.025% xylene cyanol FF, 0.025% bromophenol blue (3 µL), and incubate at 90°C for 4 min immediately before loading onto a 6% DNA sequencing gel. Electrophoresis is at 60 W constant power for about 2.5–3 h on an IBI (New Haven, CT) aluminum-block sequencer until the bromophenol blue dye elutes completely out the bottom. Reaction products are visualized for display by film autoradiography at –70°C overnight (*see* **Note 4**).

3.2. Approaches for Reamplifications, Cloning, Screening, and Sequencing of Full-Length cDNAs

Select bands by differential display, isolate, reamplify and clone. Cloned fragments are used to screen cDNA libraries for the full-length cDNA, which are then used for sequencing. Full-length cDNAs are then transfected to appropriate target cells to determine if they have an active function in observed responses to treatment.

1. Select bands of interests from the display gels, reamplify, verify for expression in the originating cells by Northern analysis, slot blotting ***(15)***, or the ribonuclease protection assay, and clone (*see* **Notes 5** and **6**).
2. Use composite anchor primers and arbitrary primers in amplifications at 0.4 µ*M* for one low-stringency round (i.e., 94°C for 1 min for denaturing; 40°C for 4 min for low-stringency annealing; 72°C for 1 min for extension), followed by 35 high-stringency rounds (94°C for 45 s for denaturing; 60°C for 2 min for high-stringency annealing; 72°C for 1 min for extension). Reamplification is with 35 high-stringency cycles.

3.3. Reamplification of cDNA

1. Dry the DNA sequencing gel on Whatman 3 MM paper and use radioactive ink and needle punches for orienting the autoradiogram to the originating gel. Precise alignment is achieved using the "border mixture" edge-defining approach

(21). The film is developed after overnight exposure at room temperature, and the autoradiogram is oriented with the landmark needle punches and ink labels. The bands of interest are located and cut out directly through the film.

2. Soak the gel and paper for 10 min in 100 µL H_2O and then boil for 15 min in a parafilm-sealed microfuge tube. After centrifuging the paper and gel down, transfer the supernatant to a fresh microfuge tube and precipitate with 5 µg glycogen, 0.3 *M* Na acetate, and 4.5 vol of ethanol at –80°C. Wash the precipitate with 85% ethanol and resuspend in 10 µL of H_2O. Alternatively, DNA can be recovered by electroelution.
3. Perform reamplification with the originating AP primer set under the originating PCR conditions in a 40-µL final volume containing: dNTP (250 µ*M*, 3.2 µL), composite T_{12}MN (10 µ*M*, 4 µL), one of the six composite AP primers (2 µ*M*, 4 µL), the cDNA template (from **step 7, Subheading 3.1.**, with appropriate matching T_{12}MN added earlier, 4 µL), PCR buffer (10X, 4 µL), Perkin-Elmer Ampli*Taq* (0.4 µL) and distilled, nuclease-free water (20.4 µL) (*see* **Note 7**). Reagents are RNAmap unless otherwise specified.
4. After the first round of PCR, use 4 µL of the PCR sample as template for the second round of PCR using identical conditions. PCR samples from both rounds (20–30 µL) are run on 1.5% agarose gels and stained with ethidium bromide. Purify the PCR reamplification products on sequencing gels, elute, extract with a QIAEX (QIAGEN, Valencia, CA) kit and store at –20°C. The sizes of the reamplified PCR products are checked on 6% sequencing gels to make sure they agree with the size of the originating bands. Since the reamplification product is often contaminated with fragments that are not visualized by or related to the originating differential display, it is prudent to screen for *bona fide* differential expression of the reamplification fragment. This screen can be accomplished by dividing [^{32}P]-cDNA recovered from the display gel and using half for reamplification, with the remainder saved as a blotting probe for the reamplification product ***(22)***.
5. Exam cDNA reamplification fragment sequence to assess fidelity of amplification/reamplification by determining the degree to which the cDNA strands are complementary.
6. Purify, reamplify, and clone cDNAs in bands of interest into a suitable plasmid vector. Plasmids obtained from several different clones are spotted onto nylon membrane and hybridized against the [^{32}P]-cDNA that was eluted and saved frozen. Autoradiography is used to reveal the plasmid clones that contain the cDNA in the differentially represented band, while negative hybridizations will occur due to DNA fragments contaminating the selected band. As little as 2000 dpm of [^{32}P]-cDNA can detect positive clones after an overnight exposure of film and membrane.
7. Alternatively, mRNA immobilized on Northern blot membranes can be used to affinity-capture radiolabeled cDNAs, which are then cloned and retested for expression in Northern analyses ***(23)***. Plasmid minipreps are run with plasmids recovered from the lysis supernatant by using QIAprep™ spin columns (QIAGEN, Valencia, CA). The plasmid pellet is resuspended in 20 µL of 10 m*M* Tris-HCl,

pH 8.0 with 1 m*M* EDTA. Solution containing plasmid (0.5 µL, 5–10 ng DNA) is spotted on a dry nylon membrane (Colony Plaque Screen™, positively charged), dipped in 0.5 *N* NaOH/1.5 *M* NaCl, and incubated for 7 min. Following two successive 3 min incubations in 0.5 *M* Tris-HCl, pH 7.5/1.5 *M* NaCl, the membrane is briefly washed twice in standard saline citrate (SSC) and dried. The membrane is then incubated for at least 1 h at 37°C in prehybridization solution (50% deionized formamide, 6X SSC, 0.5% nonfat dry milk). The [^{32}P]-cDNA is solubilized in 20 µL of 0.1 m*M* EDTA, pH 8.0 and heat-denatured for 3 min at 100°C before being quickly chilled on ice. Prehybridization solution (1.25X, 80 µL) is then added to the chilled, solubilized [^{32}P]-cDNA. The resulting hybridization buffer is added to the hybridization reaction after eliminating the prehybridization solution and allowed to hybridize for 16 h at 37°C. The membrane is then washed twice for 15 min in 0.2X SSC, 0.1% SDS and autoradiographed at –80°C with an intensifying screen for 4–16 h. After the reamplified probe has been confirmed as the originating differential-display band, the reamplification product is then verified for expression by Northern blot. Random prime labeling is used to label the cDNA probes for Northern analysis using 10 µ*M* T_{12}MN primer to improve signal. For Northern analysis, standard prehybridization and hybridization are run at 42°C. Blots are washed with 1X SSC, 0.1% sodium dodecyl sulfate (SDS) at room temperature for 15 min (two times) followed by washing with 0.25X SSC, 0.1% SDS at 58°C for 15 min. Exposure is with intensifying screens at –80°C for 12 h to 14 d (*see* **Note 8**).

8. Finally, messages that are suspected of being of very low abundance can be detected by RT-PCR ribonuclease protection assay (SNP) ***(24)***. Use two sets of sequence specific primer, one set for test and the other for control message. Perform unidirectional PCR with adapters to get the sense strand. Labeled antisense is made from RNA from cDNA clones to be tested and is hybridized to single-stranded sense cDNAs, followed by using the SNP assay for expression (*see* **Note 8**). Because extended reamplification is subject to some infidelity of the PCR enzyme and because contaminants in the PCR reaction can affect sequencing results, the reamplified PCR products should be cloned for sequencing. Selected PCR products can be blunt-end cloned ***(25)*** into the pCR-Script SK(+) vector (Stratagene, La Jolla, CA) by making use of the *Srf*I restriction enzyme to reduce nonrecombinant ligation of the vector to itself. Although it is unlikely that the rare sequence recognized by *Srf*I will be present in the PCR product, by including 5'-methyl deoxycytidine triphosphate (dCTP) in the PCR reaction, it is possible to insure that only *Srf*I sequences in the vector are cleaved. An aliquot (2 µL) of the PCR product is added to 50 ng of *Srf*I-digested pCR-Script SK(+) plasmid DNA (Stratagene) in a 10 µL volume containing 1 µL of 10X Universal buffer and ATP-ribose (rATP) (0.5 m*M* final concentration). T4 DNA ligase (4 U) and *Srf*I are then added. Ligation reactions are done at room temperature for 1 h and terminated by heating at 65°C for 10 min. An aliquot of the ligation reaction (2 µL) is used to transform 100 µL of thawed XL1-Blue competent *E. coli* cells (Stratagene) for 30 min on ice, followed by a heat pulse for 45 s at 42°C.

The cells are then placed on ice for 2 min and 1 mL of SOC media (Tryptone, 2%; yeast extract, 0.5%; NaCl, 0.05%; KCl, 2.5 m*M*; MgCl, 10 m*M*; glucose 20 m*M*) is added. The cells are aerated for 20 min at 37°C. A 100 µL aliquot of the transformed cells is then plated and colonies identified through β-galactosidase (–) white phenotype on plates spread with 20 µL of 10% X-gal (in dimethylformamide) and 20 µL of 200 m*M* isopropyl thiogalactose (IPTG). The cloned PCR product insert is recovered as *Not*I-*Bam*H1 fragments (*see* **Note 9**).

9. For Northern analysis, total RNA (10 µg) from the originating cell lines is separated by electrophoresis on 1% agarose/2.2 *M* formaldehyde gels, blotted, and immobilized onto nylon membranes (Duralon-UV, Stratagene, LaJolla, CA). The *Not*I-*Bam*H1 fragments of the cloned cDNA (or of the reamplification PCR product) are labeled with [α-^{32}P]-dCTP (NEN, Boston, MA) using a random priming kit (BRL, Gaithersburg, MD). Blots are prehybridized with 5X SSPE, 10X Denhardt's solution, 50% formamide, 0.1 mg/mL sonicated denatured salmon sperm DNA for 18 h at 42°C. Hybridization is performed in the same solution applying 2–5 × 10^7 cpm labeled probes. As a control for RNA loading and transfer, membranes are stripped and rehybridized with a labeled oligonucleotide (5'-CGGCATGTATTAGCTCTAGAATTACCACAG-3') complementary to 18S ribosomal RNA.

3.4. Obtaining Full-Length cDNAs: An Overview

Two different approaches can be used for obtaining full-length cDNAs to be used for sequencing: Partial cDNAs can be used to screen cDNA libraries containing full-length transcripts; and combination 5'- and 3'-rapid amplification of cDNA ends (RACE) can be used in extended amplifications to synthesize full-length double-stranded (ds) cDNAs.

3.4.1. Library Screening

A suitable and complete cDNA library will represent each mRNA in the cell, especially those for the cDNAs being screened for. Using an estimate of 34,000 different mRNAs in a population of 500,000 mRNA molcculcs per cell with the eight molecules comprising the rarest mRNA, there will be a 99% probability that a library of 500,000–1,000,000 independent cDNA clones should contain at least one copy of every mRNA. The mRNA source of library cDNA should originate from the cells used to obtain the differential display.

1. Ligate double-stranded cDNA (1 µg) with noncomplementary *Bst*XI ends to an equimolar plasmid-vector DNA (CDM8 from Invitrogen [San Diego, CA] or pCNTR from 5 Prime→3 Prime, Inc., Boulder, CO as appropriate), linearized at the two *Bst*XI sites and purified by electrophoresis. Equimolar vector and insert are joined with T4 DNA ligase (20 U/mL overnight at 4°C) and used to transform twenty 100-µL aliquots of competent cells (MC1061/P3 *E. coli*, Invitrogen, San Diego, CA) after verifying efficiency to at least 1 × 10^8 colonies/µg supercoiled pUC18 DNA.

2. Grow pooled transformed cells for 1 h at 37°C with shaking, and plate overnight at 37°C to 10–20 15-cm agar plates containing antibiotics for selection. Adequate library complexity, determined by manual counting, should target 0.5–2 × 10^6 colonies/µg input cDNA. The bacteria are eluted with LB medium and divided into a portion representing the unamplified library to be used after alkaline lysis/CsCl purification, if needed to regenerate the originating library, and a portion frozen in 0.2 volume of 80% glycerol as multiple aliquots to be used for inoculating liquid culture to prepare additional library DNA (*see* **Note 10**).
3. Screen a library as follows. A bacterial suspension is slowly suctioned through a level porous nitrocellulose membrane, leaving bacteria bound to the membrane surface. Transfer to an agar plate, bacteria side up, and the bacteria are allowed to form colonies into the agar. Plates are incubated with the agar side up at 37°C until colonies are ~1 mm across. Replica filters are wetted and positioned on the bacterial lawn of the originating filter and pressed hard on 3 MM paper with a glass plate. Holes for orientation are punched with a 20-gage needle. The filters are separated and placed on different agar plates bacteria side up. The replica colonies are grown up at 37°C (overnight) and the library filters stored at 25°C and then at 4°C on agar until screening is complete. Regrowth of colonies on the library filter (2–4 h, 37°C) is used for multiple replica filters. Plasmids on the replica filter are amplified to increase signal for hybridization by transferring them to an Luria Bertani (LB) plate containing 50 µg/mL chloramphenicol (4–10 h, 37°C). Two replica copies are made for hybridization to each probe. Replica filters are removed from the LB/chloramphenicol plates and placed bacteria side up on 3 MM paper soaked with 0.5 *M* NaOH (5 min) before careful transfer to 3 MM paper soaked with 1 *M* Tris-HCl, pH 7.5 (5 min), and then transfer to 0.5 *M* Tris-HCl, pH 7.5/1.25 *M* NaCl (5 min). Filters are allowed to dry on a dry sheet of 3 MM paper, baked in an oven (80°C, 90 min) before using with nick-translated probes ***(26)***.
4. Perform screening *per se* by hybridizing pre-wet filters in a sealable plastic bag. Filters are prehybridized with hybridization solution containing 50% formamide (1 h, 42°C). [^{32}]P-cDNA is obtained by radiolabeling one of the cloned inserts (at least 10^7 cpm/µg is used). Radioactive probes (1–15 ng/mL hybridization reaction volume) are boiled for 10 min in 1 mL containing 2 mg sonicated herring sperm and added to the prehybridized filters at 42°C overnight. Nonhybridized probe is washed away with low-stringency wash buffer. Starting with high-stringency wash buffer warmed to 40°C and increased at 5°C increments until background for autoradiography is low enough (estimated with a Geiger counter), filters are washed to remove mismatches and nonsequence specific interactions. Filters are mounted to used X-ray film plastic backing, covered with plastic wrap, marked with radioactive ink for alignment purposes, and then exposed using intensifying screens as needed (1–20 h). Autoradiograms are oriented to matching agar plates and positive colonies picked with a sterile toothpick that is rinsed into a microfuge tube containing 1 mL LB medium including antibiotic to which the plasmid confers resistance. The bacterial suspension is plated onto an

LB plate (+antibiotic) at a dilution that gives 25–250 clones/100 mm plate and is allowed to grow overnight. A nitrocellulose-filter replica of the bacterial lawn is made, denatured, renatured, and hybridized. The most isolated, positive colonies are selected from the secondary plate and grown up to isolate DNA for sequencing. Plasmid inserts are selected on the basis of longest length in agarose electrophoresis. Expression in the originating cell line is confirmed by Northern blot.

3.4.2. Combined 5'- and 3'-RACE

Advantages and limitations of RACE methodologies have recently been reviewed ***(27)***.

1. Cloned cDNA inserts that have been verified for expression in originating cells can be sequenced to obtain partial sequence information for use in 5'- and 3'-RACE reactions with extended PCR to amplify full-length cDNAs (*see* **Note 11**). Sequence information is used to design two gene-specific primers with at least 16-nt overlap.
2. The single-stranded cDNA selected from differential display is made double-stranded using RNase H, DNA polymerase I, and DNA ligase, and then blunt-ended with T4 DNA polymerase. T4 DNA ligase is used to add an annealed adaptor sequence. The gene-specific primers are then used with a partially double-stranded adaptor and complimentary adaptor primer to amplify both 5'- and 3'-RACE reactions. The adaptor primer is designed to be colinear with the single-stranded region of the adaptor so that a binding site for the adaptor primer is not contained in adaptor-ligated cDNA, but is introduced by extension of the gene-specific primer. In addition, an amino group is used to prevent extension of the 3' end of the adaptor-ligated cDNA in order to avoid nonspecific amplification. During the first round of PCR, only the inner gene-specific primer is extended creating an adaptor-primer site at the 5' or 3' terminus and allowing specific exponential amplification in subsequent cycles.
3. Next, the 5'- and 3'-RACE fragments are gel-purified and combined in the absence of primers so that the overlapping regions of the 5' and 3' fragments anneal and are extended by a long-distance PCR enzyme mix in 5–40 cycles of thermal cycling to generate the full-length cDNA product. To obtain full-length, fully double-stranded cDNA, long-distance PCR is performed with the adaptor primer and the cDNA synthesis primer used to make the first cDNA strand in the RT reaction. A commercially available product can be used for the 5'- and 3'-RACE reactions (Marathon cDNA amplifications, Clontech Laboratories, Palo Alto, CA). The amplified full-length product can be cloned using restriction sites in the adaptor. Verification of full-length should be done by determining the true 5' end using RNase-protection assays, primer-extension assays, and cDNA or genomic-sequencing information.

3.5. Sequencing Full-Length cDNAs: An Overview

1. Sequence determination is performed on partial fragments (<300–400 nt) of the full-length cDNA. Templates for sequencing are prepared either by heat denaturation or by the proteinase K method ***(28,29)*** after miniprep purification using QIAEX (QIAGEN, Valencia, CA) column kits.

2. Because the full-length cDNAs are too large for sequencing in a single step, two strategies for sequencing portions of the full-length cDNA can be employed. The first strategy for sequencing is "primer walking," which is well-suited to dideoxy DNA sequencing and bypasses the need for subcloning smaller pieces of DNA. With this approach, initial sequence information is obtained using a vector-based primer. As a new sequence is determined, an oligonucleotide is synthesized that hybridizes near the 3' end of the newly obtained sequence and primes synthesis in a subsequent set of dideoxy reactions ***(30)***. Alternatively, a second strategy involves subcloning by generating an ordered set of smaller DNA molecules by making progressive (nested) sets of deletions from a clone containing the entire DNA fragment. Either the protocol using exonuclease III ***(31)*** or nuclease *Bal*31 can be utilized for generating nested sets ***(32)***.

3.6. Overview of Procedures Associated with Transfection

1. cDNAs identified by differential display can be ligated into the *Hin*dIII site of pCB6. This vector permits constitutively high expression of cDNAs in most mammalian cell lines using the promoter/enhancer from cytomegalovirus (CMV) and contains a neomycin-resistance marker for selection of stably transfected cells. Cohesive *Hin*dIII ends are conferred onto PCR products of known sequence for ligation via oligonucleotide primers ***(33)***. BES (*N,N*-bis[2 hydroxyethyl]-2-aminoethane sulfonic acid)-buffered calcium phosphate can be used to gradually precipitate circular plasmids containing the selected cDNAs into the target cells ***(34,35)***. The strontium-phosphate procedure, as well as the lipofectin procedure can be explored as alternative transfection methods to be used to maximize transfection efficiency ***(36,37)***.
2. Exponentially growing test cells (~5×10^5/10 cm plate) are exposed to plasmid DNA (10, 20, or 30 μg with optimal exposure concentration forming an even, granular precipitate) in 0.1 *M* $CaCl_2$ at 35°C in a 5% CO_2 atmosphere overnight. Plasmid DNA is prepared using lysozyme-Triton lysis with CsCl isolation and phenol/chloroform extraction, followed by ethanol precipitation to avoid toxic contaminants in column prep isolations. The cells are then split 1:4 and grown overnight before initiating neomycin selection conditions. Cells that demonstrate growth after at least three passages are examined for expression of the appropriate mRNA by Northern analysis, slot blotting, or S1 nuclease protection, as already described. The functional properties of the cells are examined relative to the expression of the transfected cDNA.

4. Notes

1. Diluted RNA is very unstable, requiring that concentration be adjusted to 0.1 μg/μL just before use. Recent work has reported that the use of oligo(dT) magnetic beads to isolate mRNA can reduce the preparation time involved in performing differential display ***(38)***. We have found that the Oligotex™ Combi Kit (cat. no. 79990, QIAGEN, Valencia, CA) can be used to prepare mRNA that is suitable for differential display directly from tumors and tumor cells. For steps involving

RT and PCR amplification reactions, microcentrifuge tubes should be siliconized to minimize losses of material during transfer steps since cDNA is present in picogram quantities before reamplifications. Microcentrifuge-tube type is also a consideration with certain tubes having almost a 90% rate of failed reactions ***(39)***. Although thin-wall tubes generally have good rates of success, reaction failures of 4–10% occur so that replicate reactions should be anticipated and tubes pretested before experimental use.

2. AP primers that have been used are designated as follows:
   AP-1: 5'-AGCCAGCGAA-3', AP-2: 5'-GACCGCTTGT-3',
   AP-3: 5'-AGGTGACCGT-3', AP-4: 5'-GGTACTCCAC-3',
   AP-5: 5'-GTTGCGATCC-3', and AP-6: 5'-GCAATCGATG-3'.
3. Recent work has reported that the total time for a standard reaction of 40 cycles (~5 h) can be shortened by employing a 1 s denaturation at 94°C ***(40)***.
4. Band sharpening can be accomplished by film exposure at room temperature after drying the gel. Untreated vs treated lanes are run side-by-side so that band appearance or disappearance can be displayed readily. Gene displays are interpreted as reproducible when identical band patterns >150 nt are observed in three repeat preparations.
5. It is essential to establish expression of the selected bands since false positives can easily occur, probably due to mismatching and the low-stringency amplification conditions. We have recently examined the use of primers with extensions, which are used in a single low-stringency amplification followed by multiple high-stringency amplifications ***(41)***. These extensions include an AATT *Eco*RI restriction site that facilitates insertion during cloning and consists of the nucleotides CGGAATTCGG used with the T_{12}MN primers (i.e., 5'-CGGAATTCGGT_{12}MN-3'), and CGTGAATTCG used with the AP-x primers [i.e., CGTGAATTCG + AP-2 (5'-GACCGCTTGT-3') → 5'-CGTGAA TTCGGACCGCTTGT].
6. Comparisons utilizing primers with and without extensions indicate that different displays are obtained in the high- vs low-stringency amplifications (**Fig. 1**). Under low-stringency conditions using primers without extensions, a variability in displays occurs between preparations (**Figs. 2** and **3**: lanes 2 vs 3 vs 4; 16 vs 17 vs 18) and within preparations (**Figs. 2** and **3**: lanes 4, 9, 10, 11; 7, 12, 13, 14; 18, 23, 24, 25; 21, 26, 27, 28). Although low-activity displays can accompany displays with apparent activity in repeat amplifications (**Fig. 2**, lanes 11 and 26), some preparations demonstrate negligible activity in repeated amplifications (**Fig. 3**, lanes 4, 9–11; 18, 23–25).
7. Extended primers used under high-stringency conditions appear to be preferable in reamplification reactions. First, nonextended primers under low-stringency conditions yield less of the expected size fragments (**Fig. 4**, lanes 1–6) than extended primers under high-stringency conditions (**Fig. 4**, lanes 7–14). Smaller size fragments can be present in reamplification gels (**Fig. 4**, lanes 11, 14, 19, and 22) even when extended primers are used with high-stringency conditions. Consequently, reamplification products should be gel-purified to insure that the larg-

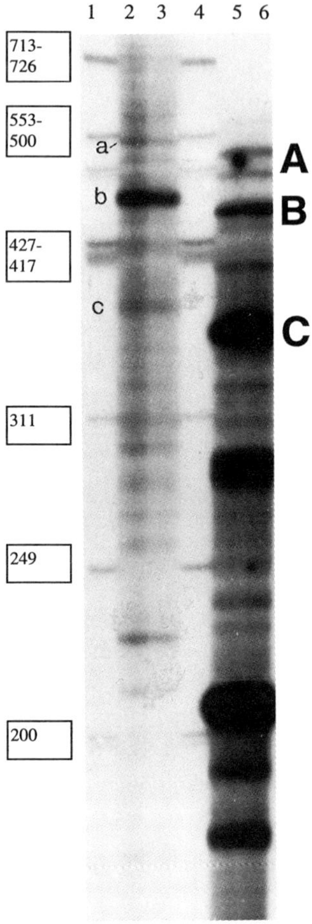

Fig. 1. Comparison between primers with and without extensions in low- and high-stringency amplifications. Total RNA was prepared from MCF-7 human breast cancer cells and used in RT reactions and differential-display amplifications with the following primers: extended T_{12}MC + extended AP-2 (lanes 2–3) or T_{12}MC + AP-2 (lanes 5–6). The ϕx174 *Hinf*I marker is in lanes 1 and 4 and fragment sizes are indicated at the left. DNA was run on a 6% acrylamide sequencing gel containing 8 *M* urea in 1X TBE at 70 W.

est fragments are used in subsequent expression tests and cloning. Second, expression signals in Northern analyses were less frequent when reamplification was with nonextended primers under low stringency (**Fig. 4**, bottom, left six lanes) than when extended primers were used at high stringency (**Fig. 4**, bottom, right six lanes). Our efforts support the contention that use of extended primers at higher stringency improves the reamplification reaction and the opportunity to assess expression. Regional alignments were made for the differential-display

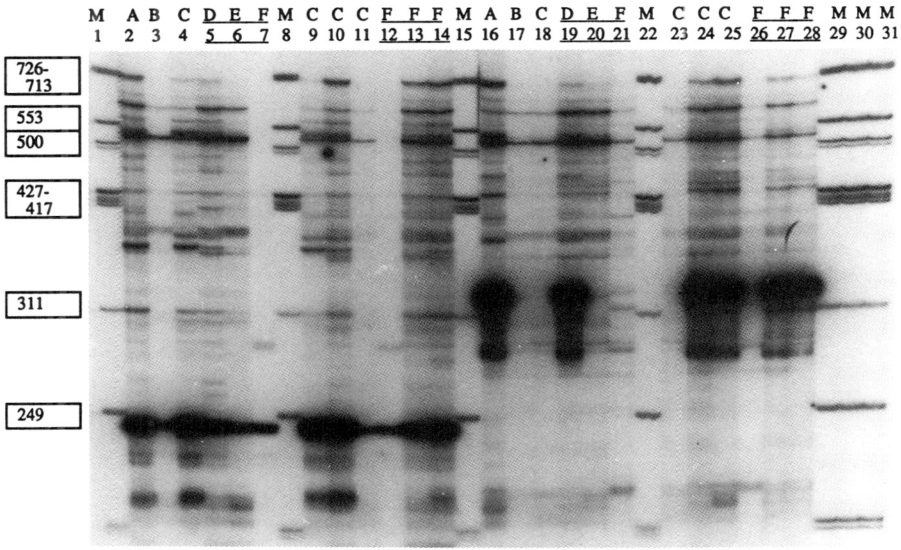

Fig. 2. Reproducibility of AP-1 and AP-2 primers without extensions in low-stringency amplifications. Total RNA was prepared from BG-1 human ovarian cancer cells and used in RT reactions and differential-display amplifications with the following primers: $T_{12}MG$ + AP-1 (lanes 2–12, 9–14) or + AP-2 (lanes 16–21, 23–28). Six different cell flask preparations (A–F) were utilized with A–C grown in the absence of estradiol and D–F grown for 3 d in the presence of 50 n*M* estradiol (underlined). Flasks C and F were subjected to replicate amplifications as shown in lanes 9–11, 12–14, 23–25, and 26–28. Sequencing gel (6%) containing 8 *M* urea was run at 70 W. The ϕx174 *Hinf*I marker is in lanes marked with a "M," and fragment sizes are indicated at the left.

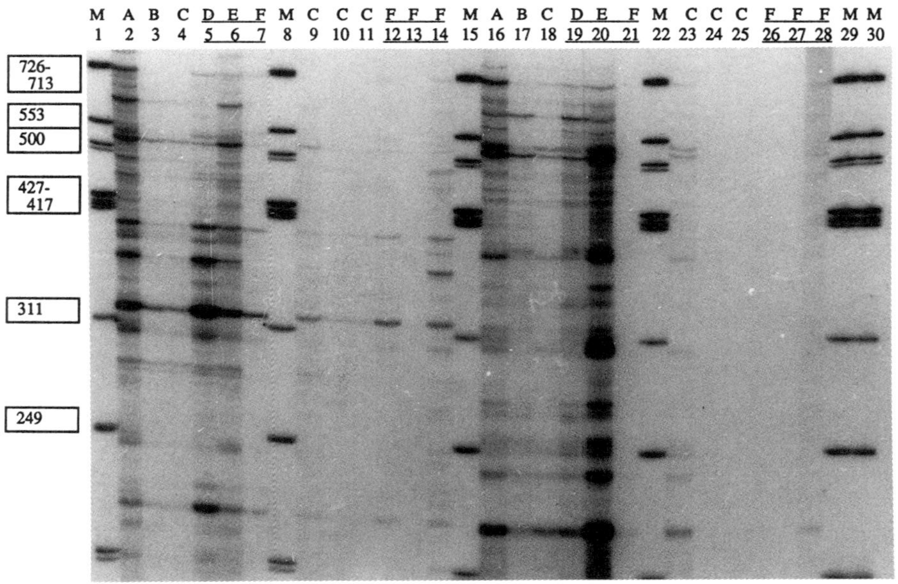

Fig. 3. Reproducibility of AP-3 and AP-4 primers without extensions in low-stringency amplifications. Total RNA was prepared from BG-1 human ovarian cancer cells and used in RT reactions and differential-display amplifications with the following primers: $T_{12}MG$ + AP-3 (lanes 2–12, 9–14) or + AP-4 (lanes 16–21, 23–28). Conditions and designations are as described in **Fig. 2**.

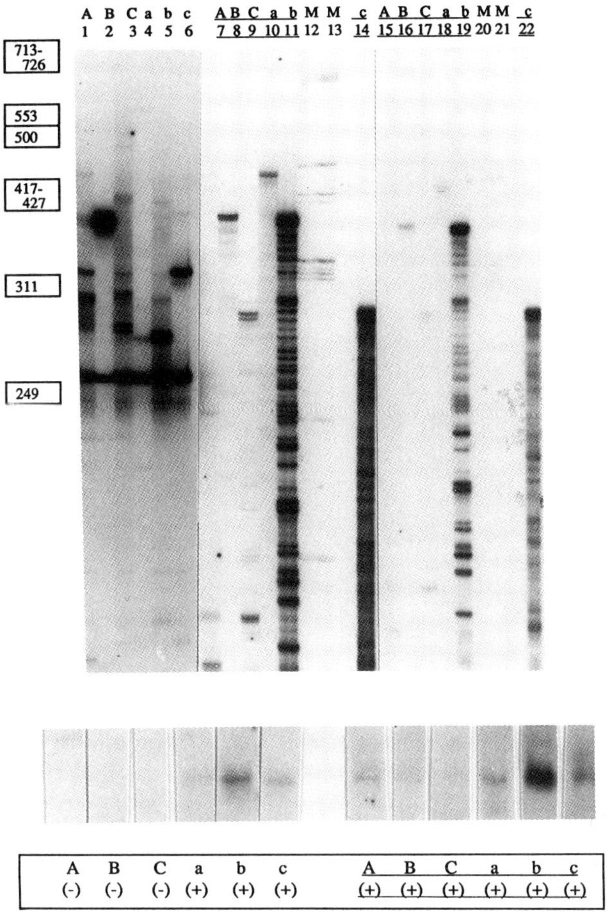

Fig. 4. Reamplifications using primers with and without extensions in low- and high-stringency amplifications. Bands identified in **Fig. 1** as A, B, C, a, b, and c were eluted from the display gel and reamplified with T_{12}MC + AP-2 (lanes 1–6) or with extended T_{12}MC + extended AP-2 (lanes 7–22). Lanes 15–22 were exposed for less time than lanes 7–14, representing the same gel. Bottom: Reamplification bands were gel-purified and used in Northern-gel expression tests. Northern gels show the result of reamplification of all six primary bands with T_{12}MC + AP-2 (left six lanes) or with extended T_{12}MC + extended AP-2 (right six lanes) and are marked positive (+) or negative (–) for expression signals on the originating film records.

band A in **Fig. 1** reamplified with T_{12}MC and are listed below in clustered order. Alignment has been made against the inverse complement of the cDNA amplified with the AP primer.

```
T12 Strand     1                          5'-tATTTAT      acacAAAACGTGC ACATGA
                                              ||||||          ||||||||| ||||||
INV. COMP      1 gggggagaattgtttttgcagtttttaacATTTATttaaaacagaAAAACGTGCaACATGA

T12 Strand    27 GCTGCCTACTCATTTTctcTCACTGCGCAccCtgGGCATTGGGGTTGGTGACTCTGATGGC
                 ||||||||||||||||   ||||||||||  |  |||||||||||||||||||||||||||
INV. COMP     62 GCTGCCTACTCATTTTtctTCACTGCGCA gCctGGCATTGGGGTTGGTGACTCTGATGGC

T12 Strand    88 CAaTTGGGCAGCTCTTTCCACGAaGGCTTTGCGGTTCTTGGAaGAAACATTGTGAaCGAaC
                 || |||||||||||||||||||| |||||||||||||||||| |||||||||||| ||| |
INV. COMP    122 CAgTTGGGCAGCTCTTTCCACGAtGGCTTTGCGGTTCTTGGAgGAAACATTGTGAgCGAtC

T12 Strand   149 TCcGCACAGTAAGAaTTGTTGCACATCAaCAaCACTTCCAaCTCCTTGACGTTGTGGACCA
                 || ||||||||||| ||||||||||||| || |||||||| ||||||||||||||||||||
INV. COMP    183 TCgGCACAGTAAGAtTTGTTGCACATCAgCAgCACTTCCAgCTCCTTGACGTTGTGGACCA

T12 Strand   210 GGAACTTCCGGAAGCCACTGGGCAaCATGTtCcTTGTTTTTTTGTTGCTTCCATAACCAAT
                 |||||||||||||||||||||||| ||||| | ||||||||||||||||||||||||||||
INV. COMP    244 GGAACTTCCGGAAGCCACTGGGCAgCATGTgCtTTGTTTTTTTGTTGCTTCCATAACCAAT

T12 Strand   271 GTTGGGCATCAAGAaCTGGCCCTTGAATCTTCTACaAACCCTGTTGTCAATGCCTCTGGGT
                 |||||||||||||| |||||||||||||||||||| |||||||||||||||||||||||||
INV. COMP    305 GTTGGGCATCAAGAtCTGGCCCTTGAATCTTCTACgAACCCTGTTGTCAATGCCTCTGGGT

T12 Strand   332 TTCCcGCCAGTTACGCcTAATTTTGACATATCGGTCTGACTGGTGCCGGATGAACTTCTTG
                 |||| ||||||||||| ||||||||||||||||||||||||||||||||||||||||||||
INV. COMP    366 TTCC GCCAGTTACGCtTAATTTTGACATATCGGTCTGACTGGTGCCGGATGAACTTCTTG
T12 Strand   393 GTTCTCTcTTTGAaaaAactTGGGcTtccaaacgcatc
                 ||||||| |||||   |   |||| |
INV. COMP    426 GTTCTCTtTTTGA cgAtggTGGGtTc
```

Using each cDNA as an inverse complement, matching was 86–92% over the entire strand sequence and 93% when unmatching strand ends were excluded. The degree of fidelity may be adequate for gaining preliminary identification of cDNA sequences. As shown in the following, the cDNA in band A of **Fig. 1** was sequenced using the AP-2 primer (451 nt) and was found to have some homology to the human mRNA for ribosomal protein L32 (GenBank locus HSRPL32, accession #: X03342, bases 1–505) ***(42)***, having a residue identity of 95% with 433 matches, 13 mismatches, 6 gaps, and no conservative substitutions.

This match was confirmed by the strand sequence using the T_{12}MC primer, showing a residue identity of 92% with 398 matches, 28 mismatches, 4 gaps, and no conservative substitutions. Thus, the possibility of observing preliminary DNA data-bank matches to reamplified cDNA prior to cloning is reasonable.

8. Very abundant messages may need to be exposed for shorter times to obtain good film images (15 min to 5 h). The time required to expose a film to obtain signal verification can be used as a relative indication that the mRNA is abundant or rare. When signals are not observed by Northern analysis or slot blots, sensitivity may be insufficient and the increased sensitivity of the S1 ribonuclease protection assay (SNP assay) can be utilized. SNP assays can be run with a kit from Ambion (SNP kit, cat. no. 1425).
9. Recently, we have utilized the PCRTrap cloning system from GenHunter and the pZErO-1 system from Invitrogen and have obtained many fewer false positives. Moreover, both the PCRTrap and pZErO-1 systems utilize PCR-based screening for insertion, which tremendously simplifies the screening process. Significant improvements in through-put and performance are associated with both systems.

```
X        10        20        30        40        50        60        70
GAA-CCCACCATCGTCAAAAAGAGAACCAAGAAGTTCATCCGGCACCAGTCAGACCGATATGTCAAAATTAA
||| ||||  ||||||||||||||||||||||||||||||||||||||||||||||||||||||||||||||
GAAGCCCAAGATCGTCAAAAAGAGAACCAAGAAGTTCATCCGGCACCAGTCAGACCGATATGTCAAAATTAA
 60        70        80        90       100       110       120

       80        90       100       110       120       130       140
GCGTAACTGGCGGAAACCCAGAGGCATTGACAACAGGGTTCGTAGAAGATTCAAGGGCCAGATCTTGATGCC
||||||||||||||||||||||||||||||||||||||||||||||||||||||||||||||||||||||||
GCGTAACTGGCGGAAACCCAGAGGCATTGACAACAGGGTTCGTAGAAGATTCAAGGGCCAGATCTTGATGCC
130       140       150       160       170       180       190       200

    150       160       170       180       190       200       210
CAACATTGGTTATGGAAGCAACAAAAAAACAAAGCACATGCTGCCCAGTGGCTTCCGGAAGTTCCTGGTCCA
||||||||||||||||||||||||||||||||||||||||||||||||||||||||||||||||||||||||
CAACATTGGTTATGGAAGCAACAAAAAAACAAAGCACATGCTGCCCAGTGGCTTCCGGAAGTTCCTGGTCCA
      210       220       230       240       250       260       270

  220       230       240       250       260       270       280
CAACGTCAAGGAGCTGGAAGTGCTGCTGATGTGCAACAAATCTTACTGTGCCGAGATCGCTCACAATGTTTC
||||||||||||||||||||||||||||||||||||||||||||||||||||||||||||||||||||||||
CAACGTCAAGGAGCTGGAAGTGCTGCTGATGTGCAACAAATCTTACTGTGCCGAGATCGCTCACAATGTTTC
    280       290       300       310       320       330       340
290       300       310       320       330       340       350
CTCCAAGAACCGCAAAGCCATCGTGGAAAGAGCTGCCCAACTGGCCATCAGAGTCACCAACCCCAATGCCAG
||||||||||||||||||||||||||||||||||||||||||||||||||||||||||||||||||||||||
CTCCAAGAACCGCAAAGCCATCGTGGAAAGAGCTGCCCAACTGGCCATCAGAGTCACCAACCCCAATGCCAG
  350       360       370       380       390       400       410
360       370       380       390       400       410       420       430
GCTGCGCAGTGAAGAAAAATGAGTAGGCAGCTCATGTTGCACGTTTTTCTGTTTTAAATAAATGTTAAAAAC
|||||||||||||| ||||||||||||||||||||| |||||| ||||||| |||||||||||| |||||||
GCTGCGCAGTGAAG-AAAATGAGTAGGCAGCTCATG-TGCACG-TTTTCTG-TTTAAATAAATG-TAAAAAC
420       430        440       450         460        470        480
      440       450
TGCAAAAACAATTCTCCCCC
||| |       ||| ||
TGCCATCTGGCATCTTCCTT
   490       500   X
```

The presence of the cloned PCR product in cell lysates should be verified by Northern blot analysis, slot blots, or the SNP assay.

10. The library is evaluated by: (a) screening a single plating with cDNA used to construct the library, expecting hybridization in at least 50% of the clones; and (b) screening for ~1-kb or larger inserts in 10/12 recombinant clones in DNA minipreps digested with *Eco*RI that are run on 1.5% agarose gels. The plasmid library approach yields for the most part full-length clones with the opportunity to efficiently identify cDNAs encoding proteins up to 150 kDa.
11. It has been reported that cDNAs can be extracted directly from the display gels, ligated into a PCR fragment cloning vector (pCRII, Invitrogen) and then directly reamplified using a T7 promoter primer and a poly dT primer adjacent to the cloning site and sequenced using either a T7 promoter primer or a pCRII polylinker primer ***(43)***. Recent work reports that uncloned cDNA can be reamplified with an extended primer set that renders the amplified DNA suitable for direct sequencing using either of the extended primers ***(44)***. Sequence information is used to design two gene-specific primers with at least 16-nt overlap.

References

1. Liang, P. and Pardee, A. B. (1995) Differential display of eukaryotic messenger RNA by means of the polymerase chain reaction. *Science* **267,** 1186,1187.
2. Liang, P., Averboukh, L., Keyomarsi, K., Sager, R., and Pardee, A. B. (1992) Differential display and cloning of messenger RNAs from human breast cancer versus mammary epithelial cells. *Cancer Res.* **52,** 6966–6968.
3. Zou, Z., Anisowicz, A., and Hendrix, M. J. (1994) Maspin, a serpin with tumor-suppressing activity in human mammary epithelial cells. *Science* **263,** 526–529.
4. Jensen, R. A., Page, D. L., and Holt, J. T. (1994) Identification of genes expressed in premalignant breast disease by microscopy-directed cloning. *Proc. Natl. Acad. Sci. USA* **91,** 9257–9261.
5. Liang, P., Averboukh, L., Zhu, W., and Pardee, A. B. (1994) Ras activation of genes: Mob-1 as a model. *Proc. Natl. Acad. Sci. USA* **91,** 12,515–12,519.
6. Mok, S. C., Wong, K. K., Chan, R. K., Lau, C. C., Tsao, S. W., Knapp, R. C., and Berkowitz, R. S. (1994) Molecular cloning of differentially expressed genes in human epithelial ovarian cancer. *Gynecol. Oncol.* **52,** 247–252.
7. Kar, S. and Carr, B. I. (1995) Differential display and cloning of messenger RNAs from the late phase of rat liver regeneration. *Biochem. Biophys. Res. Commun.* **212,** 21–26.
8. van Groningen, J. J., Bloemers, H. P., and Swart, G. W. (1995) Identification of melanoma inhibitory activity and other differentially expressed messenger RNAs in human melanoma cell lines with different metastatic capacity by messenger RNA differential display. *Cancer Res.* **55,** 6237–6243.
9. Feng, X. H. and Kung, S. D. (1994) Identification of differentially expressed members of tobacco homeobox families by differential PCR. *Biochem. Biophys. Res. Commun.* **198,** 1012–1019.
10. Zimmermann, J. W. and Schultz, R. M. (1994) Analysis of gene expression in the preimplantation mouse embryo: use of mRNA differential display. *Proc. Natl. Acad. Sci. USA* **91,** 5456–6013.
11. Liang, P., Averboukh, L., and Pardee, A. B. (1993) Distribution and cloning of eukaryotic mRNAs by means of differential display: refinements and optimization. *Nucleic Acids Res.* **21,** 3269–3275.
12. Liang, P., Zhu, W., Zhang, X., Guo, Z., O'Connell, R. P., Averboukh, L., Wang, F., and Parde, A. B. (1994) Differential display using one-base anchored oligo-dT primers. *Nucleic Acids Res.* **22,** 5763,5764.
13. Bauer, D., Muller, H., Reich, J., Riedel, H., Arenkiel, V., Warthoe, P., and Srauss, M. (1993) Identification of differentially expressed mRNA species by an improved display technique (DDRT-PCR). *Nucleic Acids Res.* **21,** 4272–4280.
14. Trentmann, S. M., van der Knaap, E., and Kende, K. (1995) Alternatives to 35S as a label for the differential display of eukaryotic messenger RNA. *Science* **267,** 1186,1187.
15. Mou, L., Miller, H., Li, J., Wang, E., and Chalifour L. (1994) Improvements to the differential display method for gene analysis. *Biochem. Biophys. Res. Commun.* **199,** 564–569.

16. Gyllensten, U. (1989) Direct sequencing of in vitro amplified DNA, in *PCR Technology* (Erlich, C. A., ed.), Stockton Press, New York, pp. 45–60.
17. Brow, M. A. D. (1989) Sequencing with *Taq* DNA polymerase, in *PCR Protocols* (Innins, M. A., Gelfand, D. H., Sninsky, J. J., and White, T. J., eds.), Academic Press, San Diego, CA, pp. 189–196.
18. Soares, M. B., Bonaldo, M. F., Jelene, P., Su, L., Lawton, L., and Efstratiadis, A. (1994) Construction and characterization of a normalized cDNA library. *Proc. Natl. Acad. Sci. USA* **91,** 9228–9232.
19. Barnes, W. M. (1994) PCR amplification of up to 35-kb DNA with high fidelity and high yield from lambda bacteriophage templates. *Proc. Natl. Acad. Sci. USA* **91,** 2216–2220.
20. Cheng, S., Fockler, C., Barnes, W. M., and Higuchi, R. (1994) Effective amplification of long targets from cloned inserts and human genomic DNA. *Proc. Natl. Acad. Sci. USA* **91,** 5695–5699.
21. Kim, A., Roffler-Tarlov, S., and Lin, C. S. (1995) New technique for precise alignment of an RNA differential display gel with its film image. *Biotechniques* **19,** 346.
22. Callard, D., Lescure, B., and Mazzolini, L. (1994) A method for the elimination of false positives generated by the mRNA differential display technique. *Biotechniques* **16,** 1096–1103.
23. Li, F., Barnathan, E. S., and Kariko, K. (1994) Rapid method for screening and cloning cDNAs generated in differential mRNA display: application of Northern blot for affinity capturing of cDNAs. *Nucleic Acids Res.* **22,** 1764,1765.
24. Woo, H. H., Brigham, L. A., and Hawes, M. C. (1995) Detection of low-abundance messages by a combination of PCR and ribonuclease protection. *Biotechniques* **18,** 778,779.
25. Scharf, S. J. (1989) Cloning with PCR, in *PCR Protocols* (Innins, M. A., Gelfand, D. H., Sninsky, J. J., and White, T. J., eds.), Academic Press, San Diego, CA, pp. 84–91.
26. Hanahan, D. and Meselson, M. (1983) Plasmid screening at high density. *Meth. Enzymol.* **100,** 333–342.
27. Schaefer, B. C. (1995) Revolutions in rapid amplification of cDNA ends: new strategies for polymerase chain reaction cloning of full-length cDNA ends. *Anal. Biochem.* **227,** 255–273.
28. Krishnan, B. R., Blakesley, R. W., and Berg, D. E. (1991) Linear amplification DNA sequencing directly from single phage plaques and bacterial colonies. *Nucleic Acids Res.* **19,** 1153.
29. Sears, L. E., Moran, L. S., Kissinger, C., Creasey, T., Perry-O'Keefe, H., Roskey, M., Sutherland, E., and Slatko, B. E. (1992) CircumVent thermal cycle sequencing and alternative manual and automated DNA sequencing protocols using the highly thermostable VentR (exo-) DNA polymerase. *Biotechniques* **13,** 626–633.
30. Straus, E., Kobori, J., Siu, G., and Hood, L. (1986) Specific-primer directed DNA sequencing. *Anal. Biochem.* **154,** 353–360.
31. Straus, N. A. and Zagurski, R. J. (1991) In vitro production of large single-stranded templates for DNA sequencing. *BioTechniques* **10,** 376–384.

32. Misra, T. K. (1985) A new strategy to create ordered deletions for rapid nucleotide sequencing. *Gene* **34,** 263–268.
33. Fenstermaker, R. A., Poptic, E., Bonfield, T. L., Knauss, T. C., Corsillo, L., Piskurich, J. F., Kaetzel, C. S., Jentoft, J. E., Gelfand, C., and DiCorleto, P. E. (1993) A cationic region of the platelet-derived growth factor (PDGF) A-chain (Arg159-Lys160-Lys161) is required for receptor binding and mitogenic activity of the PDGF-AA homodimer. *J. Biol. Chem.* **268,** 10,482–10,489.
34. Chen, C. and Okayama, H. (1987) High-efficiency transformation of mammalian cells by plasmid DNA. *Mol. Cell. Biol.* **7,** 2745–2752.
35. Chen, C. and Okayama, H. (1988) Calcium phosphate-mediated gene transfer: a highly efficient system for stably transforming cells with plasmid DNA. *Biotechniques* **6,** 632–638.
36. Neuenschwander, S., Roberts, C. T., and LeRoith, D. (1995) Growth inhibition of MCF-7 breast cancer cells by stable expression of an insulin-like growth factor I receptor antisense ribonucleic acid. *Endocrinology* **136,** 298–303.
37. VanderKuur, J. A. and Wiese, T. (1993) Influence of estrogen structure on nuclear binding and progesterone receptor induction by the receptor complex. *Biochemistry* **32,** 7002–7008.
38. McKendree, W. L., Nairn, C. J., and Bausher, M. G. (1995) Differential display from plant leaves using oligo(dT) magnetic bead mRNA isolation and hot air PCR. *Biotechniques* **19,** 715–719.
39. Chen, Z., Swisshelm, K., and Sager, R. (1994) A cautionary note on reaction tubes for differential display and cDNA amplification in thermal cycling. *Biotechniques* **16,** 1002–1006.
40. Liu, L. Z. and Shearn, A. (1995) Rapid PCR for RNA differential display in a conventional heat block thermal cycler. *Biotechniques* **19,** 44–46.
41. Zhao, S., Ooi, S. L., and Pardee, A. B. (1995) New primer strategy improves precision of differential display. *Biotechniques* **18,** 842–850.
42. Young, J. A. and Trowsdale, J. (1985) A processed pseudogene in an intron of the HLA-DP beta 1 chain gene is a member of the ribosomal protein L32 gene family. *Nucleic Acids Res.* **13,** 8883–8891.
43. Reeves, S. A., Rubio, M. P., and Louis, D. N. (1995) General method for PCR amplification and direct sequencing of mRNA differential display products. *Biotechniques* **18,** 18–20.
44. Wang, X. and Feuerstein, G. Z. (1995) Direct sequencing of DNA isolated from mRNA differential display. *Biotechniques* **18,** 448–453.

V

Miscellaneous Markers and Early Detection of Cancer

25

Transforming Growth Factor Beta

A Plasma Tumor Marker

Feng-Ming Kong, Mitchell S. Anscher, and Randy L. Jirtle

1. Introduction

Malignant tumors have been known for many years to release proteins or polypeptides into the circulation (for review, *see* **ref. *1***). Some of these molecules have biological activity, resulting in endocrinologic manifestations of malignancy referred to as paraneoplastic syndromes. In contrast, others do not cause clinical symptoms, but rather serve as markers to aid in diagnosis, monitoring of tumor response, and selection of patients for adjuvant treatment. In this latter category are prostate-specific antigen (PSA), carcinoembryonic antigen (CEA), β-human chorionic gonadotropin (β-HCG), and CA-125. Tumor markers may be specific for certain tumors, such as PSA for prostate cancer, or more generally expressed, such as CEA.

The cytokine, transforming growth factor beta (TGFβ), functions not only in an autocrine and paracrine manner but also as an endocrine growth factor ***(2–4)***. It is a potent inhibitor of epithelial cell proliferation and plays a central role in normal wound healing as well as abnormal fibrogenesis ***(5–9)***. Cells secrete TGFβ as a biologically inactive latent complex ***(5,8)***, and it also circulates in an inactive form in human plasma bound to the carrier protein α-2-macroglobulin ***(5,9)***. The proteolytic activation of TGFβ by plasmin is facilitated by the binding of the TGFβ latent complex to the mannose 6-phosphate/insulin-like growth factor 2 receptor (M6P/IGF2R) ***(10–12)***. The biological responses of TGFβ are mediated through its binding to three distinct receptors ***(13–15)***. The TGFβ types I and II serine/threonine kinase receptors form a heterodimeric complex upon TGFβ binding to the type II receptor, and a mitoinhibitory signal is then transduced into the cell ***(16)***. In contrast, the TGFβ type III receptor functions as a regulator of TGFβ access to the TGFβ type II receptor, but is not directly involved in signaling.

From: *Methods in Molecular Medicine, Vol. 14: Tumor Marker Protocols*
Edited by: M. Hanausek and Z. Walaszek © Humana Press Inc., Totowa, NJ

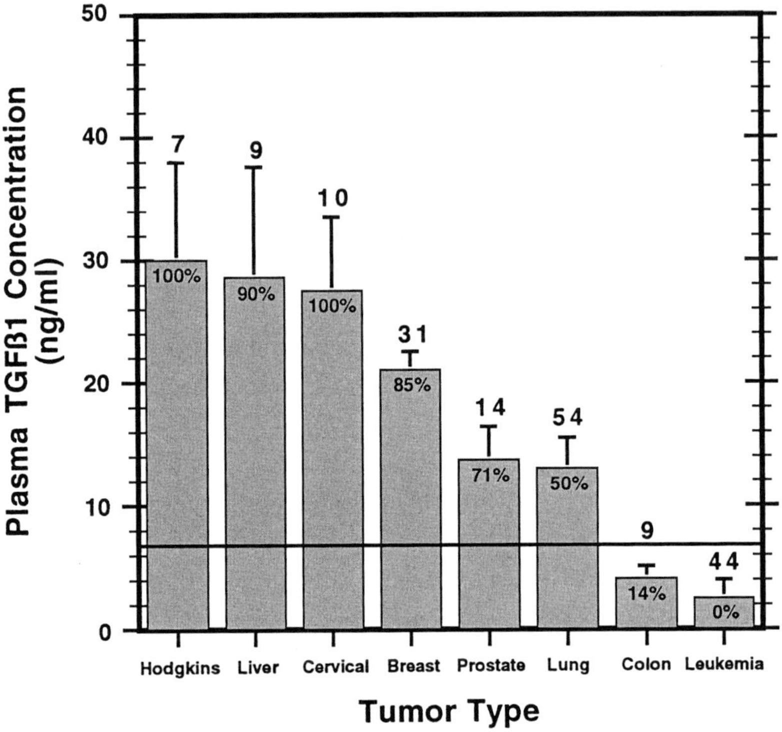

Fig. 1. Plasma TGFβ1 concentration in patients with various types of tumors. The number of patients in each tumor group is shown above the error bars (standard error of the mean). The percentage of patients whose plasma TGFβ1 level was >2 SDs above the mean plasma TGFβ1 value for control patients (horizontal line) is also provided for each tumor. The results for lung ***(27)***, liver ***(28,29)***, and breast ***(30)*** tumors and leukemia ***(31)*** have been previously published.

TGFβ1 is overexpressed locally in many tumors and is thought to play a role in tumor transformation and progression, as well as in tumor regression ***(17–20)***. However, tumor cells are often refractory to the antiproliferative effects of TGFβ1 ***(21)***. The inability of malignant cells to respond to the mitoinhibitory effects of TGFβ1 can result from a reduction or loss of the TGFβ type I or II receptor function ***(22–24)*** or an inability to effectively activate TGFβ1 because of the loss of the M6P/IGF2R gene, a recently defined tumor suppressor ***(25,26)***. In either case, this could lead to an overproduction of TGFβ1, resulting in its increase in the plasma.

The plasma TGFβ1 level is increased in patients with a variety of tumors (*see* **Fig. 1**), including hepatocellular carcinoma ***(28,29)***, prostate carcinoma

(32), and breast cancer ***(30,33)***. In newly diagnosed breast cancer patients, an elevated plasma TGFβ1 level usually decreases to normal following the complete surgical removal of the primary tumor; persistently elevated levels correlate with the presence of lymph node metastases or overt residual disease ***(30)***. Plasma TGFβ1 levels also correlate with disease status following radiation therapy for lung cancer, suggesting the possibility of its use to monitor patients for treatment response, evidence of tumor recurrence, and for the selection of patients requiring adjuvant therapy ***(27)***. Additionally, patients with a high propensity to develop normal tissue toxicity following chemo/radiotherapy can be predicted by measuring the plasma TGFβ level during treatment ***(2,34)***. In the following sections, we describe the various assays available for quantifying TGFβ plasma and tissue concentrations.

2. Materials

2.1. Acid/Ethanol Extraction of TGFβ

1. Acid/ethanol (A/E) solution: 375 mL 95% ethanol, 7.5 mL 12*N* HCl, 33 mg phenylmethylsulfonylfluoride (PMSF), and 1.9 mg pepstatin A.
2. A/E/H_2O solution: 375 mL 95% ethanol, 7.5 mL 12 *N* HCl, 105 mL distilled water.
3. 50 mL Oak Ridge centrifuge tubes (Nalgene Labware, Rochester, NY).
4. High-performance liquid chromatography (HPLC) grade glacial acetic acid (Fisher Scientific, Pittsburgh, PA).
5. Dialysis bags (3500 molecular weight cut-off, Spectrum Medical, Houston, TX).
6. 5-mL Sterile plastic tubes (Becton Dickinson Labware, Lincoln Park, NJ).
7. 3-([3-Cholamidopropyl] dimethylammonio)-1-propanesulfonate (CHAPS) buffer: 10 m*M* Na_2HPO_4, 2.0*M* NaCl, 1% CHAPS, 0.1% polyoxyethylene-sorbitan monolaurate (Tween-20) (Sigma, St. Louis, MO).
8. Phosphate-buffered saline-bovine serum albumin (PBS-BSA): 0.14*M* NaCl, 2.7 m*M* KCl, 10 m*M* Na_2HPO_4, 1.76 m*M* KH_2PO_4, and 0.1% BSA (Sigma) pH 7.4.
9. PBS-BSA/CHAPS: 1:1 mixture of PBS-BSA and CHAPS buffer.
10. Rotator/shaker.

2.2. Mink Lung TGFβ Growth Inhibitory Assay

1. Mv 1 Lu (NBL-7) mink lung cell line (American Type Culture Collection, Rockville, MD).
2. Dulbecco's modified Eagle medium (DMEM) (Gibco Life Technologies, Grand Island, NY).
3. Tissue culture plates, 24-well plastic, and T-75 plastic tissue culture flasks (Corning, Corning, NY).
4. Fetal bovine serum (FBS) (Gibco Life Technologies).
5. Mink lung-cell assay medium: DMEM/0.2% FBS, 10 m*M* N-[2-hydroxyethyl]-piperazine-*N*'-[2-ethanesulfonic acid] (HEPES) (Sigma).

6. Human recombinant TGFβ1: Purified TGFβ1 (R&D Systems, Minneapolis, MN) should be reconstituted according to manufacturer's recommendations, aliquoted, and stored at –20°C.
7. TGFβ1 neutralizing antibody: Make a stock solution (1 mg/mL) of the TGFβ1 neutralizing antibody (R&D Systems) in sterile PBS, aliquot, and store at –20°C.
8. ^{3}H-Thymidine (ICN Biomedical, Irvine, CA).
9. 1% Sodium dodecyl sulfate (SDS) (Kodak Biotech, New Haven, CT).
10. Trypsin/ethylenediamine tetra-acetic acid (EDTA): 0.25% trypsin in 1 m*M* EDTA (Gibco Life Technologies).

2.3. TGFβ Radioreceptor Assay

1. AKR-2B cell line (clone 84A) ***(35)***.
2. 6-Well plastic tissue-culture plates (Corning).
3. McCoy's modified media (Gibco Life Technologies).
4. Radioreceptor binding buffer: 128 m*M* NaCl, 5 m*M* KCl, 5 m*M* Mg_2SO_4, 1.2 m*M* $CaCl_2$, 50 m*M* HEPES, 0.2% BSA, pH 7.5.
5. ^{125}I-TGFβ1 (DuPont NEN, Boston, MA).
6. Octyl phenoxypolyethoxyethanol (Triton X-100) (Sigma).
7. Rocking platform.

2.4. TGFβ Enzyme-Linked Immunosorbent Assay (ELISA)

1. Monoclonal anti-TGFβ1 antibodies: 12H5 (nonneutralizing mouse monoclonal anti-TGFβ antibody ($IgG_{2b,\kappa}$) and 4A11 (neutralizing mouse monoclonal anti-TGFβ1 antibody ($IgG_{1,\kappa}$) ***(36)***.
2. Carbonate buffer: 13 m*M* $NaHCO_3$, 6.4 m*M* Na_2CO_3, pH 9.6.
3. PBS/Tween-20 (PT) buffer: PBS with 0.05% Tween-20.
4. PT-milk solution: PT buffer with 0.1% nonfat dried milk (1:20 dilution of milk diluent/blocking agent) (Kirkegaard and Perry, Gaithersburg, MD).
5. Human recombinant TGFβ1: Purified TGFβ1 (R&D Systems) should be reconstituted according to the manufacturer's recommendations, aliquoted, and stored at –20°C.
6. Horseradish peroxidase (HRP)-conjugated rabbit antimouse IgG1 (Zymed Laboratories, South San Francisco, CA).
7. 96-well ELISA microtiter plates (Corning Inc.).
8. 2,2'-Azino-di-[3-ethylbenzthiazoline 6-sulfonic acid] (ABTS) (Bio-Rad, Hercules, CA).
9. ELISA-plate shaker (Fisher).
10. Microtiter ELISA plate reader.

3. Methods

3.1. Plasma Sample Preparation

Since TGFβ is present at a high concentration in platelets, it is mandatory that the blood sample be processed to eliminate both platelet degranulation and platelet contamination of the plasma. The proper plasma-sample preparation involves using EDTA as the anticoagulant, handling the blood sample carefully following withdrawal, storing the blood sample at 4°C until centrifugation, and centrifuging the blood sample with sufficient force to guarantee the complete removal of the platelets from the plasma.

3.1.1. Plasma Sample Collection

1. Collect the blood sample in a 5-mL EDTA purple-top vacuum tube (*see* **Note 1**) (Becton Dickinson Labware) using a 19- or 21-gage needle (*see* **Note 2**).
2. Store the blood-collection tubes in an upright position at 4°C until sample centrifugation and avoid agitating the samples.
3. Centrifuge the blood collection tubes at 3000*g* for 25 min with the centrifuge brake off (*see* **Note 3**).
4. Remove only the top 1 mL of plasma and keep the transfer pipet tip as far away from the buffy coat as possible (*see* **Note 4**).
5. Store plasma samples at –70°C until assayed for TGFβ.

3.1.2. Acid/Ethanol Extraction of TGFβ from the Plasma

TGFβ extraction from the plasma is required for many of the presently available assays. The extraction method used is that originally described by Roberts et al. *(37)*. This procedure not only partially purifies TGFβ from the plasma sample, but also releases it from its binding proteins. Consequently, following extraction, it is no longer possible to determine whether the TGFβ was in a biologically active or inactive state while in the plasma.

3.1.2.1. Seven-Day TGFβ Extraction Procedure

This extensive extraction procedure is required to accurately measure TGFβ with the mink lung-cell growth-inhibition assay (*see* **Subheading 3.2.1.1.**).

1. Mix 1 mL plasma with 4 mL of A/E solution and 1 mL H_2O; shake the resulting mixture overnight at 4°C.
2. Centrifuge at 20,000*g* for 30 min at 4°C. Transfer the supernatant to a new tube and store at 4°C.
3. Re-extract the remaining pellet with 4 mL of A/E/H_2O. solution. Vortex the pellet until it solubilizes and shake the solution overnight at 4°C.
4. Centrifuge the pellet solution at 20,000*g* for 30 min at 4°C.
5. Combine the supernatants from **steps 2** and **4**, adjust pH to 5.2–5.3 with NH_4OH (back titrate with glacial acetic acid if necessary), and carefully note the final volume.
6. Add 1 mL of 2 *M* ammonium acetate to 8.5 mL of the combined supernatant extract.
7. Transfer this solution to a 50-mL Oak Ridge centrifuge tube, add 2 vol of cold (4°C) absolute ethanol, and precipitate at –20°C for 2 or more days.
8. Centrifuge this solution at 20,000*g* for 30 min at 4°C.
9. Resuspend the pellet in 4 mL of 1 *M* glacial acetic acid (HPLC grade).

10. Fill a dialyzing container with 4 L of 1% glacial acetic acid (HPLC grade) (*see* **Note 5**).
11. Pipet the resuspended sample (*see* **step 9**) into a dialysis tube and dialyze the sample at 4°C overnight (*see* **Note 6**).
12. Collect the dialyzed solution extraction and carefully note the final volume.
13. Freeze the dialyzed solution at –70°C for 30 min and lyophilize for 8 h or until dry (*see* **Note 7**).
14. Reconstitute the lyophilized sample in assay buffer and store at –20°C until it is used in the mink lung cell-growth inhibition assay.

3.1.2.2. Two-Day TGFβ Extraction Procedure

This shorter TGFβ partial-extraction procedure is adequate for measuring plasma TGFβ concentration by the radioreceptor assay (*see* **Subheading 3.2.1.2.**) and the ELISA assay developed in our laboratory (*see* **Subheading 3.2.2.1.**).

1. Dilute 0.2 mL of plasma with 0.8 mL of A/E solution and 0.2 mL H_2O in a 1.5 mL-screw-top-capped microcentrifuge tube. Then vortex the sample and incubate it for 4 h at 4°C.
2. Centrifuge the sample at 20,000*g* for 30 min at 4°C, transfer the supernatant to a new 1.5 mL-screw-top-capped microcentrifuge tube, and store at 4°C.
3. Re-extract the remaining pellet with 0.8 mL A/E/H_2O solution. Vortex the sample and shake it for 2 h or more at 4°C.
4. Centrifuge the re-extracted solution at 20,000*g* for 30 min at 4°C.
5. Fill the dialyzing container with 4 L of 1% glacial acetic acid (HPLC grade) (*see* **Note 5**).
6. Pipet the combined supernatant from **steps 2** and **4** into a dialysis bag and dialyze the combined sample overnight at 4°C (*see* **Note 6**).
7. Transfer the dialyzed sample to two 5-mL plastic tubes, freeze them for at least 30 min at –70°C, and lyophilize for 8 h or until dry (*see* **Note 7**).
8. Reconstitute the sample with the appropriate assay buffer PBS-BSA/CHAPS and store it at –20°C.

3.2. TGFβ Measurement

The available methods to assay TGFβs can be divided into biological assays and immunoassays. The bioassays involve measuring the growth inhibitory effects of TGFβ on cell proliferation (*see* **Subheading 3.2.1.1.**) or TGFβ binding to cell-surface receptors (*see* **Subheading 3.2.1.2.**). The main advantages of the bioassays are their high sensitivity and independence of having to produce specific TGFβs antibodies. The primary disadvantages of the bioassays are:

1. The longer preparatory procedures required to isolate TGFβ from the plasma sample;
2. The relatively large amount of variability between assays;
3. The inability to discriminate between the three isoforms of TGFβ; and

4. The inability to directly exclude the confounding effects of contamination from other positive and/or negative growth factors present in the plasma.

In contrast, the ELISA methods described (*see* **Subheading 3.2.2.**) are:

1. Highly reproducible;
2. Easy to use; and
3. Discriminate among the different TGFβ peptides with the use of isoform-specific TGFβ antibodies.

The use of an ELISA method to measure plasma TGFβ is also justified, since we found the levels determined with bioassays to be highly correlated with those measured with our ELISA method (*see* **Subheading 3.2.2.1.**) ***(38)***. Thus, even though the ELISA method has a lower sensitivity than the bioassays, it is presently the most frequently used procedure for measuring TGFβ.

3.2.1. TGFβ Biological Assays

3.2.1.1. TGFβ Mink Lung-Cell Growth Inhibition Assay

The mink lung cell growth inhibition assay was one of the first bioassays used to quantify TGFβ ***(39)***. It measures the level of active TGFβ by monitoring the growth inhibitory activity of TGFβ on proliferating mink lung epithelial cells. Briefly, cells in an exponential phase of growth are exposed to a test sample, and the amount of ^{3}H-thymidine incorporated into the DNA of these cells is compared to that observed when various concentrations of purified TGFβ1 are added to the cells to generate a standard curve. The amount of TGFβ in the test sample is determined from the standard curve and is inversely correlated with the amount of ^{3}H-thymidine incorporated into the cellular DNA.

3.2.1.1.1. Preparation of Mink Lung Cells for Assay

1. Thaw mink lung Mv 1 Lu (NBL-7) cells stored in liquid nitrogen in a 37°C water bath (1×10^6 cells/vial).
2. Dilute the 1-mL cell suspension in 10 mL of DMEM/10% FBS.
3. Centrifuge the cell suspension at 60*g* for 5 min; aspirate the supernatant and resuspend the cells in 15 mL DMEM/10% FBS.
4. Transfer the cell suspension to a T-75 cell culture flask and incubate at 37°C for 48 h.
5. Subculture these cells into four T-75 flasks following trypsinization and incubate them at 37°C for 2–3 d (*see* **Note 8**).

3.2.1.1.2. Mink Lung-Cell Assay Procedure

1. Trypsinize the subcultured mink lung cells while they are in their exponential growth phase and resuspend them in DMEM/10% FBS.
2. Centrifuge them at 60*g* for 5 min and resuspend the pellet in mink lung-cell assay medium.

3. Repeat **step 2** to completely remove trypsin and remaining dead cells.
4. Count the cells and dilute them to a concentration of 10^5 cells/mL in mink lung-cell assay medium.
5. Plate the cells into 24-well tissue culture plates (0.5 mL/well) and incubate at 37°C for 1 h to allow cell attachment.
6. Add the TGFβ standard samples (twofold serial dilutions of purified TGFβ1 from 600 to 9.4 pg/mL) and test samples in triplicate to the wells containing cells; incubate them at 37°C for 22 h (*see* **Note 9**).
7. Add 0.5 μCi of ^{3}H-thymidine/well and incubate the cells for an additional 4 h.
8. Fix the cells by adding 1 mL of methanol:glacial acetic acid (3:1) to each well for 1 h at room temperature and wash the cells twice with 1.5 mL/well of 80% methanol.
9. Remove the residual methanol, add 0.5 mL of Trypsin-EDTA/well, and incubate for 1 h at room temperature to remove cells from the well.
10. Add 0.5 mL of 1% SDS for 5 min to solubilize the cells, transfer the solution to a scintillation vial, and count the radioactivity with a β counter.
11. The TGFβ present in the test sample is determined by comparing the radioactivity in the test sample with that in the standard curve samples (*see* **Note 10**).

3.2.1.2. TGFβ Radioreceptor Assay

The TGFβ radioreceptor assay is a two-step indirect competitive assay modified from Wakefield et al. ***(40)***. In this assay, the level of active TGFβ is determined by comparing the ability of the TGFβ in the test sample to block the binding of radiolabeled TGFβ1 to the TGFβ receptors present on the surface of cells. Since only active TGFβ is detectable with this assay, TGFβ must be released from its plasma-binding proteins prior to being assayed. The amount of the radioactively labeled TGFβ bound to the cell-surface TGFβ receptors is inversely correlated with the amount of TGFβ present in the test sample.

1. Place AKR-2B cells (5×10^5 cells/well) (*see* **Note 11**) into 6-well plates and culture overnight in McCoy's modified media with 5% FBS (*see* **Note 12**).
2. Wash the cells twice with PBS-BSA and incubate them on a rocking platform for 1 h in radioreceptor-binding buffer.
3. Aspirate the radioreceptor-binding buffer, add TGFβ1 standard samples (twofold dilutions of purified TGFβ1 from 8 ng/mL to 0.03 ng/mL), and test samples diluted in 1 mL of radioreceptor-binding buffer. Both the TGFβ1 standard and test samples should be run in duplicate or triplicate.
4. Aspirate the test and TGFβ standard samples, wash the wells twice with ice-cold radioreceptor-binding buffer, add ^{125}I-TGFβ1 (0.05 μCi/well) in radioreceptor-binding buffer, and incubate on a rocking platform at room temperature for an additional 2 h.
5. Aspirate the ^{125}I-TGFβ1 containing radioreceptor-binding buffer, wash the cells three times with PBS-BSA, and incubate on a rocking platform with PBS-BSA and 1% Triton X-100 for 10 min.

6. Transfer the solubilized cellular solution to scintillation vials and count the radioactivity with a γ counter.
7. The TGFβ present in the test sample is determined by comparing the radioactivity in the test sample with those obtained for the TGFβ1 standard curve samples (*see* **Note 13**).

3.2.2. TGFβ ELISA Assays

Double-antibody sandwich ELISA methods, using either a TGFβ antibody or the TGFβ type II receptor to capture TGFβ, have recently been developed. These assays have greatly improved the ability to accurately and rapidly detect TGFβ in plasma samples.

3.2.2.1. Laboratory-Developed TGFβ ELISA Assay

1. Coat the 96 wells of microtiter plates by adding 100 μL/well of 12H5 (0.5 μg/mL in carbonate buffer) and incubate the plates overnight at 4°C.
2. Wash the microtiter-plate wells five times with PT buffer (*see* **Note 14**).
3. Block the microtiter-plate wells by adding 250 μL/well of PT-milk solution and incubate the microtiter plate for either 1 h at room temperature or overnight at 4°C on a microtiter-plate shaker.
4. Wash the microtiter plate five times with PT buffer.
5. Place 50 μL of the PBS-BSA/CHAPS buffer into the wells, add 50 μL of the extracted test samples (diluted 1:5 in PBS-BSA/CHAPS), and incubate the microtiter plate at room temperature for 3 h on a microtiter-plate shaker. A TGFβ1 standard curve must also be determined for each microtiter plate (twofold serial dilutions of purified TGFβ1 from 10 to 0.156 ng/mL). Duplicate or triplicate wells should be used for the test samples and the TGFβ1 standard curve samples (*see* **Note 15**).
6. Wash the microtiter-plate wells 10 times with PT buffer (*see* **Note 16**).
7. Incubate the microtiter plate with 100 μL/well of the 4A11 TGFβ1 monoclonal antibody (2 μg/mL dissolved in PT milk solution) for 2 h at room temperature on a microtiter-plate shaker (*see* **Note 17**).
8. Wash the microtiter-plate wells 10 times with PT buffer.
9. Incubate the microtiter plate with 100 μL/well of HRP-rabbit antimouse IgG1 (diluted to 1:3000 with PT-milk solution milk) for 1 h at room temperature on a microtiter-plate shaker.
10. Wash the microtiter-plate wells 10 times with PT buffer.
11. Add 100 μL/well of ABTS substrate and incubate the microtiter plate on a microtiter plate shaker for 30 min (*see* **Note 18**).
12. The optical density of the solution in the 96-well microtiter-plate wells is read at 405 nm using an automatic microtiter-plate spectrophotometric reader.
13. The amount of TGFβ1 in a test sample is determined by comparing its optical density with those obtained for the TGFβ1 standard curve samples (*see* **Note 19**).

3.2.2.2. Commercially Developed TGFβ ELISA Kits

TGFβ1 ELISA kits are now available commercially. We have compared the results of our TGFβ1 ELISA (*see* **Subheading 3.2.2.1.**) with those from ELISA kits obtained from both R&D Systems and Genzyme (Boston, MA). The TGFβ1 levels determined with both of these commercially available TGFβ1 ELISA kits correlate well with those obtained using our ELISA method ($r > 0.9$, $p < 0.01$) over a wide range of plasma TGFβ1 concentrations. However, at all plasma TGFβ1 concentrations tested, the R&D ELISA kit estimated a lower total plasma TGFβ1 concentration when compared to that obtained with our ELISA procedure.

4. Notes

1. We have found that the anticoagulant used for collecting blood samples influences the ability to accurately determine the plasma TGFβ1 concentration. The plasma TGFβ1 level in blood samples collected with the anticoagulant EDTA are constant for at least 12 h when the blood sample is stored at 4°C following collection. In contrast, the plasma TGFβ1 level increases rapidly when heparin is used as the anticoagulant. Therefore, EDTA is the preferred anticoagulant when the blood sample is to be stored for a period of time prior to plasma removal.
2. Platelets are the richest source of TGFβ in the blood. Thus, large-bore needles should be used to withdraw blood in order to minimize platelet lysis. Plasma samples that show evidence of red-cell hemolysis should not be used for estimating plasma-TGFβ concentration.
3. A centrifugal force of 3000*g* for 20 min is required to insure that platelets are removed from the plasma. The centrifuge brake is turned off to reduce blood-cell agitation that may cause platelet resuspension.
4. There are always some platelets in solution at the plasma/white-cell interface. Consequently, only the top 1 mL of platelet-free plasma is collected for TGFβ determination.
5. Measure the pH of this solution to ensure that it is approx 2.7 ± 0.5.
6. The dialysis tubings should be thoroughly boiled for 1 h before they are used.
7. Freezing the dialyzed samples before they are lyophilized facilitates the reconstitution of the lyophilized material in the assay buffer.
8. The mink lung cells must be kept subconfluent at all times during passage to ensure assay reproducibility.
9. A TGFβ-specific neutralizing antibody should also be added to an independent set of test samples to determine whether the cell-growth inhibition is eliminated. Radioactivity equal to or greater than that in the controls should be observed in those test-sample wells containing TGFβ-neutralizing antibodies, if the observed growth inhibition in the test-sample wells is caused by the presence of TGFβ.
10. Multiple test-sample dilutions (1:50, 1:100, and 1:250) must be used, and only the dilution that lies in the middle of the standard curve should be used to calculate the test-sample TGFβ concentration.

11. Any cell line with TGFβ receptors can be used for the radioreceptor assay. A549 human-lung carcinoma cells are also frequently used in this assay ***(40)***.
12. The cells should be confluent to reduce nonspecific binding of ^{125}I-TGFβ1 to the plate. The lower the nonspecific binding, the greater the sensitivity of this assay.
13. The nonspecific binding of ^{125}I-TGFβ1 in this assay is normally 30–50%. If it is higher than this, the assay is not reliable.
14. The pH should be 7.4 for all the ELISA buffers except the carbonate buffer.
15. Because of the presence of TGFβ binding proteins in the test samples, even after plasma extraction, the factor by which the test samples are diluted can influence the TGFβ1 value estimated by ELISA. The best test-sample dilution for human-plasma samples is 1:10 in the ELISA we have developed.
16. Removing the nonspecifically bound TGFβ1 is important because it increases assay sensitivity by decreasing the background. Less-frequent washing is used in other ELISA procedures, however, because TGFβ1 is a "sticky" molecule extensive washing is required.
17. To determine the level of TGFβ2 and TGFβ3 in test samples, monoclonal antibodies (MAbs) 3C7.14 and 1D11.6 can be substituted for 4A11 and 12H5, respectively ***(30)***.
18. The optimal development time for the substrate varies with the ELISA conditions. Consequently, the microtiter plate is read at 5-min intervals and the results used for TGFβ1 estimation are those obtained when the greatest optical-density value in any well is 2.0 at 405 nm; usually this occurs at 30 min.
19. A TGFβ1 standard curve is determined each time an ELISA is performed. A control test sample and purified TGFβ1 sample (5 ng/mL) are also run every time an ELISA is performed to check for day-to-day reproducibility. The TGFβ1 standard-curve samples and test samples are run in triplicate, and the three optical density (OD) values for each sample are averaged following the substraction of background. A standard curve is generated by plotting the average OD value for each standard-curve sample on the x-axis vs the corresponding TGFβ1 concentration on the y-axis. The data are curve-fit by least-squares analysis to the third-order polynomial $y = ax^3 + bx^2 + cx + d$. TGFβ1 concentration for each test sample is calculated by substituting its corrected averaged OD into this equation.

Acknowledgment

This research was supported by NCI grant CA25951.

References

1. Schwartz, M. K. (1993) Cancer markers, in *Cancer Principles & Practice of Oncology* (DeVita, V. T., Jr., Hellman, S., and Rosenberg, S. A., eds.), Lippincott, Philadelphia, pp. 531–542.
2. Anscher, M. S., Peters, W. P., Reisenbichler, H., Petros, W. P., and Jirtle, R. L. (1993) Transforming growth factor β as a predictor of liver and lung fibrosis after autologous bone marrow transplantation for advanced breast cancer. *N. Engl. J. Med.* **328,** 1592–1598.

3. Anscher, M. S., Kong, F.-M., Murase, T., and Jirtle, R. L. (1995) Normal tissue injury after cancer therapy is a local response exacerbated by an endocrine effect of TGFβ. *Br. J. Radiol.* **68,** 331–333.
4. Letterio, J. J., Geiser, A. G., Kulkarni, A. B., Roche, N. S., Sporn, M. B., and Roberts, A. B. (1994) Maternal rescue of transforming growth factor-β1 deficient null mice. *Science* **264,** 1936–1938.
5. Massagúe, J. (1990) The transforming growth factor-β family. *Ann. Rev. Cell Biol.* **6,** 597–641.
6. Border, W. A. and Ruoslahti, E. (1992) Transforming growth factor-β in disease: the dark side of tissue repair. *J. Clin. Invest.* **90,** 1–7.
7. Amento, E. P. and Beck, L. S. (1991) TGF-β and wound healing, in *Clinical Applications of TGFβ* (Bock, G. R. and Marsh, J., eds.), Wiley, West Sussex, UK, pp. 115–129.
8. Kankaki, T., Olofsson, A., Moren, A., Wernstedt, C., Hellman, U., Miyazono, K., Claesson-Welsh, L., and Heldin, C.-H. (1990) TGFβ1 binding protein: a component of the large latent complex of TGFβ1 with multiple repeat sequences. *Cell* **61,** 1051–1061.
9. LaMarre, J., Wollenberg, G. K., Gauldie, J., and Hayes, M. A. (1990) Alpha 2-macroglobulin and serum preferentially counteract the mitoinhibitory effect of transforming growth factor-β2 in rat hepatocytes. *Lab. Invest.* **62,** 545–551.
10. Dennis, P. A. and Rifkin, D. B. (1991) Cellular activation of latent transforming growth factor β requires binding to the cation-dependent mannose 6-phosphate/insulin-like growth factor type II receptor. *Proc. Natl. Acad. Sci. USA* **88,** 580–584.
11. De Bleser, P. J., Jannes, P., van Buul-Offers, S. C., Hoogerbrugge, C. M., van Shravendijk, C. F. H., Niki, T., Rogiers, V., Van den Brande, J. L., Wisse, E., and Geerts, A. (1995) Insulin like growth factor-II/mannose 6-phosphate receptor is expressed on CCl4-exposed rat fat-storing cells and facilitates activation of latent transforming growth factor-β in co-cultures with sinusoidal endothelial cells. *Hepatology* **21,** 1429–1437.
12. Lyons, R. M., Keski-Oja, J., and Moses, H. L. (1988) Proteolytic activation of latent transforming growth factor-β from fibroblast-conditioned medium. *J. Cell Biol.* **106,** 1659–1665.
13. Wang, X.-F., Lin, H. Y., Ng-Eaton, E., Downward, J., Lodish, H. F., and Weinberg, R. A. (1991) Expression cloning and characterization of the TGF-β type III receptor. *Cell* **67,** 797–805.
14. Lin, H. Y., Wang, X.-F., Hg-Eaton, E., Weinberg, R. A., and Lodish, H, F. (1992) Expression cloning of the TGF-β type II receptor, a functional transmembrane serine/threonine kinase. *Cell* **68,** 775–785.
15. Ebner, R. , Chen, R.-H., Shum, L., Lawler, S., Zioncheck, T. F., Lee, A., Lopez, A. R., and Derynck, R. (1993) Cloning of a type I TGF-β receptor and its effect on TGF-β binding to the type II receptor. *Science* **260,** 1344–1348.
16. Wrana, J. L., Attisano, L., Wieser, R., Ventra, F., and Massagúe, J. (1994) Mechanisms of activation of the TGF-β receptor. *Nature* **370,** 341–347.

17. Derynck, R., Goeddel, D. V., Ullrich, A., Gutterman, J. U., Williams, R. D., Bringman, T. S., and Berger, W. (1987) Synthesis of messenger RNAs for transforming growth factors α and β and the epidermal growth factor receptor by human tumors. *Cancer Res.* **47,** 707–712.
18. Kim, S.-J., Kehrl, J. H., Burton, J., Tendler, C. L., Jeang, K.-T., Danielpour, D., Thevenin, C., Kikm, K. Y., Sporn, M. B., and Roberts, A. B. (1990) Transactivation of the transforming growth factor β1 (TGF-β1) gene by human T lymphotrophic virus type 1 tax: a potential mechanism for increased production of TGF-β1 in adult T cell leukemia. *J. Exp. Med.* **172,** 121–129.
19. Jasani, B., Wyllie, F. S., Wright, P. A., Lemoine, N. R., Williams, E. D., and Wynford-Thomas, D. (1990) Immunocytochemically detectable TGF-β associated with malignancy in thyroid epithelial neoplasia. *Growth Factors* **2,** 149–155.
20. Ito, N., Kawata, S., Tamura, S., Takaishi, K., Shirai, Y., Kiso, S., Yabuuchi, I., Matsuda, Y., Nishioka, M., and Tarui, S. (1991) Elevated levels of transforming growth factor β messenger RNA and its polypeptide in human hepatocellular carcinoma. *Cancer Res.* **51,** 4080–4083.
21. Fynan, T. M. and Reiss, M. (1993) Resistance to inhibition of cell growth by transforming growth factor-β and its role in oncogenesis. *Crit. Rev. Oncol.* **4,** 493–540.
22. Sue, S. R., Chari, R. S., Kong, F.-M., Mills, J. J., Fine, R. L., Jirtle, R. L., and Meyers, W. C. (1995) Transforming growth factor beta receptors and mannose 6-phosphate/insulin-like growth factor II receptor expression in human hepatocellular carcinomas. *Ann. Surg.* **222,** 171–178.
23. Kimchi, A., Wang, S. F., Weinberg, R. A., Cheifetz, S., and Massagué, J. (1988) Absence of TGF-β receptors and growth inhibitory responses in retinoblastoma cells. *Science* **240,** 196–198.
24. Markowitz, S., Wang, J., Myeroff, L., Parsons, R., Sun, L., Lutterbaugh, J., Fan, R. S., Zborowska, E., Kinzler, K. W., Vogelstein, B., Brattain, M., and Willson, J. K. V. (1995) Inactivation of the type II TGF-β receptor in colon cancer cells with microsatellite instability. *Science* **268,** 1336–1338.
25. De Souza, A. T., Hankins, G. R., Washington, M. K., Finc, R. L., Orton, T. C., and Jirtle, R. L. (1995) Frequent loss of heterozygosity on 6q at the mannose 6-phosphate/insulin-like growth factor II receptor locus in human hepatocellular tumors. *Oncogene* **10,** 1725–1729.
26. De Souza, A. T., Hankins, G. R., Washington, M. K., Orton, T. C., and Jirtle, R. L. (1995) M6P/IGF2r gene is mutated in human hepatocellular carcinomas with LOH. *Nature Genet.* **11,** 447–449.
27. Kong, F.-M., Washington, M. K., Jirtle, R. L., and Anscher, M. S. (1996) Plasma transforming growth factor β1 reflects disease status in patients with lung cancer after radiotherapy: a possible marker. *Lung Cancer* **16,** 47–59.
28. Shirai, Y., Kawata, S., Ito, N., Tamura, S., Takaishi, K., Kiso, S., Tsushima, H., and Matsuzawa, Y. (1992) Elevated levels of plasma transforming growth factor-β in patients with hepatocellular carcinoma. *Jpn. J. Cancer Res.* **83,** 676–679.

29. Shirai, Y., Kawata, S., Tamura, S., Ito, N., Tsushima, H., Takaishi, K., Kiso, S., and Matsuzawa, Y. (1994) Plasma transforming growth factor-β1 in patients with hepatocellular carcinoma: comparison with chronic liver diseases. *Cancer* **73,** 2275–2279.
30. Kong, F.-M., Anscher, M. S., Abbot, B. D., Iglehart, J. D., and Jirtle, R. L. (1995) Elevated plasma transforming growth factor-β1 levels in breast cancer patients decrease after surgical removal of the tumor. *Ann. Surg.* **222**, 155–162.
31. Murase, T., Jirtle, R. L., and McDonald, G. B. (1994) TGF-β concentrations in the patients with lymphoma and leukemia receiving chemotherapy and marrow transplantation. *Blood* **83,** 2383.
32. Ivanovic, V., Melman, A., Davis-Joseph, B., Valcic, M., and Geliebter, J. (1995) Elevated plasma levels of TGFβ1 in patients with invasive prostate cancer. *Nature Med.* **1,** 282,283.
33. Wakefield, L. M., Letterio, J. J., Chen, T., Danielpour, D., Allison, R .S. H., Pai, L. H., Denicoff, A. M., Noone, M. H., Cowan, K. H., O'Shaugnessy, J. A., and Sporn, M. B. (1995) Transforming growth factor-β1 circulates in normal human plasma and is unchanged in advanced metastatic breast cancer. *Clin. Cancer Res.* **1,** 129–136.
34. Anscher, M. S, Kong, F.-M., Marks, L. B, Bentel, G. C., and Jirtle, R. L. (1997) Transforming growth factor β: a predictor of symptomatic radiation pneumonitis. *Int. J. Radiat. Oncol. Biol. Phys.* **37,** 253–258.
35. Tucker, R. F., Branum, E. L., Shipley, G. D., Ryan, R. J., and Moses, H. C. (1984) Specific binding to cultured cells of ^{125}I-labeled type β transforming growth factor from human platelets. *Proc. Natl. Acad. Sci. USA* **81,** 6757–6761.
36. Lucas, C., Worms, M.-M., Fendly, B. M., Figari, I. S., Patzer, E., and Palladino, M. A. (1990) The autocrine production of transforming growth factor beta 1 during lymphocyte activation: a study with a monoclonal antibody-based ELISA. *J. Immunol.* **145,** 1415–1422.
37. Roberts, A. B., Lamb, L. C., Newton, D. L., Sporn, M. B., DeLarco, J., and Todaro, G. J. (1980) Transforming growth factors: isolation of polypeptides from virally and chemically transformed cells by acid/ethanol extraction. *Proc. Natl. Acad. Sci. USA* **77,** 3494–3498.
38. Murase, T., Anscher, M. S., Petros, W. P., Peters, W. P., and Jirtle, R. L. (1995) Changes in plasma transforming growth factor beta in response to high-dose chemotherapy for stage II breast cancer: possible implications for the prevention of hepatic veno-occlusive disease and pulmonary drug toxicity. *Bone Marrow Transplant.* **15,** 173–178.
39. Danielpour, D., Dart, L. L., Flanders, K. C., Roberts, A. B., and Sporn, M. B. (1989) Immunodetection and quantitation of the two forms of transforming growth factor-beta (TGF-β1 and TGF-β2) secreted by cells in culture. *J. Cell. Physiol.* **138,** 79–86.
40. Wakefield, L. M., Smith, D. M., Masui, J., Harriss, C. C., and Sporn, M. B. (1987) Distribution and modulation of the cellular receptor for transforming growth factor-beta. *J. Cell Biol.* **105,** 965–975.

26

Anti-HMdU Autoantibodies in Human Sera as a Biomarker of Cancer Risk

Krystyna Frenkel and Jerzy Karkoszka

1. Introduction

Our laboratory has discovered that blood sera of healthy men and women contain low levels of anti-HMdU (5-hydroxymethyl-2'-deoxyuridine, an oxidized thymidine) autoantibodies (aAbs) *(1,2)*. However, patients with chronic inflammatory diseases exhibit elevated anti-HMdU aAb titers. Interestingly, people at high risk for cancer, those with cancer and, most importantly, those who were diagnosed with breast, colon, and rectal cancer 0.5–6 yr *after* donation of the blood samples we analyzed (but not those diagnosed with nonmelanoma skin cancer) had significantly elevated anti-HMdU aAb titers *(4)*. The high aAbs in apparently healthy women at the time of blood donation persisted for several years before diagnosis *(3)*. We recently also showed that these anti-HMdU aAb titers depend on subject's age and menopausal status *(5)*. Our findings point to anti-HMdU aAb titers as being a biomarker of a potential to develop cancer even in the absence of family risk factors, which constitute the majority of cases.

Our contention that anti-HMdU aAb titers can be considered as biomarkers is strengthened by the finding that HMdU levels in white blood cell DNA from women with breast cancer are elevated *(6)*. Moreover, HMdU in white blood cell DNA from women at high risk for breast cancer can be significantly decreased by an intake of a low-fat diet *(7)* and, presumably, increased intake of fruits and vegetables. Elevated levels of oxidized purines were also found in human breast tumors *(8)*. Finding that dietary intervention can cause a decrease in the levels of oxidized DNA bases *(7)* may be important to cancer prevention.

The individuals identified as being at risk for cancer because of repeatedly high anti-HMdU aAb levels in their blood sera could then be evaluated by

From: *Methods in Molecular Medicine, Vol. 14: Tumor Marker Protocols*
Edited by: M. Hanausek and Z. Walaszek © Humana Press Inc., Totowa, NJ

physicians to determine which type of cancer or immune complex disease they were at risk for. Physicians could, in turn, recommend dietary (*see* **ref.** *7*) or other (i.e., anti-inflammatory) preventive measures to be administered. The efficacy of these measures also could be tested by the same sensitive and facile assay ***(1–3)***.

Chronic inflammation has been shown to be a strong contributing factor in the development of many types of cancer ***(9–11)***. The generation of copious amounts of oxidants is an important part of the inflammatory response. These oxidants, and particularly hydrogen peroxide (H_2O_2), induce oxidative stress, which contributes to oxidation of bases in the genetic material, among other types of cellular damage ***(11)***. Cells possess extensive antioxidant defenses as well as repair enzymes, which recognize and remove oxidized bases from cellular DNA ***(11–13)***. However, chronic inflammatory conditions induced by carcinogens ***(11,14,15)*** significantly increase the burden and persistence of many oxidized bases in DNA ***(11,15)***.

In addition to the inflammatory responses, carcinogen treatment can result in the production of oxidants during the carcinogen's metabolism and, in some cases, redox cycling ***(11,16)***. For example, certain polycyclic aromatic hydrocarbons and hormones can form quinones during their metabolism. Those quinones, as well as chemotherapeutic quinones (i.e., adriamycin), are often reduced by cellular reductases (i.e., cytochrome P450 reductase) to semiquinones, which readily reduce molecular oxygen to superoxide anion radical ($O_2^{\cdot-}$) and reform the parent quinone. This process of redox cycling and $O_2^{\cdot-}$ generation can proceed for as long as there are reducing cofactors (i.e., NADPH) and molecular oxygen present. The generated $O_2^{\cdot-}$ dismutates to H_2O_2 spontaneously or when catalyzed by an enzyme superoxide dismutase (SOD). Of the many oxidants formed, H_2O_2 can migrate most readily through the cell and nuclear membranes and reach the DNA in the nucleus, where it may cause site-specific oxidation of DNA bases.

Many laboratories, including ours, have shown that treatment with antioxidants and other chemopreventive agents can inhibit oxidative stress, oxidative DNA damage, inflammatory responses, as well as carcinogenesis ***(11,17–24)***. Among the agents with which we have worked are some naturally occurring substances, such as (–)–epigallocatechin gallate ([EGCG] the main protective polyphenol in green tea and to a lesser extent in black tea), caffeic acid phenethyl ester ([CAPE] the major component in the propolis of honeybee hives), sarcophytol A (isolated from a soft marine coral), as well as tamoxifen ([TAM], an anticancer drug). Although structurally different, all of these compounds showed similar anti-inflammatory properties in vivo ***(21,23–25)***. They also suppressed formation of H_2O_2 and oxidized DNA bases (i.e., HMdU,

8-hydroxy-2'-deoxyguanosine [8-OHdG], and thymidine glycol) in cultured cells ***(22,26)*** and in animals treated with the tumor promoter 12-*O*-tetradecanoyl-phorbol-13-acetate (TPA) or in a two-stage carcinogenesis model ***(21,23–25)***. Some of the anticarcinogenic properties perhaps can be explained by inhibition of the oxidative burst by activated neutrophils, which leads to a substantial decrease in H_2O_2 generation by these cells and of H_2O_2-inflicted oxidative DNA damage ***(26–28)***.

The oxidative DNA damage may include increased DNA fragmentation occurring because of the hydroxyl radical-like attack (akin to that induced by ionizing radiation), as well as the removal of oxidized bases by the DNA excision repair system ***(11–13)***. Another consequence of the enhanced presence of oxidized bases in DNA may be to evoke an immune response in the form of elevated levels of aAb that recognize oxidized DNA bases, as we discovered several years ago ***(1–5,29,30)***. We mainly used HMdU (a chemically stable, oxidized thymidine) coupled to bovine serum albumin (BSA) as an antigen to analyze human sera for the presence and levels of anti-HMdU aAb, using the enzyme-linked immunosorbent assay (ELISA). This assay is very reproducible ***(31)***. As our preliminary results indicate, there are other antioxidized DNA base (i.e., anti-8-OHdG) aAbs present in human sera (K. Frenkel, unpublished data). In this chapter, we are presenting the methodology developed using HMdU-BSA as the antigen to coat the wells of microtiter plates.

Specifically, this chapter provides the following procedures:

1. Preparation of antigens needed for coating of plates;
2. Coating of microtiter plates with antigen and mock-antigen for the determination of specific and nonspecific binding, respectively;
3. Design of experiments, including use of positive controls;
4. ELISA; and
5. Evaluation of the results.

2. Materials

2.1. Plasticware

1. Microtiter plates, i.e., 96-well, flat-bottomed, polystyrene, with a high protein-binding capacity.
2. Seals for microtiter plates (S/P).
3. 50-mL Polypropylene tubes for reagent preparation and 1.5-mL microtubes for –80°C storage of sera samples.
4. Polyethylene solution basins long and wide enough to accommodate a multichannel pipeter needed for pipeting reagents to the wells.
5. Pipeter tips, including tips with a protective filter inside them.
6. Sterilization units with 0.2-μm filter for filtering distilled water and buffers.

2.2. Glassware

1. 125- and 250-mL Erlenmeyer flasks.
2. 20-mL Glass scintillation vials used as reaction vessels.
3. 25-cm long 1.5 cm ID Glass chromatography columns with polyethylene frits and a stopcock.
4. 5-mL Glass disposable borosilicate culture tubes.

2.3. Pipeters

1. 0–200- and 0–1000-μL pipeters.
2. One multichannel, 0–250-μL (8 channels) micropipeter.
3. 0.5–5.0-μL Micropipeter.

2.4. Instruments

1. SpeedVac concentrator (Savant Instruments, Farmingdale, NY).
2. Incubator set at 37°C.
3. Microplate reader.
4. Microbalance and pH meter.
5. UV/VIS Spectrophotometer.
6. Fraction collector.
7. UV absorbance detector (optional).
8. Fleaker® Hollow Fiber Concentration System by Spectrum, with two regenerated cellulose hollow-fiber bundles (176 fibers/bundle; 13 kDa cut-off point), and two peristaltic pumps to operate them.

2.5. Solutions

2.5.1. Preparation of Antigens

Conjugation of oxidized 5-hydroxymethyl uridine (HMU) and BSA and mock conjugation of BSA.

1. 5% K_2CO_3 (10 mL): 0.5 g K_2CO_3 + distilled H_2O to 10 mL.
2. 0.1 *M* $NaIO_4$ (5 mL): 106.95 mg + distilled H_2O to 5 mL.
3. BSA solution (6.25 mL): 175 mg BSA + distilled H_2O to 5 mL; adjust pH to 9.5 with 5% K_2CO_3, add distilled H_2O to 6.25 mL. Check pH before using.
4. $NaBH_4$ solution: 15 mg $NaBH_4$ + distilled H_2O to 10 mL. Prepare just before use.
5. 1 N formic acid: 0.495 mL 88% (20.2 *N*) formic acid + 9.505 mL distilled H_2O.
6. 1 N NH_4OH: 0.333 mL ~30% NH_4OH + 4.666 mL distilled H_2O.
7. 0.1 *M* NH_4HCO_3: 7.9 g NH_4HCO_3 in 1 L distilled H_2O.
8. P-6DG desalting gel (BioRad, Hercules, CA, cat. no. 150-0738).

2.5.2. Coating Microtiter Plates

1. Coating buffer (CB), pH 9.5: 1.59 g Na_2CO_3 + 2.93 g $NaHCO_3$ in 1 L distilled H_2O. Filter through 0.2-μm filter. Store up to 1 mo at 4°C.
2. 5X Stock Tween-20-PBS buffer (5X TP), pH 7.1: 80.0 g NaCl + 11.5 g Na_2HPO_4 + 0.4 g KH_2PO_4 + 0.4 g KCl + 5 mL Tween-20; adjust pH to 7.1 with

HCl, add distilled H_2O to 2 L. Filter through 0.2-μm filter to a sterile bottle. Store up to 1 mo at 4°C.

3. Washing buffer: Tween-20-PBS buffer (TP), pH 7.5. Dilute five times 5X TP (*see* **item 2**) with distilled H_2O. Store for only a few days at 4°C.
4. Blocking solution: 10 μg BSA/mL CB (*see* **item 1**).

2.5.3. ELISA

1. Working buffer (Tween-20-PBS-BSA [TPB], pH 7.5): Dissolve BSA in TP (*see* **Subheading 2.5.2., item 3**) at 1.0 mg BSA/mL TP. Prepare fresh daily.
2. Secondary antibody: Dilute goat antihuman IgM antibody with conjugated horse-radish peroxidase (HRPO) (Sigma, St. Louis, MO, cat. no. A-8650) with working buffer (TPB) at 1:1000. Prepare fresh and only as much as needed for use a given day. Do not use affinity-purified antibody (*see* **Note 1**).
3. Substrate buffer (SB), pH 5.0: 0.1815 mg Na_2HPO_4 + 0.1285 mg citric acid in 25 mL distilled H_2O. Prepare SB in a polypropylene tube; check pH!! **Do not use glass** for preparation of substrate buffer (*see* **item 4**).
4. Substrate: 10 μL 30% H_2O_2 + 10 mg o-phenylenediamine dihydrochloride (OPD) in 25 mL SB. Use a polypropylene tube. Prepare in the dark (under yellow light) shortly before use, cover with aluminum foil. **Do not use glass!** If glass is scratched, it can degrade H_2O_2. Be careful handling OPD, it may be carcinogenic.
5. H_2SO_4 (48%): Dilute concentrated H_2SO_4 (~90%) with distilled H_2O (add acid to the H_2O and do it under a hood).

3. Methods

3.1. Preparation of Antigens for Coating of Microtiter Plates

Conjugation of oxidized HMU and BSA results in the conjugate HMdU-BSA and mock conjugation of BSA in the M-BSA conjugate. When stored dry at –80°C, these conjugates are stable for at least one year.

3.1.1. Preparation of HMdU-BSA and M-BSA

Two 20-mL glass scintillation vials with caps labeled with different color tapes (i.e., blue for mock-coupling to BSA and red for oxidized HMU coupling to BSA) are needed.

1. Oxidation step: In this step, the ribose moiety of HMU, which is an oxidized uridine, will be oxidized by periodate and by one of the resultant aldehyde groups coupled to the ε-amino group of lysine moiety in BSA *(32)*.
   a. Place 20 mg HMU in the red vial and nothing in the blue vial.
   b. Add 1.25 mL 0.1 *M* $NaIO_4$ to the red and 1.25 mL distilled H_2O to the blue vial, stir for 20 min at room temperature in the dark.
   c. Add 75 μL ethylene glycol to each of the vials and stir for 5 min.
   d. Add 2.5 mL BSA solution, pH 9.5 to each of the vials, check pH and adjust it, if necessary, to pH 9.5 with 5% K_2CO_3 solution.
   e. Stir in the dark for 45 min, maintaining pH at 9.5 with 5% K_2CO_3 solution.

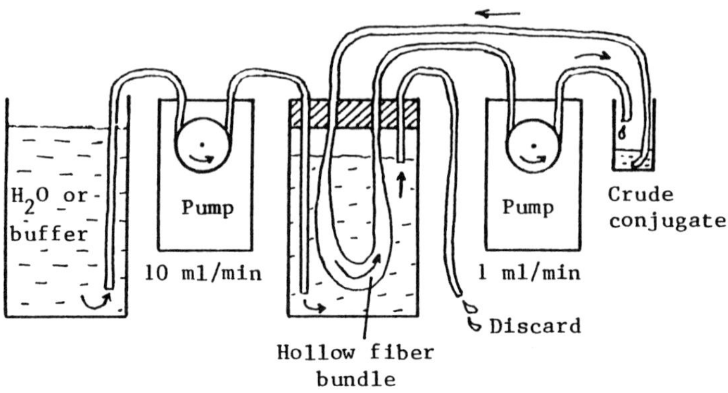

Fig. 1. Schematic representation of dialysis using hollow fiber bundle.

2. Reduction step: This procedure reduces the unstable coupling product to the stable HMdU-BSA conjugate.
   a. Slowly add 2.5 mL $NaBH_4$ solution to each of the vials (there will be foaming), and stir slowly for 18 h at room temperature in the dark.
   b. Stop the reaction by adding 1.25 mL 1 *N* formic acid to each vial, and stir for 1 h at room temperature.
   c. Adjust pH to 8.5 with ~1 mL 1 *N* NH_4OH using a pH meter.
3. Dialysis: To separate low molecular weight reagents from the high molecular weight conjugates, the contents of each vial should be dialyzed separately against ~1 L of distilled H_2O at ~10 mL/min for 1 h, then against ~1 L of 0.1 *M* NH_4HCO_3 solution using hollow fibers (bundles) with a 13 kDa cutoff. If possible, two peristaltic pumps should be used: one pumping sample through the fibers (flow rate ~ 1 mL/min), the other countercurrently pumping dialyzing solution (flow rate ~10 mL/min) (*see* **Fig. 1**). If this system is not available, regular dialyzing tubing can be utilized, but dialysis time will be much longer, and several changes of dialyzing solution should be made. Each of the samples will have a final volume of ~20 mL, which should be reduced to ~2 mL by overnight lyophilization in a SpeedVac concentrator.
4. Purification of the conjugates HMdU-BSA and M-BSA: Final removal of low molecular weight reagents and uncoupled nucleosides from the conjugates is carried out using two chromatographic glass columns containing desalting gel (P-6DG) with 6 kDa cutoff, one for HMdU-BSA and the other for M-BSA. Both columns are prepared 1 d before use. P-6DG powder (~20 g) is hydrated in distilled H_2O (~300 mL) at room temperature overnight (*see* BioRad's protocol for preparing BioGels). Each column is filled to ~24 cm in height making sure that there are no air bubbles in the packed columns. Then, 0.1 *N* NH_4CO_3 buffer is passed through the columns for 1 h. This conditioning results in some compression of the adsorbent (~2 cm). Samples obtained as in **step 3** (~2 mL each) are carefully layered on the top of each column, rinsed once with ~0.5 mL 0.1 *N*

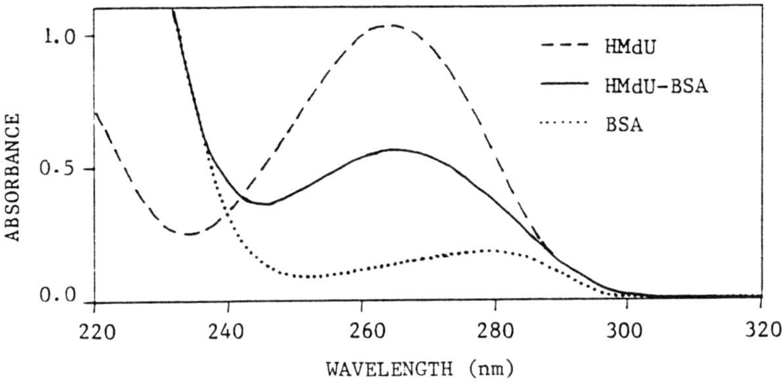

Fig. 2. UV spectra of HMdU (0.025 mg/mL), HMdU-BSA conjugate (0.25 mg/mL), and BSA (0.25 mg/mL).

NH_4HCO_3 buffer, eluted with the same buffer, and 1 mL fractions collected into glass tubes in a fraction collector. The absorbance of the fractions is either automatically monitored by a UV detector or manually in a spectrophotometer at 280 nm. The BSA conjugates elute in the void volume, while low-molecular-weight substances are retained on the column. Six to seven fractions (starting approximately with fraction 11 or 12 when the A_{280} starts to increase) are usually combined; the last several fractions with A_{280} decreasing below 1 are discarded. The combined samples of HMdU-BSA and those of M-BSA are separately scanned from 220–320 nm (retain hard copies of your UV spectra); an example of the UV spectra is shown in **Fig. 2**. Samples are aliquoted into 1-mL portions, then lyophilized overnight in a Savant SpeedVac lyophilizer, and stored at –80°C until use.

5. Determination of HMdU content in HMdU-BSA: Small, accurately weighted amounts of BSA (0.25 mg) and HMdU (0.025 mg) are dissolved in distilled H_2O, and absorbance spectra are determined. For the most accurate measurements, a full scale of 0–1 should be used, with absorbance at 265 nm (HMdU) being between 0.7 and 0.85. You can weigh out HMdU-BSA but it is not necessary because a spectrum of the conjugation product has already been taken in **step 4** above. The contribution of BSA to the 265 nm absorbance as well as the contribution of HMdU to the 280 nm absorbance are estimated by calculations based on an assumption that they are additive, since coupling of deoxyribose to the ε-NH_2 group of lysine should not change the respective chromophores. Using values shown in **Fig. 2**, the following calculations can be made:

BSA	HMdU
$A_{280} = x$	$A_{265} = y$
$A_{265} = 0.72x$	$A_{280} = 0.52y$

The 0.72 and 0.52 values are determined from spectra readings at 265 and 280 nm.

HMdU-BSA

$A_{265} = 0.72x + y = 1.10$	$A_{265}/A_{280} = 1.53$
$A_{280} = x + 0.52y = 0.72$	$(0.72x + y)/(x + 0.52y) = 1.53 \rightarrow y = 4x$
$A_{265} = 0.72x + 4x = 1.10 \rightarrow x = 0.23$	$y = 4x = 0.92$

Thus, at 265 nm 0.17 A_{265} were contributed by BSA and 0.92 A_{265} by HMdU, whereas at 280 nm 0.48 A_{280} were contributed by HMdU and 0.23 A_{280} by BSA.

Extinction coefficient (ε) for HMdU and BSA are 12,300 and 44,000, respectively. Thus,

$$\text{at } A_{265} = 1.10\text{: } \left.\begin{array}{l} 1 \times 10^6 \text{ nmol/mL} - 12{,}300 \\ z_1 \text{ nmol/mL} - 0.92 \end{array}\right\} \rightarrow z_2 = 74.8 \text{ nmol/mL HMdU;}$$

$$\text{at } A_{280} = 0.72\text{: } \left.\begin{array}{l} 1 \times 10^6 \text{ nmol/mL} - 44{,}000 \\ z_2 \text{ nmol/mL} - 0.23 \end{array}\right\} \rightarrow z = 5.23 \text{ nmol/mL BSA.}$$

$$74.5 \text{ HMdU:5.23 BSA} \quad \rightarrow 14.2 \text{ HMdU/BSA}$$

So far we have used HMdU-BSA at the ratios of 12-22 HMdU/BSA.

3.2. Coating Wells with Antigen

1. Prepare coating solution of M-BSA and HMdU-BSA (each: 10 µg antigen/mL CB, *see* **Subheading 2.5.2.**).
2. Place solutions (10 µg antigen/mL CB) in polyethylene basins and, using a multichannel pipeter, deliver 200 µL to each of the wells. Coat the left half of a microtiter plate with M-BSA and the right half with HMdU-BSA.
3. Seal the plates with sealing tape and store at 4°C for 3 d.
4. Pour the contents of the wells into the sink in one decisive movement of the hand. Wash the wells three times with TP buffer.
5. To prevent nonspecific sticking to walls, add 250 µL BSA/CB to each well, and after resealing with sealing tape, store the plates at 4°C for an additional 24 h.
6. Pour out the contents, wash wells three times with TP buffer, pat over paper towels until dry, and seal with sealing tape. Plates are now ready for use. At this state, plates can be stored at 4°C for 1–2 mo. Nonspecific binding to M-BSA-coated wells is usually very low (*see* **Subheading 3.3.2.**). Plates should not be used when nonspecific binding substantially increases.

3.3. Analysis of Human Sera for Anti-HMdU aAbs by ELISA (see Notes 2–4)

A schematic representation of ELISA using HMdU-BSA as an antigen is shown in **Fig. 3**.

3.3.1. Sample Preparation

Sera stored at –80°C retain anti-HMdU aAb activity over many years; the oldest we measured so far was stored for about 10 yr. Even when stored in the cold room, sera retain the anti-HMdU activity for several weeks. However, multiple freezing and thawing cycles eventually cause a decline in anti-HMdU aAb levels.

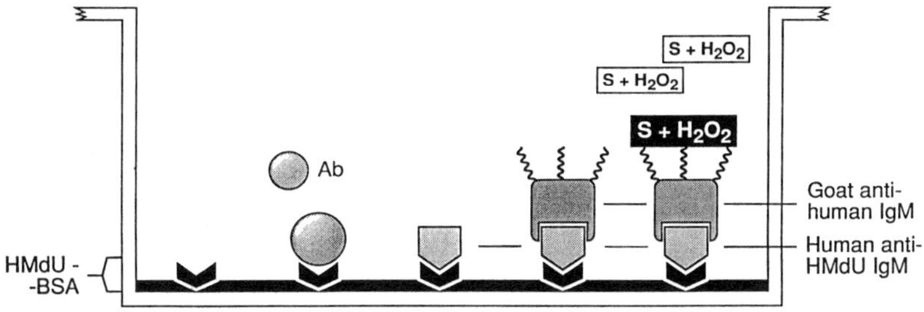

Fig. 3. Schematic representation of ELISA analysis of human sera for anti-HMdU Ab, using HMdU-BSA as the antigen. HRPO: Horseradish Px covalently bound to the secondary Ab. S: Substrate (o-phenylenediamine).

1. Split larger volumes of sera into smaller aliquots and store at –80°C, except for the vial to be used currently, which can be stored at 4°C.
2. For the results to be reproducible, take sera out of a –80°C freezer prior to use and leave at 4°C for 24 h. Remove any precipitate by centrifugation.

3.3.2. Plate Factor

1. Process the positive control serum (PCS) the same way as the unknowns. It is advisable to identify an individual with a high anti-HMdU aAb titer (i.e., 80–100 A_{492}/µL undiluted serum; 1:20,000 dilution should give A_{492} of 0.6–1.0) and secure as much serum as possible, perhaps over a period of time.
2. Combine multiple sera samples obtained from the same individual, aliquot, and store at –80°C. A tube containing PCS should be taken out of the freezer at the same time and treated the same way as other sera. Multiple analyses of the positive control should provide a reliable ("standard") aAb titer value. Knowing that "standard" value will allow control of the inherent variability in antigen coating of wells, batches of the secondary antibodies, and so on.

   We have analyzed a group of sera samples many times over a period of four years on plates coated at different times and using different batches of the secondary antibody. The use of the PCS provided us with a remarkable reproducibility, because we were able to use a plate factor (PF).
3. Calculate PF as the ratio between the net expected standard and experimental values. After subtracting the nonspecific binding value (M-BSA well) from the specific binding value (HMdU-BSA well), divide net expected standard A_{492} by net experimental A_{492}. Based on a series of assays using PCS diluted 20,000-fold, we established 0.746 as the net standard absorbance at 492 nm. This is our expected standard A_{492}. For each plate, the experimental PCS value is established based on three net readings (*see* **Fig. 4**).

   For example, if the expected PCS value is 0.746, and experimental values on different dates are 0.638 and 0.849, the plate factors are 1.17 and 0.88, respectively. To normalize the values of the unknowns they should be *multiplied* by the

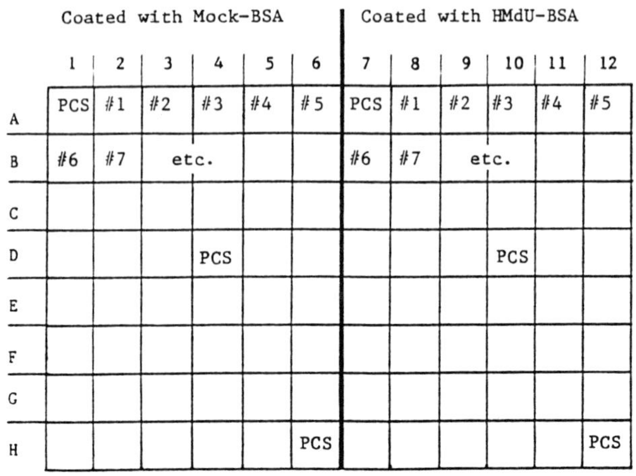

Fig. 4. Diagram showing organization of a microtiter plate. The left side (wells 1A–6H) is coated with M-BSA, while the right side (wells 7A–12H) is coated with HMdU-BSA. Positive control serum (PCS) is placed in 1A, 4D and 6H in M-BSA-coated wells, and in 7A, 10D and 12H, in HMdU-BSA-coated wells. Each of the sera samples (#1, #2, #3, etc.) is placed on both sides of the plate.

respective PF. This approach also allows comparisons among sera samples from different individuals, as well as among samples obtained from the same individual over period of time, and after dietary or medical interventions.

3.3.3. Plate Organization

1. Prepare a diagram showing where PCS will be placed on M-BSA and HMdU-BSA halves of the plate, and showing where the unknown samples will be placed.
2. Use three M-BSA (A1,D4,H6) wells and three HMdU-BSA (A7,D10,H12) wells for PCS at 20K (1:20,000) dilution (or at 10K, if 20K provides readings that are too low). **Figure 4** shows an example of such a diagram.
3. Prepare a protocol for sample setup as well as for dilution of each sample. We found that there is no need to use buffer as a negative control since binding to both M-BSA and HMdU-BSA was negligible. PCS binding to M-BSA provides a better indicator of the background (negative control) values. Although it seems tedious, this preparation helps to avoid errors, especially when many different sera samples are analyzed at different dilutions.

3.3.4. Sample Dilution

1. Dilute samples and positive controls with working buffer TPB (*see* **Subheading 2.5.3.**) to the appropriate concentration on the day of the assay. Use pipet tips with filters inside them to prevent contamination of pipet by sera (*see* **Note 2**). At the completion of ELISA, the absorbance at 492 nm (A_{492}) preferably should

be between 0.3 and 1.0 in HMdU-BSA-coated wells, but values up to ~1.7 or higher are acceptable if the linearity of the responses by the plate reader is assured.

2. To screen the unknowns, dilute sera initially 1 to 10,000 and analyze by ELISA (*see* **Subheading 3.3.5.**). The results of this screen will allow assessment of the most appropriate dilution for each of the samples. For example, if A_{492} values in HMdU-BSA-coated wells are: 0.1, 0.25, 0.7, 2.0, or "overflow" (off the scale), using Anthos II as a plate reader, we would dilute these samples with working buffer (*see* **Subheading 2.5.3.**) as follows: 1:2500, 1:5000, no change, 1:20,000, and 1:50,000 or 1:100,000. If some readings are still above the linear range of the plate reader, further dilutions should be made. Attention should be paid to the nonspecific binding levels. The serum sample should be assayed at the highest concentration on the linear portion of the HMdU-BSA binding curve that still gives low nonspecific binding to the M-BSA-coated wells.

3.3.5. ELISA for Anti-HMdU aAbs

1. Place 200 µL of each diluted sample in M-BSA (nonspecific binding)-coated and 200 µL in HMdU-BSA (specific binding)-coated wells. For dilution, use the working buffer. Start by applying PCS (*see* **Subheading 3.3.2.**), as shown in **Fig. 4**.
2. Apply all samples on a single plate within 10 min. Seal the plate with sealing tape and place it in the incubator at 37°C. Note the time. Then, if needed, apply samples to the second plate.
3. Incubate the plates at 37°C for 2 h. Remove the first plate after 2 h. Pour out solutions from the plate into a pan (*not into the sink!*) (*see below*), wash three times with washing buffer (TP) discarding the washing solution into a pan, and tap the plate on paper towels until dry. Put aside unsealed. Carry out the same procedure with the second plate. Using a multichannel pipet, apply 200 µL/well of the secondary antibody solution (goat antihuman IgM antibody) (*see* **Notes 1** and **5**) placed in a polyethylene basin, and seal the plates with sealing tape. Do not discard sera and washes into the sink. First, add concentrated bleach to the pan containing them, leave overnight, then discard and wash thoroughly. Each serum should be treated as if contaminated with HIV or hepatitis virus!
4. Incubate the plates at 37°C for 1 h. Remove all plates at the same time from the incubator. Pour liquid out into a pan, wash three times with washing buffer (TP) (*see* **Subheading 2.5.2.**), and tap plates on paper towels until dry. Use a multichannel pipet to apply 200 µL of the substrate solution per well (OPD + H_2O_2) (*see* **Subheading 2.5.3.**) placed in a polyethylene basin in the dark (under yellow lights).
5. Incubate the plates at 37°C for 0.5 h. At the end of incubation, add (in the dark, under the hood) 50 µL 48% H_2SO_4 (*see* **Subheading 2.5.5.**) per well, and seal the plates with sealing tape.
6. Incubate the plates at room temperature (in the dark) for 0.5 h. Remove the sealing tape. Place the plates successively in a plate reader and read at 492 nm versus 600 nm. This ($A_{492} - A_{600}$) provides absorbance at 492 nm after automatic subtraction of the background absorbance.

3.3.6. Evaluation of Results

1. Present the results as a net absorbance at 492 nm (A_{492}) of 1 µL undiluted serum (A_{492}/µL serum).
2. To obtain the anti-HMdU aAb titer A_{492}/µL serum, calculate two factors: plate factor (PF) and dilution factor (DF). Regardless of the dilution level, A_{492}/µL serum should be the same if the actual readings (net A_{492}) at those different dilutions of the same serum fall within linear range of the plate reader (*see* **Note 6**).

$$A_{492}/\mu L \text{ serum} = \text{net } A_{492} \times PF \times DF;$$

where

$$\text{net } A_{492} = A_{492}\text{ (HMdU-BSA-coated well)} - A_{492}\text{ (M-BSA-coated well)}.$$

a. Calculation of PF: As mentioned previously (*see* **Subheading 3.3.2.**), the PF is calculated from the results obtained for the PCS wells. Subtracting the mean ($\bar{x}$) nonspecific binding (M-BSA-coated wells) A_{492} value from the mean specific binding (HMdU-BSA-coated wells) A_{492} value of the PCS-containing wells provides net experimental A_{492} of PCS.

$$\text{Net experimental } A_{492} \text{ of PCS} =$$
$$[\bar{x} \text{ of } A_{492}\text{ (HMdU-BSA)}] - [\bar{x} \text{ of } A_{492}\text{ (M-BSA)}]$$

$$PF = \text{net expected standard } A_{492}/\text{net experimental } A_{492}.$$

Net expected standard A_{492} is established as shown in **Subheading 3.3.2.** above.

b. Calculation of DF: This factor is needed as a consequence of presenting results as $A_{492\,/\mu L}$ of serum. Serum is appropriately diluted (2500–200,000 times) and 200 µL are placed in a well for the first incubation. Therefore, DF = actual dilution of the sample/200. When all samples are initially diluted 10,000-fold (first ELISA), according to the above formula DF should be 50.

c. Calculation of anti-HMdU aAb titer: Here is an example of values we obtained. PCS diluted 20,000-fold had experimental A_{492} on HMdU-BSA-coated wells (A7,D10,H12) of 0.999, 0.925 and, 1.051, respectively. A_{492} values of the same samples placed in M-BSA-coated wells (A1,D4,H6) were 0.069, 0.077, and 0.079, respectively. The standard expected net A_{492} was 0.746. Therefore, PF = 0.81 (the calculations are shown below).

$$PF = 0.746/\{[(0.999 + 0.925 + 1.051)/3] -$$
$$[(0.069 + 0.077 + 0.079)/3]\} = 0.81.$$

Sample # 1 was diluted 10,000-fold, therefore DF = 10,000/200 = 50.
A_{492} values of the 10K-diluted sample #1 were 0.499 (A8, HMdU-BSA) and 0.099 (A2, M-BSA). Hence, net A_{492} = 0.499 – 0.099 = 0.400. Therefore the anti-HMdU aAb titer of sample #1 is:

$$A_{492}/\mu L \text{ serum} = 0.4 \times 0.81 \times 50 = 16.2.$$

d. Overall evaluation (*see* **Notes 7–9**): Each serum should be analyzed at least four times, using separately diluted samples (not from the same diluted sample!). The standard error (SE) of the results obtained using separately diluted samples is expected to be between 0.5 and 10% of the mean A_{492}/µL serum. If SE is larger than 10%, more independent assays should be carried out. Initially, to insure a good reproducibility on the same plate, duplicate analyses should be carried out from the same diluted sample. The overall number of ELISA analyses depends on the SE of the independent assays on samples of separately diluted sera.

4. Notes

1. As a secondary antibody, it is advisable to use goat antihuman IgM (µ-chain specific) IgG with conjugated HRPO, which is *not affinity-purified*. The affinity-purified antibodies have much lower specific activity in this assay. Moreover, new batches of the secondary antibodies should be analyzed against an aliquot of the antibody from a previous batch. The PF should control for this type of variability. However, if using a new batch of the secondary antibody changes PF drastically, check with the vendor. Vendors may change isolation and/or purification procedures, which could result in the need for a different dilution of that antibody.
2. All sera (plasma) should be treated as posing a significant health hazard, since some of them may harbor HIV and/or hepatitis B virus. All the procedures with sera (plasma) should be carried out in a separate location, preferably under a flow hood, using protective clothing, mouth and nose mask, gloves, and a clear face shield. Sera (plasma) samples and their solutions should first be disposed of into a pan and discarded only after an overnight treatment with concentrated bleach. Other items (gloves, pipet tips, tissues, paper towels, etc.) should be discarded into biohazard disposal bags. To dispense sera samples to the microtiter plate wells, it is advisable to use pipet tips containing a barrier inside to prevent contamination of the pipet from sera.
3. Serum or plasma can be used in these assays. We found, using plasma and serum from the same blood samples obtained from over 20 individuals, that there are no significant differences in A_{492}/µL. In this case, plasma was obtained from blood collected into EDTA- or heparin-containing vacutainer tubes. Red blood cells were separated by dextran sedimentation and white blood cells removed by centrifugation ***(28)***. The supernatant was used as plasma.
4. Sera should be taken from –80°C and placed in a refrigerator or a cold room overnight (up to 24 h) before ELISA. Aliquoting should be carried out after centrifugation at 4°C. One should avoid many freezing and thawing cycles. Rather than refreezing, it is better to keep a small (undiluted) aliquot at 4°C; anti-HMdU aAb titers are quite stable at this temperature over a period of several weeks.
5. These ELISA analyses can be carried out using secondary antibodies conjugate with alkaline phosphatase instead of HRPO. The only difference would be in preparation of the substrate for this enzyme.

6. The described assay may be utilized using different antigens. These antigens could include other oxidized nucleosides coupled to BSA or another protein or to a μ-chain of an IgM. This latter can be useful as an antigen for determination of general anti-IgM aAb levels present in human sera, of which anti-HMdU aAb is only one specific example. Assessment of both anti-HMdU and anti-μ aAb could provide a measure of anti-HMdU aAb specificity.
7. The person who is going to carry out this assay, at the end of training, should perform an "evaluation run." From one serum, take three aliquots and separately dilute each of them. These dilutions should yield A_{492} of ~0.6–0.8 in HMdU-BSA-coated wells. Apply each of these three samples to at least 10 wells coated with M-BSA and 10 coated with M-BSA. Then, run the assay. All of the readings should be very close.
8. To obtain reliable results, it is important to carry out separate ELISA analyses on sera (plasma) samples that are separately diluted. This is because the biggest error occurs by picking up a small volume of a viscous solution, which may have to be centrifuged before use. Obtaining A_{492}/μL serum (plasma) values in such separate experiments within 0.5–10% SE provides much more reliable results than assays carried out in duplicate or even in quadruplicate from the same dilution. The variability of the readings from the same diluted sample should be negligible.
9. In designing your experiments, you should take into consideration the following:
   a. Study subjects with a history of cancer among close family members (except those with nonmelanoma skin cancer) should not be used as controls for cancer cases in the case-control studies. They as a group have statistically higher anti-HMdU aAb titers than healthy controls without family history of cancer.
   b. Since anti-HMdU aAb titers can be significantly increased in individuals who will be diagnosed with cancer years later, use of those subjects as controls may confound the results of a typical case-control study. Perhaps it would be better to establish a baseline value based on anti-HMdU titers of apparently healthy controls by excluding the following: (i) those with chronic inflammatory conditions that occurred within 0.5 yr of blood donation; (ii) those who underwent surgery within a similar time-frame; and (iii) those taking potent anti-inflammatory, cytotoxic steroidal drugs. Then, the *persistently* high titers in otherwise healthy individuals may signal an increased risk of future cancer.
   c. Treatment of patients with psoriasis or immune complex diseases with anti-inflammatory, cytotoxic drugs decreases anti-HMdU aAb titers to the background value of healthy controls. It is not yet known whether those types of drugs would decrease anti-HMdU aAb levels in sera of individuals at high risk for cancer or patients with cancer. However, it is advisable to have available the information about intake of those drugs as well as of antioxidant supplements, since their intake could confound the results.
   d. The background information should also include smoking, occupational exposures, infections, etc.

References

1. Frenkel, K., Khasak, D., Karkoszka, J., Shupack, J., and Stiller, M. (1992) Enhanced antibody titers to an oxidized DNA base in inflammatory and neoplastic diseases. *Exp. Dermatol.* **1,** 242–247.
2. Frenkel, K., Karkoszka, J., Kim, E., and Taioli, E. (1993) Recognition of oxidized DNA bases by sera of patients with inflammatory disease. *Free Rad. Biol. Med.* **14,** 483–494.
3. Frenkel, K., Karkoszka, J., Glassman, T., Dubin, N., Toniolo, P., Taioli, E., Mooney, L., and Kato, I. (1997) Serum autoantibodies recognizing 5-hydroxymethyl-2'-deoxyuridine, an oxidized DNA base, as biomarkers of cancer risk in women. *Cancer Epidemiology Biomarkers and Prevention,* in press.
4. Frenkel, K., Glassman, T., Karkoszka, J., and Taioli, E. (1994) Anti-oxidized DNA base autoantibodies as a potential biomarker of high risk for the development of human breast cancer. *Proc. Am.* Assoc. *Cancer Res.* **35,** 97.
5. Frenkel, K., Karkoszka, J., Glassman, T., Powell, J., Pero, R., Toniolo, P., and Taioli, E. (1995) Autoantibodies that recognize oxidized DNA bases as biomarkers of cancer risk. *Proc. Am.* Assoc. *Cancer Res.* **36,** 284.
6. Djuric, Z., Heilbrun, L. K., Simon, M. S., Luongo, D. A., LoRusso, P. M., and Martino, S. (1996) Levels of 5-hydroxymethyl-2'-deoxyuridine in blood DNA as a marker of breast cancer risk. *Cancer* **77,** 691–696.
7. Djuric, Z., Heilbrun, L. K., Reading, B. A., Boomer, A., Valeriote, F. A., and Martino, S. (1991) Effects of a low fat diet on levels of oxidative damage to DNA in human peripheral nucleated blood cells. *J. Natl. Cancer Inst.* **83,** 766–769.
8. Malins, D. C. and Haimanot, R. (1991) Major alterations in the nucleotide structure in DNA in cancer of the female breast. *Cancer Res.* **51,** 5430–5432.
9. Dunham, L. J. (1972) Cancer in man at a site of prior benign lesion of skin or mucous membrane: a review. *Cancer Res.* **32,** 1359–1374.
10. Weitzman, S. A. and Gordon, L. I. (1990) Inflammation and cancer: role of phagocyte-generated oxidants in carcinogenesis. *Blood* **76,** 655–663.
11. Frenkel, K. (1992) Carcinogen-mediated oxidant formation and oxidative DNA damage. *Pharmac. Ther.* **53,** 127–166.
12. Breimer, L. H. (1990) Molecular mechanisms of oxygen radical carcinogenesis and mutagenesis: the role of DNA base damage. *Mol. Carcinogenesis* **3,** 188–197.
13. Teebor, G. W., Boorstein, R. J., and Cadet, J. (1988) The repairability of oxidative free radical mediated damage to DNA: a review. *Int. J. Radiat. Biol.* **54,** 131–150.
14. Trush, M. A. and Kensler, T. W. (1991) An overview of the relationship between oxidative stress and chemical carcinogenesis. *Free Rad. Biol. Med.* **10,** 201–209.
15. Frenkel, K., Wei, L., and Wei, H. (1995) 7,12-Dimethylbenz[a]anthracene induces oxidative DNA modification *in vivo*. *Free Rad. Biol. Med.* **19,** 373–380.
16, Liehr, J. G. and Roy, D. (1990) Free radical generation by redox cycling of estrogens. *Free Rad. Biol. Med.* **8,** 415–423.
17. Dipple, A., Pigott, M. A., Bigger, A. H., and Blake, D. M. (1984) 7,12-Dimethylbenz[a]anthracene-DNA binding in mouse skin: response of different mouse strains and effects of various modifiers of carcinogenesis. *Carcinogenesis* **5,** 1087–1090.

18. McCormick, D. L., Major, N., and Moon, R. C. (1984) Inhibition of 7,12-dimethyl-benz[a]anthracene-induced rat mammary carcinogenesis by concomitant or postcarcinogen antioxidant exposure. *Cancer Res.* **44,** 2858–2863.
19. Wattenberg, L. W. (1980) Inhibition of chemical carcinogenesis by antioxidants, in Carcinogenesis, Vol. 5., Modifiers of Chemical Carcinogenesis (Slaga, T. J., ed.), Raven Press, New York, pp. 85–98.
20. Perchellet, J.-P. and Perchellet, E. M. (1989) Antioxidants and multistage carcinogenesis in mouse skin. *Free Rad. Biol. Med.* **7,** 377–408.
21. Wei, H. and Frenkel, K. (1993) Relationship of oxidative events and DNA oxidation in SENCAR mice to *in vivo* promoting activity of phorbol ester-type tumor promoters. *Carcinogenesis* **14,** 1195–1201.
22. Bhimani, R. S., Troll, W., Grunberger, D., and Frenkel, K. (1993) Inhibition of oxidative stress in HeLa cells by chemopreventive agents. *Cancer Res.* **53,** 4528–4533.
23. Frenkel, K., Wei, H., Bhimani, R., Ye, J., Zadunaisky, J. A., Huang, M.-T., Ferraro, T., Conney, A. H., and Grunberger, D. (1993) Inhibition of tumor promoter-mediated processes in mouse skin and bovine lens by caffeic acid phenethyl ester. *Cancer Res.* **53,** 1255–1261.
24. Huang, M.-T., Ma, W., Yen, P., Xie, J.-Q., Han, J., Frenkel, K., Grunberger, D., and Conney, A. H. (1996) Inhibitory effects of caffeic acid phenethyl ester (CAPE) on 12-*O*-tetradecanoyl-phorbol-13-acetate-induced tumor promotion in mouse skin and the synthesis of DNA, RNA and protein in HeLa cells. *Carcinogenesis* **17,** 761–765.
25. Wei, H. and Frenkel, K. (1992) Suppression of tumor promoter-induced oxidative events and DNA damage in vivo by sarcophytol A: a possible mechanism of antipromotion. *Cancer Res.* **52,** 2298–2303.
26. Bhimani, R. S., Zhong, Z., Schleifer, E., Troll, W., and Frenkel, K. (1995) Human promyelocytic leukemia cells (HL-60), a new model to study the effects of chemopreventive agents on H_2O_2 production. *Cancer Detect. Prev.* **19,** 292–298.
27. Lim, J. S., Frenkel, K., and Troll, W. (1992) Tamoxifen suppresses tumor promoter-induced hydrogen peroxide formation by human neutrophils. *Cancer Res.* **52,** 4969–4972.
28. Frenkel, K., Chrzan, K., Ryan, C. A., Wiesner, R., and Troll, W. (1987) Chymotrypsin-specific protease inhibitors decrease H_2O_2 formation by activated human polymorphonuclear leucocytes. *Carcinogenesis* **8,** 1207–1212.
29. Frenkel, K., Karkoszka, J., Cohen, B., Baranski, B., Jakubowski, M., Cosma, G., Taioli, E., and Toniolo, P. (1994) Occupational exposures to Cd, Ni and Cr modulate titers of anti-oxidized DNA base autoantibodies. *Environ. Health Perspect.* **102(Suppl. 3),** 221–225.
30. Frenkel, K. (1996) U. S. Patent No. 5,552,285. Immunoassays methods, compositions and kits for antibodies to oxidized DNA bases.
31. Taioli, E., Kinney, P., Zhitkovich, A., Fulton, H., Voitkun, V., Cosma, G., Frenkel, K., Toniolo, P., Garte, S., and Costa, M. (1994) Application of reliability models to studies of biomarker validation. *Environ. Health Perspect.* **102,** 306–309.
32. Erlanger, B. F. and Beiser, S. M. (1964) Antibodies specific for ribonucleosides and ribonucleotides and their reaction with DNA. *Proc. Natl. Acad. Sci. USA* **52,** 68–74.

27

A Scientific Basis for Cancer Prevention

Defining the Role of Individual Cytosolic GST Isozyme

Sanjay K. Srivastava and Sanjay Awasthi

1. Introduction

Highly electrophilic functional groups of exogenous and endogenous chemicals represent a significant threat to the structural integrity of DNA because of their propensity to react with nucleophilic sites on DNA bases. The accumulation of electrophile-mediated DNA lesions in cellular-growth regulatory genes, and their expression-controlling elements with ensuing dysregulation of growth, has been proposed as a common pathway leading to neoplastic transformation *(1)*. Glutathione S-transferases (GSTs) are a large family of cytosolic and microsomal enzymes that share the common ability to catalyze the formation of thioether conjugates of glutathione (GSH) with a wide array of structurally unrelated electrophilic toxins (including several known carcinogens and mutagens), but which differ in catalytic efficiencies toward different electrophilic substrates and in other non-S-transferase catalytic activities *(2)*. Numerous animal studies showing increased tissue GST activity in response to electrophilic carcinogen exposure indicate that GSTs may function as a primary defense mechanism for protecting nucleophilic groups on DNA bases from mutagenic electrophiles *(2,3)*. A strong correlation between the ability of dietary phenolic antioxidants to preferentially induce phase II biotransformation enzymes such as GSTs, and their ability to prevent neoplasia induced by subsequent chemical carcinogen exposure, further supports the idea that GSTs serve an important role in defending DNA from electrophilic toxins *(2,3)*. These studies as well as mechanistic studies, which have linked the regulation of expression of GST isozymes with the Michael-acceptor electrophilic functional groups of carcinogens, phenolic antioxidants, and their metabolites *(3)*, have laid

From: *Methods in Molecular Medicine, Vol. 14: Tumor Marker Protocols*
Edited by: M. Hanausek and Z. Walaszek © Humana Press Inc., Totowa, NJ

a foundation for defining optimal strategies for cancer prevention by altering the expression of GST isozymes through pharmacologic and dietary means.

The substrate specificities for each class of GST isozyme are quite different and the expression of each class of GSTs appears to be controlled differentially by endogenous and exogenous chemicals in a poorly understood organ- and sex-specific manner ***(2)***. It is thus quite possible that individual GST isozymes may preferentially provide better protection against certain groups of chemically related electrophilic toxins. Pharmacologic or dietary recommendations may have to be individually tailored on the basis of sex, type of carcinogen exposure, and other individual risk factors for development of neoplasia. Identification of patterns of induction of GST isozymes associated with optimal ability to protect against the deleterious effects of individual or classes of chemical carcinogens may thus be very useful for designing cancer-preventative dietary or pharmacologic strategies. Because cancer-preventative antioxidants can function as pro-oxidants and carcinogens at high doses ***(3)***, studies designed to investigate the relationship between the dose-response curves of induction of individual GST isozymes and the cancer-preventative activity of dietary or pharmacologic antioxidants may help identify dose levels likely to provide protection against carcinogens without undue risk of promoting carcinogenesis. Such studies require standardized protocols to purify, quantify, and characterize the catalytic properties of individual GST isozymes. Since rodents are most commonly used to study the inhibition of chemical carcinogenesis by dietary antioxidants, and because GSTs of the liver are most frequently characterized, we will describe the specific protocols used in our laboratory for the purification, isolation, identification, and quantitation of individual cytosolic GST isozymes of mouse liver.

2. Materials

2.1. Purification of Cytosolic GSTs

2.1.1. Preparation of GSH-Affinity Resin

1. Epoxy-activated Sepharose 6B: 5 g (Sigma, St. Louis, MO).
2. Linking buffer: 100 mL of 44 m*M* Na/K phosphate buffer, pH 7.0, prepared by diluting 8 mL of 0.2 *M* KH_2PO_4 and 28 mL of 0.1 *M* Na_2HPO_4 to 100 mL with distilled water.
3. Glutathione (GSH) for preparation of GSH-affinity resin: 5 mL of 100 mg/mL GSH in linking buffer with pH adjusted to 7.0 with KOH solution, prepared fresh.
4. Ethanolamine: 25 mL of 1 *M* solution, pH 8.0.
5. KCl/sodium acetate: 100 mL of 0.5 *M* KCl in 0.1 *M* sodium acetate, pH 4.0.
6. KCl/sodium borate: 100 mL of 0.5 *M* KCl in 0.1 *M* sodium borate, pH 8.0.
7. Affinity buffer: 100 mL of 22 m*M* potassium phosphate, pH 7.4, containing 1.4 m*M* β-mercaptoethanol (β-ME).

2.1.2. GST Assay

1. Assay buffer: 100 m*M* potassium phosphate buffer, pH 6.5.
2. 1-Chloro-2,4-dinitrobenzene (CDNB): 20 m*M* in ethanol, prepared fresh.
3. Glutathione: 10 m*M* in assay buffer, prepared fresh.

2.1.3. Homogenate Preparation

1. Buffer A: 10 m*M* potassium phosphate buffer, pH 7.0, containing 1.4 m*M* β-ME.
2. Miracloth (Calbiochem, La Jolla, CA).
4. Dialysis tubing: Molecular weight cutoff of 6000–8000.

2.1.4. GSH-Affinity Chromatography

1. GSH-affinity resin (prepared as described in **Subheading 3.1.1.**): 1 mL bed volume of affinity resin per 30 U GST activity in the homogenate.
2. Peristaltic pump and sealed chromatography columns with 5–10-mL capacity.
4. Affinity buffer (*see* **Subheading 2.1.1.**): 500 mL.
5. Elution buffer: 10 m*M* GSH in 50 m*M* Tris-HCl, pH 9.6, containing 1.4 m*M* β-ME.

2.2. Characterization of Purified GST

2.2.1. SDS-PAGE of Affinity Purified GSTs

1. Suitable mini-gel apparatus.
2. Acrylamide/*bis*-acrylamide: 60 g acrylamide (Bio-Rad, Hercules, CA) and 1.6 g *bis*-acrylamide (BioRad) in 100 mL water.
3. Resolving gel buffer: 3 *M* Tris-HCl, pH 8.8.
4. Stacking-gel buffer: 0.5*M* Tris-HCl, pH 6.8.
5. Sodium dodecyl sulfate (SDS): 10% (w/v) solution of electrophoresis grade SDS in water.
6. β-ME: 14.3 *M* solution of electrophoresis grade β-ME (*see* **Subheading 2.1.1.**).
7. Ammonium persulfate (APS): 50 mg APS/mL water, freshly prepared.
8. *N,N,N',N'*-Tetramethylethylenediamine (TEMED): Bio-Rad.
9. Running buffer: 3 g Tris-HCl, 14.4 g glycine, 1 g SDS dissolved in water q.s. 1 L, followed by adjusting pH to 8.3.
10. Sample buffer: 1.25 mL 0.5 *M* Tris-HCl, pH 6.8, 0.15 mL 14.3 *M* β-ME, 0.2 g SDS, 0.2 mL 0.1% bromophenol blue, 1.5 mL glycerol, and water q.s. 10 mL.
11. Gel-staining solution: 0.1% Coomassie blue R (Bio-Rad), 20% methanol, and 10% acetic acid in water.

2.2.2. Raising Polyclonal Antibodies Against GST Isozymes

1. Purified GST antigens (*see* **Subheading 3.1.**).
2. New Zealand white rabbits.
3. Freund's complete and incomplete adjuvants (Sigma).
4. Buffer B: 10 m*M* potassium phosphate buffer, pH 7.0 (without β-ME).
4. DE-52 anion exchange resin, washed and equilibrated in buffer B.

5 Buffer C: 10 m*M* Tris-HCl, pH 8.0.
6. Protein A-resin (Sigma).
7. Protein A-elution buffer: 100 m*M* glycine, pH 3.0.
8. Cyanogen-bromide-activated Sepharose 4B: 5 g (Sigma)
9. 4 *M* Potassium thiocyanate in Buffer B.

2.2.3. Western Blot Analyses of GSTs

1. Blotting buffer: 20% methanol, 25 m*M* Tris and 192 m*M* glycine, pH 8.3.
2. Suitable electroblotting apparatus.
3. Tris-buffered saline (TBS): 10 m*M* Tris-HCl containing 0.9% NaCl, pH 7.4.
4. Blocking solution: 5% (w/v) low-fat milk powder in TBS.
5. Anti-GST primary antibody, specific for one class of GST isozyme (monoclonal or polyclonal).
6. Appropriate horseradish peroxidase-linked secondary antibody (Sigma).
7. Developing reagent: 100 mL of freshly prepared solution of 0.02% H_2O_2 in TBS to which 20 mL cold methanol containing 60 mg 4-chloro-1-naphthol (Bio-Rad) is added just prior to use.

2.3. Immunoaffinity Affinity Purification of GST Isozymes

1. Cyanogen bromide-activated Sepharose 4B (5 g, Sigma). It can be linked to antibodies specific for a single class of GST isozyme; GST antibody can be substituted (100 μg/mL resin) for GSH in the procedure described for preparation of GSH-affinity resin (*see* **Subheading 3.1.1.**). β-ME must be excluded from all steps.
2. 1 m*M* HCl.
3. Coupling buffer: 0.1 *M* $NaHCO_3$, pH 7.9, containing 0.5 *M* KCl.
4. 1 *M* Ethanolamine, pH 8.0.
5. KCl/sodium acetate: 100 mL 0.5 *M* KCl in 0.1 *M* sodium acetate.
6. KCl/sodium borate: 100 mL 0.5 *M* KCl in 0.1 *M* sodium borate.
7. Buffer B (*see* **Subheading 2.2.2.**).
8. 4 *M* Potassium thiocyanate.
9. GSH-affinity purified total GSTs dialyzed against buffer B.

2.4. Isolation of GSTs by Liquid Column Isoelectric-Focusing

1. Column chromatofocusing apparatus, gradient mixer, power supply, and fraction collector (Pharmacia-LKB, Uppsala, Sweden).
2. Cathode solution: 100 mL 0.25 N NaOH containing 0.6 g/mL sucrose.
3. Anode solution: 100 mL 0.15 *M* phosphoric acid in water.
4. Dense solution: 54 mL solution containing 27 g sucrose, 2.1 mL Ampholine pH, range 3.5 to 10 (Pharmacia) and dialyzed GSH-affinity purified total GST (at least 0.2 mg).
5. Light solution: 54 mL solution containing 2.7 g sucrose and 0.7 mL Ampholine, pH range 3.5 to 10.

2.5. Substrate Specificities of Purified GST Isozymes

2.5.1. Determination of S-Transferase activity of GSTs

1. Substrates: Substrates (*see* **Tables 1–3**) can be obtained from Sigma.
2. TLC plates for 9,10-epoxystearic acid (ESA) and leukotriene A_4 methyl ester (LTA_4ME)
3. [^3H]-GSH for ESA and LTA_4ME activities.

2.5.2. Determination of GSH-Peroxidase Activity of GSTs

1. Assay Buffer: 1 *M* Tris-HCl, EDTA 5 m*M*, pH 8.0.
2. 2 m*M* NADPH (Sigma) freshly prepared in water.
3. Glutathione reductase (Sigma): 10 U/mL diluted fresh from stock.
4. Cumene hydroperoxide: 10 m*M* solution, freshly diluted in water from a 5.3 *M* stock (Sigma).
5. GSH: 100 m*M* solution in assay buffer, prepared fresh.

3. Methods

3.1. Purification of Cytosolic GSTs (see Notes 1–5)

3.1.1. Preparation of GSH-Affinity Resin

Epoxy-activated Sepharose 6B linked with GSH should be prepared prior to beginning GST purification.

1. Place the resin in a Buchner funnel and wash first with approx 500 mL distilled water and then with 200 mL linking buffer.
2. Transfer the washed resin to a small flask and adjust the volume to 20 mL with linking buffer.
3. De-gas the resin by bubbling with nitrogen.
4. Add 5 mL of GSH.
5. Bubble the resin with nitrogen for an additional 5 min.
6. Allow the coupling reaction to proceed in a sealed container at 37°C on an orbital shaker for 24 h.
7. Place the GSH-Sepharose in a Buchner funnel and wash with 100 mL distilled water.
8. Block the unreacted epoxy-groups by incubating the GSH-Sepharose resin in 25 mL 1 *M* ethanolamine, pH 8.0, for 4 h at room temperature on an orbital shaker.
9. Wash the affinity resin sequentially with 100 mL each of distilled water, KCl/sodium acetate, KCl/sodium borate, and affinity buffer.

Approximately 3 mL affinity resin is obtained from each gram of epoxy-activated Sepharose 6B. The resin can be stored in a sealed dark container in affinity buffer at 4°C for up to a week. Used affinity resin can be regenerated by washing with the same three buffers sequentially (*see* **step 9**).

3.1.2. GST Assay

The substrate most useful for following GST activity during purification is 1-chloro-2,4-dinitrobenzene (CDNB). The glutathionyl-thioether of CDNB (dinitrophenyl-S-glutathione, DNPSG) is formed through a substitution of the sulfur of sulfhydryl group on GSH for the chloride at the 1-position with formation of HCl and DNPSG. At room temperature, the nonenzymatic reaction is at least an order of magnitude slower at pH 6.5 than the enzyme-catalyzed reaction. Since the extinction coefficient of DNPSG at 340 nm (9.6 mM^{-1} cm^{-1}) is significantly greater than that of CDNB (0.6 mM^{-1} cm^{-1}), the reaction at room temperature can be followed by monitoring absorbance at 340 nm of the reaction mixture against a blank containing CDNB and GSH without enzyme. The assay described is performed according to that described by Habig et al. ***(4)***.

1. To the experimental cuvet, sequentially add 20 μL appropriately diluted enzyme, 100 μL 10 m*M* GSH prepared in assay buffer, and 830 μL assay buffer.
2. Add 100 μL GSH to the blank cuvet, 850 μL assay buffer, and no enzyme.
3. Start the reaction by the addition of 50 μL 20 m*M* CDNB to both the blank and experimental cuvets.

The enzyme should be diluted sufficiently to obtain a linear increase in absorbance at 340 nm for at least 4 min. For a 20% homogenate of liver tissue, a 100-fold dilution is usually sufficient to obtain a linear increase in absorbance. One unit of GST activity is defined as 1 μmol of DnpSG formed per minute at 25°C.

3.1.3. Preparation of Homogenate

Freshly excised liver should be perfused with phosphate-buffered saline to remove the majority of blood. Purifications performed on fresh tissues give approx 10% higher yield of GSTs than tissue frozen at –20°C for up to 1 mo. Protease inhibitors are not necessary to obtain good yields of active enzyme. GSTs account for approx 3% of total cytosolic protein in the 28,000*g* supernatant of mouse liver homogenate (*see* **Note 1**). Approximately 300 to 500 μg total purified GSTs are necessary to fully and reliably quantify and characterize the individual GST isozymes of mouse liver. Purification can thus be started with as little as 500 mg tissue, though we commonly use 1 g tissue. All steps of purification should be performed at 4°C. Leaving the enzyme at room temperature for as little as 1 h can cause significant loss of yield of some GST isozymes due to differential heat stability.

1. Prepare a 20% homogenate of minced tissue using a Tekmar homogenizer in Buffer A.
2. Centrifuge the homogenate at 28,000*g* for 45 min in a refrigerated centrifuge.
3. Collect the supernatant of homogenate by filtering through Miracloth.

4. After adjusting the volume of supernatant to 20% with buffer A, take a small (<1% of total volume) aliquot for protein and activity determination (*see* **Notes 2** and **3**). The 28,000*g* supernatant of a 20% homogenate from 1 g mouse liver contains between 60 and 100 units of GST activity toward CDNB and approx 50 mg total protein.
5. Dialyze the 28,000*g* supernatant of homogenate against at least 500 volumes of buffer A over 24 h. To prevent evaporation of β-ME, the dialysis vessel should be sealed with parafilm (*see* **Note 3**).

*3.1.4. GSH-Affinity Chromatography (*see ***Note 6***)*

1. While the homogenate is being dialyzed, pack the GSH-affinity chromatography column and equilibrate with affinity buffer (*see* **Subheading 2.1.4.**) flowing at 6 mL/min. This flow rate is maintained constant for the entire affinity-chromatography. The bed volume of GSH-affinity resin should be based on total GST activity in the homogenate. We use 1 mL resin bed volume per 30 U of GST activity.
2. After taking an aliquot for protein and activity determinations, pass the dialyzed homogenate over the affinity column at a flow rate of 6 mL/min and collect the unabsorbed fraction for protein and activity determination. The level of fluid above the affinity-resin bed should be minimal in order to minimize binding of proteins to the sides of the column. After loading the protein on the affinity column, the sides of the column should first be rinsed using a Pasteur pipet with affinity buffer before washing the affinity-resin with affinity buffer. Between 20 and 30 bed volumes of affinity buffer are required to wash the resin such that the absorbance at 280 nm of the effluent reaches zero.
3. Elute GSTs with 3 to 4 bed volumes of elution buffer.
4. Dialyze the eluate against at least 500 volumes of buffer A (*see* **Subheading 2.1.3.**) for 24 h. Up to 20% of total GST activity is found in the unabsorbed fraction with the remaining activity being present in the eluate. The unbound GSTs include the θ-class GSTs, which can be purified using methods described elsewhere ***(5)***.

3.2. Characterization of Purified GST

3.2.1. SDS-PAGE of Purified Total GSTs

We have routinely performed SDS/β-mercapoethanol/polyacrylamide slab gel electrophoresis with a 7% stacking gel and 12.5% resolving gel using the buffer system of Laemmeli ***(6)***.

1. To make six resolving gels for the BioRad Protean II Mini-Gel system, prepare sufficient acrylamide/*bis*-acrylamide solution containing 14.5 mL water, 4.7 mL acrylamide/*bis*-acrylamide, 0.225 mL SDS, and 2.8 mL resolving-gel buffer.
2. De-gas the solution under vacuum for 10 min, add 4 μL β-ME and mix gently.
3. Start polymerization by adding 225 μL of freshly prepared APS followed by 22.5 μL TEMED.

4. Pour the resolving gel to a height of 5 cm with a layer of water carefully layered on top to ensure a horizontal gel interface. Polymerization requires approx 30 min.
5. To make six stacking gels, prepare sufficient acrylamide/*bis*-acrylamide solution from a solution containing 3 mL water, 0.6 mL acrylamide/*bis*-acrylamide, 0.05 mL SDS and 1.3 mL stacking-gel buffer.
6. De-gas it under vacuum for 10 min, add 1 μL of 14.3 *M* β-ME and mix the solution gently.
7. Start polymerization by addition of 50 μL APS and 5 μL TEMED.
8. After pouring off the water from the top of the resolving gel and drying carefully using blotting paper, place the combs for forming the wells on the gel polymerization apparatus and pour the stacking gel. Approximately 30 min are required for complete polymerization.
9. Dilute 2–4 μg purified lyophilized GST protein suspended in 20 μL water with 20 μL sample buffer and boil for less than 30 s before loading on the gel.
10. Submerse the gel in running buffer and run at constant 200 V for 45 min.
11. Stain the gel at room temperature overnight on an orbital shaker by submersion in staining solution. After destaining with 20% methanol and 10% acetic acid in water, the purified total GSTs of mouse liver reveal three bands at approx 25.5, 24, and 22.5 kDa, corresponding to the μ, α, and π class isozymes, respectively ***(7)***.

3.2.2. Raising Polyclonal Antibodies Specific of Each Class of GSTs

Identification of class of GST isozyme is accomplished using highly specific polyclonal or monoclonal antibodies that recognized each class of enzymes without significant cross-reactivity. The monoclonal antibodies are commercially available from Biotrin International, Dublin, Ireland. We have raised our own highly specific polyclonal antibodies in the rabbit toward the μ, α, and π class of GSTs. We have also raised highly specific polyclonal antibodies against mGST A-4, a mouse-GST isozyme belonging to the α-class that has unique substrate specificity and only 60% homology with the α-class. This subclass contains several very closely related isozymes and is also represented in the rat (rGST 8-8), human (provisionally designated hGST A4-4), chicken (cGST CL3), and cow (provisionally designated βGST 5.8). The antibodies raised against each class of mouse-GST isozymes also demonstrate a high degree of specificity for the particular class of isozymes in other species, including rat and human ***(8)***. The same is true for the antibodies raised against purified human isozymes of each class.

3.2.2.1. Raising Polyclonal Antibodies Specific for GSTs

The polyclonal antibodies used in our lab have been raised against highly purified GST isozymes separated by isoelectric focusing (*see* **Subheading 3.4.**). The protocol that can be used to raise these antibodies is presented here.

1. Dissolve an 80-μg aliquot of the purified enzyme, extensively dialyzed against water, in Freund's complete adjuvant and inject subcutaneously into rabbits.

2. Two and four weeks later, administer booster doses of 50 μg antigen dissolved in Freund's incomplete adjuvant.
3. Two weeks after the last booster, collect 30–40 mL blood by venipuncture from a vein on the posterior surface of the animal's ear.
4. Place the plasma at 4°C overnight and centrifuge the next day at 14,000*g*.
5. After heat inactivation of the supernatant at 56°C for 1 h, remove the precipitate by centrifugation at 14,000*g* for 30 min.
6. Subject the supernatant fraction to DE-52 anion-exchange chromatography in buffer B.
7. Collect the unadsorbed fraction containing immunoglobulin and dialyze against buffer C.
8. Pass the dialyzed immunoglobulin over a column containing Protein-A equilibrated with buffer C.
9. Elute the bound immunoglobulin using protein A elution buffer.

The antibodies thus obtained are then subjected to purification over cyanogen bromide-activated Sepharose 4B linked to purified GST antigens.

3.2.2.2. Purification of Polyclonal Antibodies Specific for GSTs

GST antigens are linked to the cyanogen bromide-activated Sepharose 4B using the same procedure used for preparation of the GSH-affinity resin except that use 1 m*M* HCl instead of linking buffer, and coupling buffer (*see* **Subheading 2.3.**) instead of water or affinity buffer (*see* **Subheading 3.1.1.**). Approximately 150 μg antigen is used per mL of resin for preparation of the GST-antigen-affinity resin. β-ME is specifically excluded from all steps of antibody purification and from all immuno-affinity chromatography.

1. Equilibrate the GST-antigen resins with buffer B.
2. If antibodies have been raised against GST-α, pass them first successively through GST-μ, GST-π, and mGST A4-4 antigen-affinity resins and collect the unadsorbed fractions.
3. Finally, pass the antibodies over the GST-α antigen-affinity resin.
4. Elute the bound antibodies with 4 bed volumes of 4 *M* potassium thiocyanate and dialyze against buffer B prior to use.

Antibodies thus obtained are sufficiently specific for the given GST isozyme that they will recognize only that enzyme either in a mixture of purified GSTs or in crude homogenate. We have used these antibodies for immunoprecipitation of the activity of individual GST isozymes and for immuno-affinity purification of individual GST isozymes from preparations of purified GSTs *(8)*.

3.2.3. Western Blot Analyses of Purified Total GSTs

For Western blot analyses, we perform SDS-PAGE using the Laemmeli buffer system as described above and immunoblotting that is essentially according to the procedure of Towbin et al. *(9)*.

1. After removing the polyacrylamide gel from the electrophoresis apparatus, wash it with blotting buffer.
2. Subsequently, perform blotting on to nitrocellulose membrane at 200 mA constant current for 4 h.
3. After blotting, incubate the nitrocellulose membrane with 20 mL blocking solution for 1 h to block nonspecific binding sites.
4. Add 10 μL primary antibody and incubate at room temperature overnight.
5. Wash off the primary antibody with blocking solution.
6. Add 10 μL of a horseradish peroxidase (HRP) coupled goat antirabbit secondary antibody (Sigma) in 10 mL blocking solution and incubate for an additional 4 h. If monoclonal antibodies are used, HRP-linked antimouse secondary antibodies must be used at this step.
7. Wash off the secondary antibody and milk with TBS.
8. Incubate the nitrocellulose membrane in freshly mixed developing reagent for a short period (approx 5 min) until adequate color development.
9. Terminate the color development reaction by discarding the developing solution and washing the blot with water.
10. Quantify the intensity of color of a given band by scanning densitometry. An appropriate standard curve can be used to quantify the immuno-reactive protein.

3.3. Immuno-Affinity Purification of GST Isozymes

An immuno-affinity resin can be prepared by attaching a given antibody to cyanogen bromide-activated Sepharose 4B using a procedure similar to that used for preparing GSH-affinity resin, except that we use 1 m*M* HCl instead of linking buffer, and coupling buffer instead of water or affinity buffer. The coupling between the antibody and the resin should be done at 4°C for 20–24 h. We use 200 μg antibody per mL of resin during the coupling step. As for other procedures involving antibodies, β-ME is excluded from all steps of resin preparation and immuno-affinity chromatography. The amount of immuno-affinity resin to be used for purification depends on the relative abundance of the GST isozyme to be purified. In general, 1 mL immuno-affinity resin prepared in this fashion can easily bind 100 μg antigen. To minimize loss, however, we generally use about 1.5 times the amount of resin necessary for an expected amount of a given GST isozyme. Thus, for purification of GST-π (which represents about 90% of total GSTs of mouse liver) from 100 μg of total purified GSTs, approx 1.4 mL resin is sufficient for quantitative recovery of GST-π. On the other hand, for GST-α, which represents less than 4% of total GSTs, quantitative recovery from 100 μg total purified GSTs can be expected to be from as little as 0.1 mL anti-GST-α affinity-resin. Immuno-affinity purification can be easily accomplished by batch process rather than column chromatography.

1. Incubate the affinity-resin with total purified GSTs at 4°C for 1 h in a sealed container on a shaker.

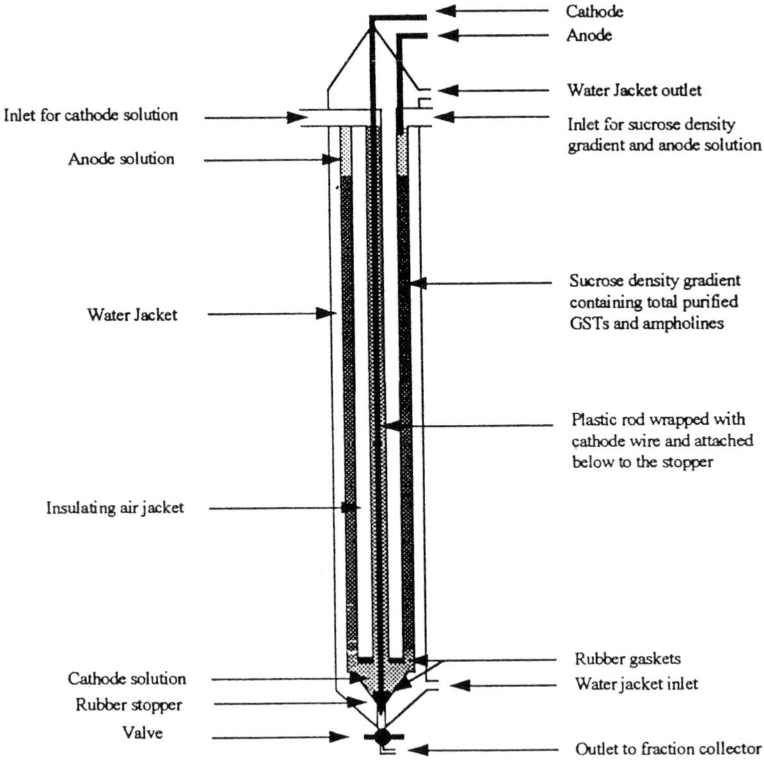

Fig. 1. A cross-sectional schematic of a liquid column isoelectric focusing column. The retractable stopcock is shown in the position in which IEF is performed. When eluting, the stopcock is retracted by pulling up on the center bar (around which the cathode wire is wrapped) to seal the column of cathode solution and open the outlet to allow elution.

2. After centrifugation at 10,000*g* for 10 min, remove the unbound protein present in the supernatant.
3. Wash the resin with buffer B repeatedly until the calculated dilution of the estimated trapped volume of fluid in the resin bed (approx 33% of bed volume) approaches at least 1×10^6-fold.
4. Elute the bound enzyme using 3–4 bed vol of 4 *M* potassium thiocyanate and dialyze immediately against cold buffer B ***(10)***. The enzyme thus obtained can be quantified by Western-blot analysis, but because of exposure to potassium thiocyanate, the enzyme loses considerable amount of activity.

3.4. Isolation of GSTs by Liquid Column Isoelectric-Focusing

We use a Pharmacia-LKB model 8100 Ampholine column for performing liquid column isoelectric focusing (IEF) (**Fig. 1**). In this apparatus, GST protein in a sucrose density gradient containing ampholines of a desired pH range are sandwiched between an alkaline cathode solution at the bottom and an acidic anode solution at the top. With application of a strong electrical field, proteins migrate and "focus" into bands dependent on their pH values. Because the column is water-jacketed and can be maintained at 4°C, IEF can be performed outside the cold room. As little as 200 μg total purified GST protein can be used.

Depending on the desired yield of isozymes and their expected concentration in the mixture of total purified GST, larger quantities of GSTs can be used as well. The recovery of activity from IEF usually exceeds 70%.

1. Load the cathode solution into the column through the designated inlet (**Fig. 1**) using a peristaltic pump.
2. Form the sucrose density gradient on top of the cathode solution using a peristaltic pump (25 mL/h) and a two-chamber gradient mixer with the "dense" solution being in the proximal chamber of the gradient mixer.
3. Layer sufficient anode solution on top of the sucrose gradient to allow contact of the liquid column to the anode wire.
4. Apply an electrical field using a power source set at 1600 V and 40 mA for 24 h.
5. Elute the column contents into glass tubes in a cooled fraction collector at a flow rate of 0.8 mL/min (120–140 fractions).
6. Check the pH of every fifth fraction with the pH meter calibrated at 4°C.
7. Assay GST activity toward CDNB (or other substrates) in alternate fractions.

A GST activity and pH profile of fractions collected from IEF performed on purified mouse-liver GSTs is presented in **Fig. 2**.

3.5. Substrate Specificities of Purified GST Isozymes

3.5.1. Determination of S-Transferase Activity of GSTs

3.5.1.1. Spectrophotometrical Determination (*see* **Note 7**)

The S-transferase activity of GSTs against several substrates can be determined spectrophotometrically and conditions used for these assays are adapted from those described by Habig et al. ***(4)*** (*see* **Table 1**).

1. Pool GSTs contained in each of the peaks from IEF separately and determine activities against several substrates. (Because of the poor water solubility of most GST substrates, they are dissolved in ethanol.)

Fig. 2. *(opposite page)* An isoelectric focusing profile of mouse liver GST isozymes. Mouse liver GSTs were purified by GSH-affinity chromatography. IEF was performed on the total purified GSTs with 15 units of activity (toward CDNB) applied to the column and 0.8-mL fractions were collected after focusing at 1600 V, 60 mA for 24 h. The GST activities towards CDNB (●) and pH (×) for each fraction are plotted vs fraction number.

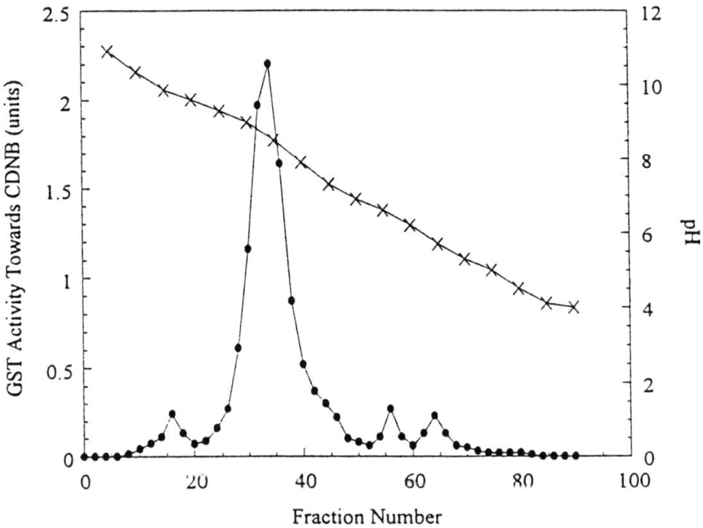

Fig. 2.

Table 1
Reaction Conditions for Spectrophotometric Assay of GST Activity Toward Electrophilic Substrates

Electrophilic substrate[a]	Final substrate concentration (m*M*)[b]		Buffer pH[c]	Wavelength (nm)	Extinction coefficient (mM^{-1} cm^{-1})
	Electrophile	GSH			
CDNB	1.00	1.0	6.5	340	9.6
DCNB	1.00	5.0	7.5	345	8.5
EPNP	5.00	5.0	6.5	360	0.5
EA	0.20	0.25	6.5	270	5
p-NBC	1.00	5.0	6.5	310	1.9
NPNO	0.20	5.0	7.0	295	7
BSP	0.03	5.0	7.5	330	4.5
TPBO	0.05	0.25	6.5	290	–24.8

[a]The stock solutions of all substrates are made in ethanol with the exception of bromosulfopthalein, which is water-soluble. The abbreviations are as follows: 1-chloro-2,4,-dinitrobenzene (CDNB), 3,4-dichloronitrobenzene (DCNB), 1,2-epoxy-3-(*p*-nitrophenoxy)propane (EPNP), ethacrynic acid (EA), *p*-nitrobenzyl chloride (*p*-NBC), 4-nitropyridine-N-oxide (NPNO), sulfobromopthalein (BSP), and *trans*-4-phenyl-3-buten-2-one (TPBO).

[b]The stock solutions of the electrophiles are at 20X the final concentration in reaction mixtures. Fresh stock solution of GSH is prepared at 10X the final concentration in reaction mixtures and kept on ice. The 1-mL reaction mixtures consist of 0.05 mL electrophilic substrate, 0.1 mL GSH, 0.02 to 0.1 mL enzyme, and buffer q.s. 1 mL.

[c]For reactions at pH 6.5 or 7.0, 100 m*M* potassium phosphate buffer has sufficient buffering capacity. For reactions at pH 7.5, 100 m*M* Tris-HCl should be used. All assays are performed at 25°C.

2. Since GST activity toward several of these substrates is near the lower range of detectability and because of problems with precipitation of substrates during assays, carefully standardize the assay procedure and carry out multiple determinations to obtain reliable results.

3.5.1.2. Nonspectrophotometrical GST Assay

For those substrates whose glutathione-conjugates are not amenable to spectrophotometric detection in the presence of the product, GST assay can be performed by directly quantifying the product.

1. For detecting the formation of leukotriene C_4 methyl ester (LTC_4ME) as a result of enzymatic conjugation of GSH with LTA_4ME, prepare a 0.1-mL reaction mixture containing 50 µL (1–2 µg) GST, 10 µL 250 m*M* potassium phosphate buffer pH 7.4, 10 µL 5 m*M* EDTA, 10 µL 20 m*M* [^{3}H]-GSH (specific activity 0.25 Ci/mol), and 10 µL water (q.s.).
2. Incubate the reaction mixture for periods up to 30 min at 30°C after adding 10 µL 200 µ*M* LTA_4ME.
3. Stop the reaction by addition of 0.1 mL cold methanol.
4. Separate the reaction mixture by TLC developed in butanol:acetic acid:water (4:2:2).
5. After spraying with ninhydrin and developing by heating at 70°C for 5–10 min, scrap the ninhydrin positive spot at R_f approx 0.45.
6. Determine the radioactivity in this spot; it is proportional to the amount of LTC_4ME formed in the given time interval ***(10)***.
7. For 9,10,-epoxy stearic acid (ESA), prepare the 0.1-mL reaction mixture containing 50 µL (1–2 µg) GST, 10 µL 20 m*M* [^{3}H]-GSH (0.25 Ci/mol), 10 µL 1 *M* potassium phosphate buffer, pH 7.4, 10 µL 2 m*M* ESA, and q.s. H_2O to 0.1 µL.
8. Incubate the reaction mixture at 37°C for 30 min after addition of ESA.
9. The remainder of the procedure is the same as for LTA_4ME ***(11)***.

GST activities of purified isozymes from mouse liver and rat pancreas are presented for reference (**Tables 2** and **3**) ***(7,12)***. Assays for other substrates including ethacrynic acid and melphalan have been described that use HPLC to separate and quantify the product ***(13,14)***.

3.5.2. Determination of the GSH-Peroxidase Activity of GSTs

In addition to their S-transferase activity, GSTs also exhibit lipid-hydroperoxide-glutathione peroxidase activity in which the lipid-hydroperoxide (LOOH) is reduced to the corresponding alcohol (LOH) with concomitant oxidation of GSH to GSSG. The most commonly used substrate for the lipid hydroperoxidase activity of GSTs is cumene hydroperoxide, though other lipid-hydroperoxides can be used as well ***(15)***. The spectrophotometric assay for this activity is based on measurement of NADPH consumption (by monitoring absorbance at 340 nm) as the GSSG formed is reduced back to GSH by glutathione reductase that is present in excess in the reaction mixture ***(16)***.

Table 2
Activity of GST Isozymes[a] of Mouse Liver Toward Electrophilic Substrates

	Specific activity (μmol/min/mg protein)[c]			
Substrates[b]	α	π	μ	mGST A4-4 (α)
CDNB	4.43	24.8	4.92	4.54
DCNB	0.33	0.035	ND	0.22
EA	0.017	1.8	ND	ND
ESA	0.064	0.078	0.203	0.26
LTA_4ME	0.024	0.232	0.478	0.306
Cu-OOH	7.67	ND	ND	ND

[a]GST-isozymes of mouse liver were purified by GSH affinity chromatography followed by isolation of individual isozymes by column IEF. The immunologic identity of the enzymes was established using Western-blot analysis against polyclonal antibodies specific for each class (or subclass) of GST. This table is adapted from *(7)*.

[b]Abbreviations for those electrophilic substrates not mentioned in **Table 1** are as follows: 9,10-epoxy stearic acid (ESA), leukotriene A_4 methyl ester (LTA_4ME) and cumene hydroperoxide (Cu-OOH).

[c]Activities toward CDNB, DCNB, and EA were determined at 25°C. Activities toward ESA and Cu-OOH activities were determined at 37°C. Activity toward LTA_4ME was determined at 30°C. ND, not detected.

1. In the experimental cuvet, prepare the reaction mixture containing 100 μL assay buffer, 20 μL 10 m*M* GSH solution, 100 μL glutathione reductase solution (*see* **Subheading 2.5.2.**), 100 μL 2 m*M* NADPH, 20 μL GST (3–4 μg protein) and water (q.s., 980 μL).
2. Incubate this reaction mixture at 37°C for 5 min.
3. Initiate the reaction by addition of 20 μL substrate. (No substrate is added to the blank cuvet. An additional blank is also used that contains the substrate but no enzyme.)
4. Subtract the rate of change of absorbance at 340 nm of the additional blank cuvet from the experimental cuvet.
5. Use the extinction coefficient for NADPH of 6.2 mM^{-1} cm^{-1} to calculate activity.

4. Notes

1. The protein biochemistry of GSTs is a relatively uncomplicated undertaking because these enzymes are relatively abundant and quite stable. They can be purified with reasonably good yields even from tissues frozen for several months at –20°C.
2. GSTs are sensitive to heat inactivation. All purification steps must be done at 4°C. All solutions (particularly dialysis buffers) should either be made using distilled water stored in a cold room or cooled to 4°C prior to use.
3. GST activity decreases rapidly in solutions not containing sulfhydryl reagents such as β-ME. Because β-ME is volatile, its concentration decreases gradually in

Table 3
Activity of GST Isozymes[a] of Rat Pancreas Toward Electrophilic Substrates

	Specific activity (μmol/min/mg protein)[c]		
Substrates[b]	α	μ	π
CDNB	3.1	14.65	9.63
DCNB	0.33	0.69	0.63
EA	ND	ND	0.29
EPNP	ND	5.2	3.75
NBC	1.45	0.81	ND
NPNO	0.40	0.30	ND
BSP	0.16	0.29	ND
t-PBO	ND	0.17	0.11
ESA	0.28	0.38	0.21
t-SO	0.10	0.47	0.21
Cu-OOH	2.1	0.12	ND

[a]GST-isozymes of rat pancreas were purified by GSH affinity chromatography followed by isolation of individual isozymes by column IEF. The immunologic identity of the enzymes was established using Western-blot analysis against polyclonal antibodies specific for each class (or subclass) of GST. This table is adapted from ***(12)***.

[b]Abbreviations for those electrophilic substrate not mentioned in **Table 1** are as follows: 9,10-epoxy stearic acid (ESA), leukotriene A_4 methyl ester (LTA_4ME), and cumene hydroperoxide (Cu-OOH).

[c]ND, not detected.

buffers if they are left open to air. Thus, it is best to add β-ME to buffers just prior to use, make only the amount necessary for immediate use, and keep the container covered with parafilm.

4. Bacterial contamination can occasionally be a problem, particularly if the purified GSTs are kept at 4°C for prolonged periods. This problem occurs can be easily remedied by using buffers filtered through a 0.2-μm filter for purification and by wearing gloves during purification.
5. All GSH solutions must be prepared fresh and preferably used within 2 to 3 h of preparation. In general, GSH solutions should be prepared in buffers rather than water because aqueous solutions of GSH are quite acidic (pH 3.0). Even when prepared in 10 m*M* phosphate buffer, depending on GSH concentration, the buffer pH can decrease significantly.
6. The purest GSTs are obtained from GSH-affinity chromatography only if the column is thoroughly washed prior to elution. Absorbance readings at 280 nm (in quartz cuvets) of the column effluent during washing must be near zero at least three times over a 4- to 6-h period before eluting. The time for washing can be minimized if care is taken to avoid a large column of homogenate above the

affinity-resin and by rinsing off the sides of the column prior to beginning the wash. Because the elution buffer contains a relatively high concentration of GSH, its pH must be adjusted to 7.0 prior to elution to avoid precipitating the enzyme due to exposure to low pH. Strict adherence to all time parameters and flow rates specified in the protocols are necessary for reproducible results.

7. During GST assays, the sequence of addition of solutions to the cuvet is important. The ethanolic substrate (*see* **Table 1**) volume should never exceed 5% and it should be the last ingredient added to a well-mixed reaction mixture (950 μL) containing GSH, GST, and the buffer. After addition of the substrate, the cuvets should be covered with parafilm and vigorously shaken. Improper mixing of the substrate can cause marked fluctuation in absorbance. If lower than expected activity is observed, the assay should be repeated with 10-fold diluted enzyme to ensure that the initial rate has not been missed due to substrate exhaustion. Results are most reproducible if *every* aspect of the assay is repeated precisely for each determination. Disposable cuvets of any kind are definitely inferior to quartz cuvets for all spectrophotometric determinations in which accuracy and precision are important.

References

1. Ames, B. N., Shigenoga, M. K., and Hagen, T. M. (1993) Oxidants, antioxidants, and the degenerative diseases of aging. *Proc. Natl. Acad. Sci.* USA **90,** 7915–7922.
2. Hayes, J. D. and Pulford, D. J. (1995) The glutathione S-transferase supergene family: regulation of GST and the contribution of the isoenzymes to cancer chemoprotection and drug resistance. *Crit. Rev. Biochem. Mol. Biol.* **30,** 445–600.
3. Awasthi, Y. C., Singhal, S. S., and Awasthi, S. (1995) Mechanisms of anticarcinogenic effects of antioxidant nutrients, in *Nutrition and Cancer Prevention* (Watson, R. and Mufti, S., eds.), CRC Press, Boca Raton, FL, pp. 141–174.
4. Habig, W. H., Pabst, M. J., and Jakoby, W. B. (1974) Glutathione S-transferases. The first enzymatic step in mercapturic acid formation. *J. Biol. Chem.* **249,** 7130–7139.
5. Hussey, A. J. and Hayes, J. D. (1992) Characterization of a human class-theta glutathione S-transferase with activity towards 1-menaphtyl sulfate. *Biochem. J.* **268,** 929–935.
6. Laemmli, U. K. (1970) Cleavage of structural proteins during the assembly of the head of bacteriophage T4. *Nature* **227,** 680–685.
7. Singhal, S. S., Saxena, M., Ahmad, H., and Awasthi, Y. C. (1992) Glutathione S-transferases of mouse liver: sex-related differences in the expression of various isozymes. *Biochim. Biophys. Acta* **1116,** 137–146.
8. Singhal, S. S., Piper, J. T., Awasthi, S., Zimniak, P., and Awasthi, Y. C. (1995) Polyclonal antibodies specific to human glutathione S-transferase 5. 8 (hGST 5. 8). *Biochem. Arch.* **11,** 189–195.
9. Towbin, H., Staehelin, T., and Gordon, J. (1979) Electrophoretic transfer of proteins from acrylamide gels to nitrocellulose sheet: procedure and some applications. *Proc. Natl. Acad. Sci.* USA **76,** 4350–4354.

10. Singhal, S. S., Ahmad, H., Sharma, R., Gupta, S., Haque, A. K., and Awasthi, Y. C. (1991) Purification and characterization of human muscle glutathione S-transferases: evidence that glutathione S-transferase corresponds to a locus distinct from GST1, GST2 and GST3. *Arch. Biochem. Biophys.* **285,** 64–73.
11. Sharma, R., Gupta, S., Singhal, S. S., Ansari, G. A. S., and Awasthi, Y. C. (1991) Glutathione S-transferase-catalyzed conjugation of 9,10-epoxy stearic acid with glutathione. *J. Biochem. Toxicol.* **6,** 147–153.
12. Singhal, S. S., Gupta, S., Saxena, M., Sharma, R., Ahmad, H., Ansari, G. A. S., and Awasthi, Y. C. (1991) Purification and characterization of glutathione S-transferases from rat pancreas. *Biochim. Biophys. Acta* **1079,** 285–292.
13. Awasthi, S., Srivastava, S. K., Ahmad, F., Ahmad, H., and Ansari, G. A. S. (1992) Interactions of glutathione S-transferase-B with ethacrynic acid and its glutathione conjugate. *Biochim. Biophys. Acta* **1164,** 173–178.
14. Awasthi, S., Bajpai, K. K., Piper, J. T., Singhal, S. S., Ballatore, A., Seifert, W. E., Awasthi, Y. C., and Ansari, G. A. S. (1996) Interactions of melphalan with glutathione and glutathione S-transferase. *Drug Metab. Disp.* **24,** 371–373.
15. Singhal, S. S., Saxena, M., Ahmad, H., Awasthi, S., Haque, A. K., and Awasthi, Y. C. (1992) Glutathione S-transferases of human lung: characterization and evaluation of the protective role of the α-class isozymes against lipid peroxidation. *Arch. Biochem. Biophys.* **299,** 232–241.
16. Beutler, E. (1975) *Red Cell Metabolism*, Grune & Stratton, New York, pp. 71–73.

28

Aberrant Crypt Foci System to Study Cancer Preventive Agents in the Colon

Principles and Guidelines

Ranjana P. Bird

1. Introduction

Knowledge of the actual active substances that act to initiate or modulate carcinogenesis in humans is limited. The past two decades have been dedicated to the discovery of environmental factors including constituents of our diet that may be carcinogenic or modulators of carcinogenic process. Recent emphasis has been on the identification of cancer preventive agents. In this regard, animal models have been most useful. They are intensively used in the identification of the initiators and modulators of carcinogenesis as well as in the elucidation of the mechanism(s) of their action.

Cancer preventive agents can be categorized into two major groups. The first group of compounds affect initiation by blocking or repairing the genotoxic effect of a carcinogen. These compounds retard the metabolic activation of a chemical into a carcinogen and may inactivate a carcinogen or increase the DNA-repair activity of the target cells *(1)*. The second group of compounds affect the postinitiation events. These compounds prevent cancer potentially by directly affecting the growth of initiated cells or indirectly by modifying the environment surrounding the initiated cells. The cellular and molecular mechanism(s) underlying their effects remain elusive. In recent years more emphasis is given to identify cancer preventive agents that retard, inhibit or regress the growth of initiated cells. Initiated cells go through a clonal expansion, each clone in turn undergoes further selection and clonal expansion. This sequential selection and clonal expansion is accompanied by enhanced growth autonomy and the acquisition of novel genotypic and

From: *Methods in Molecular Medicine, Vol. 14: Tumor Marker Protocols*
Edited by: M. Hanausek and Z. Walaszek

phenotypic features. One of the features of the advanced preneoplastic lesion is resistance to growth modulators *(2,3)*. This complexity clearly demands the need to identify cancer preventive agents capable of modulating the growth of primal and/or advanced preneoplastic lesions. The most ideal cancer preventive agent must be able to retard the growth of all preneoplastic lesions regardless of their growth stage.

The concept that colon cancer is a preventable disease is well accepted and there is an increasing thrust to actuate this concept in the population with high risk for colon cancer. In view of the preceding discussion, one of the important aspects of the cancer prevention strategies is to identify cancer preventive agents.

In order to identify agents that would inhibit, retard, or regress development of colon cancer it is important to have a system that can quantify number and growth of preneoplastic lesions. In addition, the system should be simple, inexpensive, sensitive and specific to the colon, require a short period of time, and use a small number of animals. The aberrant crypt foci (ACF)-system generally meets all the aforementioned criteria with a few exceptions (*see* **Subheading 3.1.**).

ACF are present in carcinogen-treated rodent colons. They are visualized microscopically by topographic examination of the mucosal surface of the methylene blue-stained whole mounts of colon. Aberrant crypts are identified by increased size, irregular and dilated luminal opening, thicker epithelial lining, and pericryptal zone (**Fig. 1**). The ACF are purported to be preneoplastic lesions and several studies investigating the genotypic and phenotypic features of ACF support this contention *(4)*.

A systematic examination of the induction and growth features of ACF provided strong impetus to use the system to identify modulators of colon carcinogenesis. These foci are induced in a dose-related and species-specific manner by known colon carcinogens. In addition, known inhibitors or promoters of colon carcinogenesis inhibited or promoted the number and growth of ACF.

The working premise of the ACF system is that if ACF are preneoplastic lesions, then their induction and growth should be inhibited by cancer preventive agents. Although several reports have supported the use of the ACF-system to identify modulators of colon carcinogenesis, it is important to recognize that the induction and growth characteristics of ACF are vulnerable to be affected by the same experimental variables that are known to affect tumor incidence. These variables include age, sex, strain, and species of animals, dose, route, frequency and type of carcinogen, and experimental duration. Most importantly, ACFs are in a dynamic state and there is no standard protocol that has been devised to identify cancer preventive agents employing the ACF system. However, there is an established method to visualize, identify, and quan-

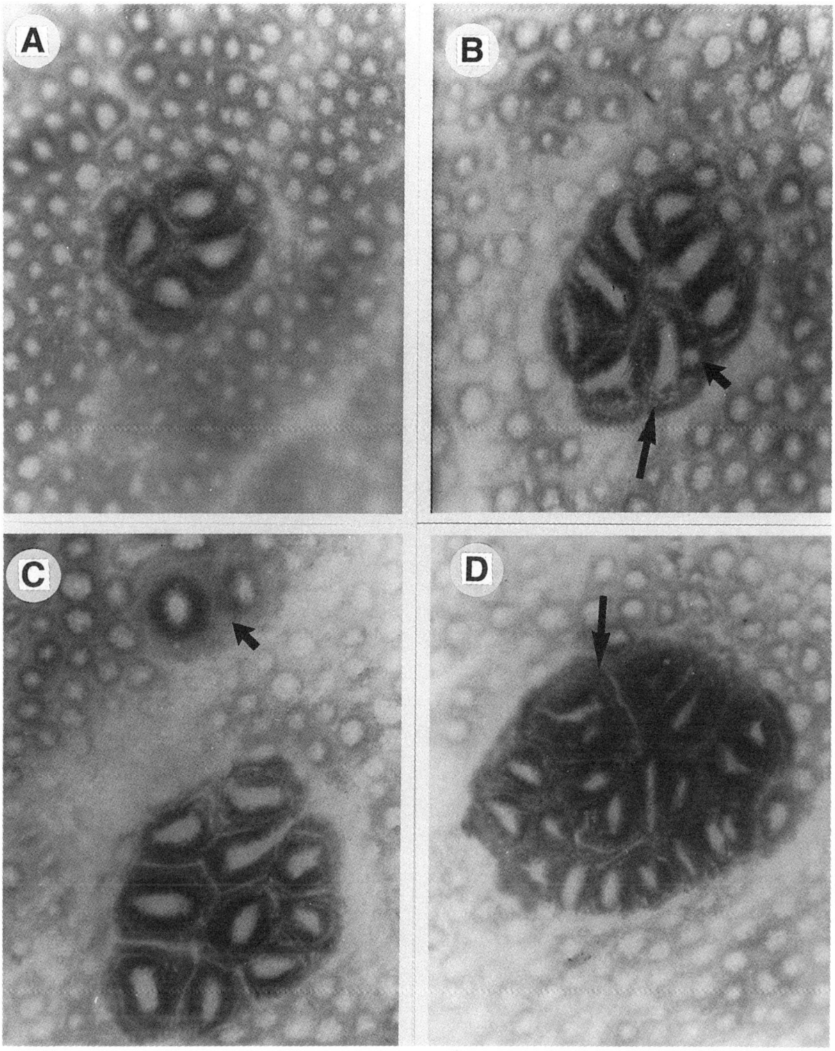

Fig. 1. A topographical microscopic view of the whole mounts of methylene blue-stained colonic mucosa exhibiting the presence of normal crypts and ACF. **(A)** A focus consisting of four crypts. **(B)** An aberrant crypt focus with eight crypts; note the presence of mucous secreting cells with depicted by white globules (long arrow) also note a branching crypt (short arrow). **(C)** Note the presence of two ACF, one consisting of one crypt (short arrow), the other consisting of ten crypts. **(D)** An advanced ACF consisting of several crypts, note one crypt exhibiting marked atypia (long arrow) compared to others (×100).

tify ACF in rodent colons. Cancer preventive agents are grouped based on their biological activities and their identification depends on the use of appropriate experimental protocol.

In view of these complexities, a brief overview is provided of the experimental protocols generally used by researchers citing their advantages and disadvantages. A section is also dedicated to the discussion of the features of ACF, the factors that may confound results, and the choice of the experimental model and procedure.

2. Materials

2.1. Materials

All materials, excluding the animals, are available through Sigma (St. Louis, MO) and are listed along with the description of the methods.

2.2. Choice and Number of Animals

Weanling male or females Sprague Dawley or F344 rats are commonly used. Rats are more sensitive than mice for developing colon cancer and require one or two injections of carcinogen. A total of 5–10 rats/group is found to be adequate to detect inhibitory effects of chemicals on ACF.

2.3. Choice of Carcinogen, Selection of Dose and Route of Administration

A variety of chemicals induce colonic tumors in rodents. The most commonly used carcinogens are azoxymethane (AOM) or 1,2-dimethylhydrazine (DMH). AOM is administered to rats at a dose level of 15–20 mg/kg body weight (intraperitoneally or subcutaneously). The sc route is preferred by several researchers. This route of administration appears to yield fewer tumors in the small intestinal tract than the ip route of administration. AOM is a liquid and diluted to the appropriate concentration in sterile saline. The volume generally injected to the rats (90–150 g body weight) is 0.2 mL.

The preparation of DMH requires a great deal of care. The DMH is dissolved in 1 m*M* ethylenediamine tetra-acetic acid (EDTA) to the appropriate concentration and the pH is adjusted to 6.5 using 2 *N* NaOH. The dose level generally administered to rats varies from 20–140 mg/kg. This carcinogen is inactivated at high pH. A limited number of researchers use this carcinogen and employ a multiple injection protocol.

2.4. Dose of the Test Agent and Route of Administration

Preventive agents include dietary constituents (nutrients and nonnutrients) and synthetic chemicals. There is no set guidelines for the selection of the dose of the test compound and its route of administration. A large number of compounds have been tested mixed in thc diet or in the drinking water. It is important to record food or water intake to estimate the daily intake of the test compound by the animals on a body weight basis. Ideally, the diet composition should be well defined. The dose level of a compound is selected based on the maximum tolerated dose (MTD) or LD_{50}. Generally the test compound is incorporated in the diet at 40 and/or 80% of the MTD. If a nutrient is being evaluated then the dose level is selected based on the level required by the experimental animals. The level of the nutrient is increased two- to fourfold in the diet.

3. Methods

3.1. Experimental Protocols

3.1.1. Choice of Experimental Protocol

As stated previously, a variety of experimental protocols are being used by investigators. Special consideration has to be given to the experimental protocol. A chemical can inhibit as well as promote cancer development depending on the time of its intervention. The most common approaches include injection (one or two) of carcinogen followed by intervention with the test agent. This approach has been quite successful in the identification of chemopreventive agents.

Ideally, the test compound should be included in the protocol once the initiation step is completed. However, it remains uncertain as to the duration required to complete the initiating events. It is estimated that the initiating effect of a carcinogen such as AOM is completed within 1 wk after its administration. Another approach would be to intervene the disease process several weeks after carcinogen treatment. This approach minimizes any interactive effect between the carcinogen and the test agent and assesses the effect of the test agent on the growth of established ACF of varying growth features. For the purpose of quick screening of the number of compounds protocols A and B are useful (**Fig. 2**).

3.1.2. Characteristics of Experimental Protocols Employed to Assess Chemopreventive Agents

Variable experimental protocols have been employed to identify cancer preventive agents (**Fig. 2**) with rats being the most common animal model.

1. In the first protocol, the animals are injected with a colon carcinogen while they are receiving the test agent (**Fig. 2A**). A few weeks later their colons are evaluated for the number and growth features of ACF. This approach is used by several researchers and has been able to identify a number of cancer preventive agents *(5,6)*. One limitation of this approach is that the test agent may interact with the biological activity of the carcinogen and the approach does not distin-

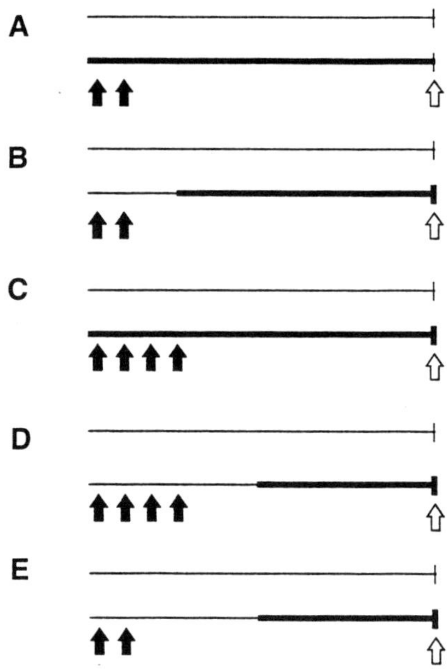

Fig. 2. A schematic presentation of experimental protocols utilizing ACF system to identify cancer preventive agents. ↑, carcinogen injection; ⇧, quantification of ACF; —, control condition; —, test agent.

guish between a cancer preventive agent which blocks or reduce the carcinogenicity of a chemical from those that affect postinitiation events.

2. In the second approach, the animals are injected with carcinogen, and 1 or 2 wk later they are treated with the test agent (**Fig. 2B**). This approach minimizes the interaction between the test agent and the carcinogen, and assumes that the test agent is affecting the postinitiating events. ACF are enumerated few weeks (4–8 wk) following the intervention with the test agent.
3. The third and fourth approaches involve giving the animals several injections while they receive the test agent (**Fig. 2C**) or the test agent is given to the animals 1 or 2 wk after the last injection of the carcinogen (**Fig. 2D**). Colons are examined for the number and growth features of ACF at one time point, before or at the appearance of tumors. This protocol is used by a limited number of researchers and prone to several limitations. For instance, exposure of the animals to a carcinogen repeatedly may obscure the effect of a test agent. A previous study has demonstrated that the growth dynamics of ACF in the colons of the animals injected four times with AOM differs a great deal from those injected with AOM once or twice ***(4)***. The effect of additional injections of AOM was to suppress the number and growth features of ACF for several weeks followed by a marked increase in both their number and growth. The value of the ACF system has not

been systematically assessed employing the multiple injection protocol. A limited number of studies have employed multiple injection protocols and failed to find a correlation between the number and growth features of ACF with tumor outcome. In these studies, ACF were assessed along with tumor outcome either in the entire colon or a segment of the colon. These studies were not designed to assess the efficacy of the ACF system to predict the tumor outcome ***(7,8)***.

4. The last protocol (**Fig. 2E**) is not commonly used. This protocol assesses the effect of cancer preventive agents on established preneoplastic lesions that are allowed to grow for several weeks before intervention with the test agent ***(9)***.

3.2. Assessment of ACF

3.2.1. Induction and Growth Characteristics of ACF

The biological features of ACF and the factors affecting induction and growth characteristics of ACF have been recently reviewed ***(4)***. The key features of ACF that may influence the usefulness of the system are as follows.

1. Aberrant crypt foci are induced by colon specific carcinogen and appear within 2 wk following a single injection of a carcinogen. The ACF are induced in a dose-related manner and their number increases, following a single injection of a carcinogen (AOM or DMH), for several weeks. There appears to be a phase during which (12–18 wk after carcinogen administration) a large number of ACF with one or two crypt multiplicity remodel or are eliminated. This phase is followed by emergence of additional ACF. ACF appear predominantly in the distal colon during early time points; as the time progresses, ACF appear in the proximal colon and a proportion of ACF start exhibiting focal expansion and may contain one to several crypts.
2. Quantification of the number and growth features of ACF in different regions of the colon, with or without intervention with the test agent, allows assessment of the cancer preventive efficacy of the test agent to the whole or to a specific region of the colon (*see* **Notes 1–3**).

3.2.2. Tissue Preparation

Tissue preparation includes the following steps:

1. Colons of the animals are excised. One end of the colon is closed using hemostatic forceps and made into a sac. A syringe filled with phosphate-buffered saline (PBS), equipped with ≥20-gage needle, is used to fill the colon with PBS. A slight pressure is exercised to stretch the wall of the colon. The clamp is removed and the contents of the colon are expelled.
2. The colon is cut at the longitudinal axis and fixed flat, mucosal side up, in a Petri dish. A number of fixatives or preservatives can be used including 10% buffered formalin, 70% ethanol, or ethanol and acetic acid (30% acetic acid in ethanol). Formalin colons can be kept for months to years without affecting the visualization of ACF. Tissues can be kept in 70% ethanol for several months. It is crucial

that tissues remain immersed in the appropriate fixative or preservative and are not allowed to air dry.

3. To visualize ACF, 3–4 cm of the colon is placed in the Petri dish and flooded with the stain. In the original method, methylene blue (0.2% in PBS or saline) was used and remains the most popular stain for visualizing ACF. With time methylene blue crystallizes and the stain should be filtered prior to use. The ACF in the tissues can also be visualized by staining the tissue with other stains such as hematoxylin. Depending on the fixative used, staining time may be 5 min or longer. Alcohol-preserved colons appear to stain faster than formalin-fixed tissue. To check whether staining is optimum colonic tissue is placed mucosal side up on a glass slide and viewed under a light microscope using ×4–10 objective. The crypt lining should appear blue. If additional staining is required, the tissue is reimmersed in the stain. If a tissue is too dark excess stain can be removed by placing the tissue in 70% ethanol. The enumeration of the ACF must be completed within a short period of time once the tissue is stained. Care has to be taken that the tissue remains wet while it is being enumerated for ACF. If the tissue starts to dry out, the mucosal surface will appear cloudy. A few drops of PBS would help keep the tissue moist.

3.2.3. Statistical Analysis

Comparison among various groups is made by an analysis of variance in combination with a test that would allow comparisons among several group means such as Duncan's Multiple Range Test.

4. Notes

1. Quantification of ACF: In order to assess the regional distribution of the ACF along the length of the colon, it is important to know the identity of the segment being enumerated. One simple approach is to divide the colon into three equal segments. Enumeration of ACF is carried out starting from the rectal end proceeding to the cecal end. The stained colonic section is viewed under an ×4 or ×10 objective of a light microscope and the number of ACF and number of crypts in each focus is quantified. Aberrant crypts appear as large crypts with dilated luminal opening and/or thicker epithelial lining than surrounding normal crypts (**Fig. 1**).
2. The number of crypts present in each focus represents the growth feature and is referred to as "crypt multiplicity." In **Fig. 1A**, the ACF has four crypts in the focus, therefore it has the crypt multiplicity of 4. The ACF shown in **Fig. 1B** has the crypt multiplicity of 8. Topographical examination of colonic mucosa also reveals the presence of goblet cells. The small white circles in the wall of ACF represent mucinous globules (**Fig. 1B**, long arrow). Any crypt or cluster of crypts quantified as ACF must be morphologically distinct from surrounding normal mucosa. A single crypt may appear to be slightly different but should not be included as aberrant crypt. In **Fig. 1C**, two foci are shown, one with a crypt multiplicity of 10 and the other of 1 (short arrow). There are several crypts in

Fig. 1B and **Fig. 1C** that appear to be slightly different from the surrounding normal crypts. Although they may be precursor to ACF they are not counted as ACF. A large ACF with a crypt multiplicity of approximately 21 is shown in **Fig. 1D**. The heterogeneity among aberrant crypts in the same focus is visualized (long arrow).

3. The number and growth features of ACF are enumerated for the entire colon and are expressed as the following parameters:
   a. Number of ACF per colon: average of the number of ACF in each rat in a group.
   b. Number of ACF in different regions of the colon: average of the number of ACF in each segment in each rat in a group.
   c. Number of aberrant crypts per focus: average of the average crypt multiplicity in each rat per group.
   d. Number of aberrant crypts per focus per group: average of the crypt multiplicity of all ACF found in a group.
   e. Distribution of ACF according to their crypt multiplicity: The ACF are grouped according to their crypt multiplicity in each colon and presented as average number of ACF with different crypt multiplicity per colon. This distribution can be used to calculate the proportion of total ACF with different growth features.

Acknowledgments

The research on ACF in the author's laboratory was supported by the Ludwig Foundation, the National Cancer Institute of Canada, and the Natural Sciences and Engineering Research Council of Canada. The author is grateful to her coworkers for their contribution in extending the knowledge on ACF, and the many researchers whose studies on the histogenesis of colon cancer provided the impetus for the development of the aberrant crypt foci system. The usefulness of the ACF system in the identification of modifiers of colon carcinogenesis would not be evident without the interest and effort of the many researchers who have assessed the ACF system in their works.

References

1. Kohlmeir, L., Simonsen, N., and Mottus, K. (1995) Dietary modifiers of carcinogenesis. *Environ. Health Persp.* **103,** 177–184.
2. Harris, C. C. (1991) Chemical and physical carcinogenesis: advances and perspectives for the 1990s. *Cancer Res.* **51,** 5023s–5044s.
3. Farber, E. and Rubin, H. (1991) Cellular adaptation in the origin and development of cancer. *Cancer Res.* **51,** 2751–2761.
4. Bird, R. P. (1995) Role of aberrant crypt foci in understanding the pathogenesis of colon cancer. *Cancer Lett.* **93,** 55–71.
5. Pereira, M. A., Barnes, L. H., Rassman, V. L., Kelloff, G. V., and Steele, V. E. (1994). Use of azoxymethane-induced foci of aberrant crypts in rat colon to identify potential cancer chemopreventive agents. *Carcinogenesis* **15,** 1049–1054.

6. Wargovich, M. J., Chen, C., Jimenez, A., Steele, V. E., Velasco, M., Stephens, L. C., Price, R., Gay, K., and Kelloff, G. J. (1996) Aberrant crypts as a biomarker for colon cancer: evaluation of potential chemopreventive agents in the rat. *Cancer Epidemiol. Biomarkers Prev.* **5,** 355–360.
7. Thorup, I., Mayer, O., and Kristiansen, E. (1994) Influence of a dietary fiber on development of dimethylhydrazine-induced aberrant crypt foci and colon tumour incidence in Wistar rats. *Nutr. Cancer* **21,** 177–182.
8. Carter, J. W., Lancaster, H. K., Hardman, W. E., and Cameron, I. L. (1994) Distribution of intestine associated lymphoid tissue, aberrant crypt foci, and tumors in the large bowel of 1,2-dimethylhydrazine treated mice. *Cancer Res.* **54,** 4304–4307.
9. Lasko, C. M. and Bird, R. P. (1995) Modulation of aberrant crypt foci by dietary fat and caloric restriction: the effects of delayed intervention. *Cancer Epidemiol. Biomarkers Prev.* **4,** 49–55.

29

Strain-Dependent Differences in the Expression of the Oncofetal Protein p65 in Mice Susceptible and Resistant to Chemical Carcinogenesis

Janusz Szemraj, Margaret Hanausek, Zbigniew Walaszek, and Alan K. Adams

1. Introduction

The 65-kDa oncofetal protein (p65), a novel tumor marker *(1–7)*, is highly conserved in different species *(2,4)*. We have identified the p65 gene as a novel member of the family of genes that encode receptors for steroid hormones, vitamin D, retinoic acid and thyroid hormone *(7)*. The p65 protein is highly homologous to estrogen receptor (ER) in its DNA binding domain but other regions of the sequence do not show similarities, indicating that p65 is a new transcription factor or receptor with an as yet unknown ligand. Using enzyme-linked immunosorbent assay (ELISA) and immunostaining, p65 was shown *(5,6)*, to be a promising marker for diagnosis and prognosis of cancer. With a view toward early detection of cancer, we have studied strain-dependent differences in the expression of p65 in different organs of mice highly susceptible (C3H/HeJ) and relatively resistant (C57BL/6N) to carcinogenesis. In this chapter, we describe a new, general procedure for detection of p65 mRNA by reverse transcription polymerase chain reaction (RT PCR). This method is sensitive enough to detect a small number of the p65-specific mRNA molecules in mammary glands and livers of young, 7–8 wk-old female mice of the C3H/HeJ strain are known to spontaneously develop mammary tumors and be sensitive to liver tumor induction. No p65-specific mRNA was detected in a control group of C57BL/6N mice known to be resistant to chemical carcinogenesis.

C3H/HeJ mice are approx 50-fold more susceptible to liver-tumor induction than are relatively resistant C57BL/6J mice *(8)*. Genetic analysis has identified

From: *Methods in Molecular Medicine, Vol. 14: Tumor Marker Protocols*
Edited by: M. Hanausek and Z. Walaszek © Humana Press Inc., Totowa, NJ

that the difference in susceptibility is a consequence of allelic differences in hepatocarcinogen sensitivity (Hcs) genes that control the growth of preneoplastic lesions in the liver ***(8–12)***. In the mouse mammary gland system, adenocarcinomas can develop by the progression of normal epithelial cells through a hyperplastic alveolar intermediate which can be visualized as a nodule of alveolar epithelial cells, i.e., hyperplastic alveolar nodule (HAN), branching from mammary ducts. Oncogenic agents which increase the frequency of mammary adenocarcinomas in mice also increase the frequency of mammary HAN ***(13)***. One such agent, mouse mammary tumor virus (MMTV) is a nonacute, transforming retrovirus which is transmitted both vertically through the germ line and horizontally in breast milk. MMTV infection of mammary glands results in proviral insertion into host DNA and activation of cellular genes ***(14)***. In a relevant study ***(15)***, however, the virus-free, transplanted C3H mammary glands maintained their susceptibility regardless of the host, indicating that the high incidence of spontaneous mammary tumors is generally predetermined for this organ in C3H mice (*see* **ref.** ***14***).

The PCR has received wide approval as a powerful method for genetic analysis. This enzyme-catalyzed chain reaction has been utilized for diverse applications including the detection of mutations, rearrangements or deletions within genes, quantitation of gene expression, cDNA amplification, the isolation and cloning of new genes and identification of specific microorganisms in clinical samples ***(16–25)***. Reverse transcription of RNA followed by polymerase chain reaction (RT-PCR) is a highly sensitive method for detection of low abundance messenger RNAs in biological samples.

We have used p65 to monitor the carcinogenic process in multistage models of rat liver carcinogenesis ***(1–3,26,27)***. The p65 protein appears to be induced very early during chemical hepatocarcinogenesis. In fact, we have successfully utilized anti-p65 antibodies for immunohistochemical detection of preneoplastic hepatic foci induced by different chemical carcinogens in the rat ***(3,26,27)***. No hepatic foci in the livers of control rats, i.e., rats not treated with chemical carcinogens were identified. We now demonstrate that using RT-PCR we are able to detect the presence of p65-specific mRNA in the susceptible organs of C3H/HeJ mice as early as at 7–8 wk of age, i.e., before tumors can be detected using any of the currently available methods. As shown in **Fig. 1** (upper panel), p65-specific mRNA was detected in the livers of 4 of 10 female C3H/HeJ mice (i.e., 40%), but not in the livers of C57BL/6N mice (lower panel). The p65-specific mRNA was present in the mammary glands of 8 of 10 C3H/HeJ mice (i.e., 80%) (**Fig. 2**, upper panel), but not in the mammary glands of C57BL/6N mice (**Fig. 2**, lower panel). The ER-specific mRNA was detected in all RNA samples. No p65-specific mRNA was detected in control organs such as kidneys, and lungs of C3H/HeJ and C57BL/6N mice (data not shown).

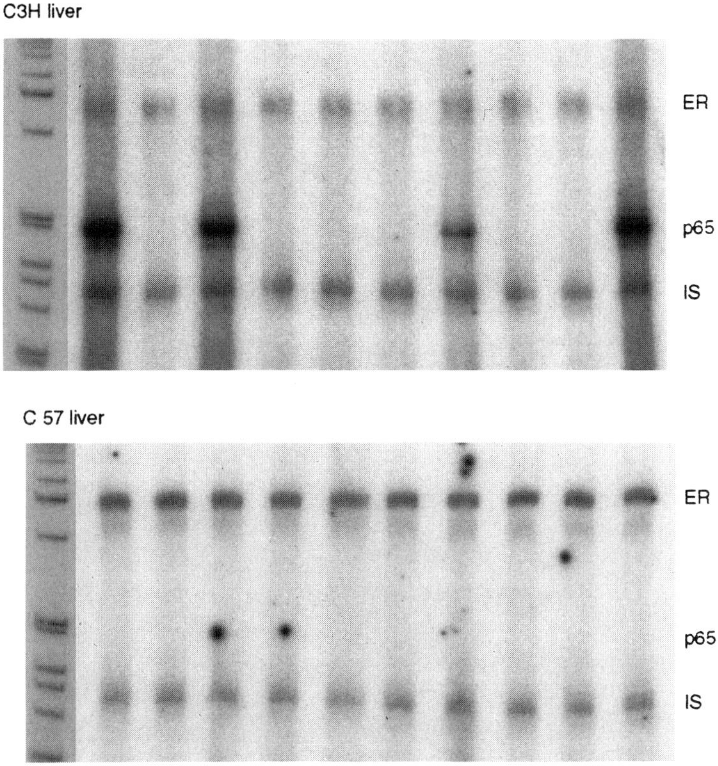

Fig. 1. RT-PCR amplification products obtained from total RNA isolated from livers of C3H/HeJ mice (**upper panel**) and C57BL/6N mice (**lower panel**). Lane 1, 1-kb DNA ladder; lanes 2–11, amplification products of RNA obtained from individual mice. IS, internal standard for p65; ER, estrogen receptor.

Using ELISA, we were not able to detect p65 in the blood serum of C3H/HeJ or C57BL/6N mice. The occurrence of p65-specific mRNA in the susceptible organs of young C3H/HeJ mice appear to correlate with the liver and mammary gland tumor incidence reported for adult mice of this strain ***(16)***. Thus, expression of p65 examined by RT-PCR may be a good diagnostic indicator for early detection of liver and mammary carcinomas.

2. Materials

2.1. Equipment

The majority of the equipment needed for this technique will be found in any well-equipped laboratory. However, the following list gives the more specialized equipment we have used.

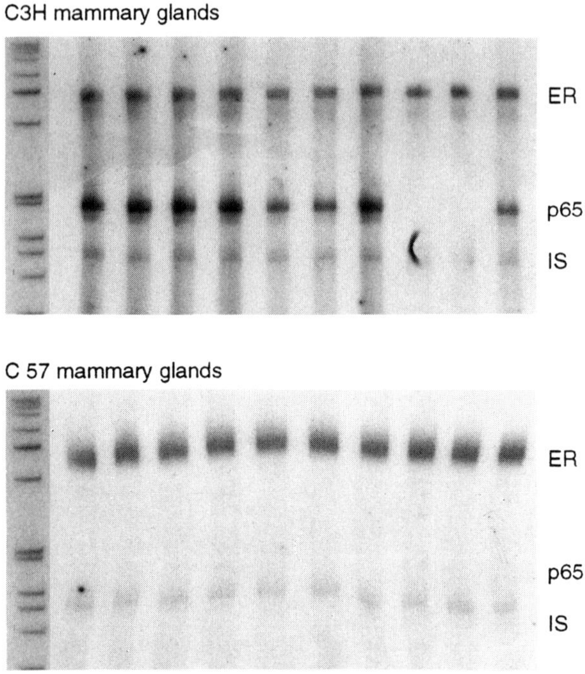

Fig. 2. RT-PCR amplification products obtained from total RNA isolated from mammary glands of C3H/HeJ mice (**upper panel**) and C57BL/6N mice (**lower panel**). Lane 1, 1-kb DNA ladder; lanes 2–11, amplification products of RNA obtained from individual mice. IS, internal standard for p65; ER, estrogen receptor.

1. Tissuemizer™ (Omni Int., Waterbury, CT) or an equivalent homogenizer.
2. 15- and 30-mL COREX centrifuge tubes. Sterilize before use.
3. Water bath.
4. Beckman J-21 and Beckman J-6 centrifuges.
5. Electrophoresis gel apparatus and power supply (Pharmacia, Piscataway, NJ).
6. Vacuum gel dryer.
7. Speed vacuum concentrator (SpeedVac, Savant Instruments Inc., Farmingdale, NY).
8. Liquid nitrogen container from Coronex Lighting Plus (Du Pont, Wilmington, DE).
9. Ultra-low freezer (–70°C).
10. Thermocycler for PCR.
11. Microcentrifuge.
12. Micropipets capable of dispensing 0.5–100 µL.
13. PCR microcentrifuge tubes.

2.2. Animals

Use C3H/HeJ and C57BL/6N virgin female mice (National Cancer Institute, Frederick, MD).

2.3. Reagents

All the chemical reagents should be of at least analytical grade, unless otherwise specified. All the solutions should be prepared using double distilled, sterile water.

2.3.1. RNA Isolation (see **Note 1**)

1. Proteinase K: Stock solution of proteinase K, 20 mg/mL (Boehringer Mannheim, Indianapolis, IN) in double-distilled water.
2. Lysis buffer: 4 *M* guanidinium isothiocyanate, 0.5% *N*-lauryl sarcosine (Sarkosyl), 25 m*M* sodium citrate, and 0.1 *M* β-mercaptoethanol (β-ME). If necessary, adjust pH to 6.0 with a citric acid solution.
3. Guanidinium solution: 6 *M* guanidinium isothiocyanate, 0.75% *N*-lauryl sarcosine (Sarkosyl), and 0.0375 *M* sodium citrate. If necessary, adjust pH to 6.0 with a citric acid solution. Make 200 mL.
4. Phenol:chloroform:isoamyl alcohol mixture (25:24:1 [v/v]). Prepare in a fume hood, just prior to use. Wear gloves. Avoid inhaling chloroform vapors. Prior to use, phenol should be redistilled under nitrogen and equilibrated with buffer containing 100 m*M* Tris-HCl, pH 6.0, 1 m*M* EDTA, 0.1 *M* β-ME, and 0.1% 8-hydroxyquinoline (w/v). Store phenol at 4°C in a dark bottle.
5. Glycogen: stock solution of glycogen 10 μg/mL (Boehringer Mannheim). Store at 4°C.
6. Diethyl pyrocarbonate (DEPC): Water and all other solutions should be treated with DEPC (0.5% [v/v]) overnight at room temperature and then autoclaved for 30 min to remove any trace of DEPC. Do not treat Tris buffers with DEPC.
7. 2 *M* Ammonium acetate.
8. Ethanol: 100 and 75% ethanol (store at –20°C).
9. Isopropanol (store at –20°C).
10. *n*-Butanol (store at room temperature).
11. 0.1 *M* β-ME: Prepare from 48.7% β-ME. **Caution:** β-ME is highly toxic. Dispense in a fume hood and wear appropriate protective equipment.
12. 3 *M* Sodium acetate, pH 5.2.

2.3.2. PCR

1. Tth Thermostable DNA polymerase: 5 U/μL (Epicentre Technologies, Madison, WI) in the reaction buffer, i.e., 10 m*M* Tris-HCl, pH 8.3, 90 mM potassium chloride, 0.005% Tween-20, 0.005% Nonidet P-40, 10 μg/mL gelatin.
2. 25 m*M* $MgCl_2$ stock solution.
3. 25 m*M* $MgSO_4$ stock solution.
4. 25 m*M* $MnCl_2$ stock solution.
5. Deoxyribonucleotide triphosphates (dNTP): 2 m*M* stock solution of each dNTP (Boehringer Mannheim).
6. 1% Tween-20.
7. 1% Nonidet P-40.

8. 1 mg/mL Gelatin.
9. Mineral oil (molecular biology grade).
10. [$\gamma^{32}P$] Adenosine triphosphate (ATP) (1000–3000 Ci/mM) from Amersham Life Sciences (Arlington Heights, IL).
11. Autoradiography films, fixer, and developer (Amersham Life Sciences).
12. Oligonucleotide probes: Oligonucleotide primers can be synthesized by any commercial DNA synthesis laboratory. The following primers were synthesized for us by Genosys (The Woodlands, TX). Because of pending patents for the p65 cDNA, the use of p65-specific primers requires negotiations with The University of Texas, M. D. Anderson Cancer Center, Office of Technology Development, Houston, TX.
    a. Reverse transcription primer specific for p65 (*see* **Note 2**): 5' ACTCGGCTC AGGTCTGGGGA 3',
    b. Reverse transcription primer specific for ER (*see* **ref. *28***): 5' ACTCCA GAATTAAGC 3'
    c. The p65-specific PCR nested primers (*see* **Note 2**):
        i. Sense: 5' AAGTGATACCCAGATTGGCC 3'.
        ii. Antisense: 5' AAGCAATGAGCCACTCCCTC 3'.
    d. The ER-specific PCR nested primers (*see* **ref. *28***):
        i. Sense: 5' CATAACGACTATATGTGTCCAGCC 3'.
        ii. Antisense: 5' AACCGAGATGATGTAGCCAGCAGC 3'.
13. Gel filtration columns (Chroma Spin Columns 10, Clontech Laboratories, Inc., Palo Alto, CA).
14. *Taq* DNA polymerase (Gibco BRL Life Technologies, Gaithersburg, MD).
15. *Taq* Polymerase reaction buffer: 100 m*M* Tricine, pH 8.4, 500 m*M* KCl, 1.5 m*M* ethylene glycol-*bis*(β-aminoethyl ether)-*N,N,N,'N'*-tetra-acetic acid (EGTA), 0.5% Tween-20, 15 m*M* $MgCl_2$, 0.1% gelatin, 1 m*M* β-ME, and 1% Thesit (polyoxyethylene-9-lauryl ether) (*see* **ref. *24***).
16. DNA ladder (Gibco BRL).
17. Restriction enzymes: *Alu*I (Boehringer Mannheim), T4 polynucleotide kinase (Gibco BRL), and appropriate buffers as recommended by the suppliers. All these reagents should be stored at –20°C.

2.3.3. Electrophoresis

1. Acrylamide/*bis*-acrylamide stock solution: 40% acrylamide, 0.66% *bis*-acrylamide in double distilled H_2O (Bio-Rad, Hercules, CA).
2. TEMED: *N,N,N'N'*-tetramethylethylenediamine (Bio-Rad).
3. Ammonium persulfate (APS) (Bio-Rad).
4. Electrophoresis loading buffer: 7*M* urea, 10% sucrose, 10 m*M* Tris-HCl, pH 7.4, 1 m*M* ethylenediamine tetra-acetic acid (EDTA), pH 7.4, and 0.05% bromophenol blue.
5. 10X TAE electrophoresis buffer: 400 m*M* Tris, 200 m*M* sodium acetate, 20 m*M* EDTA, pH 7.8.
6. Gel elution buffer: 50 m*M* Tris, pH 8.0, 0.3 *M* sodium acetate, 0.2% sodium dodecyl sulfate (SDS), and 4 m*M* EDTA, pH 8.0.

3. Methods

3.1. RNA Isolation from Different Organs of Mice

1. Excise tissues and freeze immediately in liquid nitrogen until ready for processing.
2. Place frozen tissues under liquid nitrogen, allow nitrogen to evaporate and immediately add 5 mL of lysis solution (*see* **Note 3**).
3. Homogenize lysed tissue for 15–30 s, i.e., until no visible tissue fragments remain.
4. Add 2 mg/mL of proteinase K to a final concentration of 50 μg/mL, and incubate at 55°C for 2 h; vortex gently every 30 min.
5. Immediately place cell lysates into 10-mL centrifuge tubes (RNase free) containing guanidinium solution (mix 1 vol of a lysate with 1.5 vol of a guanidinium solution (*see* **Subheading 2.3.**). Vortex solution gently but thoroughly.
6. Extract mixture twice with equal volume of phenol:chloroform:isoamyl alcohol mixture (24:24:1) (*see* **Note 4**).
7. To obtain a high yield of RNAs, add 20 μg of glycogen per each 400 μL of an aqueous phase as a carrier, add 0.25 vol of 2 *M* ammonium acetate, and precipitate RNA with an equal volume of isopropanol at –20°C for 1 h.
8. Collect the RNA by centrifuging at 15,000*g* for 20 min at 4°C. Gently aspirate the supernatant and rinse the pellet with ice cold 70% ethanol. Recentrifuge as above for 5 min, remove the supernatant, and dry the pellet under vacuum (*see* **step 8**).
9. Dissolve the RNA pellet in 200 μL of DEPC-treated water and extract again with an equal volume of anhydrous *n*-butanol. Vortex gently for 15–20 s (*see* **Note 5**).
10. Reprecipitate the aqueous phase using 0.1 vol of 3 *M* sodium acetate, pH 5.2, and 2.5 vol of 100% ethanol (–20°C, overnight).
11. Centrifuge at 15,000*g* for 20 min at 4°C, and gently wash the pellet twice with 70% ethanol (–20°C). Dry samples under vacuum. Dissolve in 100 μL of DEPC-treated water (*see* **Notes 6** and **7**).
12. Determine RNA concentration and purity by spectrophotometry readings at 230, 260, and 280 nm. The concentration of RNA in a sample is calculated using the readings at 260 nm, assuming that 1 OD = 40 μg RNA. Total RNA should be free of DNA and protein contamination and the A_{260}/A_{280} ratio should be >1.8 (*see* **Note 8**).

3.2. Reverse Transcription

1. Reverse transcribe 1 μg of total RNA by consecutively incubating it, in a 1.5-mL Eppendorf tube, at 94°C for 5 min (denaturation), at 42°C for 10 min (annealing), and at 72°C for 45 min (extension), in the presence of 5 U of Tth polymerase in the Tth polymerase reaction buffer containing 1 m*M* each dNTP and 25 pmol of the reverse transcription primer 5' ACTCGGTCAGGTCTGGGGA 3' specific for p65 (*see* **Note 9**). Perform cDNA synthesis in the total volume of 20 μL. Cover the reaction mixture with few drops of mineral oil to prevent evaporation.
2. Stop the cDNA synthesis by adding 1 μL of 5 *M* EDTA and place mixture on ice.
3. To 20-μL sample, add 2 μL (20 μg) of glycogen (carrier), 5 μL of 2 *M* sodium acetate, and 100 μL of 100% ethanol, mix gently, and keep at –20°C for 1 h.

4. Centrifuge at 15,000*g* at 4°C for 10 min. Carefully remove the supernatant without disturbing the pellet.
5. Wash the cDNA pellet with 80% ethanol, and recentrifuge at 15,000*g* at 4°C for 10 min (*see* **Note 10**).
6. Air-dry the pellet and resuspend in 10 µL DEPC-treated water.

3.3. Primer Labeling

1. Label 100 pmol of the p65-specific antisense primer with ^{32}P in a total volume of 25 µL containing 50 m*M* Tris-HCl, pH 7.6, 1 m*M* $MgCl_2$, 5 m*M* dithiothreitol (DTT), 2 µ*M* [$\gamma^{32}P$] ATP (1000–3000 Ci/mmol) and 10 U T4 polynucleotide kinase, at 37°C for 30 min.
2. Terminate reaction by heating at 65°C for 30 min.
3. Purify the labeled DNA on a gel-filtration column (Chroma Spin Column 10) to separate the primer DNA from radioactive precursors.

3.4. Polymerase Chain Reaction

1. Amplify 5 µL of cDNA solution (*see* **Subheading 3.2.**) according to the following protocol: 5 min at 85°C ("hot start"), 1 min at 60°C (annealing), 1 min at 72°C (extension), and 30 s at 94°C (denaturation) in the presence of 8 U of *Taq* polymerase in 50 µL of *Taq* polymerase reaction buffer (*see* **Subheading 2.4.**) containing 0.8 m*M* each dNTP and 25 pmol of the p65-specific nested primers, i.e., sense 5' AAGTGATACCCAGATTGGCC 3', and antisense 5' AAGCAA TGAGCCACTCCCTC 3'. Notice that you heat the tubes at 85°C for 5 min ("hot start") before adding the primers (*see* **Note 11**). Overlay the reaction mixture with mineral oil to prevent evaporation of the sample during repeated cycles of heating and cooling.
2. After 30 cycles, carry out the final extension reaction at 74°C for 10 min.
3. Analyze amplification products on 6% polyacrylamide gels in TAE buffer.
4. Carry out autoradiography at –80°C on Hyperfilm™MP (Amersham Life Sciences) with a Kodak intensifying screen.

4. Notes

1. Use only RNase-free pipets and wear gloves all the time to reduce chances of contamination with RNases.
2. The p65-specific primers were designed using the region(s) of the p65 gene nonhomologous to the estrogen receptor gene ***(7,28)***.
3. The strong chaotropic properties of the guanidinium solution completely disrupt cells and inactivate nucleases, preserving the integrity of RNA. The sample can be than processed or stored at 4°C for 1–2 wk for the later use. Isolation of not degraded total RNA from tissues is essential in studies of gene expression, therefore, pH of guanidinium solution is a very important factor. We recommend the pH for guanidinium solution to be 6.0.
4. For the first phenol extraction, the sample is mixed 30 times by inversion, while for the second extraction sample is stirred by vigorous vortex mixing for 15 s.

Residual phenol, protein, and other impurities are removed by extraction with a chloroform:isoamyl alcohol mixture (24:1). To obtain a high yield of RNA, precipitate aqueous phase using glycogen and isopropanol.

5. The *n*-butanol extraction reduces the volume and yields a colorless RNA preparation that is free from PCR inhibitors ***(23)***.
6. All RNA centrifugation steps should be conducted at 15,000*g* for 10 min at room temperature, except for isopropanol and 100% ethanol precipitation steps, which should be carried out at 15,000*g* for 20 min at 4°C.
7. Solubilization of RNA isolated from tissue using the guanidinium method usually does not pose a problem, but RNA isolated by this method is not easily solubilized. Extraction of the RNA twice with DEPC-treated water helps to dissolve the RNA. Furthermore, the RNA yield is increased when additional one or two such extractions follow.
8. A relatively low A_{260}/A_{280} ratio (i.e., <1.8) is attributed, at least in part, to protein coprecipitating with RNA. In our hands, RNA isolated by this method had the A_{260}/A_{280} ratio >1.8. If the ratio is lower, RNA samples should be extracted once more by a chloroform:isoamyl alcohol mixture (24:1) and ethanol precipitated.
9. Short sequence-specific primers enhance specificity and do not interfere with subsequent amplification because of the low-melting temperature (T_m) value. They probably allow for reduction in the amount of reverse transcriptase used in the reaction. Moreover, specific short reverse transcriptase primers are easy to design, being less empirical. We prefer to use Tth DNA polymerase because reverse transcription is taking place at high temperature. Disruption of secondary structure in the RNA template is an important factor in obtaining efficient cDNA synthesis.
10. It is important to purify the cDNA before PCR to prevent carryover of cDNA synthesis primers. cDNA solution should be placed right away on ice, if you plan to perform the PCR; otherwise, store at –70°C. We recommend reserving the portion of cDNA (–70°C) in case there is a need to repeat PCR.
11. Heating of the reaction cocktail ensure denaturation of the template if it is double stranded and melting of any stable secondary structures. For maximum amplification specificity, we performed a "hot start." All of the reaction components, except the primers, were briefly heated to 85°C. The primers were then added and thermocycling started ***(25)***.
12. We used nested primers and a high-annealing temperature (60°C), because specificity of our amplification reaction increased. The number of cycles depends on how much DNA template is present at the start, and might need to be determined empirically. The quantity of the final PCR product is determined by the concentration of dNTPs and primers assuming that a sufficient number of amplification cycles are carried out. Too many cycles will generate nonspecific products. We recommend using *Taq* DNA polymerase for PCR reactions instead of Tth polymerase. Tth polymerase is very good for reverse transcription (mediating template-dependent DNA synthesis in the presence of phenol), but in our experience is less sensitive than *Taq* DNA polymerase. The accurate quantitation of PCR

products relies on the use of standards to compensate the high number of incalculable factors affecting the yield of PCR products. In our RT-PCR assay, an internal standard specific for a p65 cDNA sequence was used. The internal standard was prepared with an amplified p65 cDNA fragment (420 bp) which was cut with endonuclease *Alu*I and then ligated. After gel elution and precipitation, we obtained a shorter p65 cDNA fragment (i.e., without the 100 bp *Alu*I-*Alu*I fragment). The construct containing this p65 cDNA fragment was amplified again using the same conditions. In each assay, 100 ag (1 ag = 10^{-16} g) of the IS probe was used.

References

1. Hanausek-Walaszek, M., Del Rio, M., and Adams, A. K. (1989) Immunohistochemical demonstration of mRNA-transport protein in rat liver putative preneoplastic foci. *Cancer Lett.* **48,** 105–108.
2. Hanausek-Walaszek, M., Del Rio, M., and Adams, A. K. (1990) Structural and immunological identity of p65 tumor-associated factors from rat and mouse hepatocarcinomas. *Progr. Clin. Biol. Res.* **331,** 109–120.
3. Mirowski, M., Sherman, U., and Hanausek, M. (1992) Purification and characterization of a 65-kDa tumor-associated phosphoprotein from rat transplantable hepatocellular carcinoma 1682C cell line. *Protein Expr. Purif.* **3,** 196–203.
4. Mirowski, M., Walaszek, Z., Sherman, U., Adams, A. K., and Hanausek, M. (1993) Comparative structural analysis of human and rat 65 kDa phosphoprotein. *Int. J. Biochem.* **25,** 1865–1871.
5. Wang, S., Mirowski, M., Sherman, U., Walaszek, Z., and Hanausek, M. (1993) Monoclonal antibodies against a 65 kDa tumor-associated phosphoprotein: development and use in cancer detection. *Hybridoma* **12,** 167–176.
6. Mirowski, M., Klijanienko, J., Wang, S., Vielh, P., Walaszek, Z., and Hanausek, M. (1994) Serological and immunohistochemical detection of a 65 kDa protein breast cancer. *Eur. J. Cancer* **30A,** 1108–1113.
7. Hanausek, M., Szemraj J., Adams, A. K., and Walaszek, Z. (1996) The oncofetal protein p65: a new member of the steroid/thyroid receptor superfamily. *Cancer Detect. Prev.* **20,** 94–102.
8. Bennet, L. M., Farnham, P. J., and Drinkwater, N. R. (1995) Strain-dependent differences in DNA synthesis and gene expression in the regenerating livers of C57Bl/6J and C3H/HeJ mice. *Mol. Carcinogenesis* **14,** 46–52.
9. Drinkwater, N. R. and Ginsler, J. J. (1986) Genetic control of hepatocarcinogenesis in C57BL/6J and C3H/HeJ inbred mice. *Carcinogenesis* **7,** 1701–1707.
10. Bennett, L. M., Winkler, M. L., and Drinkwater, N. R. (1993) A gene that determines the high susceptibility of the C3H/HeJ strain of mouse to liver tumor induction is located on chromosome one. *Proc. Am. Assoc. Cancer Res.* **33,** 144.
11. Gariboldi, M., Manenti, G., Canzian, F., Falvella, S., Pierotti, M., Della Porta, G., and Dragani, T. A. (1993) Chromosomal mapping of murine susceptibility loci to liver carcinogenesis. *Cancer Res.* **53,** 209–211.

12. Manenti, G., Binelli, G., Gariboldi, M., Canzian, F., De Gregoriol, R., Flavella, S., Dragani, T. A., and Pierotti, M. (1994) Multiple loci affect genetic predisposition to hepatocarcinogenesis in mice. *Genomics* **23,** 118–124.
13. Medina, D. (1976) Preneoplastic lesions in murine mammary cancer. *Cancer Res.* **36,** 2589–2595.
14. Schwartz, M. S., Smith, G. H., and Medina, D. (1992) The effect of parity, tumor latency and transplantation of the activation of *int* loci in MMTV-induced, transplantal C3H mammary neoplasias and their tumors. *Int. J. Cancer* **51,** 805–811.
15. Dux, A. (1981) Sensitivity of the mammary gland to tumor formation, in *Mammary Tumors in the Mouse* (Hilgers, J. and Skryser, M., eds.), Elsevier/North Holland Biomedical Press, Amsterdam, The Netherlands, pp. 516–543.
16. Festing, M. F. W. and Blackmore, D. K. (1971) Life span of specified-pathogen-free (MRC category 4) mice and rats. *Lab. Anim. (Lond.)* **5,** 179–192.
17. Burmer, G. C. and Loeb, L. A. (1989) Mutations in the K-ras 2 oncogene during progressive stages of human colon carcinoma. *Proc. Natl. Acad. Sci USA* **86,** 2403–2407.
18. Burmer, G. C., Parker, J. D., Bates, J., East, K., and Kulander, B. G. (1990) Comparative analysis of human papillomavirus detection by polymerase chain reaction and Virapap/Viratype. *Am. J. Clin. Pathol.* **94,** 554–560.
19. Burmer, G. C., Rabinovitch, P. S., and Loeb, L. A. (1989) Analysis of c-Ki-ras mutations in human colon carcinoma by cell sorting, polymerase chain reaction and DNA sequencing. *Cancer Res.* **49,** 2141–2146.
20. Frohman, M. A., Dush, M. K., and Martin, G. R. (1988) Rapid production of full-length cDNAs from rare transcripts: amplification using single gene specific oligonucleotide primer. *Proc. Natl. Acad. Sci. USA* **85,** 8998–9002.
21. Higuchi, R., von Beroldingen, C. H., Sensabaugh, G. F., and Erlich, H. A. (1988) DNA typing from single hairs. *Nature* **333,** 543–540.
22. Saiki, R. K., Gelfand, D. H., Stoffel, S., Scharf, S. J., Higuchi, R., Horn, G. T., Mullis, K. B., and Erlich, H. A. (1988) Primer directed enzymatic amplification of DNA with a thermostable DNA polymerase. *Science* **39,** 487–491.
23. Erlich, H. A. (1989) *PCR Technology: Principles and Applications for DNA Amplification*. Stockton, New York.
24. Ponce, R. P. and Micol, J. L. (1992) PCR amplification of long DNA fragments. *Nucleic Acids Res.* **20,** 3–623.
25. D'Aquaila, R. T., Bechtel, L. J., Videler, J. A., Eron, J. J., Gorczyca, P., and Kaplan, J. C. (1991) Maximizing sensitivity and specificity of PCR by pre-amplification heating. *Nucleic Acids Res.* **19,** 3749–3754.
26. Hanausek-Walaszek, M., Adams, A. K., and Sherman, U. (1991) Expression of a 65 kda tumor-associated protein and persistence of altered hepatic foci. *Progr. Clin. Biol. Res.* **369,** 91–103.
27. Hanausek, M., Sherman, U., Mirowski, M., and Walaszek, Z. Induction of a 65 kDa tumor-associated protein in altered hepatic foci of rats fed the peroxisome proliferator Wy-14,643. *Progr. Clin. Biol. Res.* **387,** 337–348.
28. Ponglikitmongkol, M., Green, S., and Chambon, P. (1988) Genomic organization of the human oestrogen receptor gene. *EMBO J.* **7,** 3385–3388.

hepatic microsomal enzyme induction by xenobiotic agents ***(14)***. The significance of this urinary excretion of GA in relation to both the metabolism of xenobiotics and the dietary intake of GA and its derivatives must be investigated (*see* **Note 1**).

At present, the physiological function of GA is unclear. Formation of the βG inhibitor 1,4-GL, from one of the products of GA's hydrolytic action could be regarded as a negative feedback mechanism ***(17–19)***. There is now growing evidence from animal models, that 1,4-GL and its precursors such as GA salts, i.e., D-glucarates may control different stages of the carcinogenic process ***(18,19)***. The mechanism, however, is not clear and needs further studies. Nevertheless, the results of recent studies on the inhibition of mammary, colon, and skin tumorigenesis in rodents clearly suggest that 1,4-GL and GA salts may exert their chemopreventive action, in part, by altering the hormonal environment and/or the proliferative status of the target organ ***(19)***. The cholesterol-lowering properties of calcium D-glucarate and potassium hydrogen D-glucarate demonstrated in our recent study ***(10)*** suggest that GA and/or 1,4-GL may be natural regulators of cholesterogenesis and steroidogenesis.

As a result of the extremely encouraging outcome of animal studies with calcium D-glucarate (*see* **refs. *18*** and ***19***), the National Cancer Institute of the National Institutes of Health has initiated a Phase I trial of calcium D-glucarate in patients at high risk for breast cancer ***(20)***. To date, no evidence of toxicity has been found, even at high doses of calcium D-glucarate. In addition, patients have tolerated the medications without complaints of any gastrointestinal distress ***(20)***. Other potential applications of D-glucarates are prevention of prostate and colon cancer as well as prevention of cardiovascular disease. In fact, potassium hydrogen D-glucarate and calcium D-glucarate have several advantages that should make them acceptable as some vitamins, to human subpopulations at risk for cardiovascular disease and cancer.

In a few earlier studies, urinary excretion of GA in cancer patients and tumor-bearing rats were found (*see* **refs. *18*** and ***19***) to be significantly lower than in healthy controls (*see* **Note 2**). In mice with experimental tumors and in cancer patients, uninvolved liver had a lower GA content ***(11)***. Cancer tissue itself lacked the GA-synthesizing system ***(11)***. We have previously reported ***(21)*** results of our preliminary study on the reduced levels of GA in the blood serum of breast cancer patients.

All the current and prospective clinical studies may be hampered by lack of a simple and accurate method to assay GA in different body fluids and tissues. At present, there are two widely accepted methods of measuring GA in biological samples. The enzyme inhibition method measures the inhibition of β-glucuronidase by 1,4-GL produced from GA during boiling of GA solutions at acid pH ***(15)***. Alternatively, the pyruvate method may be used which as origi-

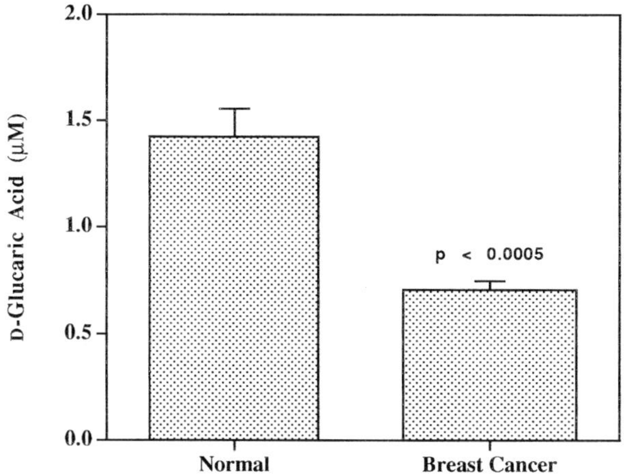

Fig. 1. GA content of serum from normal, healthy women (n = 15) and breast cancer patients (n = 19) (Mean ± SD).

nally described, is a sensitive, analytical method for urinary GA determination *(22)*. This assay employs the *Escherichia coli* catabolic enzymes that quantitatively convert GA and its salts ***(22,23)*** as well, as lactones (Z. Walaszek et al., unpublished data) to pyruvate. The pyruvate is then assayed by use of lactate dehydrogenase. The pyruvate method was also used to measure GA concentration in the blood serum ***(23)***. Currently known high-performance liquid chromatography (HPLC) methods (*see* **ref. *24***) are, in general, not sensitive enough for assaying nonradioactive GA in body fluids and tissues ***(25)***.

The goal of this chapter is to demonstrate the usefulness of the pyruvate method for determination of GA concentrations in the blood serum of breast and prostate cancer patients. Thus, using the pyruvate method we have found (*see* **Figs. 1** and **2**) normal levels of GA in the blood serum of healthy people to be 1.42 ± 0.5 μ*M* in women (n = 15) and 1.50 ± 0.29 μ*M* in men (n = 19). In cancer patients, the blood serum concentration of GA dropped significantly, i.e., to levels <1 μ*M* (*see* **Figs. 1** and **2**). Specifically, the GA concentrations in the blood serum of breast cancer patients (n = 19) and prostate cancer patients (n = 22) were 0.71 ± 0.17 μ*M* (50% reduction) and 0.95 ± 0.16 μ*M* (37% reduction), respectively (*see* **Note 3**). The values found for breast and prostate cancer patients were significantly lower ($p < 0.0005$, Student's t test) from the values found for normal, healthy women and men, respectively. We conclude that following extensive studies of a large number of cancer patients and normal, healthy people, GA may become a new and important tumor marker.

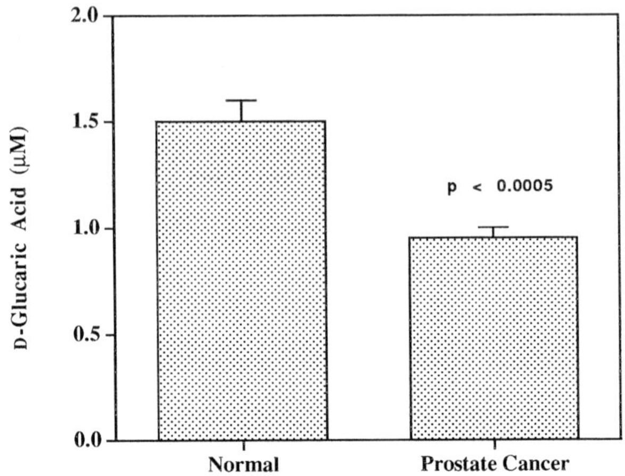

Fig. 2. GA acid content of serum from normal, healthy men ($n = 19$) and prostate cancer patients ($n = 22$) (Mean ± SD).

2. Materials

1. *E. coli* strain 8044 (American Type Culture Collection, Rockville, MD).
2. β-Nicotinamide adenine dinucleotide, reduced form (β-NADH): Disodium salt (Sigma, St. Louis, MO). 4 m*M* NADH solution in 0.01*M* NaCl should be prepared fresh and stored not longer than 1 d.
3. L-Lactate dehydrogenase (LDH): Crystalline suspension from rabbit muscle, type II (Sigma). Dilute to a concentration of 1 mg/mL with 2*M* ammonium sulfate.
4. UV/VIS Spectrophotometer.
5. 1.5-mL Cuvets (semimicro, 1-cm pathlength).
6. Trypticase-soy-agar (Difco, Fisher Scientific, Pittsburgh, PA).
7. GA monopotassium salt: potassium hydrogen D-glucarate (Sigma) (*see* **Note 4**).
8. Liquid bacterial growth medium (salt medium): Weigh out 8.0 g of GA monopotassium salt (32 m*M*), 16.5 g Na_2HPO_4, 1.5 g K_2HPO_4, 2.0 g $(NH_4)_2SO_4$, 0.2 g $Mg_2SO_4 \cdot 7\,H_2O$, 10 mg $CaCl_2 \cdot 2\,H_2O$, and 50 μg $Fe_2SO_4 \cdot 7\,H_2O$. Dissolve in double-distilled water and make volume up to 1 L. Adjust pH to 6.8 with 30% sodium hydroxide.
9. 0.4 *M* Tris-maleate buffer, pH 7.8:
   a. Solution A: 0.2 *M* Tris-maleic acid solution in double distilled water. Weigh out 24.2 g Tris and 23.2 g maleic acid and dissolve in a final volume of 1 L.
   b. Solution B: 0.2 *M* NaOH, 500 mL.

   Mix 50 mL of solution A with 58 mL of solution B and adjust volume to 200 mL.
10. 1 *M* $MgSO_4$.
11. 0.01 *M* NaCl.
12. 50 m*M* KCl.
13. 50 m*M* Tris-HCl, pH 7.5 with 3 m*M* glutathione (*see* **ref. *21***).
14. Saturated ammonium sulfate (*see* **Note 5**) neutralized with ammonium hydroxide (NAS).

15. Ethanol.
16. Bradford reagent for protein determination (Sigma) with bovine serum albumin (Sigma) as a standard (*see* **Note 6**).
17. Microwave oven.

3. Methods

3.1. Bacterial Enzyme Preparation (see Notes 7 and 8)

1. Grow *E. coli* strain 8044 overnight on trypticase soy-agar plate.
2. Inoculate starter medium (100 mL) with several loopfuls from agar plate (5% main culture [v/v]). Grow a starter culture at 37°C in salt medium containing 32 m*M* potassium hydrogen D-glucarate ***(22,23)*** as the sole carbon source for 16 h and then in a larger volume of the same medium (1–2 L), with constant shaking, for another 24 h.
3. Collect cells by centrifugation at 2000*g* for 30 min, and wash cells twice with 50 mL cold 50 m*M* KCl.
4. Resuspend the pellet in a buffer containing 50 m*M* Tris-HCl, pH 7.5 and 3 m*M* glutathione (use 10 mL of the buffer/1 g of wet pellet) (*see* **ref. *21***).
5. Sonicate cells for about 8 min using ice to cool the solution down. Keep the extract temperature below 8°C.
6. Centrifuge at 21,000*g* for 10–15 min and collect a crude cell-free extract.
7. Treat the crude extract with an equal volume of saturated ammonium sulfate solution (NAS). Add NAS dropwise with constant stirring. Incubate on ice for 20 min and then centrifuge at 21,000*g* for 10 min to remove the precipitate.
8. To 20 mL of the supernatant add 13.5 mL of NAS dropwise with constant stirring. Incubate on ice for 20 min and then centrifuge at 21,000*g* for 10 min to collect the precipitate .
9. Dissolve the pellet in 10 m*M* Tris-HCl, pH 7.5/3 m*M* glutathione buffer (use 2 mL of the buffer/1 g of packed cells).
10. Centrifuge the solution at 100,000*g* for 3 h. This step removes remaining NADH oxidase activity. The supernatant after ultracentrifugation contains approx 30–35 mg of protein/mL. Always check the protein concentration (*see* **Note 6**).
11. Dialyze the supernatant at 4°C overnight against a large volume of the 10 m*M* Tris-HCl, pH 7.5/ 3 m*M* glutathione buffer (*see* **Note 8**).

3.2. D-Glucarate Enzymatic Assay

1. Prepare five cuvets and to each cuvet add 0.25 mL 0.4 *M* Tris-maleate buffer, pH 7.3, 0.1 mL 1 *M* Mg_2SO_4, and 0.2 mL 4 m*M* NADH in 0.01*M* NaCl.
2. To all cuvets, except cuvet #2 add 50–100 µL LDH (92 U, 1 mg protein/mL in 2 *M* ammonium sulfate).
3. To all cuvets, except cuvet #1, add 0.37 mL serum.
4. Add double-distilled water to all the cuvets to a final volume of 1.5 mL. (The instrumental blank cuvet contains only 0.25 mL of 0.4 *M* Tris-maleate buffer, pH 7.3, 0.1 mL 1 *M* Mg_2SO_4, and water.)

5. Preincubate cuvets at room temperature for 1 h in order to reduce any serum pyruvate to lactate prior to the glucarate assay and decrease other sources of background oxidation of NADH by serum samples (*see* **ref.** *23*).
6. After preincubation, add 10 μL 2.5 m*M* potassium hydrogen D-glucarate to cuvet #5, 50 μL of the *E. coli* extract (*see* **Subheading 3.1.**) to all the cuvets, except # 2, and then add water to a final volume of 1.55 mL.
7. Mix well and read absorbance at 340 nm at times 0, 15, 30, and 60 min.
8. Subtract the absorbance readings for cuvets #1 and #2 from those found for cuvets #3–#5 after the 60-min assay.
9. Calculate the D-glucarate content assuming that the molar absorption coefficient of NADH is 6.22 × 10^3/*M*/cm, and that 2 μmol of NADH is oxidized/1 μmol of D-glucarate present in the solution (*see* **Note 9**).

4. Notes

1. Because urinary excretion of GA increases following exposure to xenobiotics, including different toxins and carcinogens, it was initially suggested for use as an indicator of hepatic microsomal enzymes induction by xenobiotic agents ***(13)***. There is evidence, however, that enhanced GA excretion is not always accompanied by induction of hepatic microsomal enzymes (*see* **ref.** *26*). We believe that GA synthesis and excretion is more likely to be related to β-glucuronidase enzyme induction ***(17,18)*** and to a lesser extent to dietary intake of GA ***(10)***. Nevertheless, urinary excretion of GA is considered useful as nonspecific parameter for exposure to environmental factors ***(26)***. In fact, very often, the results of the GA tests correlate well with the results of bacterial urinary assays for mutagenic activity, i.e., the Ames test ***(26)***.
2. The urinary level of GA in cancer patients was found (*see* **refs.** ***18*** and ***19***) to be approx 10 timcs lower. Pregnancy and estrogen therapy produce a nonsignificant increase in GA excretion ***(26)***. In general, smoking has been reported to cause a significant increase ***(16,26)***. Disease states known to be accompanied by increased excretion of GA include alcoholism, early stage of renal disease in children and liver diseases (*see* **ref.** *26*). Decreased values of GA excretion have been found to occur in patients with congestive heart failure, starvation, severe burns, and favism ***(26)***.
3. The blood serum concentration, found by us using the pyruvate assay, in normal, healthy men and women are in good agreement with the data found earlier ***(15)*** using the fluorometric method. Higher values are reported in **ref.** *23*. Note, however, that when we analyzed the GA content in various fruits and vegetables ***(10)*** and also in the rat urine (Z. Walaszek et al., unpublished data) by both the pyruvate and β-glucuronidase inhibition methods, our pyruvate assay gave essentially the same or only slightly higher values compared to the β-glucuronidase inhibition assay. The lowest GA concentration detectable by our pyruvate assay was ~0.2 μ*M*. We were not able to detect nonradioactive GA in the blood serum using the HPLC method described in **ref.** *25*. The lowest concentration of GA detectable by HPLC was ~2 μg/mL (i.e., ~1 μ*M*) which is comparable to the value

reported earlier *(24)*. Thus, the lowest concentration of GA detectable by the HPLC method *(24,25)* is four- to fivefold lower than the GA concentration reported for the normal human blood serum in **ref.** ***23***.

4. In earlier studies, with the large amounts of impurities in most commercial potassium hydrogen D-glucarate preparations, the D-glucarate was purified as the dicyclohexylammonium salt before being used as the substrate *(22,23)*.
5. Use purified ammonium sulfate, grade I (Sigma).
6. The protein concentration is measured in all bacterial enzyme preparations by the Bradford method *(27)*.
7. The following three *E. coli* enzymes are necessary for the D-glucarate assay: D-glucarate dehydrase (EC 4.2.4.0), α-keto-β-deoxyglucarate aldolase (EC 4.1.2.20), and tartronate semialdehyde reductase [D-glucarate:NAD(P) oxidoreductase] (EC 1.1.1.60). All three enzymes are induced in most *Enterobacteriacae (22)*. The *E. coli* strain 8044 was used in **ref.** ***23*** and in our study. The strain NCTC 10418 was used in **ref.** ***22***.
8. All enzyme preparation steps should be performed at 4°C. The enzyme extract may be used immediately after dialysis or may be stored at –20°C for more than 6 mo.
9. In **ref.** ***23*** to calculate the results the actual A_{340} of the 1.50-mL samples in cuvets 1–5 after 60-min preincubation were corrected arithmetically to 1.55 mL, and the A_{340} values between the corrected 0 and actual 60-min assay readings were then calculated from the printout of the individual readings. The amount of NADH utilized in the *E. coli* enzyme NADH oxidase control during the 60-min assay (usually <0.03 A_{340}), was then subtracted from the A_{340} for cuvets 2–5. Normal serum GA values are stable for months when sera are stored at –20°C (*see* **ref.** ***23***).
10. A number of possible serum constituents were tested which might be falsely detected as D-glucarate (*see* **ref.** ***23***). D-Glucose (up to 2 mg) or galactarate, D-glucuronate, or L-ascorbate (at a level of 2.6 μg/mL) was not detected as D-glucarate when added to serum, within the normal error of the method.

References

1. Marsh, C. A. (1963) Metabolism of D-glucuronolactone in mammalian systems. II. Conversion of D-glucuronolactone into D-glucaric acid by tissue preparation. *Biochem. J.* **87,** 82–90.
2. Gorter, M. K. (1912) Note sur les acides chlorogenique et saccharique dans le latex. *Rec. Trav. Chim.* **31,** 281–286.
3. Kessler, G., Neufeld, E., Feingold, D. S., and Hassid, W. Z. (1961) Metabolism of D-glucuronic acid and D-galacturonic acid in *Phaseolus aureus* seedlings. *J. Biol. Chem.* **236,** 308–312.
4. Dittrich, P. and Kandler, O. (1971) Biosynthesis of D-glucaric acid in needles of *Larix decidua. Z. Pflanzen Physiol.* **66,** 368–371.
5. Kingstad, R. and Nordal, A. (1975) Lactone forming acids in succulent plants. *Phytochemistry* **14,** 1868–1870.

6. Elliger, C. A., Lundin, R. E., and Haddon, W. F. (1981) Caffeyl esters of glucaric acid in *Lycopersicon esculentum* leaves. *Phytochemistry* **20,** 1133,1134.
7. Oen, H. and Vestheim, S. (1985) Detection of non-volatile acids in sweet cherry fruits. *Acta Agric. Scand.* **35,** 145–152.
8. Risch, B., Herrmann, K., and Wray, V. (1988) (*E*)-*O*-p-cumaroyl-(*E*)-*O*-feruloyl-derivatives of glucaric acid in citrus. *Phytochemistry* **27,** 3327–3329.
9. Winsnes, R. (1972) Lactonic acids in the latex of *Euphorbium canariensis* L. in relation to succulent metabolism: isolation and characterization of D-glucaric acid. *Medd. Norsk. Farm. Selskap* **34,** 1–8.
10. Walaszek, Z., Szemraj, J., Hanausek, M., Adams, A. K., and Sherman, U. (1996) D-Glucaric acid content of various fruits and vegetables and cholesterol lowering effects of dietary D-glucarate in the rat. *Nutr. Res.* **16,** 673–681.
11. Levvy, G. A. and Conchie, J. (1966) β-Glucuronidase and the hydrolysis of glucuronides, in *Glucuronic Acid: Free and Combined* (Dutton, G. J., ed.), Academic, New York, pp. 301–364.
12. Horton, D. and Walaszek, Z. (1982) Conformation of the D-glucarolactones and D-glucaric acid in solution. *Carbohydr. Res.* **105,** 95–109.
13. Marsh, C. A. (1963) Metabolism of D-glucarolactone in mammalian system: identification of D-glucaric acid as normal constituent of urine. *Biochem. J.* **86,** 77–86.
14. Dutton, G. J. (1980) *Glucuronidation of Drugs and Other Compounds*, CRC Press, Boca Raton, FL, pp. 83–89.
15. Matsui, M., Fukuo, A., Watanabe, Y., Wanibe, T., and Okada, M. (1972) Studies on the glucaric acid pathway in the metabolism of D-glucuronic acid in mammals. IV. Fluorometric method for the determination of D-glucaric acid in serum. *Chem. Pharm. Bull. (Tokyo)* **20,** 845–848.
16. Colombi, A., Maroni, M., Antonini, C., Fait, A., Zocchetti, C., and Foa, V. (1983) Influence of sex, age and smoking habits on the urinary excretion of D-glucaric acid. *Clin. Chim. Acta.* **128,** 349–358.
17. Dohrmann, R. E. (1969) *β-Glucuronidase*, Springer Verlag, Berlin-Heidelberg.
18. Walaszek, Z. (1990) Potential use of D-glucaric acid derivatives in cancer prevention. *Cancer Lett.* **54,** 1–8.
19. Walaszek, Z. (1993) Chemopreventive properties of D-glucaric acid derivatives. *Cancer Bull.* **45,** 453–457.
20. Heerdt, A. S., Young, C. W., and Borge, P. I. (1995) Calcium D-glucarate as a chemopreventive agent in breast cancer. *Isr. J. Med. Sci.* **31,** 101–105.
21. Walaszek, Z., Szemraj, J., Adams, A. K., Kordari, P., and Hanausek, M. (1996) Reduced levels of D-glucaric acid in mammary tumor-bearing hosts and the effect of its supplementation during estrogen replacement and tamoxifen therapy. *Proc. Am. Assoc. Cancer Res.* **37,** 235.
22. Marsh, C. A. (1985) An enzymatic determination of D-glucaric acid by conversion to pyruvate. *Anal. Biochem.* **145,** 266–272.
23. Blumentahl, H. J., Lucuta, V. L., and Blumentahl, D. C. (1990) Specific enzymatic assay for D-glucarate in human serum. *Anal. Biochem.* **185,** 286–293.

24. Laakso, E. I., Tokola, R. A., and Hirivisalo, E. L. (1983) Determination of D-glucaric acid by high performance liquid chromatography. J. *Chromatog.* **278,** 406–411.
25. Walaszek, Z., Szemraj, J., Narog, M., Adams, A. K., Kilgore, J., Sherman, U., and Hanausek, M. (1997) Metabolism, uptake and excretion of a D-glucaric acid salt and its potential use in cancer prevention. *Cancer Det. Prev.* **21,** 178–190.
26. Brewster, M. A. (1988) Biomarkers of xenobiotic exposure. *Ann. Clin. Lab. Sci.* **18,** 306–317.
27. Bradford, M. (1976) A rapid and sensitive method for the quantitation of microgram quantities of protein utilizing the principle of protein-dye binding. *Anal. Biochem.* **72,** 248,249.

Index

A

Aberrant crypt foci (ACF),
 assessment of colon cancer preventive agents, 465–474
 experimental protocols, 469–471
Acute lymphoblastic leukemia (ALL), 209–212
Acute myeloid leukemia (AML), 212, 213
Allelic loss,
 on chromosome 8p21, 279
 on chromosome 11q13, 281
Antigen retrieval, 169, 171
Anti-HMdU antibodies,
 biomarkers of cancer risk, 431–446
 ELISA for, 441

B

bcl-2, 75, 151
bcl-2 gene,
 detection of rearrangements in, 255–266
 PCR in analysis of, 257–263, 331–339
Binary comparisons, 14–20
Biomarker panels, 53–55
Bladder cancer biomarkers,
 classification of, 39, 40
 panels of, 53–55
 quantitative fluorescence-image analysis (QFIA), 181–205
 selection of biomarkers, 40–44
Breast cancer markers,
 ferritin, 115–125
 gene deletions, 78, 79
 oncofetal protein p65, 103–114
 oncogene products, 73–76
 oncoproteins, 95–102
  HER/*neu*, 99
  mutant p53, 98, 319
 serum vs tissue markers, 69–73
 steroid hormone receptors, 76–79
Breast cancer susceptibility genes BRCA1 and BRCA2, 82

C

Cancer susceptibility genes, 78, 79
Cathepsin D, 75
Childhood malignancies,
 cytogenetics in diagnosis of, 209–227
 immunodiagnosis of, 127–142
  indirect immunoperoxide procedure, 135, 136
  labeled streptavidin-biotin procedure, 135–137
Chromosomal *in situ* suppression (CISS), 245
Chromosome abnormalities,
 in childhood ALL, 210, 211
 in *de novo* childhood AML, 213
Chromosome staining and banding techniques, 239–243
 Giemsa banding (G-bands), 240, 241
 reverse banding (R-bands), 240–242
Colon cancer biomarkers,
 immunohistochemical evaluation of, 151–155
  p53, 152
  bcl-2, 153
 semiquantitation of, 147–149
Comparative genomic hybridization (CGH), 214, 215, 221–223, 290

Continuous markers, 27–29
Coupled amplification and ligation (CAL), 342
Cytogenetic markers, 209–266

D

Differential diagnosis, 20–23
Differential display,
  procedures, 395–404
  use of to define molecular markers, 391–414

E

Early detection of cancer, 417–495
Electron microscopy (EM), 116, 121–124
Epidermal growth factor receptor (EGF-R), 75
EWS gene,
  fusions of in sarcomas, 299–317
    RT-PCR detection of, 310–312
    Southern blot detection, 308, 309
  polymorphism in, 313

F

Factor VIII-related antigens (FAVIIIRA), 76
Ferritin,
  determination of, 115–125
    by electron microscopy, 123
    by imunocytochemical analysis, 123
    by microparticle enzyme immunoassays, 122
  increased concentration in breast cancer, 116, 117
Field disease model, 46, 47
Fluorescence *in situ* hybridization (FISH),
  preparation of probe for, 246–249
  probe hybridization to chromosomes, 246–251
  protocol for, 214, 215, 218–221
  use in cancer diagnosis, 245–254

G

D-Glucaric acid
  D-glucarate enzymatic assay, 491, 492
  reduced serum levels in cancer, 487–495
    breast cancer, 489
    prostate cancer, 490
Glutathione S-transferases (GSTs),
  assay, 452
  role of GSTs in cancer prevention, 447–464

H

Hematological disorder,
  diagnosis of, 209–227
    by CGH, 221
    by cytogenetic techniques, 216
    by FISH, 218
  recurrent chromosomal abnormalities in, 210–213
HER-2/*neu* (c-*erb*B-2) proto-oncogene, 73, 95

I

Immunocytochemical analysis (ICA), 116, 121–123
Immunodiagnosis (serum and tissue markers), 61–205
Immunofluorescence assays, 188–191
Immunohistochemical evaluation of biomarkers,
  factors affecting evaluation, 161–179
    antibodies, 162
    controls, 168
    fixation, 165, 166
    quality assurance 168
    tissue processing, 165
  methodology for, 171–176
Indirect immunoperoxidase procedure, 135–137

K

Ki-67 antibody, 72

L

Labeled streptavidin-biotin (LSAB)
procedure, 131–137
Laser capture microdissection, 293–295
Loss of heterozygosity (LOH), 78, 269–298

M

Marker data types, 12
Marker hypotheses, 13, 14
MEN1 tumor-suppressor gene
deletions, 288
Metaphase chromosomes,
harvesting of, 230–235
rapid preparation of, 230–235
use in cytogenetic analysis, 229–238
Metastasis suppressor gene (NM23),
76, 150
Microparticle enzyme immunoassay
(MEIA), 115, 121, 122
Molecular markers, 269–414
Multiple clinical tests,
predictive factors, 3–5
combining of, 6
reporting research results, 9
Multiple markers, 3–60
Mitotic index, 71
Proliferation markers, 71
Multivariable marker data analysis, 23–
25
c-*myc* proto-oncogene, 75

O

Oligonucleotide ligation assay (OLA),
341–355
Oncofetal protein p65,
breast cancer detection, 103–114
immunostaining, 106
immunoblotting, 109
differential expression in mouse
strains, 475–485
liver, 478
mammary gland, 478
RT-PCR analysis, 481, 482
Oncoprotein p53 (mutant p53),
ELISA assay for, 98, 99
immunoblotting analysis, 319–330
immunohistochemical analysis, 319–
330
Overfixation of tissue,
recovering from, 166, 167
vimentin control, 166

P

p21 (WAV-1/CIP-1), 74
p53, 74,151,319
p65, 103–114, 475–485
p185^{erbB-2}, 149
Polymerase chain reaction (PCR),
analysis of *bcl-2* gene rearrangements,
255–266, 331–339
analysis of *p65* gene expression,
475–485
detection of *ras* gene mutations, 347,
348
Proliferating cell nuclear antigen
(PCNA), 73
Prostate cancer cells,
RT-PCR detection of 357–371
in blood, 363
in bone marrow, 364
Prostate specific antigen (PSA), 357–
371
Prostate neoplasia biomarkers,
immunohistochemical evaluation of,
119–151
nm23-H1, 150
p185^{erbB-2}, 149
patterns of expression of, 144–147

Q

Quantitative, competitive RT-PCR
(QC-RT-PCR), 373–389
Quantitative fluorescence image
analysis (QFIA),

immunofluorescence assay, 188–191
instrumentation, 189
sample collection, fixation, processing, and storage, 183–188

R

ras gene mutations,
amplification of *ras* oncogenes, 346–348
detection by oligonucleotide ligation, 341–355
ras gene products, 75
Reverse transcription-polymerase chain reaction (RT-PCR), 255–266, 310–312, 357–371, 373–389, 475–485

S

Statistical considerations,
computer programs, 31–33
experimental design, 11–14
sample-size determination, 29–31
statistical analyses, 14–29

T

Tumor-associated antigens, 69, 70
Tumor-suppressor genes, 74, 151, 288
Tissue microdissection,
breast cancer, 281–284
colon cancer, 286, 287
esophageal cancer, 289, 290
melanoma, 290, 291
methodology, 270–276
neuroendocrine tumors, 287–289
PCR interpretation, 276, 277
prostate cancer, 278–281
renal cancer, 284–286
synchronous tumors, 291
visceral Kaposi sarcoma, 291, 292
Wilms' tumors, 285, 286
Transforming growth factor beta (TGFβ),
measurement of, 422–426
as plasma tumor marker, 417–430
t(14;18) Translocation in lymphoid tissue, 331–339

V

VHL gene deletion, 287